WATER RESOURCES
CENTER ARCHIVES

OCT -- 2009

UNIVERSITY OF CALIFORNIA
BERKELEY

# Environmental History of Water

# Environmental History of Water - Global views on community water supply and sanitation

Petri S. Juuti, Tapio S. Katko and Heikki S. Vuorinen

Published by IWA Publishing, Alliance House, 12 Caxton Street, London SW1H 0QS, UK
Telephone: +44 (0) 20 7654 5500; Fax: +44 (0) 20 7654 5555; Email: publications@iwap.co.uk
Web: www.iwapublishing.com

First published 2007
Reprinted 2007
© 2007 IWA Publishing

Printed by TJ International, Cornwall, UK

Apart from any fair dealing for the purposes of research or private study, or criticism or review, as permitted under the UK Copyright, Designs and Patents Act (1998), no part of this publication may be reproduced, stored or transmitted in any form or by any means, without the prior permission in writing of the publisher, or, in the case of photographic reproduction, in accordance with the terms of licences issued by the Copyright Licensing Agency in the UK, or in accordance with the terms of licenses issued by the appropriate reproduction rights organization outside the UK. Enquiries concerning reproduction outside the terms stated here should be sent to IWA Publishing at the address printed above.

The publisher makes no representation, express or implied, with regard to the accuracy of the information contained in this book and cannot accept any legal responsibility or liability for errors or omissions that may be made.

### Disclaimer

The information provided and the opinions given in this publication are not necessarily those of IWA and should not be acted upon without independent consideration and professional advice. IWA and the authors will not accept responsibility for any loss or damage suffered by any person acting or refraining from acting upon any material contained in this publication.

*British Library Cataloguing in Publication Data*
A CIP catalogue record for this book is available from the British Library

*Library of Congress Cataloging-in-Publication Data*
A catalog record for this book is available from the Library of Congress

ISBN: 1-84339-110-4
ISBN13: 978-1-84339-110-4

# CONTENTS

1 FOREWORD
Johannes Haarhoff ........................................................................................................... 1

2 "WATER IS THE BEGINNING OF ALL": GLOBAL WATER SERVICES AND CHALLENGES
Petri S. Juuti, Tapio S. Katko & Heikki S. Vuorinen ..................................................... 3

## PART I: EARLY SYSTEMS AND INNOVATIONS ............................................. 9

3 INTRODUCTION: EARLY CULTURES AND WATER
Petri S. Juuti, Tapio S. Katko & Heikki S. Vuorinen ................................................... 11

4 FIRST INNOVATIONS OF WATER SUPPLY AND SANITATION
Petri S. Juuti ................................................................................................................. 17

5 WATER AND HEALTH IN ANTIQUITY: EUROPE'S LEGACY
Heikki S. Vuorinen ....................................................................................................... 45

6 WATER SUPPLY IN THE LATE ROMAN ARMY
Ilkka Syvänne ............................................................................................................... 69

7 CONCLUSIONS
Petri S. Juuti, Tapio S. Katko & Heikki S. Vuorinen ................................................... 93

## PART II: PERIOD OF SLOW DEVELOPMENT ............................................... 97

8 INTRODUCTION
Petri S. Juuti, Tapio S. Katko & Heikki S. Vuorinen ................................................... 99

9 THE EMERGENCE OF THE IDEA OF WATER-BORNE DISEASES
Heikki S. Vuorinen ..................................................................................................... 103

10 BIRTH AND EXPANSION OF PUBLIC WATER SUPPLY AND SANITATION IN FINLAND UNTIL WORLD WAR II
Petri S. Juuti & Tapio S. Katko .................................................................................. 117

11 COLONIAL MANAGEMENT OF A SCARCE RESOURCE: ISSUES IN WATER ALLOTMENT IN 19TH CENTURY GIBRALTAR
Lawrence A. Sawchuk & Janet Padiak ...................................................................... 131

| 12 | COPING WITH DISEASE IN THE FRENCH EMPIRE: THE PROVISION OF WATERWORKS IN SAINT-LOUIS-DU-SENEGAL, 1860–1914 |
|---|---|
| | Kalala J. Ngalamulume ..................................................................................................147 |
| 13 | WATER SUPPLY IN THE CAPE SETTLEMENT FROM THE MID-17TH TO THE MID-19TH CENTURIES |
| | Petri S. Juuti, Harri R.J. Mäki & Kevin Wall ..................................................................165 |
| 14 | DEVELOPMENT OF THE SUPPLY AND ACQUISITION OF WATER IN SOUTH AFRICAN TOWNS IN 1850-1920 |
| | Harri R.J. Mäki .............................................................................................................. 173 |
| 15 | WATER, LIFELINE OF THE CITY OF GHAYL BA WAZIR, YEMEN |
| | Ingrid Hehmeyer .............................................................................................................197 |
| 16 | HISTORY AND PRESENT CONDITION OF URBAN WATER SUPPLY SYSTEM OF TASHKENT CITY, UZBEKISTAN |
| | Dilshod R. Bazarov, Jusipbek S. Kazbekov & Shavkat A. Rakhmatullaev ........................ 213 |
| 17 | PHILADELPHIA WATER INFRASTRUCTURE 1700-1910 |
| | Arthur Holst ...................................................................................................................221 |
| 18 | PRIVATISATION OF WATER SERVICES IN HISTORICAL CONTEXT, MID-1800S TO 2004 |
| | Petri S. Juuti, Tapio S. Katko & Jarmo J. Hukka ............................................................235 |
| 19 | CONCLUSIONS |
| | Petri S. Juuti, Tapio S. Katko & Heikki S. Vuorinen .......................................................259 |

PART III: MODERN URBAN INFRASTRUCTURE ..................................................................263

| 20 | INTRODUCTION |
|---|---|
| | Petri S. Juuti, Tapio S. Katko & Heikki S. Vuorinen .......................................................265 |
| 21 | HISTORY OF WATER SUPPLY AND SANITATION IN KENYA, 1895 – 2002 |
| | Ezekiel Nyangeri Nyanchaga & Kenneth S. Ombongi .....................................................271 |
| 22 | THE HISTORY OF WATER CONSERVATION AND DEVELOPMENT IN THE MWAMASHIMBA AREA IN THE BUHUNGUKIRA CHIEFDOM AND IN RUNERE VILLAGE, TANZANIA |
| | Jan-Olof Drangert ..........................................................................................................321 |
| 23 | PROVISION AND MANAGEMENT OF WATER SERVICES IN LAGOS, NIGERIA, 1915-2000 |
| | Ayodeji Olukoju..............................................................................................................343 |

24 EXPANDING RURAL WATER SUPPLIES IN HISTORICAL PERSPECTIVE:
SIX CASES FROM FINLAND AND SOUTH AFRICA
Petri S. Juuti, Tapio S. Katko, Harri R. Mäki & Hilja K. Toivio ............................................. 355

25 SISTER TOWNS OF INDUSTRY: WATER SUPPLY AND SANITATION IN MISKOLC
AND TAMPERE FROM THE LATE 1800S TO THE 2000S
Petri S. Juuti & Viktor Pál ........................................................................................................ 381

26 WATER SUPPLY AND SANITATION IN RIGA: DEVELOPMENT, PRESENT, AND FUTURE
Gunta Springe & Talis Juhna ................................................................................................... 401

27 WATER AND ENVIRONMENT IN ONE INDIGENOUS REGION OF MEXICO
Patricia Avila García ................................................................................................................ 411

28 THE HISTORICAL DEVELOPMENT OF WATER AND SANITATION
IN BRAZIL AND ARGENTINA
José Esteban Castro & Leo Heller ............................................................................................ 429

29 THE GEOPOLITICS OF THIRST IN CHILE
–NEW WATER CODE IN OPPOSITION TO OLD INDIAN WAYS
Isabel Maria Madaleno .............................................................................................................. 447

30 CASE OF TOKYO, JAPAN
Yurina Otaki ............................................................................................................................. 463

31 HEALTHY WATER FROM AN INDIGENOUS MAORI PERSPECTIVE
Ngāhuia Dixon .......................................................................................................................... 475

32 THE MEDICAL IDENTIFICATION OF NEW HEALTH HAZARDS TRANSMITTED BY WATER
Heikki S. Vuorinen .................................................................................................................... 489

33 CONCLUSIONS
Petri S. Juuti, Tapio S. Katko & Heikki S. Vuorinen ................................................................ 501

## PART IV: FUTURE CHALLENGES IN WATER SUPPLY AND SANITATION SERVICES AND ENVIRONMENTAL HEALTH ...........................507

34  INTRODUCTION
    Petri S. Juuti, Tapio S. Katko & Heikki S. Vuorinen ..................................509

35  AGING AMERICAN URBAN WATER INFRASTRUCTURE
    Laurel E. Phoenix ..................................................................................511

36  FROM WATER RESTRICTIONS TO WATER DEMAND MANAGEMENT:
    RAND WATER AND WATER SHORTAGES ON THE SOUTH AFRICAN
    URBAN LANDSCAPE (1983-2003)
    Johann W. N. Tempelhoff ......................................................................531

37  WATER AND NEPAL – AN IMPRESSION
    Sanna-Leena Rautanen ..........................................................................563

38  IMPROVEMENT OF THE LIVES OF SLUM DWELLERS - SAFE WATER
    AND SANITATION FOR THE FUTURE CITIES OF THE FOURTH WORLD
    Jarmo J. Hukka & Osmo T. Seppälä ......................................................575

39  CONCLUSIONS:
    DOES HISTORY MATTER? PRESENT WATER GOVERNANCE CHALLENGES
    AND FUTURE IMPLICATIONS
    Petri S. Juuti, Tapio S. Katko & Heikki S. Vuorinen ..................................589

EPILOGUE: LOCAL SOLUTIONS BASED ON LOCAL CONDITIONS
Petri S. Juuti, Tapio S. Katko & Heikki S. Vuorinen ..........................................593

ABOUT AUTHORS ......................................................................................599
INDEX OF PERSONS ..................................................................................609
INDEX BY PLACES ......................................................................................613
INDEX BY SUBJECT ....................................................................................621

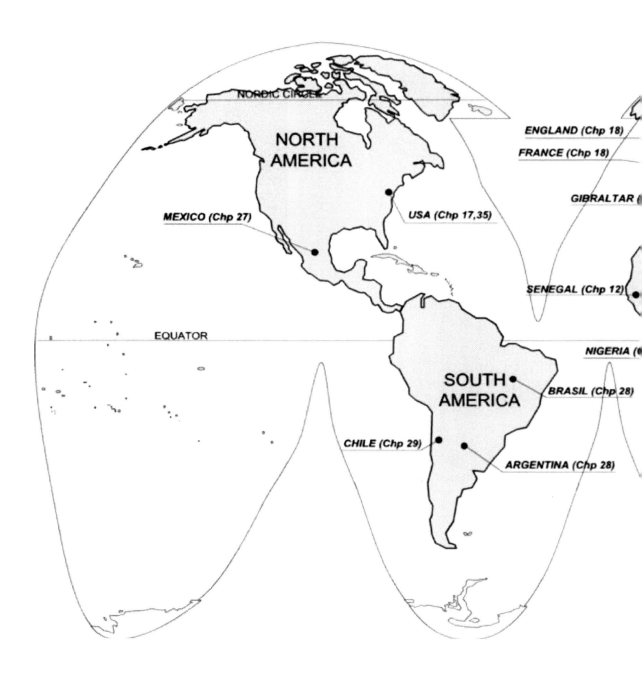

# 1
# FOREWORD

Water supply and sanitation are the primary points where our human activities connect and interact with the larger hydrological cycle of our planet. During the early days of human existence, these human connections were small and insignificant; human activities were to a large extent dictated by the constraints posed by the natural environment. With the explosive growth of the human race during the past century, the opposite became apparent; human activities began to strain and damage the environment at those points where water is abstracted and wastes are disposed. At the same time, there is an increasing awareness that water supply and sanitation are flip sides of the same coin, which demand a tightly integrated approach. Moreover, both are vital determinants of our quality of life. In the words of a popular slogan adopted by the South African Department of Water Affairs: *Water is Life; Sanitation is Dignity.*

This collection of expert contributions provides a wealth of compelling examples of the delicate and universal interaction between our human needs and our environment. Early man had to adapt his methods of war, medicine, agriculture and economic production to make best use of what nature provided, while modern man had to invent and apply new technologies to allow life at increasing urban concentrations that otherwise would be stifling. The case studies presented cut a broad swathe through history. They come from different periods — from the dawn of the first human settlements in Mesopotamia and extend up to the present day. They include examples from almost all regions on earth — from the arid parts of Botswana to the seasonally snowy hinterland of Finland. A wide variety of communities are covered — from small, specialised communities such as the Roman Army or the British contingent at Gibraltar, to the scale of large cities such as Riga and Lagos. A rich variety of natural and political factors is demonstrated at different places and times — from the tsetse fly in East Africa a hundred years ago to the most recent privatisation policies. The resulting tapestry of water-related histories woven by this volume is a pioneering benchmark for the contextualisation of other existing and future "histories" of water and sanitation. The many individual case studies on water and sanitation scattered throughout the literature come to new life when examined in the light of the broader patterns and historical process that are brought to the fore in this volume.

The authors and editors have filled an important gap and there is no question that this volume will both stimulate and guide future studies of our water history.

*Johannes Haarhoff*
Professor
Department of Civil Engineering Science
University of Johannesburg, South Africa
March 2006

Figure 1.1 Sammy Marks Fountain, Zoological Gardens, Pretoria, South Africa. This monumental fountain with cast-iron figures imported from Ireland to South Africa by Samuel Marks (1844 - 1920). It was installed in Church Square of Pretoria in 1906 and moved to present position in Zoological Park in 1910. The entrepreneur and senator Marks was born in Lithuania and arrived in South Africa in 1868. (Photo: Petri Juuti)

# 2
## "WATER IS THE BEGINNING OF ALL": GLOBAL WATER SERVICES AND CHALLENGES

*Petri S. Juuti, Tapio S. Katko & Heikki S. Vuorinen*

Figure 2.1 The Bembo Fountain, Kornarou Square, Iraklion, Crete, Greece from 1588 is a mixture of Venetian and Roman antiquities decorated with coats of arms and scenes. (Photo: Petri Juuti)

Water is life – and all life on earth is linked to water. Our existence is dependant on water – or the lack of it – in many ways and one could say our whole civilization is built on the use of water. Approximately 50 000 years ago modern man began to inhabit every corner of the world and people where constantly on the move. Occasionally people were troubled by pathogens transmitted by contaminated water, but all in all aversion for water that tasted revolting, stunk and was odious looking must have developed quite early on the evolutionary history of mankind. But 10 000 years ago mankind started to settle down and became sedentary agriculturists and this changed the relationship with water – settlements concentrated near lakes, rivers and wells. People and their cattle produced excreta, which started to pile around permanent settlements and eventually contaminated water. Earlier the problems arising from water which was polluted by human wastes had been easily avoided by moving the camp to another place. Farming, settling down and building villages and towns also meant the start of the problems mankind suffers this very day – how to get drinkable water for humans and cattle and how to manage waste we produce.

Acquisition of water from rivers, lakes, wells or springs has always been a daily task for human beings. The first evidence of water supply network emerged with cities of first ancient civilizations (Egypt, Sumerian and other civilizations in Middle East, Indus River Civilization, China). The ancient civilizations of New World (e.g. Maya, Aztec, Inca) also developed sophisticated water technologies. The Romans organised centralized system of aqueducts and collection of used water while they also used various types of small systems such as wells and fountains. In the Middle Ages water distribution was largely based on carried water by private water carriers or vendors. They fetched water mainly from well or fountains. Such systems played the key role until industrialisation when piped water and sewerage systems were established for growing communities. However, small systems still play a role in the 21$^{st}$ century in a modern society alongside with larger systems.

## 2.1. CHALLENGES

The World Water Development Report (WWDR) from 2003 noted several challenges for water management in the years to come.

*The challenges to life and well-being* cover the issues related to the ways we use water and the increasing demands we are placing on the resource. Signs of stress and strain are apparent across every sector: ecosystems, cities, food, industry and energy. With population growth and continuing pollution, these pressures are likely to increase. (http://www.unesco.org/water/wwap/targets/index.shtml)

Respectively, the *management challenges* - stewardship and governance – cover the issues how water's competing needs, uses and demands can be met. These challenges focus on the tools available or still to develop and to encourage an efficient and equitable use of the resource. As the Report pointed out: "Sadly, the tragedy of the water crisis is not simply a result of lack of water but is, essentially, one of poor water governance."

The WWDR report further states that "by 2030, over 60 percent (nearly 5 billion people) of the world's population will be living in urban areas. As a result, competing demands from domestic, commercial, industrial and peri-urban agriculture are putting enormous pressure on freshwater resources". (http://www.unesco.org/water/wwap/targets/index.shtml) Interestingly enough altogether 23 of the United Nations departments or agencies were involved in preparing the report, in other words almost all of them. This shows in a very pragmatic way how water management is very inter- and multidisciplinary by its nature. The next respective report for 2006 involves 24 of the UN bodies.

Since water management is as long as the history of mankind and water services as long as urban history, it is obvious that intra- and international historical comparisons and analyses could provide potential lessons to be learnt and utilised. These would also help us to understand better the present crisis and available future options. Such interest towards water history has indeed risen as can be seen through the establishment of the International Water History Association and international conferences such as the one in Crete in October 2006 to be organised by International Water Association, the home of the publisher of this book.

## 2.2. PRIORITY OF WATER SUPPLY AND SANITATION SERVICES

Water supply and sanitation services are also one of the ways to reach the millennium development goals (MDGs): reducing the number of people struggling with poverty and lacking several basic services by half by 2020. VISION 21 of the Water Supply and Sanitation Collaborative Council (WSSCC) aims at achieving "a world by 2025 in which each person knows the importance of hygiene, and enjoys safe and adequate water and sanitation".

Water management can be categorised by three major needs - water for people, water for food, and water for nature. The international discussion in the early 21$^{st}$ century points out, among others, the need for Integrated Water Resources Management (IWRM). Intepretations of the IWRM approach as good as it may be have, however, largely ignored various water uses purposes, their obvious priorities and potential conflicts. A recent study (Katko & Rajala 2005) shows how community water supply has practically in any culture the highest priority among various water use purposes. It is true that in many countries some 70 percent of fresh water is used for irrigation, often in inefficient ways. Yet, in addition to quantity even more important is the question of water quality. Thus water management is balancing water quantity, water quality and water use purposes and their priorities in varying local conditions. (See for example Juuti & Katko 2005)

Another water related debated issue is the global climate change (e.g. rising temperature, changing rainfall). In many connections it has been considered merely as an energy issue. Yet, the consequences of this cause changes in hydrological cycle and rainfall patterns, being therefore also a fundamental water question.

The early 21st century have experienced international water hazards in various parts of the world. We cannot overstress our consolation for all those innocent people that faced these tragedies - lost their lives, family members, shelters and/or ways of livelihood. However, it is also a fact that some 10 to 20 thousand people die every day in the global village due to water-related diseases – largely due to the lack of safe water supply and proper sanitation. Thus, the lack of these water supply and sanitation services is in fact the biggest catastrophe of the world.

## 2.3. FOCUS AND STRUCTURE OF THE BOOK

In this book the main focus is on water and sanitation services seen from various angles and disciplines while there are also chapters dealing with water resources management. The book reveals that history of water and sanitation services is strongly linked to current water management and policy issues - even its futures.

The authors of individual chapters were requested to view the issues of population growth, health, water consumption, technological choices and governance when appropriate. Therefore, the overall approach of this book can be summarised by figure 2.2. Pasts, presents and futures – indeed, there are many of these – and various interpretations of them, are interlinked with each other. Especially in water and sanitation systems we may have several decisions that through path dependence limit the available future options. The form of the pasts, presents and let alone the futures are symbolically more like ameba than rectangular shape. (Juuti & Katko 2005)

This book is by no means any final word on the history of water services and management, but perhaps one of first of its kind. It hopefully promotes the importance of history for present and future water management and governance as well as will have several other studies to be produced in the future.

The book is divided into four chronological parts although some of the chapters may cover several time periods. After the introduction of the book, the first part describes the water supply and sanitation services during the first urbanization in the ancient civilizations and focussing on ancient Greece and Rome. The second part of the book concentrates on the long 19th century (roughly 1700–1910 A.D.). The third part of the book deals with the modern urban infrastructure in the 20th century and the fourth part with the future challenges of water supply and sanitation services. In the end of each part the editors compile the key findings of the period. Finally some more general findings are presented by the editors.

The book has 29 peer reviewed chapters produced individually or jointly by altogether 33 authors representing several disciplines and research traditions. The book is based on voluntary contributions of individual authors and thus has a bottom-up approach. The editors have by purpose given only thematic guidelines while individual authors have used their approaches and ways of writing. By this way the book hopefully gives a better overview of the variety of issues and ways for managing water and sanitation services.

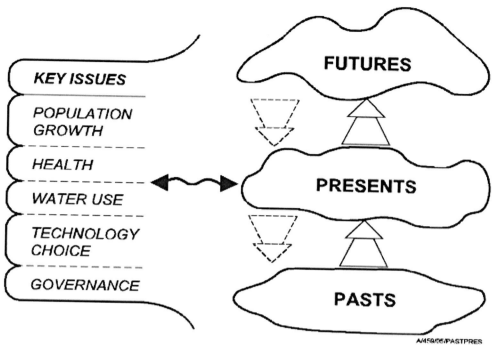

Figure 2.2 Approach for the book: interaction between the pasts, presents and futures and the related key issues.

We have also tried to have a good and balanced geographical coverage of water supply and sanitation services in the global village. By purpose the cases do hardly present capitals and major European cities but rather such cases which have been less written about, including smaller cities and townships. The majority of the cases are on cities and urban centres while some of the chapters also deal with rural areas. Besides, in water and sanitation at least by water sources there are often urban-rural interactions.

The length of a chapter varies depending on the subject. Although such a variety can be criticized we hope that this variety will serve the actual purpose of the book: through various cases showing the diversity of ways and issues in managing water supply and sanitation services and promote the interest towards long-term thinking. It hopefully also shows how several current ideas argued to be "new innovations" may have ancient roots. If we really wish to be interested in and have impact on our alternative futures, we have to know where we are and where we are coming from.

In addition to all the individual authors and their great interest we wish to acknowledge the interest and support from the IWA publishers. We also wish to thank the Academy of Finland (grand number 210816) for their financial support. Special thanks go to Ian Borthwick, Joan Löfgren, Sari Merontausta, Riikka Rajala and Katri Wallenius for their assistance.

In Tampere, Finland
*In the middle of crystal clear waters,*
World Water Day 23rd of March, 2006
*Editors*

## 2.4. REFERENCES

LITTERATURE:

Katko T.S. & Rajala R.P. (2005) Priorities for fresh water use purposes in selected countries with policy implications. *IJ on Water Resources Development.* Routledge. Vol. 21, no. 2. pp. 311-323.

Juuti P.S. & Katko T.S. (eds.) (2005) Water, Time and European Cities. History matters for the Future. www.watertime.net

INTERNET:

http://www.unesco.org/water/wwap/targets/index.shtml (accessed 6.12.2005)

# PART I

## EARLY SYSTEMS AND INNOVATIONS

Miserum est opus, igitur demum fodere puteum, ubi sitis fauces tedet.
[It is wretched business to be digging a well just as thirst is mastering you.]
– Plautus, Mostellaria (II, 1, 32) –

The round water basin in kings chamber, Knossos, Crete, Greece. (Photo: Elsi Maijala-Juuti)

# 3

# INTRODUCTION: EARLY CULTURES AND WATER

*Petri S. Juuti, Tapio S. Katko & Heikki S. Vuorinen*

Figure 3.1 Moros fountain from Iraklion, Crete, from 1600s. (Photo: Petri Juuti)

# INTRODUCTION: EARLY CULTURES AND WATER

Water shapes the environment, revising its history and future – thus to a very great extent it shapes our lives too. Humans have dwelled on earth some 200 000 years and during major part of it has lived as a hunter-gatherer, gradually growing in number and inhabiting every corner of the world. But archaeological findings or written sources concerning water and sanitation can be found only concerning relatively recent times. Thus in reconstructing the history of water and sanitation of this hunter-gatherer phase, we have to rely on analogies from later societies. The modern anthropological studies and recorded mythologies of indigenous people play an important role in forming these analogies and also observing primates and other more evolved mammals can give us useful information.

The earliest sites where a safe supply of water was found were springs and freshwater streams such as small creeks. Not just humans but also other mammals prefer flowing water. Some mammals even dig water holes for themselves. For example, elephants can dig quite deep pits for water in dry areas and seasons. The earliest type of well, the pit well or a deep water hole without any fortified walls, is the forerunner of the first properly constructed dug or sunk well. Water was lifted from the well using the means available at that time, first with a bucket or a similar vessel, or possibly with the help of a rope or other tools.

Humankind established permanent settlements about 10 000 years ago, when people adopted an agrarian way of life. This new type of livelihood spread everywhere and the population began to expand faster than ever before. This sedentary agricultural life made it possible to construct villages, cities and eventually states and for all these first settlements water was very essential part of life.

The oldest known written sources date back to about 5000 years ago, whilst archeological records extend roughly to the same era when the first great civilizations – Mesopotamia, Egypt, Indus culture, and China – also appeared. The history of water supply and sanitation in early civilizations is presented by Petri S. Juuti in his chapter titled, "First innovations of water supply and sanitation". In his chapter, he discusses various traditions and myths linked with water. For instance, people have prayed for rain, fertility, waterfowl and fish from Mother Water etc. A well or spring was considered to be a living creature and its spirit was believed to be frightened by noise or whistling. It got irritated if it was mucked up or dishonoured and that could happen if water was drawn from a well with a dish washing bucket. This is also well illustrated in the chapter "Water and Indigenous People in a White Settler Society: The Case of the Maori of New Zealand" by Ngâhuia Dixon in the third part of this book.

Religious beliefs connected to water served a good purpose even though the reasoning behind them was not what we would now call scientific. Nevertheless, they taught people to respect pure water and to adopt safe customs.

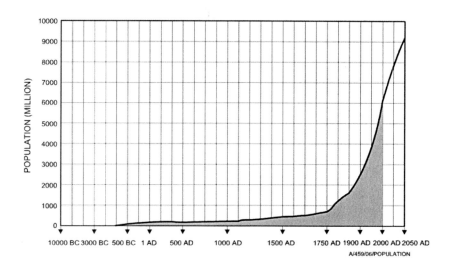

Figure 3.2 Human population growth 10,000 years ago to the present and estimation until year 2050.

Our knowledge of actual water supply and sewage dates back to the same era as of the early urban settlements (Figure 3.2). The earliest known permanent settlement, which can be classified as urban, is Jericho from 8000–7000 BC, located near springs and other bodies of water. In Egypt there are traces of wells and in Mesopotamia there are traces of stone rainwater channels from 3000 BC. It is estimated that Ur, one of the first and best known cities of Mesopotamia, had already in 2000 BC rainwater and drainage systems and the water closet was quite common in private houses. From the early Bronze Age city of Mohenjo-Daro, located in modern Pakistan, archaeologists have found hundreds of ancient wells and water pipes.

In the New World there was a wholly independent development of civilisation: the Inca, Mayan and Mexican cultures developed urban centres with relatively sophisticated systems.

Around the year 3000 BC the draw well with counterpoise lift was invented in Babylonia and for over 2000 years it was almost the only effective means of drawing water. The counterpoise lift was spread by the Greeks and Romans from the Near East to Central and Western Europe, Africa and eventually also to the North. In Egypt it was called a *shaduf* and was used to lift water from the river Nile. A windlass or winch was used when the well

Figure 3.3 Presumably the first European road from Knossos, Crete. (Photo: Petri Juuti)

was very deep and the counterpoise lift mainly in less deep draw wells or irrigation channels. The windlass and counterpoise lift were followed by windmill engines, crank reels and hand pumps. The first tube wells were built at the end of the 19th century. Besides technological development, ideas and philosophies about water also evolved.

In Europe, already in Minoan culture, water was considered to be holy. Minoans started the building of the first palace city around 4000 years ago. It evolved further and become presumably the oldest European city culture with advanced water and sewage systems and other infrastructure (Figures 3.1, 3.3 & 3.4). Later, the early Greek philosopher Thales (app. 624–546 BC) said: "Arkhen de ton panton hydor - Water is the beginning of all". It has been suggested that already the Etruscans had a highly developed water supply in their cities. Later the Romans used this knowledge in their own systems. From antiquity the best known are the gravitational water pipes or aqueducts built by the Romans. By the heyday of the Roman Empire, the water supply system of the city of Rome consisted of 19 aqueducts with a total length of 600 kilometres. Water was led through aqueducts to cities, where it was distributed mostly by a network of lead pipes. Private consumers were charged by the diameter of their own pipes. There were also numerous public toilets in Rome.

The importance of water for the health of people was a widely held view of ancient Greek and Roman writers. The role of water in health is described in the chapter "Water and health in antiquity: Europe's legacy" by Heikki S. Vuorinen. In the Roman Empire the destiny of the state relied heavily on the shoulders of soldiers. The army used water for many purposes. The chapter titled "Water Supply in Late Roman Army" by Ilkka Syvänne describes these uses circa AD 300-640.

Figure 3.4 Minoans considered water holy - sealife is natural object of art for sailor nation. Figure with doplhins originally from the Queen's Megaron, Knossos.

# 4

# FIRST INNOVATIONS OF WATER SUPPLY AND SANITATION

*Petri S. Juuti*

Figure 4.1 Key places mentioned in the chapter.

# FIRST INNOVATIONS OF WATER SUPPLY AND SANITATION

## 4.1. GENERAL BACKGROUND

The first major innovations of water supply and sanitation were probably the well and the toilet. Without these two simple but necessary innovations, human life and wellbeing face constant risk and nature is under serious stress. If these simple basic facilities are in good order, health problems and environmental risks can be avoided. Of course one should keep in mind that different kinds of water-lifting devices in the use of irrigation are also thousands of years old. The history of wells and toilets is as long as the history of permanent human settlements. What makes these innovations so important is the fact that wells and latrines are still in use and will certainly remain so also in the future. The well and the toilet are still the most common technical systems in the service of mankind.

Water is a necessity for human life and thus it also plays a vital role in many religions; springs and wells are frequently mentioned in their scriptures. For example, the first few books of the Old Testament have several descriptions of building, using, owning and securing wells. (Spier 1989, 34)

> 19. Isaac's servants dug in the valley and discovered a well of fresh water there.
> 20. But the herdsmen of Gerar quarreled with Isaac's herdsmen and said, "The water is ours!" So he named the well Esek, [dispute] because they disputed with him.
> 21. Then they dug another well, but they quarreled over that one also; so he named it Sitnah. [opposition]
> 22. He moved on from there and dug another well, and no one quarreled over it. He named it Rehoboth, [room] saying, "Now the Lord has given us room and we will flourish in the land." (Genesis 26:19–22)

There were disputes over wells and the right to use them, especially in the dry areas, where the well has been of vital importance. Toilets haven't aroused the interest of artists, philosophers or folklorists as much as wells. But it is not surprising that there were disputes over toilets, especially if they were too near a neighbor's border or house. Toilets have been found repugnant and in many countries and cultures talking about bodily functions is still a taboo. The caretakers of latrines are often seen as the lowest of all people. In poor living conditions the social status of tenants was visible in their location – the further one lived from the toilets the better status. Toilets have never been as appreciated a meeting place as wells, however, public toilets were a place for meeting people in ancient times. And although the well-tended outhouse with several adjacent seats in many bigger houses or factories offered a moment of rest in between heavy work. (Juuti 2003, 47–48; Juuti & Wallenius 2005, 25, 87.)

One of the oldest remaining references to the disposal of human waste is found in the Old Testament:

> 12. Designate a place outside the camp where you can go to relieve yourself.
> 13. As part of your equipment have something to dig with, and when you relieve yourself, dig a hole and cover up your excrement. (Deuteronomy 23:12-13)

Figure 4.2 Water carriers in Cape Town, South Africa in early 1800's. Detail from the drawing by Anne Barnard. (Source: NAR/AG 15691)

Emptying toilets and carrying water (Figure 4.2) were—and still are—considered to be unpleasant tasks. These tasks were usually assigned to the lowest social group available—women, children and slaves. The task of water carrier was also used as a punishment, for example: *"They continued, "Let them live, but let them be woodcutters and water carriers for the entire community." So the leaders' promise to them was kept."* (Joshua 9:21)

Until the birth of water and sewage works, toilets and wells were the main technological innovations in the service of the humankind. These simple solutions have been in general use for thousands of years and are today not radically different from those used in the past. For example, most people in Europe and the USA took their drinking water from private wells until the last quarter of the 19th century. It must be noted that also public fountains were used in many countries till the mid 1800s and even later in some cases. (Figure 4.3) The needs of fire fighting, businesses and industries, real estate owners, and health authorities hastened the birth of water works. (Keating 1989; Juuti 2001; Juuti & Katko 2005)

This chapter describes the general development of wells and toilets up to the early 20th century.

## 4.2. WATER HOLES AND FIRST WELLS

It is impossible to say for sure where the first human-made well was. It is easier to search for wells that still exist. There are several remains of wells from the Neolithic era. For example, two very old wells have been found in Israel and Cyprus. In northern Israel, a well that is roughly 10,000 years old was found in the Pre-Pottery Neolithic B settlement of Atlit Yam. The well was constructed by dry-stone walling, and its diameter is 1.5 meters with a depth of 5.5 meters. In western Cyprus, an old well has been found at Kissonerga-Mylouthkia. It is about 7-8 meters deep and from the same era as the Atlit Yams well. (arcl.ed.ac.uk/arch/annrept/report99/; witwib.com/Atlit_Yam, accessed Aug 24, 2005)

Figure 4.3 Old public well from Vaasa, Finland and author. Well was built in late 1800s. (Photo: Tapio Katko)

While wells were built already 10 000 years ago, permanent places for drawing water are even older than that. The oldest constructions relating to the use of water were used for irrigation. Springs and wells have not just been places to get water but also important meeting points for thousands of years. They have also served as boundary markers, places of worship and many other purposes.

The earliest sites where safe supplies of water were found were springs and freshwater streams such as small creeks. Not just humans but also other mammals prefer flowing water and some mammals even dig water holes themselves. For example, elephants dig quite deep well-like holes in dry areas. An average elephant needs approximately 160 liters per day; therefore the need for an adequate water source is obvious. If there is no water available, for example, in the river, they dig wells in the dried riverbeds with their trunks and bring the water up. In the dry seasons also other mammals use elephant wells. So the pit well, a deep water hole without any fortified walls, is the forerunner of the dug well. Water has been taken from this sort of well by whatever means were accessible, usually just using simple vessels. One possibility was to form a chain of water carriers—this enabled the drawing of

Figure 4.4 A draw well with counterpoise lift from Juupajoki, Finland. (Photo: Petri Juuti)

water from deep in the ground without advanced technology. In this way it was possible to reach water lying tens of meters deep. Water has been lifted from the dug well using the means available at that time, first with a bucket or a similar vessel, or possibly with the help of a rope or other tools.

Approximately 3000 BC the draw well with counterpoise lift (Figure 4.4) was invented in Babylonia and it was for over 2000 years almost the only effective means of drawing water. The counterpoise lift was spread by the Greeks and Romans from the Near East to Central and Western Europe and eventually also to Scandinavia.(Toivonen et al. 1981, 42; Katko 1996, 26.) In Egypt it was called a *shaduf* and was used to lift water from Nile. Traditionally a draw well was built from wood, but some iron fortification might also have been used. However, the column, the counterpoise lift, the bucket pole and the bucket were wooden. If there was a need for extra weight to counterweight the bucket, it was usually made of a heavier material.

## 4.3. MONEY DOESN'T STINK – EARLY TOILETS AND WELLS

The earliest known permanent city-like settlement was Jericho, dating from 8000–7000 BC, which was near springs and other bodies of water. In Egypt there are traces of wells from 3000 BC and in Mesopotamia there are traces of stone rainwater channels from same era. It is estimated that one of the first and best known cities of Mesopotamia, Ur, had already in 2000 BC rainwater drainage systems and the water closet was quite common in private houses.

The Indus civilization, from its emergence in the early third millennium BC to its collapse in the early second, is best known from the excavations of Mohenjo-Daro and Harappa, located in modern Pakistan. Between these two cities lie several hills where the remnants of other cities of the same culture can be found. In the Bronze Age city of Mohenjo-Daro, there are still to be seen ancient wells and water pipes. The Indus culture was highly developed and similar to Mesopotamia and Egypt. Houses were well built and spacious, as were baths, which had brick sewers. (Gray 1940, 939–946)

Many houses in Mohenjo-Daro had a well inside the building. Wells were usually round, sometimes oval shaped, the floor level was stone or brick and the brick lining went deep down. There were altogether some 700 wells and the average distance to the nearest well was only 17 meters. This is an absolutely unique example in the history of water supply. The main innovation was nevertheless storing the water in advance where it was needed in the city for easy access and use.

Almost every house in Mohenjo-Daro had a toilet, placed on the street side of the house, so that wastewater was easy to dispose of into street gutters. Washing places were right next to the toilets. Bath and household water, wastewater from toilets and rainwater were usually led through brick-lined pits and then through the outlet to the street gutters instead of leading them there directly. In some cases the gutters were situated too close to the wells and it is possible that wastewater contaminated the drinking water. The ruins of Mohenjo-Daro depict a community in which private and public hygiene were maintained effectively and the water supply of which was adequately protected from contamination. (Foil et al. 1993, 1-7; Gray 1940, 939-946; Wijmer 1992, 12; Jansen 1993, 17) The Indus culture is a prime example of a functional solution to sanitation problems in early times.

Figure 4.5 Minoans had separate sewer systems in Knossos. Rainwater was collected with this kind of "stone drains". (Photo: Elsi Maijala-Juuti)

Figure 4.6 Old and new technology side by side in Knossos. Clay water pipes from Minoan time some 4000 years ago and bottle top of water bottle on the left upper side of pipe. This is part of of water system that channelled drinking water from a source in the vicinity of the palace (nearby mountain). This took form of a series of clay pipes with one end narrower than the other so that they could fit together, thereby increasing the water pressure and allowing it to flow more easily. (Photo: Petri Juuti)

Figure 4.7 Minoan toilet (left) from Knossos and drain (right). (Photo: Petri Juuti)

In Crete, the Bronze Age Minoan culture (circa 2600-1100 BC) was highly advanced. In the capital city Knossos there were three different water systems; one for drainage, one for rain water and one for drinking water. (Figure 4.5) Drinking water to palace was taken from nearby mountain via 1,6 kilometers long clay pipes. These water pipes are probably the oldest still remaining ones and approxmately 4000 years old. (Figure 4.6) There were also several toilets (Figure 4.7, 4.9) using water to flush wastes to the river. In Crete other city Phaistos relied on rain water because there was no natural springs. Rain water was collected to several cisterns. Water was also in central role in Minoan mythology. (Kakudakis 17.10.2005; Juuti, P. October 2005; Vasilakis, 69-70) Read more about this systems from next chapter by Vuorinen.

Good insight into ancient wells is offered by the over 90 meter-deep Bir Yusuf, also known as (aka) the Spiral Well (aka Joseph's Well, aka Well of Saladin) well, in Cairo, where the water was lifted and hauled up with a long line of carriers and later with a wooden water wheel. The well was dug in the limestone rock during the Graeco-Roman era, but when Sultan Salah el Din (Saladin, aka Jusuf, aka Joseph 1171-1193 AD), ordered his engineer Bahaaeddin Ibn Qaraqosh to construct the Citadel of Cairo, the place where the well was situated was within its walls. The well was also enlarged and it consisted of two levels

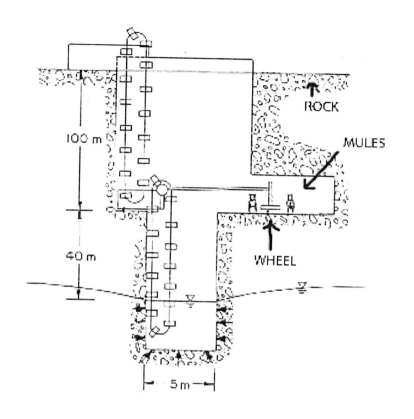

Figure 4.8 Sketch of the Joseph's well. (Source: Hendricks 1991. Unpublished manuscript.)

connected via a spiral staircase. A narrow path begins in the upper part and continues down to 50 meters deep, intended for mules to be able to go all the way up and down the ramp to the cistern. Within the shaft is a platform on which a water wheel, *saqiya*, stood. The water wheel was powered by animals, mainly by oxen, and brought the water to the surface. In figure 4.8 is a rough sketch of the well. (http://www.warlinks.com/cairo/cairo.shtml; http://weekly.ahram.org.eg/2003/640/heritage.htm)

There are several descriptions of the well in the memoirs of travelers and also in novels. An American, J. L. Stephens, described the well early in the 19th century as follows:

> From hence, passing around outside the walls, I entered by the Gate of the Citadel, where I saw what goes by the name of Joseph's Well, perhaps better known as the Well of Saladin. It is

Figure 4.9 Drains from Minoan Knossos, Crete. (Photo: Petri Juuti)

forty-five feet wide at the mouth, and cut two hundred and seventy feet deep through the solid rock to a spring of saltish water, on a level with the Nile, whence the water is raised in buckets on a wheel, turned by a buffalo. (Stephens 1837)

Also Mark Twain wrote in 1869:

[...] I shall not tell of Joseph's well which he dug in the solid rock of the citadel hill and which is still as good as new, nor how the same mules he bought to draw up the water (with an endless chain) are still at it yet and are getting tired of it, too [...] (Twain 1869)

The best known infrastructure constructions from antiquity are the gravitation water pipes or aqueducts constructed by the Romans. In the sixth century A.D. in the system of the city of Rome there were altogether 19 aqueducts and their total length was 600 kilometers. Water was led through aqueducts to cities, where it was distributed into a network of lead pipes. Private consumers were charged by the diameter of their pipes. It is estimated that around

100 AD the citizens of Rome got approximately 67 liters of water per day. (Bruun 1991) These systems are described later in this book. (See, for example, chapter 5 by Vuorinen and chapter 6 by Syvänne.)

The first toilets did not require technical construction; they were just holes in the ground. There were private and public toilets already in ancient towns. Some of the public toilets were free of charge, some not. A story about Roman Emperor Vespasian (ruled 69–79 AD) tells how he ordered that there be a charge for using public toilets. When he was criticized, he answered:

*"Pecunia non olet"* – *"money doesn't stink"*. Another version of the story tells how Vespasian's son Titus (Emperor in 79–81 AD) criticized his father for issuing a urine tax. Vespasian waved money from the first payment under his nose, asking him whether he found the smell offensive. When he said he didn't, Vespasian replied, "And yet it comes from urine." Hence the saying, "The smell of profit is good no matter what the source".

There were numerous public toilets in Rome. Under the seats was running water and it took the excrement through the sewer network into the Tiber. For wiping, people most likely used a sponge on a stick. At home most people probably used chamber pots or commodes and emptied them down the drain or onto night soil wagons. Night soil wagons went to surrounding agricultural areas to dump the excrement on the fields.

After the Roman Empire, all of society in Western Europe changed, including systems of water supply and sanitation. In medieval cities the water supply was based on wells, some of which were located outside the cities. In Tallinn, Estonia, the services of water carriers were paid by the town council in the 1330s. Tallinn had adopted the so-called Lübeck Act in 1257. It stated how toilets and pigsties were to be located in relation to streets and neighboring bounds (Kaljundi 1997). Castles and fortresses usually had their own wells. There were some occasional water pipe projects even in this era. It was also a period in which private water enterprise developed, when the vendor took water to his customers with a donkey cart. In the 18th century the water supply of Paris was taken care of by thousands of water vendors using buckets. In Cairo there might have been even more water vendors in medieval times. (Ponting 1991, 349; Coffey & Reid 1976, 133) They are still seen on the streets of Cairo, mainly to attract tourists. However, for the slums of the developing countries, it's still a reality even in the new millennium.

Although neglected in Europe, toilets were not forgotten: the basic need remained. Outside Europe they still developed further. For example, in Japan the earliest known toilets were in use in the 700s A.D. They were pits, as well as more advanced toilets consisting of a ditch carrying water through part of the house to convey the waste outdoors. Toilets were also built over running streams. Pit toilets came into widespread use over the following centuries. In the thirteenth century the Japanese began to use night soil from the toilets as fertilizer. Traditional Japanese toilets consist of a hole or basin in the floor and are not made to sit on but to squat over. Sewerages and seated toilets were introduced in Japan around the beginning of the twentieth century, but it was only after World War II that Western-style toilets began to spread on a major scale. Using human night soil as fertilizer was banned for sanitary reasons, and flush toilets became common. Still, Western-style toilets did not fully replace Japanese-style ones, and even now the majority of toilets at train stations and other public facilities are the squatting type. This is probably largely because many Japanese prefer outside of the home not to sit directly on toilets, which they suspect may not be hygienic. (http://web-japan.org/kidsweb/techno/toilet/history.html) See more about this matter from chapter by Otaki later in this book.

## 4.4. TYPES OF WELLS AND TOILETS

Well water is groundwater, developed by the absorption of rain and melt waters into the ground, where it sinks through different porous layers of soil and the cracks of the rocky foundation under the influence of gravity. This percolation purifies the water and minerals dissolve into it. Groundwater is usually best in the core of the ridges, while in the rim areas high levels of iron and manganese may occur. Thus it is important that the filtering layers are clean. The level of groundwater varies by season depending on the amount of rainwater and the melting of the snow and soil frost. The best groundwater resources exist in sand and gravel deposits. Virgin groundwater is pure, for it is naturally quite well protected from any pollution. However, it is a slowly renewing natural resource, so it should be protected carefully.

There are several ways to categorize wells; one way is to sort them in types in chronological order, taking into account the method of lifting water (Table 4.1).

The simplest well is the bottomless barrel or a vessel that is sunk into a natural spring. These so-called spring wells (Table 4.1, type A I) use the natural overflow of groundwater. There are, of course, also different kinds of springs. However, they are not within people's reach everywhere and compared to the dug well, the spring well is much less widely used. Building a well like this doesn't require much work, but the construction is very vulnerable and if it is the only source of water, much harm can result if it gets ruined.

A pit well (type A II) is an evolved version of a watering hole, which can even be dug and used by some animals. Sometimes a drinking hole is enlarged over decades or even centuries when its users dig it deeper because the groundwater level gets lower, thus turning it into a pit well. This type isn't really planned or built.

Table 4.1 Types of wells in chronological order.

| TYPE OF WELL | METHOD OF LIFTING | TECHNICAL REALIZATION |
|---|---|---|
| A I Natural Spring & bottomless barrel in spring | Hands, scoop, bucket | No construction or very simple construction |
| A II Pit well | Hands, scoop, bucket | Pit in the ground, no construction or very simple construction |
| A III Dug well | Well & rope, bucket pole, hand pump | Place and construction planned, built shaft, place carefully chosen by observing the terrain |
| A IV Draw well, counterpoise lift | Counterweigh | As above, extra planning to lifting |
| A V Windlass well | Winch or reel | As above, extra planning to lifting |
| A VI Tube (a) & drill well (b) | Pressure of the groundwater formation or pump | Pipe is pushed into the ground (tube well) or pipe is drilled on rock foundation (drill well), place and construction planned, requires precise knowledge of groundwater location |
| A VII Wells with wind engine | Wind power, rotor | Lifting power directly from the site of well |
| A VIII Wells operated with engine | Combustion engine, electricity | Power from outside source |

Table 4.2 Types of toilets in chronological order.

| TYPE OF TOILET | TIME PERIOD OF GENERAL USE (NOT FIRST TIME) | METHOD | CONSEQUENCES |
|---|---|---|---|
| B I Hole in the ground "pit toilet" | All periods of human existence | None or covered with soil | Waste won't compost |
| B II Water toilets, no flush | Ancient times and cultures | Running water | Wastes to bodies of water |
| B III "Privies" | Castles in the Middle Ages | No waste treatment | Leakage in ground or into body of water, environmental hazards, wells and watercourses endangered |
| B IV Outhouses | From 1800s to present | a) no waste treatment<br><br>b) centralized collection of waste | a) leakage in ground or into body of water, environmental hazards, wells and watercourses endangered<br>b) depends on further treatment |
| B V WC | From 1800s to present | a) waste flushed and led into watercourse<br><br>b) waste collected into precipitation tank (one- or multiple-piece), heavier matter sinks to the bottom<br>c) waste flushed into closed tank, then collected and transported to the network of sewer works<br>d) WC with filtering on the ground, often with precipitation tank<br>e) WC with small, local treatment plant<br>f) WC connected to the sewer network and a wastewater treatment plant | a) catastrophic; watercourse and in worst case drinking water contaminated and polluted<br>b) refinement only partial<br><br><br>c) good<br><br><br><br>d) result varies<br><br>e) result varies<br><br>f) advantages of bigger units: better treatment result |
| B VI Dry toilets & ECO-toilets | 1800s and again from late 1900s to present | Composting | Controlled recycling of nutrients, separation of urine |
| B VII Other type of toilets | from late 1900s to present | Waste treated with chemicals, burned, frozen etc | Depends on further treatment, nutrients wasted |

Figure 4.10 Medieval, wooden toilet seat from Turku, Finland. (Photo: Petri Juuti)

## WELLS AND TOILETS IN THE CASTLES

Some of the oldest still remaining wells and toilets in the world were constructed in castles. In most cases superterranean parts of wells and toilets were destroyed in fires, but the remaining ones were mainly built from stone. They are usually quite similar around the world in terms of their location and construction. For instance, a well was placed usually in the main castle, either in the inner yard or in the basement, to protect it from the enemy. The elite used privies found in the castles. For example, in English castles in the 15th and 16th centuries, privies were placed over a vertical shaft leading to the moat. Common soldiers and servants used much more modest places for their daily needs. (http://www.bottlebooks.com/privyto.htm)

The old castle of Calmar (Sweden) has a well, with a depth of 15.5 metres and a diameter of 2.5 meters and a potential water mass of 60 cubic meters. (Sinirand 1992, 7-25; Hörberg 1997, 17) Public wells (aka joint wells) are mentioned in Stockholm and Calmar in the 15th century, but they may have existed even earlier. In Stockholm there were at least three joint wells in the 15th century. (Bjur 1988, 15-16)

In Finland, which was part of the kingdom of Sweden until the year 1809, Häme castle was founded at the end of the 13th century and it is one of Finland's medieval royal castles. Its first well was built at the same time as the castle and it was 12 metres deep and lined with stones. (Kilkki 1973) The impressive old well can still be seen near the east corner of the fortress and it serves as a "wishing well" for visitors. However, already soon after its construction waste and rainwater leaked into the well, polluting the water and it could be used only for firefighting purposes. A new well was built but its location is unknown.

# FIRST INNOVATIONS OF WATER SUPPLY AND SANITATION

Figure 4.11 King's privy from Turku Castle, Finland. (Photo: Petri Juuti)

Naturally a toilet was needed too in Häme castle and presumably it was attached to the encircling wall on the lake side. The so-called "himmelhuusi" ("sky privy"), built outside the walls was mentioned in the inventory from 1687, but hasn't survived to our day. (Ailio 1917, 182–183; Kilkki 1973, 14; Katko 1996, 27) The only indoor toilet of the main castle, which still exists, is located near the clerk's office on the middle floor. (Vilkuna 2001, 52) Only a few privileged people had access to this indoor toilet with a wooden seat—such as the clerk, who was among the highest ranks in the castle.

In the Turku castle, founded at the mouth of the Aurajoki River in southwestern Finland in the 1280s, were several toilets. From the 15th century onwards there was a privy in the gatekeeper's chamber located in the corner of the gate tower. One was also needed in the prison and yet another was located in the north wing. These toilets were connected to the same toilet drainage system and formed an independent system. The king, other noble residents and high officials had their own privies. One is still left in the medieval great hall. (Puhakka & Grönros 1995, 40, 48, 57) These privies were usually constructed on top of the corbels or supportive beams, as sort of a bay toilet, being located partially outside the wall. In the Turku castle, most of these privies in the quarters of the aristocracy were built completely inside the walls (Figure 4.10, 4.11). In the 1540s near the king's hall the chambers of the young noblemen were built with access to the bay toilet built on beams. Similar privies were attached to the castellan's chamber and the queen's hall. (Puhakka & Grönros 1995, 59, 63, 75) None of this type of wooden privies has survived to the present day; fires and time have taken their toll, but they can be seen in the scale model of the castle. However, other types of privies have been preserved quite well in this castle.

Figure 4.12 Bay privies attached to towers (Olavinlinna castle, Finland), entrance, the scenery downwards and the seat (constructed later). (Photo: Petri Juuti)

The massive Olavinlinna castle (Figure 4.12), nowadays located in the city of Savonlinna, southeastern Finland, was founded on a steep, rocky islet in the Kyrönsalmi strait. Work on the building site was started in 1475 and by 1483 the main castle reached its form, although the construction work still continued for a long time thereafter. (Sinisalo 1986, 7–8; Laamanen & Prusi 2004, 5–10) Olavinlinna castle is unique for not having a single well in it – it was impossible to build one in the location. Instead the surrounding waters of the strait were used.

Olavinlinna castle with its five towers was built from the point of view of defence. The towers of the main castle were extended upward during the second main building era in the mid-16th century. These round artillery towers, or rondels, were originally used as living quarters for the aristocracy. (Sinisalo 1986, 8–9; Laamanen & Prusi 2004, 5–10.) The "bay privies" (Figure 4.13) made of stone are attached to the walls of rondels. Originally there was only a horizontal log for seating, but later on proper wooden seats were built. At times, ventilation might have worked even too effectively. An old anecdote describes these privies as the first water toilets in Finland—bay privies were above the water and the height of the drop was great: the towers of the main castle were over 20 metres high.

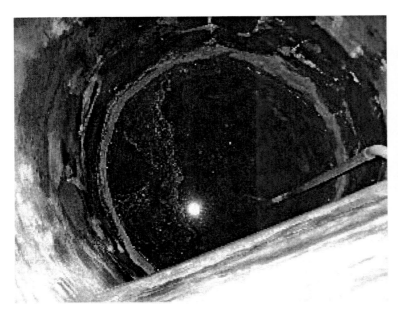 

Figure 4.13 Dug well and reel, oldest still existing in Cape Town, South Africa. (Photo: Petri Juuti)

Three of these privies still remain. Walls and slope roofs were built of grey granite on supporting console beams. (Laamanen & Prusi 2004, 58) The choice of building materials has preserved these examples to our day, although usually they have been destroyed in fires—the frequent threat of castles.

At the opposite side of the globe, in South Africa, the castle of Good Hope in Cape Town provided water from its wells. Basically this castle has one big dug well from the 1600s and a couple of smaller ones, the biggest one with a reel (Figure 4.13). The walls of the big well were made from stone, which was quite typical also elsewhere around the world. (Juuti & Wallenius 2005, 12-15)

This well is the oldest well in existence in South Africa, dating back to the year 1682. The well was situated in the centre of the castle and it is still accessible to the interested. In 1691, a wall was constructed across the castle courtyard and the well was left inside it. The well is about ten metres deep and two metres across. (Werz 2002, 95-97) Later, around the beginning of the 1700s, it was operated by hand pumps. Water was lifted by these hand pumps to an iron water tank with a capacity of about three cubic metres and from there via pipes to several places where water was needed most. The smaller wells were in the kitchen and in the yard. (Juuti 2004)

## 4.5 TOILETS AFTER ROMAN ERA

The lack of proper sanitation increased the effects of epidemics in medieval towns. Sewers deteriorated and it wasn't until the mid-19th century when scientific knowledge and civil engineering coincided with the needs of public health care and modern wastewater treatment and management were born. The population of the megapolis Rome collapsed in the 500s to 30,000. The situation was just as bad elsewhere too. In a gathering called together by the Emperor of the Holy Roman Empire in 1183, the floor of the (main) great hall collapsed and the noblemen and knights dropped into the cesspit, where many even died. Emperor Frederick I (Barbarossa) himself was barely saved. (Foil et al. 1993, 1-7; Gray 1940, 939-946)

Usually towns built few modest public latrines for the inhabitants, but these were mostly inadequate for the size of the population. For example, London had only sixteen public latrines in a population of 25-30,000. Quite often towns built these latrines near bodies of water, for example projecting out over a river. In Helsingör, Denmark, the emptying of the cesspits belonged to executioners. The executioner apparently didn't do his job well, for a Dutchman living there annoyed the locals by emptying his own toilet, when the city officials were unable to do it after many efforts. In any case private toilets were needed, and they were dug in the backyards and even under the apartments. The contents were dumped in a local lake or river or used as manure in the farming areas. (Foil et al. 1993, 1-7; Gray 1940, 939-946; http://the-orb.net/medieval_terms.html)

The streets of medieval London and Paris were almost full of waste. Residents emptied their potties out of the windows and only the lucky and fast could avoid the spillage. Common and upper-class people had equally as bad manners. For example, on the 6th of August 1606 all the residents of St. Germain Palace were forbidden from urinating inside and the very next day the king's son urinated on the wall of his room. A regulation from the year 1531 in Paris required landlords to build a toilet in every house, but it wasn't really enforced. In the era of the great French Revolution, the number of public toilets was great, but they were so disgusting that people avoided using them. Instead, the terrace of Tuileriers was favoured and eventually it became so filthy too that a public toilet was built in the area charging a fee of two *sou*. People were enraged at the high price and started to use the grounds of the royal palace and thus forced the Duke of Orleans to build a dozen new lavatories, which were kept in better condition.

In Berlin refuse was piling up in front of St. Peter's church, until the law from the year 1617 required that each peasant visiting the city had to take a load of refuse when leaving. The little existing information on using clay pipes reveals that they were mainly used in the sewage systems of castles and the manors of local noblemen and gentry. (Foil et al. 1993, 1-7; Gray 1940, 939-946; http://the-orb.net/medieval_terms.html) Emptying one's bowels on the street is not just a problem nowadays – and not done only by dogs, but also by people.

# FIRST INNOVATIONS OF WATER SUPPLY AND SANITATION

Figure 4.14 Chamborg Castle and well from Loire Valley, France. Attached to walls cherubs and gargoyles spilled out rain water. (Photo: Petri Juuti)

Development was not straightforward in this matter, and some innovations were made even earlier. Already Leornardo da Vinci (1452–1519) made a sketch for a flushing water closet for the castle of Francis I at Ambrose. His drafts include flushing channels inside the walls, and also a ventilating system. These plans were never realized. Leonardo worked at the time also with staircase plans, which were finally built in Chamborg Castle (Figure 4.14, 4.15) and the spiral staircase is considered as one of Leonardo's masterpieces. Castle itself was completed around 1547 when Francis I died. A few decades later an Englishman, Sir John Harrington, wrote in 1596 an article titled "Plan Plots of a Privy of Perfection" in which he described a flushing water closet and shortly after that erected one in Kelston, England. The system was working technically well and thus Queen Elizabeth I allowed Harrington to install a water closet in the Royal Palace. Elizabeth was very happy with her new accessory. (http://www.masterplumbers.com/plumbviews/1999/toilet_tribute2.asp)

An Englishman, Joseph Bramah, is often named as the developer of the first actual water closet. He got a patent in 1778 but also several other names are mentioned as a developer of the first toilet, e.g. Thomas Crapper, Alexander Cummings and Henry Moule. All these and several others developed similar systems in parallel.

It took several centuries before the toilet became a common component in ordinary houses. From England this innovation gradually spread to the continent of Europe and elsewhere. For example, in the mid-1800s the water closet came to Hungary. (Péter 2005)

In Sweden a debate about installing and connecting water closets to water and sewerage systems began in the year 1880 and permission was granted to install WCs in 1883. Finally, in 1904, general permission for flush toilets using piped water was given, and in 1909 it was allowed to connect water closets to the municipal sewerage system. (Katko & Stenroos 2005) In Finland WCs were allowed around the year 1900. Some unofficial installations already existed in both countries before these dates.

The water closet was seen as a victory for public health care, because excrement was moved away immediately from houses and yards. (Harjula 2003, 41) But not everybody rushed to exploit this new invention. For example, some wealthy burghers could hire people cheaply from the countryside and when there was a servant to empty the pots in the morning; there was no hurry to get a water closet. In these circles the WC and other facilities that eased everyday life weren't acquired until the lady of the house was alone or there was only one servant taking care of the household. (Brunow-Ruola 2001, 234) But as the WC was starting to spread, in workers' quarters it was still considered progress just to get an outhouse. (Jutikkala 1979, 147)

The WC gained popularity in the United States after World War I, when American troops came home from England. They were talking about a "mighty slick invention called the crapper." However, one of the first flushing water closets had been installed at Harvard University already in 1735. (http://www.masterplumbers.com/plumbviews/1999/toilet_tribute2.asp)

It is surprising from environmental point of view to see that some authors still see the WC as the great improvement that even prevented diseases:

> The flush toilet is an invention of which humanity can be very proud. Without this marvelous contraption, disease would still be rampant and water supplies throughout the world would be undrinkable. (http://www.masterplumbers.com/plumbviews/1999/toilet_tribute2.asp)

In fact, this so-called English toilet is the most dangerous type of the toilet for the environment and health. This water closet, WC, is connected to a sewer without wastewater treatment facilities. This kind of system has caused several fatal epidemics and the pollution of lakes and rivers. (See, for example, Koskinen 1999; Juuti 2001; Juuti et al. 2003a; Juuti et al. 2003b)

The most environmentally friendly type of toilet is the compost toilet (Table 4.2, B VI), especially the dry compost model in which urine is collected separately. Urine diluted with

# FIRST INNOVATIONS OF WATER SUPPLY AND SANITATION

Figure 4.15 Medieval well with reel (type A V) and tower, Castle Chenonceau, Loire Valley, France. (Photo: Petri Juuti)

water can be used as fertilizer and composted solid waste can be used for soil improvement. The annual amount of urine of one individual could be used to produce 200 kilograms of grain. This method not only recycles the nutrients in the urine but it also prevents them from getting into the groundwater and watercourses. The whole process is manageable by people themselves. The separating model also has a less distinctive smell. It is notable that in the 19th century there was already dry compost and compost toilets in cities combined with different transport systems. Choosing the water closet as the primary system in the late 19th and early 20th centuries ended the product development of dry compost and compost toilets for over a hundred years. (Juuti & Wallenius 2005; Mattila 2005)

Throughout history in the rural areas a separate toilet lacked or it was a modest structure, which is still true to many people in many areas of the world. For instance, in coastal areas and northern parts of Finland, a compost heap was used for relieving oneself, whereas in inland areas it was the field and in Ostrobothnia the manure shed. (Katko 1996, 34) It was quite common to do one's business just behind the back door. (Kangasalan historia 1, 495.) In general it was considered proper to go further from the house, well and away from other people's sight.

Later on the toilets were often placed near the cowshed. They were also built above the dung pit of the cowshed. Sometimes a similar method was used in a town, for example in Hamina in Finland. (Juuti, P. July 2004; Jokinen 1994, 121) This kind of location was natural since the amount of manure was much greater than human excrement. According to calculations from the 1920s by engineer Ben Mitro, the annual accumulation of solid excrement in Tampere, Finland was 44 kilograms per person. (Juuti 2001, 200-201) Depending on the way of making calculations and the conditions, several other estimations have been made both earlier and later. A good rule of thumb to estimate the annual amount of waste per person is approximately 50 litres of excrement and 500 litres of urine.

In the 18th century manors and parsonages in Finland and Sweden started to have toilets in separate structures (type B IV). In some parsonages toilets were also built in the entrance hall. (Katko 1996, 34) It was exceptional in that era and must have bewildered the locals, for in the countryside still in the mid-19th century it astonished people to "do their business" indoors. The Farmer's Handbook from the year 1863 recommended to build a separate toilet, but this custom spread slowly. (Katko 1996, 35)

Manure as well as human excrement were the most important fertilizer substances before artificial fertilizers were introduced. For that reason it was necessary to deposit excrement on the field for further use as a fertilizer. (Maamiehen käsikirja 1945; Juuti & Wallenius 2005, 50; Maijala 25.8.2004) It was important to make sure the fertilizing substances went to the right place.

Cattle needed a lot of water, therefore the consumption of water was high in farms. At the end of the 19th century there was still some livestock in city houses in the least industrialized countries (excluding the biggest cities) and even as late as the early 20th century in small towns. In some parts of the world it is still common to have domestic animals in townships. As the old phrase goes: "*A cow is the best insurance*". (Maamiehen käsikirja 1945, 339, 348, 353-355, 361)[8]

It's hard to estimate the precise figures on how much water the cattle needed, for the amount is dependent on several factors such as nutrition, the size of the animal, how hard the animal worked, etc. The Farmer's Handbook from the 1940s says: "There are different opinions on watering the cattle. Some say that for a bull, 30–40 litres a day is enough, for excessive drinking makes a bull pot-bellied and listless. Others let the bull drink as much as it wants and don't see any harm coming." The handbook defined exactly what kind of water was best for cows. "Cows must have plenty of good drinking water. It should be clear, good-tasting, odour-free and at a temperature of 8–10°C." (Maamiehen käsikirja 1945, 353-355)

A common estimation of the water that was needed daily for cows is 50 litres. Water was needed also to clean the cowsheds and wash the cows. "Every now and then cows have to be washed all over with a brush, soap and warm water, which should be used as little as possible." (Maamiehen käsikirja 1945, 355-360)

When living with the cattle in the towns and countryside, watering the livestock formed the major part of the water consumption. Thus a well was placed closer to the cowshed than the house itself. (Katko 1988, 8–11; Paulaharju 1958, 32–33; Paulaharju 1906, 7)

In today's world the pit toilet is still the most popular model of toilet. And it should be noted that in many areas toilets are not used at all. In some rural areas it is still common that people go to the bushes or directly in the field. For example, in some areas in the Nepal countryside, the so-called "pig-toilet" is in use. A pig toilet is not really a toilet at all; it just means that pigs eat human excreta and later pigs are used for human victuals. (Rautanen 2005) It sounds like the perfect nutrient cycle; but, unfortunately, it is very risky to the health of humans.

## 4.6. CONCLUSIONS

In 2006 hundreds of millions of people have their own wells and on-site sewage system. For hundreds of millions of people, a properly constructed well and toilet would even nowadays mean a remarkable improvement in their living conditions. Wells and toilets don't represent "yesterday's world" but rather are functional solutions for sparsely populated areas. Especially wells are continuously built in such areas but toilets are also needed simultaneously—one does not work properly without the other. Ancient inventions aren't necessarily outdated, but are in all their simplicity well functioning.

Even in a highly developed country such as Finland the sparsely populated areas are still mainly without a municipal sewer network. Although there is no municipal sewerage, there is no risk to human lives with on-site system. The majority of the households are connected to a water system and get their water from waterworks, in cooperation with each other or directly from their own wells. This system is prone to risks, for the consumption of the water and amount of waste water may increase by even ten times—there's not the trouble of lifting and carrying the water inside and spontaneous use is easier. It is harder to follow the amount of consumption compared to checking the level inside the bucket.

Outhouses and wells remained in use in cities for some time after the establishment of central water supply systems in many countries –if not even in all countries. Water and sanitation services reached the suburban areas slowly and in some areas of scattered settlements they still are not available. At the end of the 1800s the WC was considered an improvement that saved people from unpleasant tasks—such as emptying a chamber pot in the morning.

The water closet was seen as a victory for public health care, because excrement was moved away immediately from houses and yards. But where was the human excreta going through the sewer pipes? That is yet another question. It went to bodies of water. When it comes to the health of humans and the state of the environment, the WC was not an improvement, it was major step backwards. Nutrients were getting out of the use of agriculture and nature's metabolism into the bodies of water. This unfortunate development is described in many later chapters in this book.

Improved techniques in these basic innovations can help also to promote not only sustainability, but also equality. Still, wells and toilets play a key role as sounds solutions for the future.

## 4.7. REFERENCES

Ailio J. (1917) Hämeen linnan esi- ja rakennushistoria. Hämeenlinnan kaupungin historia osa 1. Hämeenlinna

Bible, the New International version.

Bjur H. (1988) Vattenbyggnadkonst i Göteborg under 200 år.

Brunow-Ruola M., Helene & Augusta. (2001) Porvariston elämää Turussa 1870-1920. Keuruu.

Bruun C. (1991) The water supply of ancient Rome. A study of Roman imperial administration. Societas Scientiarum Fennica. Commentationeres Humanarum Litterarum 93/1991.

Coffey K. & Reid G. (1976) Historical implications for developing countries of the developed countries water and wastewater technology. The University of Oklahoma.

Ellis B. (2002) 'White Settler Impact on the Environment of Durban, 1845-1870', in *South Africa's Environmental History: Cases & Comparisons*, ed. by Stephen Dovers et al. Cape Town.

Fehr, W. (1955) The Town House, Its Place in the History of Cape Town. Cape Town.

Foil J.L., Cerwick J.A. and White J.E. (1993) Collection Systems Past and Present. Operations Forum, 10(12).

Gardberg C.J. (1959) Åbo slott under den äldre Vasatiden. Helsinki.

Gardberg C.J. (1961) Turun linna ja sen restaurointi. Turku.

Gray H.F. (1940) Sewerage in Ancient and Mediaeval Times. Sewage Works Journal.

Harjula M. (2003) *Tehdaskaupungin takapihat*. Tampere.

Hattersley A.F. (1956) More Annals of Natal. London.

Hattersley A.F. (1973) An Illustrated Social History of South Africa. Cape Town.

Hendricks D. W. 1991. Water and Wastewater Practices And Institutions. Ancient To Modern. Notes As Developed Through. Colorado State University. Unpublished manuscript.

Hörberg I. (1997) Vårt välsignade vatten. Kalmars vattenförsörjning 1897-1997. Kalmar Vatten och Renhållning Ab.

Jansen M. (1993) Mohenjo-Daro – Stadt der Brunnen und Kanäle – Wasserluxus vor 4500 Jahren. Mohenjo-Daro – City of Wells and Drains –Water splendour 4500 years ago. Frontius-Geschällschaft e.V. Supplementary Volume II. Bonn.

Jokinen P. (1994) *Kangasalalaista huumoria*. Kangasala.

Jutikkala E. (1979) *Tampereen historia III. Vuodesta 1905 vuoteen 1945*.

Juuti P. (2001) Kaupunki ja vesi. Doctoral dissertation. Acta Electronica Universitatis Tamperensis 141. Tampere.

Juuti P. (2003) "Elämää ja hieman historiaakin opiskelemassa 1985-1991", Heikkinen et al., eds: Patinan neljä vuosikymmentä. Tampere, 45-50.

Juuti P., Rajala R. & Katko T. (2003a) *Aqua Borgo ensis*. Porvoo.

Juuti P., Äikäs K. & Katko T. (2003b) *Luonnollisesti vettä. Kangasala, Vesilaitos 1952–2002*. Saarijärvi.

Juuti P.S. and Wallenius K.J. (2005) Brief History of Wells and Toilets. Pieksämäki.

Kaljundi J. (1997) 660 years of water business in Tallinn, 1337-1997. Tallinna Vesi.

*Kangasalan historia 1*. (1949) Virkkala K., Luho V. & Suvanto S. *Längelmäveden seudun historia I.Kangasalan historia I*. Forssa.

Katko T. (1988) Maaseudun vesihuollon kehittyminen Suomessa: suuntaviivoja kehitysmaille? Taustaselvitys. TTKK, VYT B 35.

Katko T. (1996) Vettä! – Suomen vesihuollon kehitys kaupungeissa ja maaseudulla. Tampere.

Keating A.D. (1989) Public-private partnerships in public works: A bibliographic essay. pp. 78-108. In: APWA. 1989. Public-private partnerships: privatization in historical perspective. Essays in Public Works History. No. 16. 108 p.

Kilkki P. (ed.) (1973) Hämeen linna. Mikkeli.

Koskinen M. (1995) *Saastunut Näsijärvi terveydellisenä riskinä – Kulkutaudit, kuolema ja puhdasvesikysymys Tampereella 1908–1921.* Master thesis, University of Tampere.

Laakkonen S. (2001) Vesiensuojelun synty. Helsingin ja sen merialueen ympäristöhistoriaa 1878–1928. Tampere.

Laamanen M. & Prusi H-L (2004) *Olavinlinnan opaskirja.* Museovirasto.

Leyds G.A. (1964) A History of Johannesburg. Johannesburg.

*Maamiehen käsikirja (1945)* Otava.

Mattila H. (2005) Appropriate Management of On-Site Sanitation. Tampere University of Technology, Publication 537. Tampere.

Paulaharju S. (1906) *Asuinrakennuksista Uudella kirkolla Viipurin läänissä.* SKS, Helsinki.

Paulaharju S. (1958) *Kainuun mailta. Kansantietoutta Kajaanin kulmilta.* Porvoo.

Péter J. (2005) Hungary. In Juuti, P.S. and Katko, T.S. Water, Time and European Cities – History Matters for the Futures, WaterTime/EU.

Ponting C. (1991) A Green History of the World. Penguin Books.

Puhakka M. & Grönros J. (1995) Turun linna. Jyväskylä.

Sinirand I. (1992) Tallinna veevarustuse ja kanalisatsiooni minevik ja tänäpäev. Tallinn Valgus.

Sinisalo A. (1986)Olavinlinnan rakentamisen vaiheet suuresta pohjan sodasta nykypäiviin. Savonlinna.

Spier P. (1989) *Joonaan kirja*. Karas-Sana.

Stephens J.L. (1837) Incidents of travel in Egypt, Arabia Petraea, And The Holy Land By An American (J. L. Stephens). Based on the 2nd edition New York: Harper & Brothers. http://chass.colostate-pueblo.edu/history/seminar/stephens/stephens1-3.htm.

Toivonen R., Mäki-Kuutti T. & Bonsdorff M. (ed.) (1981) TEK keksintöjen kirja. WSOY.

Twain, M. (1869) Innocents Abroad. http://www.mtwain.com/Innocents_Abroad/59.html.

Vasilakis A (sa.) Phaistos. V.Kouvdis - V.Manouras Editions. Iraklion.

Vilkuna A.-M. (2001) Hämeen linnan opaskirja. Museovirasto.

Werz, Bruno E.J.S. (2002) Survey and excavation of the main well in Cape Town Castle, South Africa: fieldwork and structural aspects, *The International Journal of National Archaeology*, 31, 1.

Wijmer S. (1992) Water om te drinken. VEWIN, the Netherlands.

INTERNET:
http://www. arcl.ed.ac.uk/arch/annrept/report99/
http://www.bottlebooks.com/privyto.htm
http://bridge.ecn.purdue.edu/~alleman/w3-articles/swallow-sanit-1911/swallow-1911.html
http://www.masterplumbers.com/plumbviews/1999/toilet_tribute.asp
http://www.masterplumbers.com/plumbviews/1999/toilet_tribute2.asp
http://the-orb.net/medieval_terms.html
http://nautarch.tamu.edu/portroyal/CHAMBER/Textfile.htm
http://www.warlinks.com/cairo/cairo.shtml;
http://web-japan.org/kidsweb/techno/toilet/history.html
http://weekly.ahram.org.eg/2003/640/heritage.htm)
http://www.whazo.com/seniorproject.html
http://www. witwib.com/Atlit_Yam, read Aug 24,2005

INTERVIEWS & PERSONAL COMMUNICATION
Kostas K. 17.10.2005.
Maijala M. 25.8.2004.
Rautanen SL. 2.6.2005.

FIELD RESEARCH TRIPS
Juuti P. – July 2004 (Castles of Finland.)
Juuti P. –November 2004 (History of Water supply and sanitation in South Africa).
Juuti P. – October 2005 (Environmental history of Crete).

# 5

## WATER AND HEALTH IN ANTIQUITY: EUROPE'S LEGACY

*Heikki S. Vuorinen*

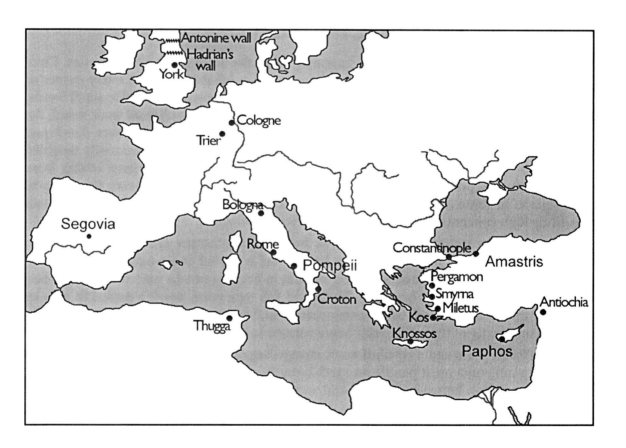

Figure 5.1 Key places mentioned in the chapter.

## 5.1. GENERAL BACKGROUND

The importance of water for the health of people was a widely held opinion of ancient Greek and Roman writers. The purpose of this article is to examine: i) ideas of salubrity of water, ii) the organization of water supply in towns, and iii) the influence of water on public health during antiquity (circa 500 B.C. – 500 A.D.). The material for the study consists primarily of the works of ancient Greek and Roman medical authors, especially the Hippocratic writings. Several Roman authors outside medicine proper are also invaluable for studying water and health, especially Vitruvius, Pliny the Elder and Frontinus. Special studies relevant for understanding public health during antiquity are also used. Translations of ancient texts are from the Loeb Classical Library editions, as listed in the references.

## 5.2. PROLOGUE

Modern human beings (*Homo sapiens*) have existed on the earth some 200,000 years. Circa 50,000 years ago a rapid cultural evolution began and we inhabited almost every corner of the world. People got there living by hunting, fishing and gathering, and lived in small bands, constantly on the move. The waterborne health risks of hunter-gatherers were small. An aversion to water that tasted revolting, stunk and was odious-looking must have developed quite early on in the evolutionary history of mankind. People were occasionally troubled by pathogens transmitted by contaminated water, but generally the problems arising from water, for example, polluted by human wastes, could be avoided by moving the camp to another place. In some restricted areas of the world people inevitably drank water that had unhealthily high concentrations of, e.g., arsenicals and fluorides.

The most serious waterborne health risks for hunter-gatherers must have been some parasites, which besides living off human beings had a period in an aquatic organism to complete their lifecycle. Plasmodia causing malaria in humans need a period in mosquitoes, which in turn need water for breeding. The broad fish tapeworm needs a period in man or some other mammal and fresh water crustaceans and fish(s) to have its lifecycle completed. The mere complexity of these lifecycles bears witness to the antiquity of these parasites.

Roughly 10,000 years ago a revolutionary change began to shape people's living. During the following millennia most people on earth became sedentary agriculturists. This created a brand new relation between humans and water (Garbrecht 1983). People concentrated around wells, lakes, rivers and other sources of water. The sedentary population did not move from their houses when the filth piled up around them. Human excreta contaminated drinking water, while the domesticated animals must have been even a more serious risk for potable water.

Pathogens transmitted by contaminated water became a very serious health risk for the sedentary agriculturists. In different areas and eras, pathogens causing cholera, dysentery, diarrhoeas etc. began to take their toll on human beings. Serious waterborne pathogens must

Figure 5.2a and 5.2b Restored clay tub (a) and drains from the supposed water toilet (b) near the Queen's hall in the palace of Knossos, Crete, Greece. (Photo: Marjatta Vuorinen)

have emerged especially in the first urban centres containing several thousand people.

The first urbanization in Europe occurred during antiquity (500 B.C. – 500 A.D.) around the Mediterranean region. The urban population reached some 10–20 % in the days of glory of the Greco–Roman world in the centuries around the birth of Christ. The majority of townspeople lived in quite modest-sized towns (5 000–10 000 people). The most urbanized areas were the Eastern Mediterranean, Egypt, North Africa (modern Tunisia), the Apennine Peninsula (modern Italy), and the southern part of the Iberian Peninsula, most of which were areas of quite modest rainfall. This was a world in which the problem of guaranteeing pure water for people became a prerequisite for successful urbanization.

The first evidence of the purposeful construction of the water supply, bathrooms, toilets and drainage in Europe comes from Bronze Age Minoan (and Mycenaean) Crete in the second millennium B.C. (Graham 1987: 99–110, 219–221, 255–269). The inhabitants in the palace of Knossos were supplied with running water by clay pipes. These pipes were constructed in a way interpreted to make a pressure system feasible. (Graham 1987: 219) Minoans built bathrooms and toilets in their houses, the most elegant ones being in the palaces, especially in the palace of Knossos (Figures 5.2a, 5.2b). The clay tubs in the bathrooms of Minoan (Mycenaean) palaces and houses must have been filled and emptied by hand. The toilets were also probably flushed by hand in most cases. Elaborate drainage was discovered in the palace of Knossos and rainwater was probably used to flush the toilet near the Queen's Hall (Figure 5.2b). A. Trevor Hodge expressed, however, severe doubts about the efficiency of the drain of the toilet. (Hodge 1992, 477 n17)

## 5.3. WATER AND HEALTH ACCORDING TO ANCIENT GREEK AND ROMAN AUTHORS

The realization of the importance of pure water for the health of people is evident in the myths of ancient people. Religious cleanness and water were important in various ancient cults. The first known Greek philosophical thinker in the early 6$^{th}$ century B.C. also recognized the importance of water: "*Thales of Miletus, one of the Seven Wise Men, affirmed that the principle of all things is water [...]*" (Vitruvius *De Architectura*. VIII. preface. 1) Alcmaeon of Croton (floruit c. 470 B.C.) was the first Greek doctor to give a definition of health and sickness. He also stated that the quality of water may influence the health of people: "*Disease may come about from external causes, from the quality of water, local environment or toil or torture.*" (Aëtius, *On the opinions of the philosophers* V.30.1)

It was, however, in the so-called Hippocratic writings, mostly written around 400 B.C., that we have the first surviving texts concerning comprehensively water and the health of people. Classical Greek medicine adopted a sincere confidence in rationality and logic and in the rejection of the supernatural causation of diseases during antiquity. The confidence in rational thinking is clearly expressed in those treatises that deal with the health aspects of water.

Especially one treatise among the Hippocratic writings has attracted attention. An unknown Greek physician wrote in the second half of the fifth century B.C. (Jouanna 1999, 375) the famous treatise called *Airs, Waters, Places*, which starts by describing what circumstances an itinerant physician must consider when he comes to an unfamiliar town:

> Whoever wishes to pursue properly the science of medicine must proceed thus. First he ought to consider what effects each season of the year can produce; for the seasons are not at all alike, but differ widely both in themselves and at their changes. The next point is the hot winds and the cold, especially those that are universal, but also those that are peculiar to each particular region. He must also consider the properties of waters; for as these differ in taste and in weight, so the property of each is far different from that of any other. Therefore, on arrival at a town with which he is unfamiliar, a physician should examine its position with respect to the winds and to the risings of the sun. For a northern, a southern, an eastern, and a western aspect has each its own individual property. He must consider with the greatest care both these things and how the natives are off for water, whether they use marshy, soft waters, or such as are hard and come from rocky heights, or brackish and harsh. The soil too, whether bare and dry or wooded and watered, hollow and hot or high and cold. The mode of life also of the inhabitants that is pleasing to them, whether they are heavy drinkers, taking lunch, and inactive, or athletic, industrious, eating much and drinking little. (Airs, Waters, Places. 1)

The treatise later on contains several passages about the health effects of water, and the ancient Greek author deals with the different sources, qualities and health effects of water in length.

> I wish now to treat of waters, those that bring disease or very good health, and of the ill or good that is likely to arise from water. For the influence of water upon health is very great. Such as are marshy, standing and stagnant must in summer be hot, thick and stinking, because there is no outflow; and as fresh rain-water is always flowing in and the sun heats them, they must be of bad colour, unhealthy, bilious. [...] Such waters I hold to be absolutely bad. The next worst will be those whose springs are from rocks – for they must be hard – or from earth where there are hot waters, or iron to be found, or copper, or silver, or gold, or sulphur, or alum, or bitumen, or soda. [...] So from such earth good waters cannot come, [...] The best are those that flow from high places and earthy hills. [...] (Airs, Waters, Places. 7)

> Rain waters are the lightest, sweetest, finest and clearest. [...] Such waters are naturally the best. But they need to be boiled and purified from foulness. [...] [...] Waters from snow and ice are all bad. [...] I am of opinion that such waters derived from snow or ice, and waters similar to these, are the worst for all purpose. (Airs, Waters, Places. 8)

> Stone, kidney disease, strangury and sciatica are very apt to attack people, and ruptures occur, when they drink water of very many different kinds, or from large rivers, into which other rivers flow, or from a lake fed by many streams of various sorts, and whenever they use foreign waters coming from a great, not a short, distance. [...] Such waters then must leave a sediment of mud and sand in the vessels, and drinking them causes the diseases mentioned before. (Airs, Waters, Places. 9)

Ideas expressed in the *Airs, Waters, Places* had a definite influence during antiquity and later. The philosopher Plato (428/427–348/347 B.C.) most probably knew the legendary Hippocrates and was familiar with Hippocratic medicine and the ideas of environmental influences on health. He advised a lawmaker who was about to found a city to take into account several factors, among them water: *"Some sites are suitable or unsuitable because of varying winds or periods of heat, others because of the quality of the water."* (Plato *Laws*. 5.747d) The ideas on climate, place and water expressed in *Airs, Waters, Places* are seen in the manual of architecture *De Architectura* by Vitruvius in the late first century B.C. *"[...] he must know the art of medicine in its relation to the regions of the earth (which the Greeks call climata); and to the characters of the atmosphere, of localities (wholesome or pestilential), of water-supply. For apart from these considerations, no dwelling can be regarded as healthy."* (Vitruvius *De Architectura*. I.i.10.) Columella (1st century A.D.) and Palladius (5th century A.D.), who both wrote about agriculture and Roman officer and administrator Pliny the Elder (23/24–79 A.D.), who dedicated his encyclopaedic *Natural History* to the emperor Titus in 77, were also clearly influenced by the same ideas. The most famous physician in antiquity (second after the mythic Hippocrates), Galen, (129–199/200/216 A.D.) wrote a commentary on *Airs, Waters, Places*, and it was among the few Hippocratic treatises to be translated into Latin in late antiquity. (Jouanna 1999, 356, 361, 375, 478.)

Various other Hippocratic treatises contain short comments on the influence of water on the health of people (*Internal Affections*. 6, 21, 23, 26, 34, 45, 47; *Diseases I*. 24; *Epidemics II*. 2.11; *Epidemics VI*. 4.8, 4.17; *Aphorisms*. 5.26; *Humours*. 12; *Regimen IV or Dreams*. 93) One Hippocratic treatise besides *Airs, Waters, Places*, and almost contemporaneous with it, deals quite largely with the influence of the environment on disease. However, the author of this Hippocratic treatise, called *Humours*, which is dated at the end of the fifth century B.C. or the beginning of the fourth (Jouanna 1999: 389, 394), had quite little to say about the influence of water on health of people: *"Some [diseases] spring from the smells of mud or marshes, others from waters, stone, for example, and diseases of the spleen; of this kind are waters because of winds good or bad."* (Humours. 12) As a whole *Humours* is quite an inspiring piece of work and it was not a surprise that Galen was to write also commentary on this treatise. (Jouanna 1999, 356)

A surviving fragment of the writings of fourth century B.C. Greek doctor Diocles of Carystus consists of, besides a general opinion on the good effects of water and the variability of the quality of water by locality, a testimony that opinions among ancient doctors differed as to what kind of water was considered the best:

> Diocles says that water promotes digestion and causes no flatulence, and that it is moderately cooling and clears sight and least makes the head heavy, and stimulates both the soul and the body. Praxagoras also says this, and he recommends rain-water, while Evenor recommends water from cisterns. He also says that water from [the sanctuary of] Amphiaraos is good compared to water in Eretria. (Diocles of Carystus. 235)

In the late first century B.C. Vitruvius had in his *De Architectura* a lot to say about waters:

> Therefore, inasmuch as physicists, philosophers and the clergy judge that everything consists of the principle of water, I thought fit that, having explained in the pervious seven volumes the methods of building, I should write in the present volume about the discovery of water, the qualities of its special sources, the methods of water supply and of the testing water before using it. (Vitruvius *De Architectura*. VIII.Preface.4)

According to Viruvius (*De Architectura*. I.iv.1), marshy areas must be avoided when the site of a city is chosen:

> First, the choice of the most healthy site. Now this will be high and free from clouds and hoar frost, with an aspect neither hot nor cold but temperate. Besides, in this way a marshy neighbourhood shall be avoided. For when the morning breezes come with the rising sun to a town, and clouds rising from these shall be conjoined, and, with their blast, shall sprinkle on the bodies of the inhabitants the poisoned breaths of marsh animals, they will make the site pestilential.

Pliny the Elder in the first century A.D. had in his works a long section concerning the different opinions on what kind of water is the best:

> It is a question debated by the physicians what kinds of water are most beneficial. They rightly condemn stagnant and sluggish waters, holding that running water is more beneficial, as it is made finer and more healthy by the mere agitation of the current. For this reason I am surprised that some physicians recommend highly water from cisterns. […] Rain-water, it is agreed, becomes putrid very quickly, and it is the worst water to stand a voyage. […] But cistern water even physicians admit is harmful to the bowels and throat because of its hardness, and no other water contains more slime or disgusting insects. Yet it must be admitted, they hold, that river water is not ipso facto the most wholesome, nor yet that of any torrent whatsoever, while there are very many lakes that are wholesome. What water then, and of what kind, is the best? It varies with the locality. […]Slime in water is bad. […] From what source then shall we obtain the most commendable water? From wells surely, as I see they are generally used in towns, but they should be those the water of which by frequent withdrawals is kept in constant motion, and those where due thinness is obtained by filtering through the earth. […] One point above all must be observed – and this is also important for a continuous flow – well water should issue from the bottom, not the sides. (Plinius NH, XXXI, xxi–xxiii)

Galen summarises the preferable qualities of water in his *De Sanitate Tuenda*:

> I do not consider, as some do, that such children should be wholly enjoined from cold beverage; but I permit its use mostly after meals and in the hottest seasons, when they incline to cold, especially, if possible, from a fresh spring which has no acquired harmful quality […] But one should be on guard against stagnant, muddy, malodorous, and salty pools, and in short those which display any unpalatable quality. For the purest water should seem to them the best, not only to the taste but also to the smell. (Galen. De Sanitate Tuenda. I.xi)

The general attitude in the ancient dietetics was to avoid sudden changes in one's way of life, including the water that one was used to. Ancient authors expressed a healthy caution towards drinking unfamiliar waters, although the author of *Airs, Waters, Places* was of the opinion that "*A man in health and strength can drink any water that is at hand without distinction, [...]*". (*Airs, Waters, Places.* 7) However, in the Hippocratic treatise called *Internal Affections* there is a statement that the change of water could cause disease: "*[...] it* (erysipelas) *may also attack as the result of eating meat, or from a change of water.*" (Internal Affections. 6) The dangers of change in one's habits is clearly expressed by Diocles of Carystus: "*One should not drink water one is not used to as it comes to hand, for this is bad and risky [...]*" (Diocles of Carystus. 182.10)

Galen expresses the same caution towards waters that the subject has no previous experience of, but affords then some clues for assessing the quality of unfamiliar water:

> It is safest, then, to judge such water by experience. But if any one should wish to prejudge its property by signs, those whose sources, springing from rocks, flow towards north, having the sun turned away from them, are all to be considered slow of digestion and passage, and they will warm and cool slowly. But those whose sources face the sunrise, and are strained through a clean opening of pure earth and warm and cool most quickly, these one must expect to be best for all ages. (Galen. De Sanitate Tuenda. I.xi)

The ancient qualifications for good drinking water would have primarily protected people against those hazards that we would now consider as biological (bacteria, viruses, protozoa etc.). The health of people drinking stagnant, marshy water is discussed at length in *Airs, Waters, Places*, and includes a fine description of malarial cachexia:

> In winter they (waters) must be frosty, cold and turbid through the snow and frosts, so as to be very conducive to phlegm and sore throats. Those who drink it have always large, stiff spleens, and hard, thin, hot stomachs, while their shoulders, collar-bones and faces are emaciated [...] This malady is endemic both in summer and in winter. In addition the dropsies that occur are very numerous and very fatal. For in the summer there are epidemics of dysentery, diarrhoea and long quartan fever, which disease when prolonged cause constitutions such as I have described to develop dropsies that result in death. These are their maladies in summer. [...] (Airs, Waters, Places. 7)

The author of *Internal Affections* had the same warning against stagnant and marshy waters:

> The next dropsy arises in the following way: if, in summer, a person on a long journey happens upon some stagnant rain water, and drinks a large amount of it at one draught [...]

> Another ileus: this one occurs mainly in summer, in swampy areas; in most cases, it comes on as result of drinking water [...] (Internal Affections. 26, 45)

The ancient Greeks and Romans were, however, also quite aware of some of the risks, which we now regard as caused by poisonous chemical elements in the water. The dangers of water coming from hills and mountains where mining was carried out did not escape the notice of the keen ancient observer. The Hippocratic author of *Airs, Waters, Places* stated:

> The next worst will be those whose springs are from rocks – for they must be hard – or from earth where there are hot waters, or iron to be found, or copper, or silver, or gold, or sulphur, or alum, or bitumen, or soda. For all these result from the violence of heat. So from such earth good waters cannot come, but hard, heating waters, difficult to pass and causing constipation. (Airs, Waters, Places. 7)

Educated people outside proper medicine were aware of these dangers as evidenced by Vitruvius:

> But when gold, silver, iron, copper, lead and the like are mined, abundant springs are found, but mostly impure. They have the impurities of hot springs, sulphur, alum, bitumen; and when the water is taken into the body and, flowing through the vessels, reaches the muscles and joints, it hardens them by expansion. Therefore the muscles swelling with expansion are contracted in length. In this way men suffer from cramp or gout, because they have the pores of the vessels saturated with hard, thick and cold particles. (Vitruvius De Architectura. VIII,iii,5)

## 5.4. WATER TESTING AND PURIFICATION

Different means were used to find good drinking water. Springs or flowing water was valued, but out of necessity wells and cisterns were also used. Limpid, tasteless, odourless and cool water was regarded well for drinking but more elaborate methods to evaluate the salubrity of water were also proposed. Vitruvius suggested quite an ingenious method to examine the water or food sources in an area of an intended town:

> Therefore emphatically I vote for the revival of the old method. For the ancients sacrificed the beasts which were feeding in those places where towns or fixed camps were being placed, and they used to inspect the livers, which if at the first trial they were livid and faulty, they went on to sacrifice others, doubting whether they were injured by disease or faulty diet. When they had made trial of many, and had tested the entire and solid nature of the livers in accordance with the water and pasture they established there the fortifications; if, however, they found them faulty, by analogy they judged: that the supply of food and water which was to be found in these places would be pestilential in the case of human bodies. [...] Hence we may know by food and water whether the properties of places are pestilential or salubrious. (Vitruvius De Architectura. I,iv,9,10)

Later in his manual Vitruvius further explores the methods to be used when testing the suitability of water:

> The discovery and testing of springs is to be pursued in the following manner. When they are abundant and in the open, we are to observe and consider, before we begin to lay the water on, what is the physique of those who live in the neighbourhood. If they are strong, of clear complexion, free from distortion and from inflamed eyes, the water will pass. Again, if a fresh spring be dug, and the water, being sprinkled over a vessel of Corinthian, or any other good bronze, leave no trace, the water is very good. Or if water is boiled in a copper vessel and is allowed to stand and then poured off, it will also pass the test, if no

sand or mud is found in the bottom of the copper vessel. Again, if vegetables being put in the vessel with water and boiled, are soon cooked, they will show that the water is good and wholesome. Likewise, if the water itself in the spring is limpid and transparent and if wherever it comes or passes, neither moss nor reeds grow nor is the place defiled by any filth, but maintains a clear appearance, the water is indicated by these signs to be light and most wholesome. (Vitruvius De Architectura. VIII, iv,1,2).

We have no way to evaluate whether or not these thorough and sound recommendations of Vitruvius were ever followed in practice in antiquity. Nowadays we follow these recommendations almost in every aspect; of course using different methods: i) we examine the water source and water obtained from it with our senses in the same way as Vitruvius; ii) we make chemical and physical laboratory tests (the sprinkling over bronze, boiling in a copper vessel and boiling of vegetables as Vitruvius described); iii) we make epidemiological studies (observing the health status of people in the neighbourhood of spring as Vitruvius described); and iv) we experiment on animals in order to test the health effects of various substances (like the sacrificing of animals in Vitruvius). The only thing that was totally missing in Vitruvius' writing is any statement that could be interpreted to allude to microbiological studies, which did not appear before the late 19th century after the improvement of microscopy and the discovery of disease-causing microbes.

The Greeks and Romans used different methods to improve the quality of the water if it did not satisfy their quality requirements. From written sources and archaeological excavations, we know that using settling tanks, sieves, filters and the boiling of water were methods used during antiquity. At least boiling water, which was widely recommended by the medical authors during antiquity, would have diminished the biological risks of poor quality water. Unfortunately the available sources do not permit the estimation of how often boiling or other methods were used or how effectively they improved the quality of the water. As a consequence, it is not possible to evaluate the effects of these methods on public health. Although the boiling of water might have been feasible from a hygienic point of view, it was ecologically and economically not feasible in extensive use since firewood and other combustibles would sooner or later have become a scarce resource around the Mediterranean.

The author of *Airs, Waters, Places* already recommended boiling for purifying rain-water. There is also a short and confusing passage on boiled water in another Hippocratic treatise (*Epidemics* VI. 4.8). Pliny the Elder in the first century A.D. had a more detailed recommendation about boiling: *"At any rate it is agreed that all water is more serviceable when boiled, ... It purifies bad water to boil it down to one half."* (Pliny *NH*, XXXI, xxiii) Besides boiling, the fourth century B.C. Greek doctor Diocles of Carystus recommended even more elaborate methods for improving water:

> You will make water weakest by boiling off a third of it; you should boil off whitish water in the same way, and add to it lumps of dry potter's earth until they are soaked, about a twelfth of a measure per amphora; when you have boiled it, drink it. You can expel the hot

smell of the water in the following way, by beating it with the hand against the wind and exposing it to sun and air in a wide-mouthed vessel and by pouring it into many vessels, in small portions." (Diocles of Carystus, 236)

Paulus Aeginata in the 7th century A.D. also considered that marshy, stinking water could be made drinkable by boiling (Paulus Aeginata I.20).

Vitruvius recommended the use of several cisterns for the clarification of rain water collected from roofs or higher grounds: *"If the cisterns are double or treble, so that they can be changed by percolation, they will make the supply of water much more wholesome. For when the sediment has a place to settle in, the water will be more limpid and will keep a flavour unaccompanied by smell."* (Vitruvius De Architectura. VIII,vi,14) Filtering through sand was also used to purify water, as discussed by Ilkka Syvänne in Chapter 6.

## 5.5. WATER SUPPLY AND SEWERAGE IN TOWNS

The ancient Greek and Roman medical authors did not have much to say about the actual organisation of the water supply besides general recommendations as to which kind of water should be the best for drinking, etc. The most detailed descriptions of the water supply come from the Roman authors Vitruvius and Frontinus. Columella, Pliny and Palladius also refer to the water supply in their writings but sometimes their descriptions resemble very much those of Vitruvius.

Much of the evidence concerning water supply and sewerage comes from archaeological excavations. The Romans have commonly been celebrated for their aqueducts, which brought vast amounts of water to towns. The building of an aqueduct seems to have been a cultural necessity for a decent town life in Roman times. Clear evidence of this is provided by certain towns in Britain, where many remains of aqueducts have been found in archaeological excavations (see Wacher 1975 and Burnham & Wacher 1990); the main problem in Britain, however, must have been to get water out of a town, not into it. The best-known aqueducts are in Rome (Figure 5.3), but they existed in towns everywhere in the Roman world (Figure 5.4.).

The amount of water distributed to the towns by the aqueducts is not easy to estimate; even so, the estimated figures are so high that only a part of the presented amounts would have been enough to ensure sufficient volume for lavish daily use. The estimate of water that every person received per day was for: Roman Bologna approx. 1400 litres (Giorgetti 1988); Trier approx. 1000 litres (Grewe 1988); Pergamon approx. 250 litres (Garbrecht 1987a); and Pompeii approx. 800 litres (Eschebach 1987). Christer Bruun estimated that the amount of water that every person received was only 67 litres per day in imperial Rome (Bruun 1991: 103), but for Rome, higher figures than Bruun's have been presented: e.g. 520–635 litres per person per day (Garbrecht 1987b).

# WATER AND HEALTH IN ANTIQUITY: EUROPE'S LEGACY

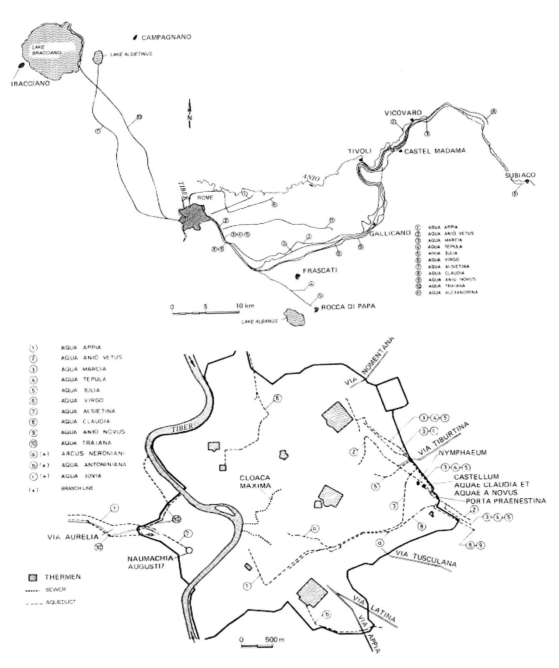

Figure 5.3 The aqueducts supplying imperial Rome, the main sewer, Cloaca Maxima, is also shown. (Source: modified from Garbrecht 1986, 34–35).

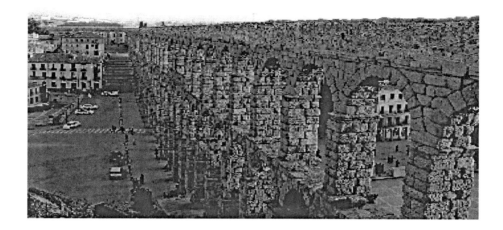

Figure 5.4 The archaeological remains of many Roman aqueducts are still quite impressive, like this in Segovia, Spain included in The World Heritage List since 1985. (Photo: Tapio Katko)

It is justified to argue that the basic reason for building the aqueducts was luxury, the baths, not the acquisition of good drinking water. (Ward-Perkins 1984: 125–126) However, good quality drinking water seems to have been highly esteemed by Romans as evidenced in the writings of Vitruvius and Frontinus. (Vitruvius *De Architectura*. VIII; Frontinus 88–92) The Roman senator and high military and civil officer Frontinus (circa 30–103/4), who wrote *De Aqueductu urbis Romae* probably in the year 98 A.D. (Rodgers 2004: 1–8), classified the aqueducts of Rome according to the suitability of their water for drinking. It has also been argued that, e.g., in Pompeii, the aqueduct was built in the Augustan age for the sake of getting good drinking water to the town, rather than to supply water to baths. (Laurence 1996, 44)

For their water supply, ancient cities had relied on wells and cisterns before the aqueducts were built and many of them continued to rely on wells and cisterns throughout antiquity. For instance, the Roman towns of Italy were firmly established and many of them prospered before the aqueducts were introduced into them in the first century A.D. (Ward-Perkins 1984, 121–122). Pliny the Elder even recommended the use of water from wells in towns. (Pliny *NH*, XXXI, xxiii) In rural areas wells and cisterns seem to be considered as a common solution for water supply (Columella *Rei Rusticae*. I.5; Palladius *Opus Agriculturae*. I, 17). Several ancient authors discussed the construction of wells and the danger of noxious gases to the well-diggers (Vitruvius *De Architectura*. VIII, vi, 12–13; Pliny *NH*, XXXI, xxviii, Palladius *Opus Agriculturae*. IX, 9).

Figure 5.5 The public fountains in Pompeii, Italy, were mostly located on street corners. (Photo: Marjatta Vuorinen)

Old wells might fall into disuse and fill with rubbish as is indicated to have happened in Pompeii after the building of an aqueduct. (Laurence 1996: 44) When circumstances demanded, a town could return to the use of wells even after an aqueduct was built. This happened in Pompeii after an earthquake damaged the town in 62 A.D.; the aqueduct was not connected to the water supply network in the town before the eruption of Vesuvius finally destroyed the town in 79 A.D. (Laurence 1996: 36) In the sixth century A.D., during the Gothic Wars, Neapolitans and Romans successfully relied on wells and springs after the aqueducts were cut off by the besiegers. (Ward-Perkins 1984: 122–123)

The water supply was considered important for the professional Roman army (see the chapter by Syvänne in this book). Roman forts had typically a good water supply, aqueducts and baths. A modern visitor may witness this when observing the archaeological remains in the excavated forts of Hadrian's and Antonine walls in the periphery of Roman Britain. (Robertson 1990; Bédoyère 1998) Army veterans were well accustomed to baths and to an ample water supply during their active service, and they may have been a quite important pressure group for building an aqueduct and bath(s) in a town.

Baths consumed only a part of the water that the aqueduct(s) brought to the town. Besides baths, water was distributed to the public street fountains, to the taverns and other small businesses and private houses. Most of townspeople must have fetched their daily drinking and other domestic water from the public fountains (Figure 5.5). There were approximately 50 public fountains in Pompeii, which meant about 160 people per fountain (Eschebach 1988). It has been calculated that most people in Pompeii lived within 80 metres of a public fountain, which was nearly always located at a street junction (Figure 5.5). (Laurence 1996, 46, 47 [map], 50)

In contrast to the water supply, the sewer systems of ancient towns are quite poorly known. In some towns like Rome, York, Cologne and Trier, the main sewer could have been quite vast (Hodge 1992: 333–344; Addyman 1989: 244–264; Watermann 1974: 277-294). Although sewers in Rome are highly praised (Garbrecht 1983), many Roman towns had no sewers or they covered only part of the town's area (Hodge 1992, 334). Hence streets and rivers must have worked as sewers in many towns in the same way as evidenced by literary sources in ancient Smyrna (Strabon. *Geografia*. XIV, i, 37) and Amastris. (Plinius. *Epistulae*. X, 98). Also in the poorer areas of Rome, streets must have worked as sewers. Against this background, it probably was salubrious that the abundant waters of aqueducts were flushed to the streets and sewers. (Frontinus. *De aquaeductu urbis Romae*. 88, 111) The Romans did not use taps like modern people, but kept their water running all the time. The constant flow of the water assured that in towns equipped with aqueduct(s) there was plenty of "waste" water to flush the toilets, sewers and streets.

Good drainage was a necessity for a bath, where huge amounts of water were consumed. Although the Romans are considered to have been good engineers, they sometimes failed due to problems in drainage, as evidenced by the abandonment of a large bath in Britain after only a couple of decades in use (Burnham & Wacher 1990, 300).

A public toilet was often built in proximity to or inside a bath so that it was easily entered from both inside and outside of the bath (Figures 5.6 a and b). (Brödner 1983: 116-118, 153-154; Yegül 1992: 411–413) This solution for the location of public toilets had several advantages: 1) the abundance of water that was conducted to the bath could also be used to flush the toilet; and 2) the distance to the sewer needed for the bath was short. Although the archaeological evidence of private toilets is scant when compared with public toilets, it can be concluded that at least the inhabitants of Pompeii and Ostia seemed to consider a toilet to be natural to have in their house. (Hermansen 1981; Jansen 1997) The private toilets were either flushed with water or they had a vertical drainpipe that hardly needed flushing with water. Piped water for flushing private toilets seems to have been a rarity. In many cases the private toilet was located near the kitchen, most probably because of the convenience of building a common drain for them.

Figure 5.6 a Roman toilet in the bath of Cyclopes in ancient Thugga (modern Dougga in Tunisia). (Photo: Heikki S. Vuorinen)
Figure 5.6 b The water channels running in front of and under the seats are typical remains of toilets found in archaeological excavations of ancient baths; Roman bath in Paphos, Cyprus. (Photo: Marjatta Vuorinen)

## 5.6. WATER AND PUBLIC HEALTH

Ancient authors have made some comments about the influence of different kinds of water on the health of people, but this data is hard to interpret. The archaeological evidence is also quite meagre on the connection between the health of people and their water supply. Thus, because of the inadequacy of sources, it is practically impossible to evaluate the health of ancient town populations and the role of water in it. It is, however, quite safe to conclude that despite the impressive measures used to obtain pure potable water, urban centres had serious health problems. (Morley 2005; Scobie 1986)

The poor level of waste management, including wastewater, most probably involved a major risk for public health during antiquity. For instance, toilet hygiene must have been quite poor. The Romans lacked our toilet paper. They probably commonly used sponges or moss or something similar, which was moistened in the conduit in front of the seat and then used to rinse their bottoms. In public toilets facilities were common to all; they were cramped, without any privacy, and had no decent way to wash one's hands. The private toilets lacked running water and they were commonly located near the kitchens. All this created an excellent opportunity for the spreading of intestinal pathogens.

Water-borne infections must have been among the main causes of death. (Grmek 1989: 16) Dysentery and different kinds of diarrhoeas must have played havoc with the populations. Although the ancient medical writers described different kinds of intestinal diseases, the retrospective diagnoses are difficult and the causative agents cannot be identified. (Stannard 1993) Summer and early autumn, when water resources were meagre in the Mediterranean world, must have been a time when drinking water was easily contaminated, and intestinal diseases were rife as presented in several passages in the Hippocratic writings. (e.g. *Airs, Waters, Places*. 7; *Aphorisms*. III, 11, 21, 22; *Internal Affections*. 26, 45) The mortality of children, especially recently weaned, must have been high, which is probably echoed in the following passage of a Hippocratic author: "*It is mostly children of five years that die from this disease* (dysenteries), *and also older ones up to ten years; other ages less.*" (*Prorrhetic II.* 22)

It should also be kept in mind that the salubriousness of the water supply must have differed markedly in accordance with the social status of people in the Roman towns. The rich had running water; the poor had to fetch their water from public fountains. The rich had their own baths and toilets, the poor had to use public toilets and baths. All this must have caused differences in the health of rich and poor people.

Written sources have very little to say about the water supply inside private dwellings. The comments made by the author of the Hippocratic treatise *Regimen in Acute Diseases*, dated from the end of the fifth century B.C. (Jouanna 1999: 410), suggest that the water supply in a typical Greek dwelling was far from ideal. "*[...] few houses have suitable apparatus and attendants to manage the bath properly. [...] The necessary things include a covered place free from smoke, and an abundant supply of water [...]*" (*Regimen in Acute Diseases.* 65)

As previously stated, a lot of the water in a Roman town was consumed in bath(s) connected to the aqueduct(s). Ideally shining marble walls and limpid water were considered a feature of a bath in Rome, the cleanliness of which was watched over by aediles (Seneca. *Ad Lucilium epistulae morales.* 86). Baths were probably also beneficial for public health in towns where there was an abundance and rapid turnover of water. However, in towns where water was in short supply, cisterns had to be used and the turnover of water was slow, the role of baths was probably negative for public health.

The relatively extensive urbanization during antiquity may to a significant extent be attributable to the importance assigned to the transportation of sufficient amounts of good quality water to the towns. However, during antiquity the indirect effects of water on public health might have been greater than the direct effects. Agriculture depended on the proper amount of available water. Too much or too little were bad things. Droughts and floods led to food shortages and famines. Two important diseases caused by parasites were intimately connected with water and the ways water was managed during antiquity: malaria and schistosomiasis.

The breeding of mosquitoes depended on water and mosquitoes spread malaria, which was a serious and widespread health problem around the Mediterranean during antiquity. (Sallares 1991: 271–281, 467–469; Sallares 2002; Grmek 1989: 275–283) Malaria was well documented by Greek and Roman medical authors from the Hippocratic writings onwards. Among the cases in *Epidemics I* and *III*, a serious complication of chronic malaria, blackwater fever, has been identified by Mirko D. Grmek at least in one patient, Philiscus, but probably also in another, Python. (*Epidemics I* fourteen cases, case 1; *Epidemics III*, sixteen cases, case 3; Grmek 1989: 295–304) A fine description of malarial cachexia is to be found in *Airs, Waters, Places* (*Airs, Waters, Places*. 7; Grmek 1989, 281).

Not only did the Romans lead water into the towns by aqueducts, but they also drained water out of towns. Most of the drainage (e.g. lakes and marshes) occurred, however in rural areas. Drainage in antiquity, as in later times, was a complicated business. It has even been argued that many drainage schemes in antiquity were failures. (Sallares 2002: 75) Sometimes the efforts to drain a lake or marsh might have favoured the spread of malaria in the area.

Schistosomiasis (bilharzias) has been for millennia a scourge in Egypt. The parasite (blood-vessel inhabiting worms) has an intricate relationship between the human host and a snail intermediate host. The disease must have been very common in villages of the Nile Delta and other areas where humans had constant contact with fresh water. The type of agriculture (irrigation, flooding of the Nile) must have spread the disease. Although the evidence from ancient Egyptian medical papyri remains hard to interpret, there is strong paleopathological evidence of schistosomiasis in human remains from ancient Egypt (Sandison & Tapp 1998, 39–40; Millet et. al. 1998, 99–101; Nunn & Tapp 2000).

Water had an important indirect effect on the health of people through water transportation, which played a definite role during antiquity in the spread of disease. Like food and people, also pathogens were easily transported by water during antiquity. Maritime trade was especially vigorous around the Mediterranean in the period 200 B.C.– 200 A.D. This meant that the Mediterranean world became more or less a common pool of infectious diseases. (McNeill 1979: 78–140)

The contamination of water by lead has been a topic in the discussions concerning the health of people in Roman cities. Roman authors expressed doubts concerning the use of lead pipes and recommended the use of ceramic pipes (Vitruvius. *De Architectura*. 8.6.10–11; Palladius. *Opus Agriculturae*. 9.11; Columella. *Rei Rusticae*. 1.5.2; Plinius. *NH*. XXXI.31.57). However, in practice it seems that although ceramic pipes were used (Figure 5.7), water was in many situations routinely distributed by lead pipes, as revealed by both written sources (Vitruvius *De Architectura*. 8.6.1, 4–6; Frontinus. 25.2, 27.3, 29.1, 30.1, 39–63, 105.5, 106.3,

Figure 5.7 Various degrees of calcium carbonate incrustation seen in sections of Roman water pipes piled up against the wall of a Roman bath in the town of Kos, island of Kos, Greece. (Photo: Marjatta Vuorinen)

115.3, 118.4, 129.4–6) and archaeological remains (Bruun 1991: 124–127; Hodge 1992: 307–315). So in theory there was a possibility for lead poisoning to occur through the water. Yet, there are two reasons to believe that exposure through water was quite minimal, as pointed out by A. Trevor Hodge (Hodge 1981 and 1992: 308). Firstly, as a consequence of the quality of the water, a calcium carbonate coating separated the lead and the water in most cases (Figure 5.7). Secondly, because of the constant flow, the contact time of water in the pipe was too short for contamination by lead.

## 5.7. EPILOGY

Throughout antiquity tasty or tasteless, cool, odourless and colourless water was considered the best, and stagnant, marshy water was avoided. These ideas were held until the end of antiquity as expressed by the fifth century author Palladius (*Opus Agriculturae*. I, 4). Paulus Aeginata in the 7th century A.D. referred to Hippocrates, and favored rainwater and considered best water that is limpid, cool, odorless and tasteless. (Paulus Aeginata I.50)

Frontinus expressed clearly that a water system needed constant maintenance to function efficiently. (Frontinus 116–123) For instance, calcium carbonate incrustation that formed inside the conduits (Figure 5.7) needed constant removal, otherwise the flow of water would eventually stop. (Hodge 1992: 227–232) In Italy aqueducts and baths seem to have been maintained even after other monumental buildings in the towns, with the exception of town walls and palaces, fell into disuse in late antiquity. (Ward-Perkins 1984: 31, 128) In Antioch and other Near Eastern towns, at least part of the ancient water system was maintained into the Byzantine period and possibly up to the Era of Islam. (Kennedy 1992) Although there were continuities from antiquity to the Middle Ages, the water supply was more limited and the Christian water patronage replaced the classical one: it was a move from *luxuria* to *necessitas*. (Ward-Perkins 1984: 152)

## 5.8. CONCLUSIONS

Written sources indicate that the idea of the salubrity of water is connected to the general scientific level of the society. The quality of the water was examined by the senses: taste, smell, appearance and temperature. Also the health of the people and animals using a water source was considered. Tasty, cool, odourless and colourless water was considered the best. Stagnant, marshy water was avoided.

Settling tanks, sieves, filters and the boiling of water were methods used to improve the quality of water during antiquity. The available sources do not permit the estimation of the public health effects of these methods.

Taking care of transporting a sufficient amount of good quality water to a town might have been the fact that made it possible to reach a relatively high level of urbanization during antiquity. The role of baths was probably positive for the public health in towns where an abundant amount of water was available. In towns where water was scarce, the role of baths was probably negative for the health of people.

The poor quality of waste management (including wastewater) was most probably a major danger for public health during antiquity. Consequently, waterborne infections must have been one of the main causes of death during antiquity. In ancient towns there were, however, vast social differences in the salubrity of the water supply.

The indirect public health effects of water might have been greater than the direct effects during antiquity. Malaria was widespread around the Mediterranean and schistosomiasis was common in Egypt. Droughts and floods led to food shortages and famines. Food, people and pathogens moved most easily by water during antiquity.

## 5.9. REFERENCES

PRIMARY SOURCES

Aëtius, On the opinions of the philosophers V.30.1, in Longrigg J. Greek Medicine. From the Heroic to the Hellenistic Age. A Source Book. Duckworth, London 1998: 31.

Airs, Waters, Places. In Hippocrates Volume I, with an English translation by W. H. S. Jones. The Loeb Classical Library.

Aphorisms. In Hippocrates Volume IV, with an English translation by W. H. S. Jones. The Loeb Classical Library.

Columella. Rei Rusticae. Lucius Junius Moderatus Columella on Agriculture in three volumes; with a recension of the text and an English translation by Harrison Boyd Ash. The Loeb Classical Library.

Diocles of Carystus. A collection of the fragments with translation and commentary. Volume one, text and translation by Philip J. van der Eijk. Brill, Leiden 2000.

Diseases I. In Hippocrates Volume V, with an English translation by Paul Potter. The Loeb Classical Library.

Epidemics I. In Hippocrates Volume I, with an English translation by W. H. S. Jones. The Loeb Classical Library.

Epidemics II. In Hippocrates Volume VII, edited and translated by Wesley D. Smith. The Loeb Classical Library.

Epidemics III. In Hippocrates Volume I, with an English translation by W. H. S. Jones. The Loeb Classical Library.

Epidemics VI. In Hippocrates Volume VII, edited and translated by Wesley D. Smith. The Loeb Classical Library.

Frontinus Sex. Iulius. De aquaeductu urbis Romae. The stratagems and The aqueducts of Rome; with an English translation by Charles E. Bennett; edited and prepared for the press by Mary B. McElwain, The Loeb Classical Library.

Galen. De Sanitate Tuenda. A translation of Galen's Hygiene (De Sanitate Tuenda) by Robert Montraville Green. With an introduction by Henry E. Sigerist. Charles C. Thomas, Springfield 1951.

Humours. In Hippocrates Volume IV, with an English translation by W. H. S. Jones. The Loeb Classical Library.

Internal Affections. In Hippocrates Volume VI, with an English translation by Paul Potter. The Loeb Classical Library .

Palladius. Opus Agriculturae. Palladii Rutilii Tauri Aemilian Vri Inlustris. Opus Agriculturae, De Veterinaria Medicina, De Insitione. Bibliotheca Scriptorum Graecorum et Romanorum Teubneriana. B. G. Teubner Verlagsgesellschaft, Leipzig, 1975.

Paulus Aeginata. Paulos von Aegina des besten Arztes Sieben Bücher. Übersetzt und mit Erläuterungen versehen von I. Berendes. E.J. Brill, Leiden 1914.

Plato. Laws. The Loeb Classical Library.

Pliny (the Elder). Natural History; in ten volumes; with an English translation by H. Rackham, The Loeb Classical Library.

Pliny (the Younger). Plinius Secundus Caecilius minor. Letters; with an English translation by William Melmoth; rev. by W. M. L. Hutchinson, The Loeb Classical Library.

Prorrhetic II. In Hippocrates Volume VIII, edited and translated by Paul Potter. The Loeb Classical Library.

Regimen IV or Dreams. In Hippocrates Volume IV, with an English translation by W. H. S. Jones. The Loeb Classical Library.

Regimen in Acute Diseases. In Hippocrates Volume II, with an English translation by W.H.S. Jones. The Loeb Classical Library.

Seneca L. Annaeus minor. Ad Lucilium epistulae morales; in three volumes; with an English translation by Richard M.Gummere, The Loeb Classical Library.

Strabo. The Geography of Strabo; in eght volumes, with an English translation by Horace Leonard Jones, The Loeb Classical Library.

Vitruvius. On Architecture; in two volumes; translated into English by Frank Granger, The Loeb Classical Library.

## LITERATURE

Addyman P.V. (1989) The archaeology of public health at York, England. World Archaeology 21, 244–264.

Bédoyère de la G. (1998) Hadrian's Wall: History and Guide. Tempus Publishing, Stroud.

Bruun C. (1991) The Water Supply of Ancient Rome. A study of Roman Imperial Administration. Societas Scientiarum Fennica. Commentationes Humanarum Litterarum 93, Helsinki.

Brödner E. (1983) Die römischen Thermen und das antike Badewesen, eine kulturhistoriche Betrachtung. Wissenschaftliche Buchgesellschaft, Darmsstadt

Burnham B.C. & Wacher, J. (1990) The `small towns' of Roman Britain. Batsford. London.

Eschebach L. (1987) Pompeji. In Die Wasserversorgung Antiker Städte. Band 2. pp. 202-207. Verlag Philipp von Zabern, Mainz am Rhein.

Garbrecht G. (1983) Ancient water works – lessons from history. Impact of science on society, No. 1, 5–16.

Garbrecht G. (1986) Wasserversorgungstechnik in römischen Zeit. In Die Wasserversorgung Antiken Rom. 3. Auflage. pp. 9–43, R. Oldenbourg Verlag, Munchen – Wien.

Garbrecht G. (1987a) Die Wasserversorgung des antiken Pergamon. In Die Wasserversorgung Antiker Städte. Band 2. pp. 11–47, Verlag Philipp von Zabern, Mainz am Rhein.

Garbrecht G. (1987b) Rom. In Die Wasserversorgung Antiker Städte. Band 2. pp. 208–213. Verlag Philipp von Zabern, Mainz am Rhein.

Giorgetti D. G. (1988) Bologna. In Die Wasserversorgung Antiker Städte. Band 3. pp. 180–185, Verlag Philipp von Zabern, Mainz am Rhein.

Graham J.W. (1987) The Palaces of Crete. Revised edition. Princeton University Press, Princeton.

Grewe K. (1988) Römische Wasserleitungen nördlich der Alpen. In Die Wasserversorgung Antiker Städte. Band 3. pp. 43–97. Verlag Philipp von Zabern, Mainz am Rhein.

Grmek M. D. (1989) Diseases in the ancient Greek world. Translated by Muellner M and L. The Johns Hopkins University Press. Baltimore.

Hermansen G. (1981) Ostia: Aspects of Roman *City Life*. The University of Alberta Press, Edmonton.

Hodge A.T. (1981) Vitruvius, lead pipes and lead poisoning. American Journal of Archaeology, 85, 486–491.

Hodge A.T. (1992) Roman Aqueducts & Water Supply. Duckworth, London.

Jansen G. (1997) Private toilets at Pompeii: appearance and operation. In Bon, S.E. and Jones, R. (Edited by). Sequence and Space in Pompeii. Oxbow Monograph 77, pp. 121–134, Oxford.

Jouanna J. (1999) Hippocrates. Translated by M. B. De Bevoise. The Johns Hopkins University Press, Baltimore and London.

Kennedy H. (1992) Antioch: from Byzantium to Islam and back again. In Rich, J. (Edited by). The City in Late Antiquity. pp. 181–198, Routledge, London

Laurence R. (1996) Roman Pompeii, space and society. Routledge, London.

McNeill W.H. (1979) Plagues and Peoples. Penguin Books, Harmondsworth.

Millet N.B., Hart G.D., Reyman, T.A., Zimmerman, M.R. and Lewin P.K. (1998) ROM I: mummification for the common people. In Cockburn, A., Cockburn, E. & Reyman, T.A. (Edited by). Mummies, Disease & Ancient Cultures. Second edition. pp. 91–105. Cambridge University Press, Cambridge.

Morley N. (2005) The salubriousness of the Roman city. In King, H. (Edited by). Health in Antiquity. pp. 192–204. Routledge, London.

Nunn J.F. & Tapp, E. (2000) Tropical diseases in Ancient Egypt. Transactions of the Royal Society of Tropical Medicine and Hygiene 94, 147–153.

Robertson A.S. (1990) The Antonine Wall: a handbook to the surviving remains. Revised and edited by Lawrence Keppie. 4th edition. Glasgow Archaeological Society.

Rogers R.H. (2004) Introduction. In Frontinus De Aqueductu Urbis Romae. Edited with introduction and commentary by R. H. Rogers. pp. 1–61. Cambridge University Press, Cambridge.

Sallares R. (1991) The Ecology of the Ancient Greek World. Duckworth, London.

Sallares R. (2002) Malaria and Rome. A history of malaria in ancient Italy. Oxford University Press, Oxford.

Sandison A.T. & Tapp, E. (1998) Disease in ancient Egypt. In Cockburn, A., Cockburn, E. & Reyman, T.A. (Edited by). Mummies, Disease & Ancient Cultures. Second edition. pp. 38–58. Cambridge University Press, Cambridge.

Scobie A. (1986) Slums, sanitation, and mortality in the Roman world. KLIO, 68, 399–433.

Stannard J. (1993) Diseases of Western Antiquity. In Kiple, K.F. (Edited by). The Cambridge World History of Human Disease. pp. 262-270, Cambridge University Press, Cambridge

Wacher J. (1975) The towns of Roman Britain. Batsford, London.

Ward-Perkins B. (1984) From Classical Antiquity to the Middle Ages. Urban Public Building in Northern and Central Italy AD 300–850. Oxford University Press, Oxford.

Watermann R. A. (1974) Medizinisches und Hygienisches aus Germania inferior. Ein Beitrag zur Geschichte der Medizin und Hygiene der römischen Provinzen. Habitationsschrift. Universität Köln.

Yegül F. (1992) Baths and Bathing in Classical Antiquity. The Architectural History Foundation. The MIT Press, Cambridge (Massachusetts) and London (England).

# 6
# WATER SUPPLY IN THE LATE ROMAN ARMY
*Ilkka Syvänne*

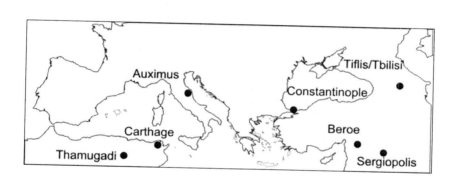

Figure 6.1 Key places mentioned in the chapter.

## 6.1. GENERAL BACKGROUND

The aim of this chapter is to trace the systems of water supply and the usages of water in the late Roman army and their historical importance. The source material for the study consists of period military treatises such as the Strategicon of Maurice, Peri Strategikes of Syrianus the Magister, Excerpts of Polyaenus, and the Epitome of Vegetius, and period histories by such authors as Ammianus Marcellinus, Procopius, and Theophylact. Furthermore, the period sources are compared with earlier and later developments in the field of water supply in the Roman and Byzantine empires (see the primary sources listed in the references).

The Romans were fully aware of the importance of the water supply for the success of any military operation. For example, Syrianus the Magister (6th century) stated in no uncertain terms that making provisions for food and water for the army and for the civilian population was both the beginning and the end of any plan of defense. (Syrianus 6.30-2) The whole military doctrine was based on the assumption that the Roman soldiers would have an adequate and safe water supply available to them wherever their forces were operating.

## 6.2. CITIES AND FORTS

During the early empire, cities often grew up on the places where the Romans had built permanent camps. In other words, when the civilians (and the families of the soldiers) serving the needs of the soldiers came to live next to the army base, the army camp gradually grew into a city. The role of the army was important in the urbanization of the more remote areas, and the increasing civilian population put additional stress on the availability of water. However, during the late empire the process was different. In peacetime the army was billeted usually in already existing cities and forts, each of which already had its own sources of water and storage facilities for other liquids such as wine. This meant that now the cities also had to cater to the needs of the army too. However, new cities and forts were also founded to meet the new needs of defence. Not surprisingly, Syrianus noted that even smaller border forts required a good supply of food and water to withstand possible sieges. Furthermore, on the list of requirements for the founding of new cities, right after the defensibility of the site, was the second requirement as noted by Syrianus: the availability of safe local drinking water (usually springs and wells) to the population of the city as well as to the likely refugees when the enemy had invaded. If the source of water was outside the walls, the building of the site was to be abandoned or a way was to be found for the water bearers to go there even in the presence of enemies. (Syrianus 9.37-9, 10; note also Procopius, Wars, 6.27)

Even earlier, Vitruvius (1.4) had stressed the importance of choosing a healthy site that had good and healthy water source(s). According to him, when the ancient Romans chose to build a town or an army post, they first conducted a series of tests at predetermined intervals by sacrificing some of their cattle so that they could examine their livers to see if there were any signs of abnormality resulting from disease, food or water. No defensive works were built until the Romans had satisfied themselves that the site had good and healthy water and food sources. Vitruvius (8.4) also included some other common-sense instructions for the testing of water springs before using them as water sources for the cities via aqueducts and pipes, such as observing the physique of the people dwelling there, and by conducting tests to see whether there was any sand or mud residue in the water or any "filth" in the water bed. He (8.3) also classified different types of water according to their beneficial or detrimental qualities for health.[1] Vitruvius's text (8.1-6) also includes methods used for the finding of water as well as for the building of aqueducts and wells. Vitruvius recommended the use of clay pipes instead of lead pipes, because he considered lead detrimental for human health (for additional information about the dangers of lead etc., see Bruun 1991, 123-130 and the chapter by Heikki Vuorinen in this book). At this juncture it should be noted that the criticism shown by Zonaras (III.157) towards the emperor Justinian (527-565) for the destroying of the conduit made of lead is beside the point. Justinian's actions in this case were entirely justified. The lead was and is harmful to the human health.

Despite the fact that the Romans wanted to avoid springs or rivers that had sandy and muddy water, they were occasionally forced to use them as a result of the population growth in the cities. However, the Romans had the necessary technical know-how to resolve this problem. They simply used large settling tanks (piscinae) near the water source and near the city. (See Frontinus Book 1; Klein 2001, 31; Blackman et al. 2001, 36-8.) Regardless, this solution was still found ineffective against clay particles (1ìm across in size) and the quality of the water was found unsatisfying. The Romans knew that the water could be cleared by using alum (naturally occurring aluminum sulphates), but its use was impractical for the continuous flow system in use in the Roman aqueduct systems and therefore the water was classified according to its quality for different purposes. (Blackman et al. 2001, 37; Frontinus, Aqueducts 1.15) In other words, the availability of safe water was very high on the list of requirements.

---

[1] Note also the classification of the types of water according to their supposed effects on health in the Pseudo-Aristotelian text Secret of Secrets (Secretum secretorum, Arabic version in Opera hactenus inedita Rogeri Baconi Fasc. V, Secretum secretorum cum glossis et notulis, Robert Steele, Oxford (1920), 176-266). In this text, (Pseudo-)Aristoteles supposedly gives Alexander the Great general advice on every facet of life the ruler should know. The text was originally probably of Byzantine or Syrian origin and was then translated into Arabic and then later into Latin and other European languages. In the surviving and oldest version in Arabic: water pp. 205-6 (English translation); wine 206-9; bath 209-212.

Then again, there were exceptions to the above recommendation. In practice, many of the cities had grown in size and population after their founding, which meant that the local water supply was often insufficient on its own. The best example of this development is the city of Constantinople (for the water supply of Constantinople, see: Mango 1995; with Procopius, Buildings Book 1, water supply esp. 1.11.10-15). It did not have an adequate local water supply in the form of wells or springs, which meant that other solutions had to be sought out. In the traditional Roman manner, the first of the solutions had been the building of aqueducts (Hadrian's aqueduct and later Valens' aqueduct, which was really a network of water pipes both above ground and below). However, the flow of water from these pipes was still insufficient for the needs of this late antiquity megalopolis (included hospices, gardens, fountains, parks, lawns, sewers, public baths; note Syrianus 11.25ff) especially during the summer months. The most important of these were the needs of the public baths, which required huge amounts of water to function.

The solution that the Romans adopted was based on the use of water cisterns (100 known examples) for storing the water for the drier summer months. There were both covered and open-air cisterns. As is obvious, besides the water from the aqueducts, the open-air cisterns received additional water intake from the rainwater. The size of the cisterns varied from the smaller ones intended for private use and other use (mansions, hospices, monasteries, churches etc.) to the larger and huge ones for public and military use. The building of new cisterns reflected the ever-increasing size of the population and the building of new public baths etc. However, the building of new cisterns did not solely reflect the needs of the population; it also reflected the needs of the military and the court. For example, the imperial palace and the accompanying military barracks had their own enormous open-air cistern (130m x 80m, max. capacity approx. 125,000 cubic m).

The city needed a safe source of water especially during sieges, when it was very likely that the enemy would cut off the water supply by severing the aqueducts leading into the city, hence the building projects of cisterns by the emperors Phocas (602–610) and Heraclius (610–641). The availability of water for the civilians during siege(s) was of utmost importance for the successful defence of the city. Firstly, the civilians defended the walls as militia; secondly, they supported the military; and, thirdly, they could have been a source of serious unrest if their needs were not taken into account. It is well known that, besides the faction riots, there were also food riots that required the use of military force. A lack of water would have resulted in even worse problems. For additional information about the water supply of the cities, especially from the point of view of the civilians, see the chapter by Heikki S. Vuorinen in this book.

When discussing the rebuilding, restoration and building projects of the emperor Justinian (527–565) outside the city of Constantinople, in his *De aedificiis* (Buildings), Procopius concentrated his attention on the improvements made in the defences (fortifications, towers, firing loopholes, moats, walls etc.) as well as on the availability of safe water for drinking and bathing purposes. The principal areas facing a water shortage lay in the east but there are also some instances elsewhere. According to him, there were many means of obtaining additional water supplies for both drinking and bathing purposes. When there was a river near the site, the water could be obtained by means of transferring the water channel to run beside the city or through it or an additional new channel was made or tunnel cisterns were dug from the city to the river. On occasion, when there were problems with flooding river(s), dams with sluices or new water channels were also built to control the flow. In some cases, when the water sources (springs, rivers) were further away, Justinian had conduits and aqueducts constructed on the sites. In some cases, the city was simply placed on an island. In practice, most of the sites also received new water cisterns for the purpose of storing water from the aqueducts or rivers for the drier summer months. In addition, cisterns were also built for the purpose of obtaining rainwater where there were no local water sources at all (some of the desert fortresses and cities) or when the other sources were inadequate for the purpose. Most importantly, Procopius praised the construction of new public baths or the restoration of former ones as a sign of urban development under Justinian. The baths were considered a necessity for the military garrisons as well as for other public officials such as the postmen. In short, the availability of safe water was very high among the state priorities for both military and cultural reasons. Though, since Procopius noted the construction of wine cellars to satisfy the needs of the army, it is clear that there were also other needs involved.[2]

The sources do not tell how the healthiness and good quality of the water was maintained when it was in storage in the cisterns. Perhaps the best guess is that the constant flow of water was maintained by leading some of the water to the distribution pipes while additional water was brought from the aqueduct. It is also known that the Romans knew how to purify the water through the use of alum and sand filtering (see later) and if all else failed through the boiling of the water. (Julius 2.6, esp. 2.6.17-20, For additional information about the methods of water purification, see especially the chapter 5 in this book) The use of any of these techniques is possible.

As is obvious, the size and type of military component in garrison duty as well as the size of the civilian population depended upon the perceived availability of water supplies in the locality. For example, in a desert environment the size of the garrison was severely limited

---

[2] Procopius, Buildings 2.2, 2.3.16-26, 2.4.12-13, 2.4.22-5.1, 2.5.9-11, 2.6.1-11, 2.7.1-16, 2.8.8-25, 2.9.1-20, 2.10.2-25, 2.11.2-12, 3.3.7-8, 3.4.15-20, 3.5.13-15, 3.7.1, 3.7.19-25, 4.1.19-27, 4.1.39-42, 4.2.1-15 (esp.4.2.6, 14), 4.2.18, 4.3.1-5, 4.3.8, 4.3.27-8, 4.8.18, 4.9.14-6, 4.10.20-23, 4.11.11-3, 5.2.1-5.4.6, 5.4.15-18, 5.5.4, 5.5.8, 5.5.14-20, 5.8.1, 6.2.5-11, 6.4.6-11, 5.5.1-11, 6.5.8-16, 6.7.1-11. Of particular interest are the following: 2.6.11 (the bath restored for the enjoyment of the garrison troops); 2.9.12-17 (spring and a lake probably cleaned from pollution by unknown means but lacunae in the text prevents one from reaching a final conclusion on this subject); 2.10.22-5 (water-channels, fountains, sewers, theatres, baths, public buildings, hospices); 3.5.13-15 (Bizana transferred to a new place called Tzumina 3 miles away because the former had been situated on a plain that had unhealthy standing pools of water); 4.10.20-23 (baths, and storehouses for grain and wine built to satisfy the needs of the soldiers); 5.3.3 (bath at the lodgings of the veredarii/couriers of the public post restored); 6.5.8-16 (Caputveda founded because a spring was found).

by the availability of water in the locality (usually an oasis), which, on the plus side, naturally meant similar difficulties for any enemy force in question that had to carry with them all of the supplies of water to even begin any operations against the fort. In other words, in a desert environment even small forts, giving the Romans control of the water sources, were often all that was needed in containing the threat of the desert tribes (such as the Bedouins and Berbers). The siege of Sergiopolis in 542 A.D. shows well what the difficulties were facing any invasion force in the desert terrain. In that case, the 6,000-strong Persian army was forced to retreat after a short siege because the Persians had exhausted their supplies.

The desert terrain was seen as a defensive barrier bound to cause troubles for any invader. However, since the desert was also less populated and less well guarded, the armies sometimes attempted to cross the deserts in order to achieve strategic surprise. If this failed—and it usually did, the invaders were in serious trouble because they now had to retreat through the waterless tracts of land. Therefore, these desperate gambles were rarely taken by any others than by the desert tribes themselves. (Procopius, Wars 2.20.1-16, Buildings 2.2.3-9; Syvänne 2004, 67, 445-7, 462-4, 503)

Even though there is contemporary proof from the period under study, 300–640 A.D., the best evidence for the actual construction methods in use for the water pipes, baths and latrines in the Roman army dates from an earlier period. This is because the state of the archeological research of the army sites is most advanced in Great Britain. These studies have shown that every permanent fort had its bathhouse (a changing room, a cold bath, a warm bath, a sauna and hot bath). Clean water was led by gravity through aqueducts or conduits into settling tanks from which it was flowed through pipes to different parts of the fort. Parts of the water were directed to the drain for the flushing of the latrines. Part of this water was also channelled for the cleaning of the sponges (used instead of toilet paper) in front of the toilet seats. The sewers discharged into rivers at points well below the watering points of the animals and when this was not possible the Romans used large soak-ways that resemble the modern septic tank. However, when the site was on a hill, the Romans constructed water tanks below ground or used open-air cisterns for the collecting of the rainwater, as discussed above. As Webster noted, the tanks were probably kept fresh and clean by fatigue parties. In this case, the latrines were deep slots cut into the ground with buckets underneath that were emptied by those assigned to this unenviable duty. (Bohec 2000, 160-161; Goldsworthy 2003, 106-107; Webster 1998, 269-260)

Archaeology of late Roman and Byzantine military sites has also shown that the water could be stored in special water towers or even in tunnel cisterns (confirmed by Procopius and military treatises) that were connected to a nearby river. Excavations in Syria and Africa (5-7$^{th}$ centuries) have also revealed the presence of rows of latrine chutes in tower-houses and also small chambers corbelled out from the wall (= corbelled latrines). The dearth of water in drier areas had clearly forced the Romans to adapt their designs to the prevailing conditions. These lavatories are clear predecessors to the typical Italian medieval toilet-towers. Ironically, the excavations at Thamugadi in North Africa have also revealed the presence of baths that are nowadays called Turkish (sic!) baths rather than Roman baths. (Foss & Winfield 1986, 7-23, 151, 154, 226-7; Pringle 1981, 164-166) All in all, the late Roman military adopted a great many different solutions to the same problems facing them depending on the circumstances and thereby paved the way for later developments in their successor states.

The evidence for the laundry services in the military garrisons is less clear or entirely neglected by the historical sources. However, it is (still) quite clear that such services were required and used. Whether the Romans used their bath facilities for this or whether they used some other solution is not known with certainty and probably depended upon the circumstances. However, from the civilian side of life it is known that the Romans knew how to clean their clothes with detergents. In short, the availability of safe drinking water (and wine and food and medical care), good personal hygiene and effective sanitation ensured better fighting effectiveness for the Roman army than their enemies had.

## 6.3. ESTIMATING THE WATER SUPPLY

How did the Romans know that they would have an adequate supply of water for their population and military establishment? To answer this question, they firstly needed to know what the water consumption was (desert environment vs. temperate climate; cultural requirements, i.e. parks, sewers and baths) and what the bare necessities were when the city was besieged (rationing of water supplies). This should not have posed them any great problems because the Romans knew from practice what the physical needs were of the individuals in different environments as well as what the needs were of the public establishments (i.e. baths etc. required a known amount of water to function). In fact, the army and navy had a logistical handbook (not extant) that could be used when the army prepared to campaign in different environments.

The principal problem is to know what actual methods of calculation were used in quantifying the supply of water that was needed to satisfy the known needs. Generally speaking, most of the modern historians are of the opinion that the ancient Romans (c.100 AD) were incapable of calculating the volume of flowing water exactly. However, all agree that the Romans possessed the means to obtain rough estimates that were in fact adequate to their needs. It was based on trial and error. (See Bruun 1991, 385-8; Blackman et al. 2001, 7-24; Kleijn, 2001 44-74.)

Mainly after Heron of Alexandria (2-1 cent. BC), Heron of Byzantium (c. 950) demonstrated the mathematical methods of calculation needed to measure the quantity and velocity of the water discharge of a spring. (The following is based on: Heron of Byzantium, Geodesia 1, 9-10; Sullivan, 281) First of all, he noted that the rates of gush varied according to the weather. When it rained there was excess water in the mountains and when there was a drought the volume diminished. This meant that during winters there was a plentiful supply whilst during summers not. Consequently, it was necessary to enclose the water of the spring and to attach quadrangular lead pipe to the lower part of the spring so that a certain amount

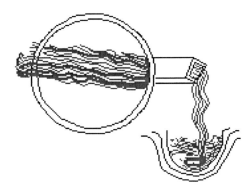

Figure 6.2 Measuring the discharge from a spring according to Heron. (Source: Vaticanus graecus 1605, folio 54v/Sullivan 2000)

of consistency could be achieved. The pipe was made much larger at the receiving point in comparison with the discharging point (height 5.85 cm, width 11.7 cm = flat pipe 35.1 cm, figure very close to one of the specifications used in ancient Rome for round pipes) and the sloping of the pipe was calculated by using the dioptra (a measuring device ). (See the commentary of the editor and translator Sullivan 249ff. See also: Heron of Alexandria, Dioptra; Julius Africanus 7.15.) The speed of the water flow was calculated by digging a ditch beneath the flow of the water. Thereafter, one simply calculated the amount of water discharged in an hour's time by using a sundial to calculate the daily output. Adjustments to the total daily output were naturally made according to the time of the year (day shorter during the winter etc.; water flow higher during the winter) in order to achieve a yearly average. In his example, Heron estimated that the hourly flowing into the ditch was on average approximately 20 kadoi (approximately 540 litres). Kados usually meant a shipboard water amphora, 18 kg, 27 litres; see Pryor 2003, 89) comprising a total of 480 kadoi (12,960 litres) per 24-hour cycle. It is clear from this why the Romans had to store up the extra flow of water in cisterns for later consumption and why they needed to tap several springs simultaneously. It also shows that the measurements were not standardized and depended upon many variables achieved through trial and error, as the previous studies have already demonstrated for the earlier period.  However, by using the above method of calculation, the Romans were (still) able to obtain relatively accurate measurements for both the volume and velocity for each of their water pipes separately for different time periods.

Heron of Byzantium also included methods of calculation for the volume of water reservoirs. It was a simple matter of calculating the dimensions of the cistern and then calculating how many measures keramia of water it held (according to Heron of Byzantium, the keramion= amphora equaled 80 Italian litres = one Roman cubic podes/foot = 80 pounds of wine). As an example he calculated the volume of the cistern of Aspar (constructed circa 459) rather than the cistern of Aetius (constructed circa 421) because the former had equal dimensions at its base. However, his values are at significant variance with the archaeological evidence (Heron 130.9 square metres vs. archaeology approx. 152m square) and should in this instance be seen only as mathematical examples.

In short, the Romans possessed the necessary tools to estimate the velocity and volume of water even if these were only roughly accurate. However, it is still quite clear that the Romans did not have the necessary tools to estimate the amount of water available to them from water deposits and wells. A case in point is the siege of Beroe by the Persians in 540. In that instance, the water supply from the well was exhausted in the middle of the siege because the soldiers had brought their animals and horses with them to the citadel. (Siege in Procopius, Wars 2.7.1-37)

## 6.4. MILITARY CAMPAIGNS

The late Roman military campaigns were thoroughly planned affairs that involved all of the state structures but in particular the logistical services under the prefects. Besides bread and meat, the logistical services also provided wine and safe water for consumption (Syrianus 14.31-2). The water supplies were transported in jars, barrels, flasks that were placed on camels, horses, wagons, and ships depending upon the circumstances. However, on campaign, the drinking water was usually obtained locally (at least when travelling light) rather than carried, whereas some of the wine (it retained its quality better) was always kept on hand for emergencies. For example, according to Theophylact (7.5.6-10), when in 594 the scouts of the army led by Peter, the brother of the emperor Maurice, failed to find water, the soldiers were forced to resort to using wine (what a terrible fate!). Obviously, a disaster would have ensued had this condition lasted, since the horses needed copious amounts of drinking water. In fact, it is quite safe to assume that the Roman armies usually carried only small amounts of water with them that were replenished periodically from some local water sources along their marching route.

It was the duty of the scouts and quartering parties to ensure that the next marching camp would have adequate water supplies. According to Syrianus (26), when the quartering parties rode ahead, they were required to test the water; if from streams, whether it was salty, bitter, or potable; if from standing pools it was tested to see if it might be harmful as a result of nature or because the enemy had tampered with it. Unfortunately, he doesn't state how the testing

Figure 6.3 Measurement of the capacity of a cistern according to Heron: The Cistern of Aspar. (Source: Vat. gr. 1605, fol. 53/Sullivan 2000)

was done, but a good guess is that the inspectors used their eyesight, nose, taste, common sense, and some other tests (how fast the water boiled; did it leave marks on the pot? etc.). One could visually inspect the site for any suspicious signs like poisonous herbs lying about, dead animals or birds near the locale, or some irregularity in the look and smell of the water. On the other hand, if there were local civilians using the water, it was probably safe. Finally the water could be tasted either by the persons in question or by using animals (or prisoners) when one suspected foul play or any real natural danger.

Indeed according to the Strategicon, it was the duty of the scouts to select a healthy and clean site (with adequate local supplies of food, water and fodder if possible) so that there would not be any outbreaks of diseases. The soldiers were also forbidden to address their sanitary needs inside the camp (i.e. latrines were dug outside the camp) especially when the army remained long in one place. Similarly, the cavalry was kept outside of the camp until the army was near the enemy. If the river in the locale was wide and swift or there was a lake, it was used to protect one of the flanks of the camp. The horses were always watered downstream from the camp, except when the water source was only a small stream. In that case the horses were watered from buckets. It was very important to maintain the quality of the water for drinking and cleaning purposes. In critical situations, it was acceptable to choose a site with a small river flowing through the camp. The treatise also noted that sometimes it was not necessarily recommendable to place the camp near water before they got near the enemy since this could cause the soldiers to become used to the habit of drinking a great deal of water. However, when the army prepared to engage the enemy in combat, the general was still required to pay special attention to the defense of his water supply in case the camp was needed as a place of refuge against the enemy. Before the battle, the general was to see that his soldiers did not drink copious amounts of wine. The military doctrine recommended the use of water (carried by the servants to the soldiers during the lulls in combat) before and during the battle to quench the soldiers' thirst so that they would maintain their battle

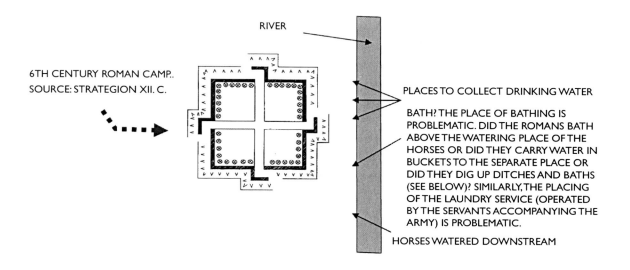

Figure 6.4 The sixth century Roman marching camp. (Source: Strategicon xii.C)

readiness. The rest of the late Roman military treatises are not as detailed as the Strategicon, but in what they portray, they resemble the former in all the major points. (Vegetius 1.21-5, 2.9-11, 3.8; Syrianus 26-29; Strategicon 5.3-4, 7.1.13, 7.2.9-10, 12.B.17, 12.B.22-3, 12.C.)

Unfortunately, the treatises do not mention in detail how the soldiers actually bathed during the campaigns or how their latrines were constructed. However, certain educated guesses can be made. When building latrines, the soldiers undoubtedly dug a deep ditch over which they placed their ersatz wooden toilet seats. If they used their sponges for wiping their rear ends, they undoubtedly used buckets of water to clean them or they dug a water channel from the river for the same purpose. Alternatively, instead of the sponges, they could have used leaves, moss or whatever vegetation was available there. During long stays, the ditches were certainly covered with dirt and new ones dug. They did not do their business into the river, which was also used for the watering of their horses downstream.

The Romans also saw to it that the bowels of their soldiers would work regularly. The simple but nourishing diet was eaten in small portions during the whole day to keep the digestive system operational throughout the daily cycle. They also used herbal concoctions or herbs such as absinth in their drinking water or sour wine to promote the good functioning of the digestive system. This solution ensured that the soldiers always had enough energy in their veins for whatever they were doing whilst it also ensured that the soldiers had very low amounts of body fat. In other words, they were lean and mean fighting machines. See Julius Africanus 2.9.

One may presume that in general the soldiers simply bathed in the rivers, streams and lakes, or the water was carried to them in buckets, or separate ditches were dug to make baths, but there were also other alternatives available to them. In desert conditions, the Romans undoubtedly had to put up with quite unsanitary conditions (note also the use of corbelled latrines in desert fortresses). There simply were not sufficient amounts of water available for all purposes. They undoubtedly rubbed their sponges in sand to clean them or used sand in the left hand as a substitute for the sponge. In short, besides the danger of dehydration, the lack of water in deserts also affected the health of the soldiery in other ways.

Similarly, when on scouting duty or taking part in other small-scale guerrilla operations, it is clear that the soldiers had fewer chances of taking care of their sanitary needs. For example, it is clear that they had to dig small holes for their excrement so that the enemy would not detect them from their smell and use moss or other undergrowth as a substitute for the sponge, etc.

Occasionally when there was enough time and the commander wanted to reward a group of soldiers, they could also build a travelling-bath. The use of the travelling baths by Roman armies was an age-old practice. For example, Velleius Paterculus (2.104.2) already tells us how the future emperor Tiberius in AD 6 loaned his bathing equipment to wounded soldiers. We are fortunate in that one of Sidonius Apollinaris' letters (c. AD 461-467) contains a description of how to build an ersatz bath.[3] According Sidonius, one first dug a trench close to the spring or river and poured a pile of heated stones into it. Then the ditch was roofed with a dome constructed of pliant hazel twigs twined into a hemispherical shape, and rugs of hair-cloth were thrown over this roof to keep the rising steam inside when the water was poured on the hot stones. After the participants had sweated sufficiently, they plunged into the hot water (presumably the ditch with the stones after it had cooled down or alternatively a separate ditch where the hot water was led after being warmed up by the stones) and after that they plunged into the cold water of the spring or the well or the river. (Sidonius, Letters, 2.9 To Donidius, esp. 2.9.8-9) In short, even in campaigning conditions the Roman soldiers did not necessarily have to suffer unsanitary conditions and could even afford some comfort.

From the point of view of the water supply, the Strategicon also contains a very important and interesting chapter dealing with the process of building a border fort by stealth. (The following discussion is based on the Strategicon 10.4. See also Syvänne, 2004, 310-311.) Firstly, one was to choose a suitable defensible site with a local water supply if possible. Secondly, one was to collect a sufficient force of artisans, building material, machines, infantry and provisions to last for three to four months. The enemy was to be diverted to another direction, and when this happened, the assembly of men, equipment and provisions was sent to the locale. From the viewpoint of the water supply, the most interesting section deals with

---
[3] Note also Herodotus 4.74-5. The Scythians also used quite a similar sauna.

## THE PROBABLE STRUCTURE OF A ROMAN TRAVELING-BATH

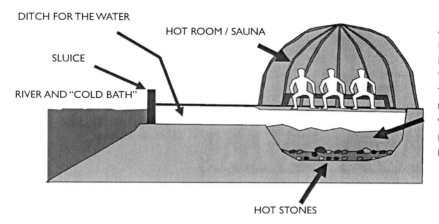

Figure 6.5 The probable structure of a Roman travelling-bath.

the methods being used when the chosen site had no local water supply, no streams or wells. Then the Romans carried their water at least 3-4 months' supply – or enough to last until winter rains arrived! – to the site in large earthenware jars or well-built barrels. The Romans filled the bottom of the water containers with clean gravel from a riverbed and then poured the water in. When they reached the building site, the water containers were stored and peg holes were drilled in them and receptacles placed below. When the water receptacles were full, they were emptied back into the water containers. By using this method, the water was filtered through a sand filter one drop at a time and also aerated in the process. The Strategicon also recommended the pouring of vinegar into the water to lessen the odour if the water had already begun to turn bad. The third century treatise of Julius Africanus also mentions the use of a series of water filters in jars that were used in the clearing of olive oil (and a wine substitute) and one may presume that a similar process could also have been adopted occasionally for the purification of the water supplies of the military establishment. However, the filtering of the water was feasible only when the army was stationary. It was not feasible when the army was constantly moving and therefore one can safely say that the army could not carry all its water supplies in a war of maneuvers. [Julius Africanus (3rd cent.) 1.19.27-33, 1.19.93-8 = Apparatus bellicus (c.950), 27 (cols.943-4), 29 (col.946)]

# WATER SUPPLY IN LATE ROMAN ARMY

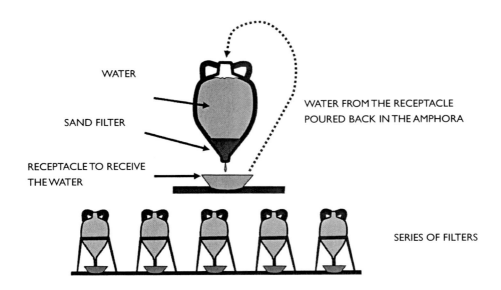

Figure 6.6 Sand filter according to Maurice.

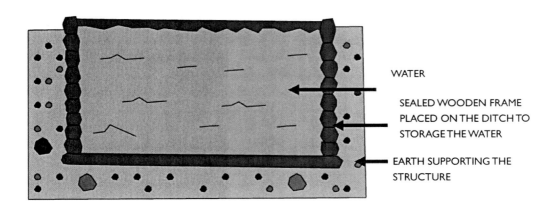

Figure 6.7 Open air ersatz cistern according to Maurice.

Maurice also recommended the building of one or more moderate-sized ersatz cisterns (6m x 3m wide and 3m deep) before cement cisterns could be built in the fort because the water usually kept its quality better in larger containers. The ersatz cisterns were made first by digging a trench into which was placed a wooden box made of thick planks. The seams and joints were sealed with pitch and tow or wicker. Again, the Strategicon does not mention how the Romans ensured that the water maintained its healthiness in the ersatz or cement cisterns. Yet again, one may presume that the water was constantly kept in motion because of the constant use of the water and the refilling of the containers through the rainfall or other means. It is also notable that at least in 343 near Singara, the Sasanian Persians also stored their water supplies in cisterns when they built a fortified camp. (Julian, Or. I.19D) It is also plausible to suggest that the Romans also followed a similar procedure when they stayed in one place for longer periods, for example, during offensive sieges – as in fact the use of cisterns when building border forts already shows.

Water supplies were also transported in ships, but the ability of the war galleys to transport adequate supplies of water to last more than three to four days was very limited. The oarsmen needed a litre per hour just for drinking to prevent dehydration when rowing and even more to compensate for the salty foods. According to John H. Pryor's calculation, a ship crew of 108 men (= an ousia, a later standard complement of men in a dromon), marines and their officers required at least 1,000 litres or one ton of water per ship per day (c. 8 litres per man). With bigger crews the figures obviously rise and with horse transports the figures rise even further. Apparently during this period the Romans still usually used the standard water amphorae (weight 18kg) that had a capacity of 27 litres rather than the more efficient barrels, but this did not make much difference in the overall capacity of the ships. The smaller dromons could not carry much more than approximately 5 tons of water (100 amphorae weighed around 4.5 tons) whilst the upper limit for the large dromons can be deduced from the later 14th and 15th century Genoese figures according to which the large galleys carried between 4 and 8 tons of water. This meant that under normal circumstances and in favourable winds (3–4 knots while at sea), the fleet could advance no more than approximately 170 miles (=275 km) by rowing in a three-day period after which it was necessary to refill the water containers for the next stage of the voyage. This meant that a great deal of pre-campaign planning was a necessity for any admiral and general. The loading of the water from a single well was too time-consuming, which meant that large fresh-water rivers or other similar voluminous water sources were sought after – and most of these were already protected by city walls. (Pryor 2003, 87-95).

However, even if the fleets usually replenished their water supplies every fourth day, it was still possible to sail even further by using sails in favourable winds or by using transport ships. For example, in 533 when Belisarius was about to disembark his troops on African soil, one argument put forth by the admiral against this was that it was difficult to supply the army on land from the ships when there were no safe harbours between there and Carthage and whilst the terrain was waterless. In other words, even if the fleet had previously replenished its water supplies regularly, it now carried enough food, water and wine to last at least about

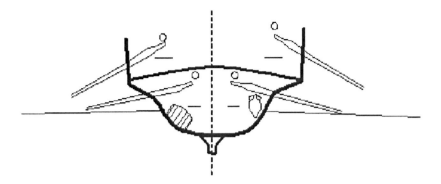

Figure 6.8 Storage of barrels or amphorae according to John H. Pryor, 2003, 91. There were 245-litre (54 gallon), 40-litre (8.8 gallon), 27-litre (6 gallon) barrels, and 27-litre (6 gallon) amphorae. The barrels or amphorae were placed next to the feet of the rowers.

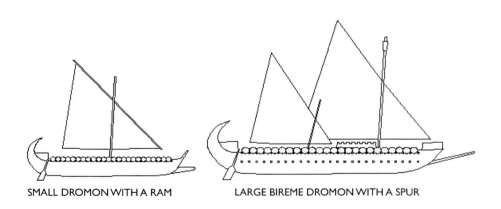

Figure 6.9 Byzantine dromon warships: Syvänne, 2004, 319. It is important to remember that these reconstructions of dromons as well as others in other studies are highly conjectural and differ from each other because we do not possess detailed pictures of the dromons. Marine archaeology will hopefully change this in due course.

19–23 days (for a force of approx. 50,000 men)—a huge figure by any standards.[4] However, in this case the Romans did not need additional water supplies from their fleet because they could still obtain them from local sources. Firstly, luck smiled on them. They found a spring of water at their landing site when the soldiers were digging the ditch. And, secondly, the land forces were subsequently able to use local resources because these lay undefended.

As far as hygiene on board a galley is concerned, one can state that it was very rudimentary. The smell on board the rowing ships was undoubtedly often horrific, especially on horse transports. The washing was done by using sea water and the latrines were basically only buckets that were periodically emptied or the men simply relieved themselves overboard.

## 6.5. WATER IN WARFARE

The use of water in warfare can be divided into three categories: 1) Denying the enemy access to water sources; 2); The use of water as a weapon; 3) The use of water as a means of transmitting poisons and diseases (In general for the spread of diseases and their treatment in antiquity, see: Vuorinen 1996; Vuorinen 1997; Vuorinen 2002). We have already touched upon the first usage above, namely the importance of the availability of water in sieges and in campaigning. In the field of defensive strategy, the Romans positioned their armies and forts and fortresses so that they always had a safe supply of water, whilst their enemies faced difficulties in obtaining drinking water. Similarly, military campaigns and marching routes were always planned with the availability of water in mind. When the Romans failed to follow this dictum or chose to take a risk, they often faced serious troubles. However, the availability of drinking water was at its most important in sieges. In defensive sieges, it was naturally of prime importance to secure the local water supply. Similarly in offensive sieges, it was always important to cut off the water supply to the city if possible. This could be done by blocking or by cutting off the water conduits and aqueducts; redirecting the river channel; using dams; blocking the wells; or poisoning the wells, etc.

This leads naturally to the question of the use of water as a weapon. The power of the flow of a river lent itself naturally also to its use as a weapon when the river flowed through or by a city. There were several variations. Firstly, one could dam the river upstream (and downstream) and then open up the dikes to crush the city walls. Secondly, the besiegers could build up dams below the city to flood it and then proceed to open up the sluices so that the walls would collapse. Thirdly, the besiegers could direct the flow of the river against a section of the wall to collapse it. During the fourth century, the Persians were particularly

---

[4] Belisarius' campaign against the Vandals: Procopius, Wars, 3.11.1ff, esp. 3.15; Syvänne, 2004, 63, 376, 434-7. The length of the land march and the dates in Pringle, 1981, 18-21. The troops landed at Caputveda on 27 August and the last troops from the fleet disembarked on 15 September and arrived at Carthage. When one adds to this figure the four days it took to sail into Caputveda, the total amount of water supplies carried by the fleet was meant to last about 23 days for 30,000 sailors, 2,000 rowers/marines, 10,000 infantry, 6,000 cavalry plus unknown numbers of Belisarius' bodyguard cavalry (approx. 2,000-5,000 men) and other people in 92 dromons and 500 other ships—a remarkable feat indeed.

fond of using rivers against Roman cities when the locale allowed this but the Romans were no less enthusiastic if they were presented with a similar opportunity, as the siege of Tiflis in 627 shows. Fourthly, when there were dams, these could also be used defensively by opening up the sluices to block the advance of the enemy as, for example, happened in 635. (Sources of the 4th century sieges translated in Dodgeon & Lieu 1991, 165-210; Tiflis: Syvänne 2004, 501-502; Dams opened: Syvänne 2004, p.456/Fihl.)

Thirdly, and possibly most importantly, the Romans also knew how to use water to spread diseases and poisons so that their enemies would succumb. Water was used in delivering poisons or disease-causing agents to the stomach of the enemy soldiers. The most commonly used method was the poisoning of wells, pools, cisterns and reservoirs with herbs, poisons, or dead animals, compost. Occasionally, the poisons, dead animals and compost were ground into powder form for ease of use. (For additional details, see the latest partially revised Syvänne 2004, 287-8, 491). A similarly devious tactic was to force the enemy to occupy unhealthy terrain such as swamps or places where the water was naturally contaminated. In other words, ancient peoples knew how to exploit the presence of unhealthy water sources for their own benefit. (For these ancient practices, see: Mayor 2003, 99-169; Grmek 1979.)

On occasion, when the enemy refrained from using water, the Romans could either give the enemy an abundance of wine in the form of false booty or alternatively they could deny the enemy the availability of wine to diminish their spirits. However, the treatises also recommended the spreading of poisons via the water pipes to a besieged city or via rivers to the enemy encampments at their meal times. When besieging a city, the Romans usually used poisons rather than diseases because they intended to occupy the target. The sole exception to this rule of which I am aware is the instance of Belisarius using dead animals in conjunction with poisonous herbs and lime (all dropped into the spring outside the city walls) during the siege of Auximus in 539.[5] However, he did this only after he had failed to block the well and the enemy was well aware of his actions. Consequently, the defenders did not drink the water and therefore there did not exist any real danger that the Romans would catch any disease from the defenders after the latter had surrendered. On the other hand, when the enemy had invaded, the Romans did often resort to using biological contaminants (in modern terms approximately the equivalent of germ warfare) against the invaders. Julius Africanus and some tenth century Byzantine military treatises recommended this tactic alongside the use of poisons and scorched earth tactics (the latter used also during invasions). The Romans may also have known how to spread a sprayed mix of water and hellebore by using large hand-held syringes. Period historians do not actually mention this usage, but Julius Africanus recommended the use of truly large syringes to spray poison

---

[5] Procopius, Wars, 6.27.21. I have mistakenly claimed in my doctoral thesis (p.286) that Belisarius' action in this instance only involved the use of poisons (asbestos stone = probably lime) but in actuality he did use dead animals, poisonous herbs, and asbestos. I do not know how this error came about, since my notes on the text indicate all of the above with an exclamation mark. However, in the most recent printing, this error has been corrected.

on the advancing enemy horses in pitched battles. Unfortunately, it is not known with any certainty if this advice was ever followed in practice (in high winds it would have been a very dangerous practice) and it may also be a later Byzantine addition to the text from a period when the Byzantines were already using hand-held flame-throwers to spread their Greek liquid fire.[6]

Since modern studies have shown that the spreading of diseases is very difficult, and the level of understanding amongst the Romans of how the diseases spread was very rudimentary, it is not known how effective the methods were. From time to time it may have worked and now and again it may have been counterproductive. Unfortunately, the sources are very sparse when it comes to the question of spreading disease-causing agents and their after-effects. It is quite possible that in some cases the Romans caught a disease spread, for example, by the Persians and vice versa. It is also possible that some of the period epidemics or pandemics resulted from the intentional spreading of the diseases although it must be said that whenever the enemy armies were stricken down by diseases, the late Roman armies appear to have been unaffected. Even after the ailing invaders had returned home, the sicknesses appear not to have spread back to their place of origin. This suggests that the Roman countermeasures (antidotes, medicines, good hygiene, avoidance of contact etc.) were usually successful, if not merely fortunate.

## 6.6. CONCLUSIONS

Period military treatises and histories are a testament to the importance of the water supply to the late Roman army. Every general had to plan his operations according to the availability of water. The Romans did know how to obtain adequate amounts of drinking water for their garrisons, cities and armies in the field even if their methods of calculation were rudimentary by modern standards. However, the narrative sources also show that there were instances of the Romans taking calculated risks in this matter, but these occasions were rare. Most importantly, the Romans understood clearly the importance of clean and healthy water for the health of their soldiers. Good hygienic conditions were very high on the agenda. Unfortunately, we do not know all of the details but clearly a fair amount of attention was paid to the question. Similarly, we do not know how all of the details of the methods used in water purification and storage worked. This question is in great need of further archaeological research. The availability of safe drinking water to meet the needs of the army was not only important for the maintenance of the effectiveness of the Roman army. It also featured in the actual military planning. Most importantly, the above discussion has demonstrated that for the

---

[6] Julius Africanus 1.2, 2.2; Frontinus, Stratagems, 3.7.6; PE 54.20; Apparatus bellicus 2-4; DOT 82-3 (p.59.9-16); Sylloge tacticorum 61, 64. See also the notes of Vieiellefond in his edition of Julianus' Africanus' Kestoi and the important summary by E.L. Wheeler (1997) of Julius Africanus' strategy of magic. However, note also that some of the so-called magic did have a basis in practice (note, for example, Julius 1.9), just as the use of poisons or herbs did.

enterprising crafty Roman general water could also be a means to an end. One can summarise these ingenious military usages of water as follows: 1) The denial of access to water supplies; 2) The harnessing of the power of the flowing water against the walls; 3) The use of water to block invasion routes through the opening up of the sluices in dams; and 4) The spreading of poisons or diseases via water. In short, the availability of a clean water supply was the single most important factor in warfare. Everything depended upon it.

## 6.7. REFERENCES

PRIMARY SOURCES

A fuller discussion of the military manuals and other sources can be found in my doctoral dissertation "The Age of Hippotoxotai" and Dain, Alphonse, 1967, "Les strategists byzantins", Travaux et Memoires 2, 1967, 317-363.

Apparatus bellicus (10th century), partial edition and French tr. in, Zuckerman, Constantin, Chapitres peu connus de L'Apparatus bellicus, TM 12 (1994), 359-389. Complete edition J. Lami, Joannis Meursii opera omnia (=Jan van Meurs) VII, Florence 1746, cols 899-980.

De obsiditione toleranda (=DOT) (c.950), Anonymous, "A Byzantine Instructional Manual on Siege Defense: The De obsidione toleranda," Introduction, English Translation and Annotations by Denis F. Sullivan (with a reprint of the Greek text edited by Hilda van den Berg, Leyden 1947), in Byzantine Authors: Literary Activities and Preoccupations, Text and Translations dedicated to the Memory of Nicolas Oikonomides, ed John W. Nesbitt, Leiden/Boston (2003).

Dodgeon, M.H. and Lieu, S.N.C. (Compiled and edited) (1991) The Roman Eastern Frontier and the Persian (AD 226-363). A Documentary History (contains partial translations of several primary sources), London and New York (1991).

Excerpts of Polyaenus, in Polyaenus, Stratagems of War, 2 Vols., ed. and tr. by Peter Krenz and Everett L. Wheeler, (Chicago, 1994), Vol. II, 851-1003.

Frontinus (Frontin, c.AD102), The Stratagems and the Aqueducts of Rome, tr. by Charles E Bennett, Loeb ed. Cambridge Mass.-London (1925/1980).

Herodotus, Histories, available at www.perseus.tufts.edu.

Heron of Alexandria, Dioptra, ed. H. Schöne, Heronis Alexandrini opera quae supersunt omnia III, (Leipzig, 1903), 141-315.

Heron of Byzantium=Anon, Parangelmata Poliorcetica in Siegecraft. Two Tenth-Century Instructional Manuals by "Heron of Byzantium", ed. and tr. By Dennis F. Sullivan, Dumbarton Oaks Studies XXXVI, Washington D.C., 2000 (includes Geodesia).

Julian, The Works of the Emperor Julian, tr. by Wilmer Cave Wright, 3 vols., Loeb ed., London-Cambridge Mass. (1913/1954).

Julius Africanus (c.AD 230), Kestoi, Les "Cestes" de Julius Africanus. Ètude sur l'ensemble des fragments avec edition, traduction et commentaries, Jean-René Vieillefond, Firenze, Sorbonne-Paris (1970), also in J. Lami, Joannis Meursii opera omnia (=Jan van Meurs) VII, Florence 1746, Juli Africani Cesti, cols 899-980.

Procopius, Opera (History of the Wars, The Anecdota or Secret History, Buildings) ed. and tr. by H.B. Dewing, Cambridge Mass., Loeb Edition reprints, 7 Vols.

Secretum secretorum, Arabic version in Opera hactenus inedita Rogeri Baconi Fasc. V, Secretum secretorum cum glossis et notulis, Robert Steele, Oxford (1920), 176-266.

Sidonius Apollinaris, Poems and Letters, Loeb ed. 2 vols, tr. By W.B. Anderson (1965).

Strategicon, The Strategicon, Das Strategikon des Maurikios, CFHB XVII, Vienna 1981, edited by G.T. Dennis and German tr. by Ernst Gamillscheg. Maurice's Strategicon. Handbook of Byzantine Military Strategy, English tr. by G.T.Dennis, Philadelphia (1984).

Sylloge tacticorum quae olim "Inedita Leonis tactica" dicebatur (c.950), ed. Alphonse Dain, Paris (1938).

Syrianus/Peri strategikes, G.T. Dennis, Three Byzantine Military Treatises, CFHB XXV, Washington 1985,

The introduction, text and tr. of the Anonymous Byzantine Treatise on Strategy, 1-135.

Theophylact, Theophylacti Simocattae, Historiae, ed Carolus de Boor, Lipsiae (1887) Teubner. Theophylact Simocattes, The History of Theophylact Simocatta, tr. by Michael and Mary Whitby, Oxford (1986).

Vegetius, Flavius Vegetius Renatus, Epitoma Rei Militaris, ed. with an English tr. by L.F. Stelten, New York, Bern, Frankfurt, Paris (1990). Vegetius: Epitome of Military Science. Translated with notes and introduction by N.P.Milner. TTH 16, Liverpool, 1993, 2nd ed. (1996).

Velleius Paterculus, Compedium of Roman History, Loeb ed. (1924), 2.104.2.

Vitruvius, The Ten Books on Architecture, tr. by M.H. Morgan, New York.

Zonaras, Ioannes (11th century), Epitome historiarum, ed. L. Dindorf, Teubner (Leipzig, 1868-1874) and also Patrologia Graeca.

## LITERATURE

Blackman D.R. and Hodge A.T. with contributions from Grewe, K., Leveau, Ph. and Smith, N.A.F. (2001) Frontinus' Legacy, Essays on Frontinus' de aquis urbis Romae. Ann Arbor, The University of Michigan Press.

Bohec Y. Le (2000) The Imperial Roman Army, tr. Raphael Bate, London New York (1994/2000).

Bruun C. (1991) The Water Supply of Ancient Rome, A Study of Roman Imperial Administration, Series: Commentationes Humanorum Litterarum 93, 1991, Societas Scientiarum Fennica/The Finnish Society of Sciences and Letters, Helsinki.

Foss C. and Winfield D. (1986) Byzantine Fortifications: An Introduction. Pretoria.

Goldsworthy A. (2003) The Complete Roman Army. London.

Grmek M. (1979) Les ruses de guerre biologiques, Revue des etudes grecques 91, 141-163.

Klein G. de, (2001) The Water Supply of Ancient Rome. City Area, Water, and Population, Series: Dutch Monographs on Ancient History and Archaeology, editors H.W. Pleket and F.J.A.M. Meijer vol. XXII, Amsterdam.

Mango C. (1995) The water supply of Constantinople. In Constantinople and its Hinterland, Papers from the Twenty-seventh Spring Symposium of Byzantine Studies, Oxford, April 1993, Edited by Cyril Mango and Gilbert Dagron with the assistance of Geoffrey Greatrex, Society for the Promotion of Byzantine Studies Publications 3, Variorum Aldershot and Brookfield.

Mann John (2002), Murha, taikuus ja lääkintä (Originally Murder, Magic and Medicine, Oxford UP 1992) Tr. Tiina Onttonen, Arthouse Oy. Helsinki.

Mayor A. (2003) Greek Fire, Poison Arrows & Scorpion Bombs. Biological and Chemical Warfare in the Ancient World. Woodstock, New York, London.

Pringle D. (1981) The Defence of Byzantines Africa from Justinian to the Arab Conquest, An account of the military history and archaeology of the African provinces in the sixth and seventh centuries, 2 vols., BAR International Series 99(i).

Pryor J. H. (2003) "Byzantium and the Sea: Byzantine Fleets and the History of the Empire in the Age of the Macedonian Emperors, c.900-1025 CE". In War at Sea in the Middle Ages and the Renaissance, Hattendorf, J.B. and B. Unger, R.B. (eds.), Woodbridge, pp. 83-104.

Syvänne I. (2004) The Age of Hippotoxotai, Roman Art of War in Military Revival and Disaster (491-636), Acta Universitatis Tamperensis 994, Tampere 2004, 67, 445-7, 462-4, 503; Partially revised edition 67, 287-8, 463-5, 491, 504.

Vuorinen H. S. (1996) Terveyden markkinat ja lääketieteen etiikka antiikin aikana, Hippokrates 13, 104-123.

Vuorinen H. S. (1997) Taudeista antiikin aikana, Hippokrates 14, 74-97.

Vuorinen H. S. (2002) Tautinen historia. Vastapaino, Tampere.

Webster G. (1998) The Roman Imperial Army of the First and Second Centuries A.D, Third Edition. University of Oklahoma Press, Oklahoma.

Wheeler E. L. (1997) Why the Romans Can't Defeat the Parthians: Julius Africanus and the Strategy of Magic. In W. Groenman van Waateringe et al. (eds.) Roman Frontier Studies 1995, Oxford.

# 7 CONCLUSIONS

*Petri S. Juuti, Tapio S. Katko & Heikki S. Vuorinen*

Figure 7.1 Some fountains from Crete. (Photos: Petri Juuti)

The experience of humankind from the very beginning testifies to the importance and safety of groundwater as a water source, particularly springs and wells. Groundwater is developed by the absorption of rain and melt waters into the ground, where it infiltrates through different porous layers of soil, mainly sand and gravel, and the cracks in the bedrock under the influence of gravity. This percolation purifies the water and minerals dissolve into it. Although ancient philosophers speculated on the hydrological cycle its real nature wasn't understood until the 17$^{th}$ century.

The early evidence from agricultural and nomadic societies shows that there were disputes over wells and springs and the right to use them, especially in the dry areas, where these water sources were of vital importance. The way in which water supply and waste management were organized was essential for early agricultural societies. If wells and toilets were in good shape, health problems and environmental risks could be avoided.

It has been postulated that the waterborne health risks of hunter-gatherers were small, while pathogens transmitted by contaminated water became a very serious health risk for sedentary agriculturists starting around 10 000 BC. Consequently, the first urbanization of Europe during antiquity was not possible without transporting or supplying a sufficient amount of good quality water to the town.

The realization of the importance of pure water for people is evident already in the myths of ancient cultures. Religious cleanness and water were important in various ancient cults. The first known Greek philosophical thinkers and medical writers also recognized the importance of water. Ideas of the salubrity of water were connected to the general scientific level of the society. Cool, tasteless or tasty, odourless and colourless water was considered the best, while stagnant and marshy water was to be avoided.

These ideas remained until the end of antiquity. In fact, the ancient qualifications for good drinking water would have primarily protected people against those dangers which we would now consider to be biological. The ancient Greeks and Romans were, however, also quite aware of some of the risks, which we now consider to be caused by poisonous chemical compounds in the water.

Simple settling tanks, sieves, filters and the boiling of water were methods for improving the quality of water already in antiquity. Available sources do not permit the estimation of the public health effects of these treatment methods. Yet, the poor level of waste management, including waste water, was most probably a major danger for public health during antiquity. Consequently, water borne infections, such as dysentery and various diarrhoeal diseases, must have been one of the main causes of death at that time.

The indirect public health effects of water might have been greater than the direct effects during antiquity. Malaria was widespread during antiquity and schistosomiasis was well established in Egypt. Both are parasites, which besides growing in human beings, have a period in an aquatic organism to complete their life-cycle: malaria in mosquitoes and snails in schistosomiasis in snails.

The availability of safe drinking water, wine, food and medical care, good personal hygiene and effective sanitation ensured better fighting effectiveness for the Roman army than their enemies had. In short, the availability of a clean water supply was the single most important factor in warfare and the significance of clean and safe water was evident in ensuring that the health of their soldiers wasn't undermined. The Romans did know how to obtain adequate amounts of drinking water for their garrisons, cities and troops in the field and thus successfully planned their operations according to the availability of water.

Toilets have been despised and in many countries and cultures talking about bodily functions is still a taboo. Until the birth of water and sewage works, toilets and wells were the main technological innovations in the service of humankind. Local materials have traditionally been used for constructing them, and they were applied to local conditions. Most people in the world took their drinking water from private wells until the last quarter of the 19th century. In case of cattle farming, watering the livestock formed the major part of the water consumption. Thus, a well was placed closer to the cowshed than the house itself. The needs of fire fighting, businesses and industries, real estate owners, and health authorities hastened the birth of water works.

The well and the toilet are still the most common technical systems in the service of humankind. In today's world the pit toilet is still the most popular model of toilet. However, the development related to wells and toilets has not been deterministic and straightforward, as development hardly is in any field. Thus, it has been, and still is, possible to influence preferable development paths. What makes these innovations so important is the fact that wells and latrines are still in use and will certainly remain so also in the future. Improved techniques in these basic innovations can help also to promote equality, in a concrete way.

Traditional techniques can be useful and mean improvements to many if used properly. But old as such does not have to mean outdated or bad. We can even claim that in antiquity water supply and sanitation was better than nowadays in many countries. Time-honoured and good solutions were usually very durable and without fancy and easily breakable parts. The local materials, methods and cultural aspects of human behaviour are to be considered carefully together with population and water consumption growth when assessing future solutions. Traditional success stories of water supply and sanitation provide good sources for this process.

PART I – CONCLUSIONS

Figure 7.2 Fountain of Fontano di Trevi in Rome, Italy, constructed in 1762 (Photo: Tapio Katko)

# PART II

# PERIOD OF SLOW DEVELOPMENT

Expect poison from the standing water.
- William Blake

The rising world of waters dark and deep.
- John Milton, Paradise Lost (bk. III, l. 11)

In 2003 technologies of various levels and different periods could still be seen in Alexandria, Egypt. The public phone on the left is gradually being replaced by mobile phones, in the middle are water jars for public use, and on the right a Lada taxi imported from the former Soviet Union. (Photo: Petri Juuti)

# 8

# INTRODUCTION
*Petri S. Juuti, Tapio S. Katko & Heikki S. Vuorinen*

Figure 8.1 Old water main on Table Mountain, Cape Town. Camps Bay and Atlantic Ocean on the back. (Photo: Harri Mäki)

After the fall of the Roman Empire, the centre of "western" science and knowledge was in the Byzantine Empire (c. 330-1453) and in the Islamic world (after c. 620). Daily life and political systems changed and many things were managed in different ways compared to those in antiquity. After the fall of the Roman Empire, also water supply experienced changes. In medieval cities it was largely based on wells and fountains instead of aqueducts. Castles, fortresses and monasteries usually had their own wells or cisterns. Yet, there were some water pipe projects already during this era.

The lack of proper sanitation increased the effects of epidemics in medieval towns. Usually towns built a few modest public latrines for the inhabitants, but these were mostly inadequate for the size of the population. For example, the streets of medieval London and Paris were almost full of waste (see chapter 4 by Petri Juuti). At the same time, in Tokyo the situation was relatively much better (see chapter 30 by Yurina Otaki). Remarkable changes occurred in the late medieval period when the socio-economic forefront moved to the northern shores of the Mediterranean. At this time science and knowledge got their institutionalised base for the first time in history when the development of modern universities started in Europe.

An Englishman, Joseph Bramah, is often named as the developer of the first actual water closet. He got a patent in 1778 but also several other names are mentioned as a developer of the first toilet, e.g. Thomas Crapper, Alexander Cummings and Henry Moule. All these and several others developed the system. Yet, it took centuries before the toilet became a common component in ordinary houses. From England this innovation gradually spread to the continent of Europe and elsewhere. By 1900, the water closet became generally accepted in many European countries and elsewhere.

While the change in the global population from 400 to 1700 was only threefold it increased 1.5 times in just one hundred years, from 1700 to 1800. Thereafter the growth became even faster; from 1800 to 1900 it already increased 1.8 times and the growth was still accelerating ; see Figure 3.2, page 22.

This section deals mainly with how some two hundred years ago the agricultural world set out to industrialize and how it affected water supply and sanitation. Yet, we do not focus only on the "Western countries", although the sanitary movement started in England. Instead, we follow this development also on other continents with focus on different environments and socio-political systems.

Industrialization started in Britain in the 18th century. This was also a period of rapid population increase in Europe, which led to a large emigration overseas. Together with industrialization and urbanization in the Western world, the enlightened people were filled with the idea of progress. Ever since the 18th century, science and reason were considered to be able to lead humankind towards an ever-happier future.

The start of the industrial revolution and the related growth of cities gradually created the need for centralized water and sanitation. In many respects, England was the forerunner of modern water supply and sanitation systems, but the innovations soon spread to Germany, other parts of Europe, USA and later also elsewhere. As the cities grew, sanitary and environmental problems overwhelmed city governments to a greater degree than before, and modern technology was often seen as the solution.

It was also the time of early private water enterprising, where vendors took water to their customers with a donkey cart. In the 18th century the water supply of Paris was taken care of by approximately 20,000 water vendors using a bucket and in Cairo there might have been even more water vendors in medieval times (see chapter 4 by Juuti). They are still seen on the streets of Cairo, though mainly to attract tourists.

Several case studies in this section deal with the solutions of water supply and sanitation in arid or semiarid environments. Chapter 15 by Ingrid Hehmeyer describes a case study in Yemen, in the southern part of the Arabian peninsula, where humans have taken an environmentally hostile situation and turned it into a source of economic success.

L. A. Sawchuk and Janet Padiak in chapter 11 point out that in the 19th century Gibraltar supported a population of around 20,000 civilians on a peninsula of under five square kilometres with no source of standing surface water. This work reviews the issues in water management Gibraltar faced during the 19th century, such as the collection and allotment of ground and stored water.

The semi-arid, dry climate of Uzbekistan in Central Asia has forced local people to bring water from distant sources for survival. The first stages of the water supply system comprised mainly a conveyance network of small earthen ditches. The water supply system of Tashkent city, the capital of Uzbekistan, has a long history of development and has been modified in various ways under different political and economic systems as described by Bazarov R. Dilshod et al. in chapter 16.

In early 19th century Europe, France was considered to lead progress in public health, while in the 1830s it lost this leadership to Great Britain. Drinking water was connected to serious public health problems in rapidly industrializing and urbanizing Britain in the early 19th century. The role of water in the spread of diseases was considered anew, which is described by Heikki S. Vuorinen in chapter 9.

In Finland by the late 1800s towns in Finland had become more densely populated, sharing the space sometimes even with a large number of livestock. Securing a supply of water was of utmost importance when carrying out town planning, even in a country with relatively rich water resources. The development of water supply and sanitation in Finland until World War II is dealt with by Petri Juuti and Tapio Katko in chapter 10.

The expansion of Europeans into the African continent south of Sahara and their solutions in water supply and sanitation are described in three chapters. Cape Town is the oldest European-style city in South Africa. In sanitary matters Cape Town was in a lamentable state in the late 1850s. The development of water supply and sanitation in Cape Town is dealt with

# INTRODUCTION

by Petri Juuti, Harri Mäki and Kevin Wall in chapter 13. The development of water supply in four South African towns — Cape Town, Grahamstown, Durban and Johannesburg — is outlined by Harri Mäki in chapter 14. The history of infrastructural building or improvement in colonial cities in Africa is still in its embryonic stage. As for West Africa, the study by Kalala Ngalamulume (chapter 12) contributes to the literature on urban Senegal. It focuses on the losing battle to provide fresh drinking water to Saint-Louis, the first capital of French Senegal.

Further to North America, chapter 17 by Arthur Holst takes a look at the water supply system in Philadelphia, Pennsylvania, USA, from its beginnings in 1700 when the city's population was 4,400 up to 1910, when it served 1.5 million people. It highlights technical innovations adopted by the city and improvements made to its distribution system during this period.

In the last chapter (18) of this section of the book, Petri S. Juuti, Tapio S. Katko and Jarmo Hukka write about the privatisation of water services in the historical context of the last 150 years. The start of the Industrial Revolution and the related growth of cities gradually created the need for centralized water and sanitation.

Figure 8.2 View from Table Mountain. Woodhead Reservoir (completed in 1897) in right and Hely-Hutchinson (completed in 1904) in left at the background. (Photo: Petri Juuti)

# 9
# THE EMERGENCE OF THE IDEA OF WATER-BORNE DISEASES

*Heikki S. Vuorinen*

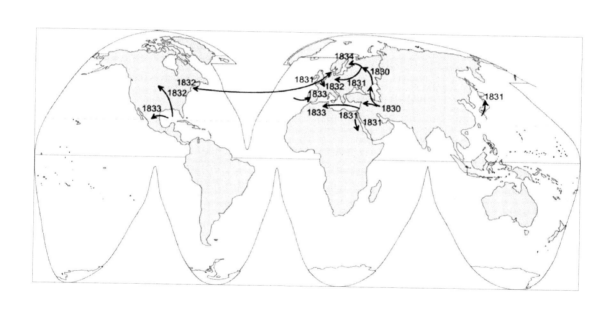

Figure 9.1 The spread of cholera in the early 1830s during the second pandemic, constructed from the data in Pollitzer 1954.

## 9.1. GENERAL BACKGROUND

A fundamental socio-economic change started in the world over two hundred years ago: an agricultural world set out to industrialize. Industrialization started in earnest in Britain in the 18th century. At the same time, the number of population began to increase in European countries one after another. Mass emigration started and millions of Europeans moved to the Americas. All over Europe people also moved from the countryside to the unrestrainedly growing towns, which soon faced serious public health problems.

Simultaneously with industrialization and urbanization in the Western world, enlightened people were filled with the idea of progress. Ever since the 18th century, science and reason have been considered to be able to lead humankind towards an ever happier future. The ruling "enlightened" elite considered old truths and unfamiliar ideas prevailing among uneducated people and in foreign cultures to be superstition and in need of being cast away. The idea of progress also overwhelmed Western medicine and public health during the 19th century. Because of the genuine confidence in rationality and logic and the rejection of supernatural causation of diseases, which classical Greek medicine had adopted already during antiquity, Western medicine was wide open to the strong emphasis on reason.

The first international public health congress took place in Paris in 1851. France has been considered to have led the progress of public health in the early 19th century but starting in the 1830s lost the leadership in public health to Great Britain. (La Berge 1992, 315; Ramsey 1994) Britain was at the vanguard of industrialization from the 18th century onwards. Drinking water was connected to the serious public health problems in rapidly industrializing and urbanizing Britain in the early 19th century. The prevailing ideas concerning the role of water and the measures taken for the protection of people's health were keenly followed in the rest of Europe, and the main focus of this chapter is on the experience of Britain. However, Germany began to play the primary role in science in the late 19th century and the role of Germany was growing in medicine and public health. Thus, for example, when a new type of institution for the teaching of public health – the public health school – was created in the USA in the 1910s, the German idea of science and the laboratory as the basis of teaching public health was influential.

## 9.2. SANITATION MOVEMENT

Contemporaries in early 19th century Britain fully realized that part of the public health problems originated from polluted drinking water and poor drainage in towns. The highly influential Edwin Chadwick's *Report on the Sanitary Condition of the Labouring Population of Gt. Britain* appeared in 1842, and the Public Health Act was passed in Britain in 1848. Later in the 19th century Chadwick's report was considered to be famous and the British public health legislation to be the most explicit and comprehensive in Europe (Palmberg 1889, 1–4) However, effective legislation to protect the water sources, especially the rivers, was slow to develop even in Britain (Luckin 1986, 158–176).

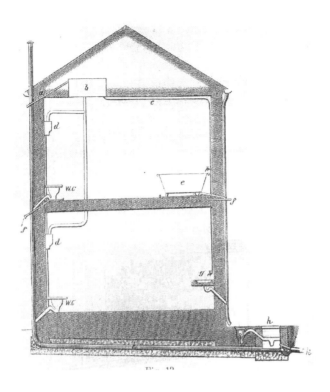

Figure 9.2 Piped water supply and sewers were introduced into the town houses during the 19th century. (Source: Palmberg 1889, 80.)

Lawyer Edwin Chadwick (1800–1890) was a vital person in the 1830s and 1840s when the public health movement was formed in Britain. (Hamlin 1998) To state it bluntly; his suggested solution was a system in which a constant abundant flow of good quality water flushes away diseases, moral deficiencies like criminality and drunkenness, and other possible problems in English cities through correctly constructed sewers, and the sludge was then to be used to fertilize agricultural land. The success of the sanitation movement was manifested in the construction of piped water supply systems and sewers in towns all over Europe in the late 19th century (Figures 9.2 and 9.3). (Juuti & Katko 2005) Although the Chadwickian ideas were also criticized, they had a significant influence even on the periphery of Europe, as in Finland in the 1870s when the first public health act was formulated (see Hjelt 1875).

## 9.3. MIASMA AND SANITATION

It has been argued that medicine played only a secondary part in the process of resolving the problems of public health in the middle of the 19th century (Rosen 1958, 224–225). In hindsight it has been argued that the great sanitary reform in Britain in the middle of the 19th century was based on erroneous medical theories but led to the right solution (Rosen 1958, 225). The role of medicine might, however, be larger than customarily realized. For contemporaries it was a well-established medical fact that filth caused diseases through the noxious miasma originating from decaying organic matter (Flinn 1965, 62). The large-scale construction of sewers, which occurred around Europe in the late 19th century, was justified as a solution for the transfer of filth away from living quarters in towns (Figure 9.3).

For many (or most) contemporaries, the miasmatic origin of epidemic diseases was the most rational theory of the causes of diseases until the latter part of the 19th century. Air that had become disease-producing (miasma) was something that people could observe with their senses. The concept of malaria – a term of Italian origin signifying bad air – and miasma were often used indiscriminately and synonymously by contemporaries (Barker 1863, 1–3). Stagnant, putrefying pools of water could be smelled; rotten carcasses of animals and vegetable remains together with the vapours arising from them could be seen. It is easy to understand how this was logically connected to the epidemic diseases, which were rife at the same time. The opinions of the contemporaries varied on the actual disease-generating agent of the miasma or malaria: was it inorganic or organic? (Barker 1863, 8–49, 67–69) It was not until the middle of the 19th century that microscopes (and related technology) developed sufficiently so that minor structures like cells and bacteria started to be reliably and reproducibly seen.

Edwin Chadwick's theoretical reliance on a very comprehensive miasma theory, which guided his plans for sanitary reform, is evident. According to Chadwick's way of thinking, there was miasma both inside and outside the human body and it influenced a human being in a multitude of ways: "*Here, then, we have from the one agent, a close and polluted atmosphere, two different sets of effects; the one set here noticed engendering improvidence, expense, and waste, – the other, the depressing effects of external and internal miasma on the nervous system, tending to incite the habitual use of ardent spirits; both tending to precipitate this population into disease and misery.*" (Report 1842, 197–198)

Chadwick used the miasma theory also in a more narrow sense to justify his sanitary proposals. In this narrow meaning he used miasma to designate the noxious air originating from putrefying vegetable and animal remains and rising from sewers and drains: "*On the subject of the escapes of gas from the sewers there is no one point on which medical men are so clearly agreed, as on the connexion of exposure of persons to the miasma from sewers, and of fever as a consequence.*" (Report 1842, 371), and  "*[...] of instances of disease and death occasioned by miasma from badly made and sluggish or stagnant drains that pervade whole towns [...]*" (Report 1842, 375).

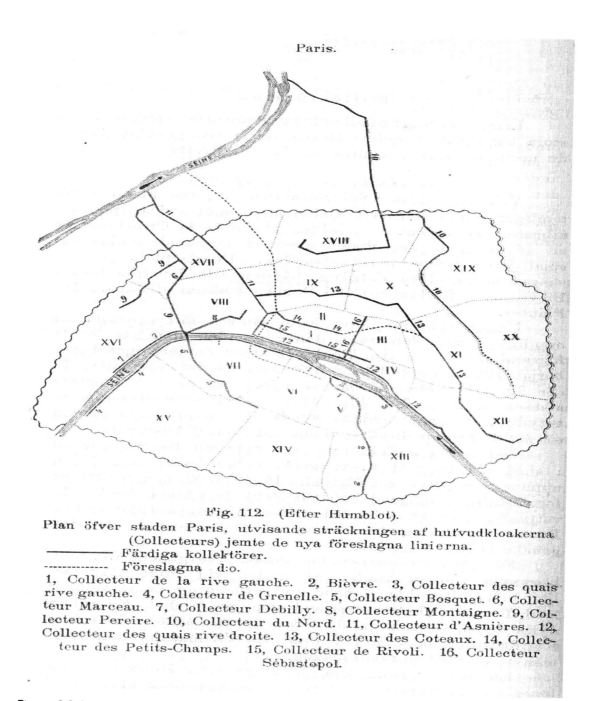

Figure 9.3 Sewers were built in European towns during the 19th century, for example in Paris as illustrated by Palmberg 1889, 506.

In the conclusions of his report, Chadwick concisely presents the way of thinking that was to lead the sanitation movement first in Britain and later on all over Europe. Although Chadwick is not using the word miasma in his conclusions, he eloquently describes the theory of atmospheric alteration, in other words the disease-causing miasmas, and the steps to be taken to combat them. He argued:

> "That the various forms of epidemic, endemic, and other disease caused, or aggravated chiefly amongst the labouring classes by atmospheric impurities produced by decomposing animal and vegetable substances, by damp and filth, and close and overcrowded dwellings prevail amongst the population in every part of the kingdom, [...]
>
> [...] That such disease, wherever its attacks are frequent, is always found in connexion with the physical circumstances above specified, and that where those circumstances are removed by drainage, proper cleansing, better ventilation, and other means of diminishing atmospheric impurity, the frequency and intensity of such disease is abated; [...]
>
> That the formation of all habits of cleanliness is obstructed by defective supplies of water.
>
> [...] The primary and most important measures, and at the same time the most practicable, and within the recognized province of public administration, are drainage, the removal of all refuse of habitations, streets, and roads, and the improvement of the supplies of water. [...]" (Report 1842, 422–423)

## 9.4. CHOLERA

A new dreadful disease emerged in the awareness of Europeans in the 19th century – cholera. Cholera was the disease that served as a justification for the sanitation movement in Europe in the 19th century. The four cholera epidemics (1831–1832, 1848–1849, 1853–1854 and 1866) that afflicted Britain were to have a profound influence on the way of thinking about the role of water in the transmission of diseases (Luckin 1986, 69–99). It is informative that Edwin Chadwick referred to cholera in several places of his report (Report 1842, 89, 95, 111, 163–164, 360, 365, 372, 396–397). Chadwick clearly states that physicians recommended sewerage for the prevention of cholera: "*This had been so far observed by medical men that there was, perhaps, no point on which they were more anxious and urgent than that increased sewerage and cleansing should be adopted as preventives of the cholera.*"(Report 1842, 372)

Cholera started to wreak its havoc in India in 1817. Although there is some disagreement among historians about the exact years assigned to different pandemics, there is a common opinion of about six pandemics during the 19th century (Table 9.1). (Hirsch 1881, 278–348, Haeser 1882, 793–943, Pollitzer 1954) (Figure 9.1). Cholera attracted enormous attention in the societies where it appeared and made an important contribution to the idea that each communicable disease has a specific cause.

Although there were diverging opinions, it was quite widely realized in the medical profession during its first approach towards Western Europe in the beginning of the 1830s (Figure 9.1) that cholera was a contagious disease and an epidemic. (Swederus 1831, 36–43) Swede G. Swederus' book on cholera was published already in 1831 in Stockholm, Sweden. It was based on publications in Russia and Germany. Swederus, among other things, described in detail the origin and spread of cholera from India towards Europe.

G. Swederus' opinion that cholera was contagious and spread via trade routes was followed by, for example, the famous English doctor John Snow (1813–1858, Snow was a medical practitioner in London). In the words of John Snow in 1855:

> It travels along the great tracks of human intercourse, never going faster than people travel, and generally much more slowly. In extending to a fresh island or continent, it always appears first at a seaport. [...] Its exact progress from town to town cannot always be traced; but it has never appeared except where there has been ample opportunity for it to be conveyed by human intercourse. (Snow 1855, 2)

John Snow developed the comprehension of the epidemiology of cholera in a very important way. Earlier ideas as to how cholera was communicated were based on the idea *"that a person should be very near a cholera patient in order to take the disease"* and that *"it must spread by effluvia given off from the patient into the surrounding air, and inhaled by others into the lungs".* (Snow 1855, 6) For instance, in 1831 Swederus held the explicit opinion that cholera was transmitted by air and required proximity to be communicated further (Swederus 1831, 38, 41). In contrast to this view, Snow held the well-justified view that the poison of cholera can be transmitted through contaminated water, a view which he actively propagated, for instance, in articles and letters in various medical journals in the 1850s. Although not everyone in Britain agreed with Snow on the mode of cholera transmission in the 1850s (Eyler 2001), it is quite evident that the way of thinking was changing from the times of Swederus, who explicitly rejected the idea that water had something to do with the spread of cholera (Swederus 1831, 39). Snow argued that it was through the alimentary tract that cholera "poison" was spread:

> Diseases which are communicated from person to person are caused by some material which passes from the sick to the healthy, and which has the property of increasing and multiplying in the systems of persons it attacks. [...] the morbid material producing cholera must be introduced into the alimentary canal – must, in fact, be swallowed accidentally, [...] (Snow 1855, 9)

John Snow eloquently expresses his ideas about the role of water in the transmission of cholera, for instance, in his *On the Mode of Communication of Cholera* in 1855:

> The great prevalence of cholera along the course of rivers has been well known for a quarter of a century; and it meets with a satisfactory explanation from the mode of communication of the disease which I am inculcating. Rivers always receive the refuse of those living on the banks, and they nearly always supply, at the same time, the drinking water of the community so situated. [...] The principles I have laid down afford a satisfactory explanation of the circumstances, that

Table 9.1 The pandemics of cholera in the 19th century.

| 1st | 1817–1823 | (or 1817–1837) |
|---|---|---|
| 2nd | 1827–1837 | (or 1826–1837 or 1840–1850 |
| 3rd | 1844–1855 | (or 1840–1864 or 1841–1862 or 1852–1860) |
| 4th | 1863–1874 | (or 1863–1873 or 1864–1875 |
| 5th | 1881–1896 | (or 1882–1896) |
| 6th | 1899–1923 | |

absence of drainage promotes the prevalence of cholera, […] Without drainage, the refuse of the population permeates the ground, and gains access to the pump-wells. (Snow 1855, 78)

Snow had a clear idea about the organic nature of the agent causing cholera although he was fully aware that the contemporary methods of microscopy did not allow the demonstration of it: *"For the morbid matter of cholera having the property of reproducing its own kind, must necessarily have some sort of structure, most likely that of a cell. It is not objection of this view that the structure of the cholera poison cannot be recognized by the microscope […]"* (Snow 1855, 10). Snow's idea that the poison of cholera was a specific living being, which entered the alimentary canal by ingestion, and the crucial role of water in disseminating it, was considered ingenious by many (Barker 1863, 23, 86; Eyler 2001), but it was proved to be correct only in 1883 when Koch identified *Vibrio cholerae* as the cause of cholera. (Rosen 1958, 285–286)

## 9.5. MICROBES

Turbid water that tasted and smelled badly, at least after it had been allowed to stand for some time, had been considered unsuitable for drinking since antiquity. General acceptance of the idea that in such waters there was a multitude of living organisms capable of producing diseases in human beings had to wait until the so-called bacteriological revolution in the late 19th century.

The idea of an organic agent capable of reproducing itself inside the human body as a cause of epidemic diseases was shared by quite many doctors already in the middle of the 19th century (Barker 1863, 23–49, 68, 84–89). However, the prevailing idea was that the agents causing diseases were predominantly spread by air – miasma – and the pathological process inside the body was some sort of chemical process resembling fermentation. For instance, the most prominent English vital statistician and "epidemiologist" of the era, William Farr, supported this idea in the 1850s although he was sympathetic towards the ideas of Snow (Eyler 2001). In the 1860s the general attitude had become more favourable towards John Snow's ideas and e.g. William Farr supported the role of water in the transmission of cholera

(Eyler 2001), and by the 1870s the following diseases were considered to be transmitted by water: typhoid fever, cholera, dysentery and diarrhoeas.

Besides cholera, typhoid fever was quite early on recognized as being spread by water (Budd 1873; Luckin 1986, 118–138). Already at the same time as John Snow, William Budd (1811–1880) in Bristol propagated similar ideas about the role of water in the transmission of cholera and he recognized that these views also applied to typhoid fever (Rosen 1958, 287; Moorhead 2002). In the words of Budd: "*Drinking-water, whether mixed or unmixed with other things, is a frequent and very deadly vehicle of the typhoid poison. That its condition should always be looked to is of the first importance in fever outbreaks, and, where there is any suspicion of impurity, even, it should be provided against.*" (Budd 1873, 90–91) In 1881 in his monumental handbook of geographical pathology, August Hirsch stated that the surest thing in the aetiology of typhoid fever is that its poison is spread by drinking water (Hirsch 1881, 475).

One after another, the causative organisms of infectious diseases transmitted by contaminated water were identified in the late 19th century. The causative microbe of typhoid fever (*Salmonella typhi*) was found in 1880 by a German, Carl Eberth, cholera (*Vibrio cholerae*) in 1883 also by a German, Robert Koch, and bacillary dysentery (*Shigella dysenteriae*) by a Japanese doctor, Kiyoshi Shiga, in 1898. Summer diarrhoea, which in Britain in the 1850s and 1860s was also believed to be transmitted by water, was a more complicated issue and the finding of causative agents and their means of transmission had to wait until the 20th century (Luckin 1986, 100–117).

The medical discipline of hygiene was especially developed in Germany from the middle of the 19th century onwards, and the name of professor Max von Pettenkofer (1818–1901) arose above the other contemporaries (Locher 2001). He became a full professor of organic chemistry in the year 1853, and hygiene in 1865 in München. Pettenkofer developed his theories of the origin of diseases in cities around the concept of polluted soil and changing ground water levels. Pettenkofer did not assign the primary role in the causation of cholera and typhoid fever to drinking water but to the domestic and industrial wastewater that contaminated the soil and created the disease poisons there. Water had through ground water, however, a very prominent place in his theories of epidemic diseases in cities. In contrast to Pettenkofer, who consistently advocated his cholera theories still in the 1890s, Robert Koch advocated the view that cholera epidemics were due to drinking water contaminated by cholera vibrios.

Medicine, including the new discipline of hygiene, was highly international already in the 19th century. Ideas were rapidly exchanged among the leading scientists. Internationally the best known Finnish hygienist, MD Albert Palmberg, wrote a handbook on the European sanitary situation in Swedish in 1889 (Palmberg 1889) and it was published in French in 1891, in Spanish in 1892 and in English in 1893 and again in 1895. In his handbook, Palmberg gave information about public health in Britain, Belgium, France, Germany, Austria, Sweden and Finland. Water supply, water closets and sewers have a quite eminent place in his book.

The Vitruvian way of examining water quality (see chapter five by Vuorinen in this book) was in the late 19th century undergoing an important modification. Palmberg still represents the old guidelines for potable water based on senses but also presents the chemical and especially the bacteriological examination of water in the evaluation of its quality (Palmberg 1889, 163–166). However, he points out that the bacteriological study of water then was as yet unreliable.

## 9.6. THE IMPROVEMENT OF THE QUALITY OF DRINKING WATER

Piped water supply systems, which became common in many European towns during the late 19th century, could be quite dangerous for the health of populations without an efficient mean of water purification, especially if surface water was used. This was proved in a ghastly way in Hamburg during the cholera epidemic in 1892 (Evans 1990, 292).

The rapid spread of water closets (Figure 9.4) in various European countries was a controversial blessing. On the one hand, it efficiently washed human excrements away from the immediate surroundings of people. This might protect people who had a bathroom and water closet in their apartment, as the experience in Hamburg during the cholera epidemic in 1892 shows (Evans 1990, 425–426). On the other hand, in crowded conditions when a lot of people shared a common water closet and bathing facilities, diseases spread easily as also evidenced by Hamburg in 1892 (Evans 1990, 424–426). Water closets also efficiently increased the pollution of surface waters.

The controversy connected to the water closets is seen in the opinion of John Snow:

> It follows from what I have said above that I should recommend the discontinuance of water-closets, or at least their diminution, instead of the continued increase of their numbers. A complete and well-regulated water-closet is so great a convenience that one cannot expect it to be discontinued in the better class of houses; but the so-called water-closet used by the working classes, who form the great bulk of population, is according to the experience I have had, a worse nuisance than an open privy over a cesspool; [...] the greatest evil of water-closets is the inordinate demand for water they occasion, and thus prevent most large towns being supplied otherwise than from polluted rivers. If the general use of water-closets is to continue, and to increase, it will be desirable to have two supplies of water in large towns, one for the water-closets, and another, of soft, spring, or well water from a distance, to be used by meter, like the gas. (Snow 1858)

Already in antiquity it was known that the filtering and boiling of poor quality water makes it more salubrious to drink (see chapter five by Vuorinen earlier in this book). In the 19th century, e.g., John Snow followed these ancient recommendations when he presented measures that should be adopted in the presence of cholera:

> 3rd. Care should be taken that the water employed for drinking and preparing food (whether it come from a pump-well, or be conveyed in pipes) is not contaminated with the contents of cesspools, house-drains, or sewers or, in the event that water free from suspicion cannot be obtained, is (sic) should be well boiled, and, if possible, also filtered. (Snow 1855, 84)

Figure 9.4 The water closet was introduced to solve the problems of sanitation in European towns during the late 19th century. (Source: Palmberg 1889, 204.)

To these recommendations, he added:

8th. To effect good and perfect drainage. 9th. To provide an ample supply of water quite free from contamination with contents of sewers, cesspools, and house-drains, or the refuse of people who navigate the rivers. (Snow 1855, 85)

During the 19th century the filtering of water supply of whole urban populations was introduced. The effectiveness of filtration was dramatically demonstrated during the 1892 cholera epidemic. The town of Altona, which had a sand-filtered water supply, did not suffer from cholera but neighbouring Hamburg, which used unfiltered surface water from the river Elbe, experienced a terrible cholera epidemic (Evans 1990, 291–292). The mostly well to do citizens of Hamburg who had the habit of letting their drinking water to be boiled had some advantage over their fellow citizens who did not boil their water even during this catastrophic epidemic of 1892, when the water supply was heavily polluted (Evans 1990, 424).

The bacteriological "revolution" brought with it an understanding of how the filtration and boiling of water actually worked. New ways to bring about a rational improvement of drinking water were developed after the transmission of disease-causing microbes through water was discovered. Disinfectant properties of chlorine compounds were already noticed in the middle of the 19th century. After the discovery of the causative agent of cholera, the disinfection of drinking water gained impetus. Chlorination of drinking water started during the 1890s and rapidly spread around the world during the first decades of the 20th century, for example the chlorination of drinking water was introduced systematically in London in 1915 (Luckin 1986, 134) and on the periphery of Europe, in Finland, in Helsinki in 1916 (Lillja 1938, 154) and in Tampere in 1917 (Juuti 2001, 190).

Water was distributed through leaden pipes in the urban centres as in antiquity. Concerns about lead exposure from leaden plumbing systems continued to be expressed since the 16$^{th}$ century. Despite several reports of waterborne plumbism (chronic lead poisoning), especially in the 19$^{th}$ century the use of lead in plumbing systems continued. (Nriagu 1985)

The role of water in the epidemiology of some very important epidemic diseases was realized in the last decades of the 19$^{th}$ century and at the turn of the 20$^{th}$ century: malaria and yellow fever were both discovered to be transmitted by mosquitoes. The prevention of both diseases could now be more rational and systematic. One branch of the prevention strategies was to be directed to the watery breeding sites of the mosquitoes: marshes were drained and oil spread on the water pools.

## 9.7. CONCLUSIONS

The sanitation of towns around Europe was one of the great achievements of the 19$^{th}$ century. First sanitation was based on the miasmatic theory of the origin of diseases, but during the 19$^{th}$ century the role of water in the transmission of several important diseases – cholera, dysentery, and typhoid fever – was realized. The final proof of the role of water came when the microbes causing these diseases were discovered in the late 19$^{th}$ century.

The chemical and microbiological examination of water was added to the evaluation of water quality based on the senses. During the 19$^{th}$ century, the sand filtering of an entire water supply of a town was introduced and the systematic chlorination of drinking water was started in the early 20$^{th}$ century. The discovery of microbes and the introduction of efficient ways in which to treat large amounts of water paved the way to an era in which the public health problems caused by polluted water seemed to belong to the history of mankind.

## 9.8. REFERENCES

Barker T. B. (1863) On Malaria and Miasmata and their Influence in the Production of Typhus and Typhoid Fevers, Cholera, and the Exanthemata: Founded on The Futhergillian Prize Essay for 1859. John W. Davies, London.

Budd W. (1873) Typhoid Fever. Its Nature, Mode of Spreading, and Prevention. Referred page numbers are from the book in www.deltaomega.org/typhoid.pdf.

Evans R. J. (1990) Death in Hamburg, Society and Politics in the Cholera Years 1830–1910. Penguin Books, London.

Eyler J. M. (2001) The changing assessments of John Snow's and William Farr's cholera studies. Soz.-Präventivmed. 46: 225–232.

Flinn M. W. (1965) Introduction. In Report on the Sanitary Condition of the Labouring Population of Gt.Britain by Edwin Chadwick 1842. pp. 1–73. Edinburgh University Press, Edinburgh.

Haeser H. (1882) Lehrbuch der Geschichte der Medicin und der epidemischen Krankheiten. Dritter Band. Geschichte der epidemischen Krankheiten. Gustav Fischer, Jena.

Hamlin C. (1998) Public Health and Social Justice in the Age of Chadwick: Britain 1800–1854. Cambridge University Press, Cambridge.

Hirsch A. (1881) Handbuch der historisch-geographischen Pathologie. Zweite vollständig neue Bearbeitung. Erste Abtheilung: Die allgemeinen acuten Infectionskrankheiten vom historisch-geographischen Standpunkte und mit besonderer Berücksichtigung der Ætiologie. Ferdinand Enke, Stuttgart.

Hjelt O. E. A. (1875) Bidrag till sundhetslagstiftningen i Finland. J. Simelii arfvingar, Helsingfors.

Juuti P. (2001) Kaupunki ja vesi, Tampereen vesihuollon ympäristöhistoria 1835–1921. PhD Thesis, Petri Juuti & KehräMedia.

Juuti P. & Katko T.(Eds.) (2005) Water, Time and European Cities, History matters for the Futures. Printed in EU.

La Berge A. F. (1992) Mission and Method: the Early Nineteenth-century French Public Health Movement. Cambridge University Press, Cambridge.

Lillja J. L. W. (1938) Helsingfors stads vattenledningsverk 1876–1936. Otava, Helsingfors.

Locher W. G. (2001) Max von Pettenkofer – Life stations of a genius. On the 100th anniversary of his death (February 9, 1901). International Journal of Hygiene and Environmental Health: 203, 379–391.

Luckin B. (1986) Pollution and control: a social history of the Thames in the nineteenth century. Adam Hilger, Bristol.

Moorhead R. (2002) William Budd and typhoid fever. Journal of the Royal Society of Medicine 95: 561–564.

Nriagu J. O. (1985) Historical perspective on the contamination of food and beverages with lead. In Mahaffey, K.R. (Edited by). Dietary and environmental lead: human health effects. pp. 1–41. Elsevier, Amsterdam.

Palmberg A. (1889) Allmän helsovårdslära på grund av dess tillämpning i olika länder. Werner Söderström, Borgå. [In English A treatise on public health and its application in different European countries (England, France, Belgium, Germany, Austria, Sweden, and Finland). Swan Sonnenschein, London 1893.]

Pollitzer R. (1954) Cholera studies: 1. History of the disease. Bulletin of the World Health Organization 10, 421–461.

Ramsey M. (1994) Public health in France. In Porter, D. (Edited by). The History of Public Health and the Modern State. pp. 45–118. Rodopi, Amsterdam – Atlanta.

Report on the Sanitary Condition of the Labouring Population of Gt.Britain by Edwin Chadwick 1842; edited with an introduction by M.W. Flinn. pp. 74–425. Edinburgh University Press, Edinburgh 1965.

Rosen G. (1958) A History of Public Health. MD Publications, New York.

Snow J. (1855) On the Mode of Communication of Cholera. Page numbers referred to in the text are from the book at www.epi.msu.edu/johnsnow/publishedworks.htm. Snow's book is also available at www.ph.ucla.edu/epi/snow/snowbook.htlm but without page numbers.

Snow J. (1858) Drainage and Water Supply in Connexion with the Public Health. Medical Times and Gazette 16 (20 February 1858): 188–191.

Swederus G. (1831) Cholera Morbus. Uppkomst, Härjningar, Kurmethod och Preservativ, efter Skrifter utgifna i Tyskland och Moskwa år 1831. P. A. Nordtedt & Söner, Stockholm.

# 10
# BIRTH AND EXPANSION OF PUBLIC WATER SUPPLY AND SANITATION IN FINLAND UNTIL WORLD WAR II

*Petri S. Juuti & Tapio S. Katko*

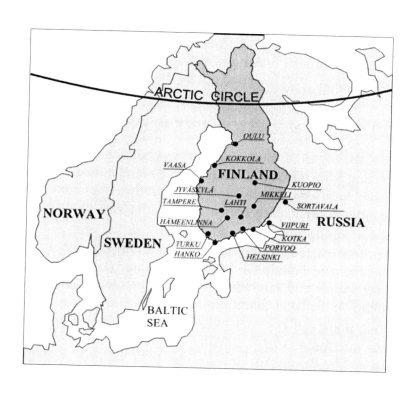

Figure 10.1. Location of the first Finnish cities with water and sewage works established by 1917.

## 10.1. GENERAL BACKGROUND

By the late 1800s towns in Finland had become more densely populated, sharing the space sometimes even with a large number of livestock. Securing a supply of water was of utmost importance when carrying out town planning, even in a country with relatively rich water resources. A location near water also provided a good means of transportation.

The overall development of Finnish water supply and sewerage services can be divided into the following five phases:

(i) Early discussions and proposals for private concessions until 1880;

(ii) Establishment of the first water supply and sewerage systems in the biggest cities, 1880 to 1917, as municipal departments and works;

(iii) Expansion of the systems and establishment of new ones, 1920 to 1940;

(iv) Reconstruction followed by major expansion of systems including stronger water pollution control measures, 1950 to 1980;

(v) Increasing autonomy, inter-municipal cooperation and outsourcing of non-core operations, 1980 to 2000. (Juuti & Katko 2005, 61)

This chapter focuses on the first three phases – from the mid-1800s until WW II. The chapter is largely based on several case studies (Juuti & Katko 1998, Juuti et al. 2000, 2003) on the evolution of individual Finnish water and sewage undertakings and a few overall studies (Juuti & Katko 2004, Katko 1997), carried out by the authors.

## 10.2. EARLY URBANIZATION CREATING DEMAND

When continental Europeans began to think about urban planning, they took high-density apartment living as a starting point. On the other hand, in Finland industrialisation occurred mainly in newly established locations next to water bodies rather than around the big cities (e.g., Hietala 1987). The latter is particularly true of forest industries.

In Finland the first water supply and sewerage systems of urban centres in the 1870s to 1890s were in most cases constructed around the same time, although often under separate organisations. There was demand mainly for fire-fighting water (Hietala 2002; Juuti 1993, 2001), but drinking water supply and sanitation, and in some cases industrial needs, also played their role depending on the local conditions and needs.

Nelson & Rogers (1994) point out the background and birth of the First Public Health Law in Sweden that came into force in 1874. Initially it was clearly influenced by the British Public Health Act of 1848. The committee drafting the 1874 Swedish Act considered the promotion of preventive health care of utmost importance. Along with the Act, for instance, public health boards became compulsory in each town. The Swedish Act also served as a model for the Health Decree of 1879 in Finland (Nygård 2004b).

In Finland, fire insurance companies contributed significantly towards the development of water services. Water was needed for putting out fires as well as for domestic and other community purposes. At first, Finnish houses were insured, if at all, with the General Fire Insurance Fund in Stockholm. The "semi-official" Finnish Fire Insurance Bureau was established in 1809 with state support. The issue of fire insurance became increasingly topical immediately following the Great Fire of Turku in 1827.

The General Fire Assistance Company of the Grand Duchy of Finland was established in 1832 (Nikula 1972, Nuoreva 1980). Later on cities received funding from this company on good terms for establishing water works. The company operated under the Superintendent's Office with its domicile in Helsinki. It was a government body, not owned by cities. In 1858 the company was renamed the General Fire Assurance Company of Finnish Cities.

The Finnish Rural Fire Assurance Company was founded in 1857, while in 1871 the Finnish Cities' Fire Assurance Company was set up to insure chattels. In 1873 fire services became a municipal responsibility for good. In 1882 the Fennia Fire Insurance Company opened up for business and was the first in Finland to underwrite industrial fire insurance. The above companies supported the acquisition of fire-fighting water and equipment in different ways.

The quite advantageous loans from the fire insurance company considering the prevailing interest rates (on average about 6% in the second half of the $19^{th}$ century) played a significant role in financing and establishing the undertakings. The taxes from spirits distilleries were also of great significance. In each city, a company was given the exclusive right to distil spirits against the payment of a liquor tax.

Normally a small initial amount of capital was raised over time for the establishment of a water works: about 10 percent of the total required – through taxes and quite substantial donations and wills. Loans were also taken from local banks when necessary. A loan from the fire insurance company was nevertheless generally the largest single source of financing, and its interest rate was clearly lower than with those of other creditors. (Juuti & Katko 2005, 61)

House owners were solely responsible for maintaining open ditches and simple sewers within their own plots until public sewage works were set up. Before that, in exceptional cases, a city could have constructed some minor sewers in the core area. No wonder that house owners eagerly supported the establishment of sewage works. They also bore the financial responsibility for street maintenance, which further made them support putting in sewers. Waste disposal was also the duty of house owners, making them favour municipal waste collection and disposal (Juuti 2001).

## 10.3. THE FIRST URBAN WATER AND SEWAGE WORKS

In Finland, a total of 16 urban water supply and sewerage systems were established by 1917 (Figure 10.1) when the country gained full independence, after having been an autonomous Grand Duchy of Russia since 1809. The first one was established in the capital of Helsinki in 1876. In most cases water supply and sewerage systems were created simultaneously – or sewerage even a few years earlier (Table 10.1).

After the decision for municipal ownership and responsibility, some technology-related selections were made, including metering-based billing, a ban on lead pipes, and the acceptance of flush toilets. The first one was taken as a normal procedure in Finland after the first, and so far the only private concession in Helsinki, was eventually bought back by the city. In this case, the company Neptun operated the waterworks and was paid by the town. The company also got a monopoly to build house connections, and together with three other companies, a monopoly for plumbing installations. In 1879 the company established a special plumbing unit, which was the beginning of the Huber Plumbing Company. Yet, in July 1880, the town of Helsinki assumed the operational responsibility for the water pipes which ended the company's monopoly and the agreement. The operation of the works was now given to Robert Huber, the former Neptun local director, for an agreed annual payment. Huber took care of the works operations until December 1882, but in the beginning of 1883 the town started to operate the system (Juuti et al. 2005; Herranen, 2001; Lillja, 1938).

The second technology-related selection - ban on lead pipes - was decided upon after tests carried out in Helsinki showed that excess lead became diluted in the raw water. The third one - the acceptance of flush toilets - was heavily debated in several cities around the turn of the century.

Extensive groundwater inventories were made in Vaasa from 1896 to 1897, for the first time in Finland, and again at the turn of the 1800s and 1900s and during the first years of the 1900s. Even the first artificial groundwater experiments in Finland were carried out there. The knowledge and expertise used were international, mainly from Sweden and Germany, while also domestic experts were used. Also in Tampere in the surroundings of the city a quite rich groundwater source was found in Vuohenoja. The matter was considered and explored for a decade and active discussions were held at times. The result was, however, that in 1920 the city council finally abandoned the plans for establishing a groundwater intake in Vuohenoja. It was thought that the groundwater would not suffice for the needs of the growing city. There were different, partly contradictory schools of thought how to assess the safe yield of groundwater deposits. On the other hand, around the same time chemical treatment methods had come on the market, thus making surface water treatment a realistic option. (Vaasa Health board 1901; Vaasa Water committee 1918; Juuti & Katko 1998, 101-108)

A few Finnish cities started wastewater treatment in the 1910s while the actual boom in modern wastewater treatment happened in the 1960s and 1970s, mainly due to the Water Act that came into force in 1962. For the first time, this Act had the necessary legal

Table 10.1 Years of establishing the first Finnish urban water and sewage works with their water sources, from 1876 to 1917. (Source: Juuti 2001)

| City | Waterworks (year) | Water source | Sewerage (year) |
|---|---|---|---|
| Helsinki | 1876 | river | 1880 |
| Viipuri | 1892 | groundwater | 1873 |
| *Tampere | 1882 | lake | 1894 |
| Tampere | 1898 | lake | 1894 |
| *Oulu | 1902 | river | 1897 |
| Oulu | 1927 | river | 1897 |
| Turku | 1903 | groundwater | 1896 |
| Hanko | 1909 | groundwater | 1906 |
| Lahti | 1910 | spring | 1910 |
| Hämeenlinna | 1910 | spring | 1910 |
| Jyväskylä | 1910 | groundwater | 1911 |
| Mikkeli | 1911 | groundwater | 1911 |
| Porvoo | 1913 | groundwater | 1894 |
| Kuopio | 1914 | lake | 1906 |
| Sortavala | 1914 | lake | 1907 |
| Vaasa | 1915 | groundwater | 1904 |
| Kotka | 1916 | river | 1890 |
| Kokkola | 1917 | groundwater | 1923 |

*In Tampere and Oulu the first systems were protosystems.

enforcement and permit mechanisms to make communities start modern wastewater treatment and management. This was preceded by the introduction of separate sewers that made it technically feasible to treat wastewaters. (Katko et al. 2005)

Important social and political reforms such as municipal reforms and universal suffrage also influenced positively the development of the sector. Private companies providing the sector with goods and services have emerged gradually based on demand (Hukka & Katko 2003, 120).

In the 1860s there were plans to organize water supply by private entrepreneurs in Helsinki and Tampere. Finally, they ended with the municipal systems. This development is described in chapter 18 by Juuti, Katko & Hukka. Health reasons and, for instance, the requirements of fire protection, led to the laying of a high-pressure water pipe in Tampere in 1898 and also later in Oulu in 1927. At the same time, a water charge based on consumption was established in 1898. The measurement of consumption is not as self-evident as one might imagine: earlier water was charged for a flat rate in Tampere while in Oulu during the first periods of the waterworks a charge was collected according to the method with which the water was fetched. (Juuti 2001, 140-150; Katko 1996)

It is interesting that Tampere initially chose to use surface water while many other cities such as Hanko, Hämeenlinna, Lahti, Turku and Viipuri (Vyborg) went for groundwater (Table 10.1). In some cities, the establishment of a waterworks was postponed far into the 20$^{th}$ century - in Savonlinna until 1951. (Katko 1996, 45, 102)

## 10.4. MANY FACTORS CREATED DEMAND FOR WATER SERVICES

The need for fire-fighting water promoted indirectly also the improvement of hygienic conditions along with sewerage systems. In spite of the incorrect scientific theory of miasma, the solutions made, however, advocated the right causes, i.e., improvement of the environment and safety of the city.

Contemporaries were searching for a solution to water acquisition for waterworks, and for drainage and environmental pollution from sewerage. Thus, water supply and sewerage were seen as solutions to the water issue, along with fires. When contemporaries in the various Finnish cities spoke about these problems, they were commonly using the term "water question". (Juuti 2001, 12, 138-164) This problem of water supply was solved only after prolonged planning and transitional periods. The transition from the so-called bucket system - based on wells, carrying the bucket - to the protosystem, based on low pressure - and finally modern water supply, based on high pressure - was a demanding process for municipal administration: many decisions requiring special knowledge had to be made. (Juuti 2001)

Probably the first municipal "water pumping installation" in Finland was founded in Tampere in 1835. (Voionmaa 1929, 481) The system was a very simple one based on gravity and was not operational for long. The first water-protection regulation in Tampere concerned this system. (Tampere city record office 1838) The rapid growth period in Tampere started a few years later. (Rasila 1984, 131)

A low-pressure system was constructed in 1882-84 in Tampere, while the city's high-pressure facility was completed in 1898, though not on the scale of the original plan. Since slow sand filtration was rejected and the outlets of the sewers were too close to intake pipes, the efficiency of the new facility was also its weakness: typhoid fever spread fast over a wide area aided by the water pipe network. In 1916 the death of hundreds of people finally prompted the necessary decisions to be made. The threat of typhoid fever and other diseases spreading through water was removed in 1917 when chlorination of water was started. There have been no typhoid epidemics in Tampere since then. (Juuti & Katko 1998, 9-72; Juuti 2001, 182-190) In 1919 infant mortality was lower in the cities than in the countryside; earlier the situation was the reverse. (Ylppö 1922) At least in this respect, the cities had become healthier places to live than the countryside. Later phases of the Tampere case are described in chapter 25 by Juuti & Pál.

The evolution of sewerage began usually from free-flowing ditches and as years went by and towns grew, the ditches were straightened, opened and covered. These measures, however, proved to be insufficient and the dirt and filth continued to spread. The exacerbated problem forced the decision-makers of municipalities to work out a plan for underground sewerage following the hygienic reform started in England and personified by E. Chadwick. (See Hamlin 1998)

When the growth of the city accelerated due to industrialisation, problems began to accumulate: there was not enough water and what little there was, was of poor quality. A discussion about changing this bad situation started. For example, the first most visible measure in Tampere was the founding of the "Sundhetskommittén I Tammerfors" (Public health committee: Swedish being used as the administrative language at that time) called by contemporaries the "temperance committee". The committee started its work in 1866, inspired, for instance, by the example of London: members knew closely the reform started in England and aimed to adapt its doctrines in Finland. (Tampere city record office 1866; Juuti 2001, 67-70)

The local newspaper *Tampereen Sanomat* followed the work of the committee closely and considered its progress. (Tampereen Sanomat 1867) The first aim of the committee was to organise drainage in parts of the city. It proposed the building of a sewer network as a remedy and appealed to the fact that the typhoid problem was worst in the least drained area of the city. The committee's report clearly shows that the members' beliefs about the causes of disease were consistent with the miasma theory. According to this theory, diseases were born in wet and contaminated soil as organic material was getting digested. The construction of sewers was a way to get rid of this. Thus, the model came directly from England, not from any other city in Finland (Tampere city record office1867; Porter 1999, 79-82; Juuti 2001, 70-71)

The building of the sewers in Tampere started after four decades of discussion in 1876. (Tampere city record office 1875; Juuti 2001, 71) The 1879 public health decree obliged the city to prepare a plan for a sewer system commensurate with the estimated population within 10 years. The city administrators took seriously the deficiencies in the sewerage and the demands of the government: starting in the early 1880s, the municipal health board repeatedly exhorted the city to expand and upgrade their sewerage system.

At the end of the 1880s, the Finnish people followed closely the development of the bacteriological revolution and hygienic reform. Slowly the miasma theory began to lose ground. (K.F.M 1885a, 66-73; K.F.M 1885b, 92-131; Terveydenhoitolehti 1897a, 1897b, 1897c, 1897d; Juuti 2001, 85-87) In this phase, discussion about the water question also started to become livelier and the first initiatives to modern systems were made, but usually without success. (Tampere city record office 1880; Tampereen Sanomat 1880)

## 10.5. SOLUTIONS AND THEIR EFFECTS

The new high-pressure waterworks provided safety and comfort in Finnish towns. Security was also essentially increased when a regular fire brigade was founded. The lack of water pipe had also caused various other difficulties, extra work and trouble. After the founding of the waterworks, it was a great relief for the city's inhabitants to get rid of the duty to carry and transport water. In Finnish towns this quite long process also improved, after some setbacks, the sanitary situation and the appearance of the city area. (Juuti 2001, 140-164, 238-240)

In cities sufficient water for fire fighting became available only after the emergence of high-pressure waterworks and professional fire brigades. This was the case both in Tampere and Oulu, since both cities had initially low-pressure waterworks. (Table 10.1) It is probable that the decisions in Tampere were known well in Oulu as the two cities followed closely developments in each other's water supply and sewerage. In addition, Tampere and Oulu used the same external experts, like Mr. Hausen from Helsinki. (Katko 1996, 52; Juuti 2001, 141-164) Networking of the experts in the Finnish water sector was quite advanced already in the last years of the 19th century. Besides, Finnish experts and civil servants went on numerous fact-finding tours abroad to Sweden, England and Central Europe to familiarise themselves with foreign solutions. (See Hietala 1987) This happened in many ways: visits were paid to foreign cities and also many foreigners were invited as experts to several Finnish cities.

Problems with water quality were also largely solved only after the introduction of high-pressure waterworks, although Tampere needed a severe typhoid epidemic before cost minimisation-minded decision-makers realised the necessity of efficient water treatment. There had been knowledge of proper treatment and the dangers of not having it for years as a result of the domestic expert network and the active foreign connections. (Koskinen 1995, 54-55, 64, 77; Juuti 2001, 140-164, 182-185)

In 1916, one year before national independence, hundreds of people died, finally prompting the necessary decision to be made that *Aamulehti* had been determinedly advocating. In Helsinki, Hämeenlinna and Lahti related problems were not as great, because they did not use untreated surface water. Lahti was using good quality groundwater from the Laune spring, Hämeenlinna used groundwater from Ahvenisto and Helsinki used from the beginning surface water treated with slow sand filters. These modern systems were thus safer than the one in Tampere. In addition, the other cities were taking care of their wastewaters in a modern way compared to the protosystem in Tampere: in Lahti the wastewaters from the entire planned city area were treated already in 1910. The facility in Lahti was the most advanced in Finland then. On an international level it was quite advanced, too. The systems in Hämeenlinna and Turku also surpassed the one in Tampere in most areas since they were using safer groundwater. (Aamulehti 1915; Koskinen 1995, 54-55; Juuti 2001, 140-164, 182-185)

Water shortages were not totally over in Finland after modern waterworks were introduced. For example, in dry summers there were problems with water quantity in the 1920s in several towns, such as Turku, Porvoo and Vaasa. Also dry summers in 1941-1942 caused troubles. (Stenroos et al. 1998, 89-98; Juuti et al. 2005; Juuti & Katko 2006.; http://www.fmi.fi/saa/tilastot_99.html#6)

Apparently economic interests also stirred up dispute since some people were afraid that the costs were going to be shared by everyone while only a few could enjoy the advantages. In Tampere there was no opposition to the waterworks at any point, only some details aroused criticism in *Aamulehti*. In Hämeenlinna the committee preparing the plan for the waterworks followed the principle of not forcing the facility on the public. It thought that the importance and necessity of the facility were so well known that no discussion was needed. This nearly destroyed the whole plan. With hindsight it can be said that the importance and necessity of the waterworks were not a big enough factor to sell it, at least, to the local newspaper *Hämeen Sanomat*. (Juuti 2001, 155-164)

## 10.6. SEWERAGE

The other side of the water question, i.e., sewerage, also had to be solved. The public health decree of 1879 obliged cities to do so since the act required that levelling of the city areas was to be carried out. Several plans for sewer systems were made and some were also constructed in the late 1800s. Although the wettest areas of the towns were drained and hygiene improved, lakes were still being polluted due to untreated wastewater discharges. The bucket was replaced by a drainpipe, and the problems were flushed out of sight, untreated, to the nearest water systems as is typical of protosystems. Luckily wastewaters were not used for irrigation like in Germany and France (Reid 1991) at that time. This kept the groundwater unpolluted.

By the beginning of the 20$^{th}$ century, the raw water basins of several towns became polluted. For example in Tampere, Lake Näsijärvi was in bad shape, and typhoid epidemics were plaguing the inhabitants. The threat of typhoid fever and other diseases spreading through water was removed in 1917, the year of Finland's independence, when the chlorination of water was started. Since then there have been no typhoid epidemics in Tampere. The system did not, however, include collection and treatment of wastewater. (Juuti 2001, 182-185)

The typhoid epidemic on its part made local decision-makers examine the question of community and industrial wastewaters. For various economic reasons it was finally decided not to do anything about the wastewater at that time: it was assumed that the Tammerkoski Rapids could purify it sufficiently. Yet, the situation in Tampere was considerably better than, e.g., in certain cities in Germany; in Tampere the amount of wastewater was only a fraction of the amount of supplied water. (Koskinen 1995, 68-69; Juuti 2001, 204-210.) Concerning water and epidemics, see also chapter 9 by Heikki S. Vuorinen.

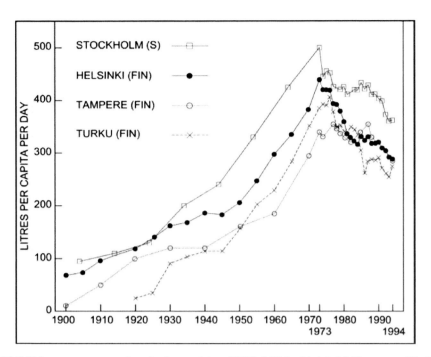

Figure 10.2 Water consumption in four cities 1900-1994: Helsinki, Tampere, Turku (Finland) and Stockholm (Sweden).

Although water and sewerage systems were established in these 16 cities or towns already during the first two phases there are still challenges remaining – old and new ones. For example, the discussion about the use of ground water versus surface water is still as active in 2006 as it was in the first decades of the 1900s. How far is it feasible to go for fetching good quality ground water or possibly introduce artificial recharge instead of treating surface water typically located closer to the urban areas? Also the dry periods are still a problem to some extent as they were in the end of 1800s, early 1900s and 1941-1942. The most recent very dry summers in Finland occurred in 2002 and 2003 and they were problematic in many cities, for example in Porvoo. That year, 2002, the typical seasonal autumn rains did not occur, and therefore ground water deposits did not fill up before winter as usually happens. The decrease in water levels over the year was several tens of centimeters. At the beginning of 2003 the water levels in southern and central Finland were exceptionally low for the time of year. The water level increase in spring was only normal or even below normal. Towards the end of the year water levels increased and in the end of 2003 lake water levels in southern and central Finland were tens of centimetres above the level of a year earlier. (http://www.ymparisto.fi/default.asp?node=12760&lan=en)

Contrary to the continuous growth of systems and water consumption during the period and later, some new challenges have also emerged; water consumption (Figure 10.2) decreased significantly after the 1973 oil crisis and the introduction of a wastewater charge in 1974.

126

(Juuti et al. 2005) This was followed by various reactions by water suppliers and customers, thus leading to tariff structure modifications and other ways of safeguarding adequate income. (Katko et al. 1998; Rajala & Katko 2004).

While we in water services are in constant connection with the environment, surprises are always possible. For example, in 2005 in southern Finland the squirrel population increased. In November a sudden campylobacter epidemic occurred in the town of Nummela in Vihti municipality when about 10 squirrels managed to get inside the water tower and were drowned to death. Over 600 people became ill and the whole network had to be disinfected with chlorine. (Helsingin Sanomat 2005) This reminds us of the need for continuous control, check-ups and preventive maintenance of these systems.

## 10.7. CONCLUSIONS

In Finnish towns many decisions have been made on water and wastewater management that have affected their development and the environment: First, the waterworks were born as a solution to the water question after long discussions - often after various, inadequate and temporary solutions. In terms of quantity, there was enough water, and the selected technological, administrative and economic solutions were also successful. The well-being of people and equality improved compared to the earlier situation as waterworks expanded and better quality water slowly reached also the working class people.

Second, the waterworks was excellently suited for the needs of fire fighting. There were no great fires in the cities after the founding of waterworks and fire departments. The choice of the pressure level, the charging system based on metered consumption, the selection of materials and machinery, the working methods and the financing of the systems were largely successful in the 16 Finnish case cities by 1917.

Third, on a national scale, the health situation improved after the founding of the waterworks; especially typhoid fever cases decreased with the exception of a few epidemics. In 1919 infant mortality was lower in the cities than in the countryside; earlier the situation was the reverse. At least in this respect, the cities had become healthier places to live than the countryside.

Fourth, the urban water and sewage works have been owned by cities except for the capital Helsinki for its first years. Yet, from the very beginning the private sector has supplied goods and services based on competition.

Fifth, in Tampere a groundwater project was abandoned in 1920, since there was not proper means or knowledge to estimate safe yields that time. This choice probably led also other cities to use surface water while the large-scale use of groundwater was not reintroduced until the end of WW II.

Sixth, wastewater treatment was introduced to two cities in 1910 and studied in others around 1920, but was taken seriously only some 30 to 40 years later after proper legislation and enforcement were enacted.

This chapter has concentrated on the birth and early development of community water supply and sanitation in Finland. It shows that cities developed their water supply systems gradually based on demand. There was, and even in 2006 still is, a debate on the use of surface versus ground water although the first ground water inventories were made already in 1896-97. Also artificial ground water, introduced in Vaasa a couple years later, is still heavily debated.

While sewerage systems were mostly constructed jointly with urban water supply, wastewater treatment was for many years used only in a few cities. Indeed, it took a long time for the society to adopt this necessity.

So far the systems have expanded continuously but by the early 21$^{st}$ century, national discussion had started on how far these systems can be expanded.

## 10.8. REFERENCES

PRIMARY SOURCES:

Tampere city record office (1838) Minutes of city administrative court 3.4.1838.

Tampere city record office (1866) Minutes of city administrative court 27.11.1866.

Tampere city record office (1867) Minutes of the city administrative court 30.1.1867.

Tampere city record office (1875) Minutes of city council 18.8.1875.

Tampere city record office (1880) Minutes of city council 21.4.1880.

Vaasa Health board (1901) Annual report. Vaasa City archieves.

Vaasa Water committee (1918) Report: "Katsaus vesijohtokysymyksen aikaisempiin vaiheisiin", 3.3.1918. Vaasa City archieves.

LITERATURE:

Aamulehti (1915) 16.4.1915

Hamlin C. (1998) Public health and social justice in the age of Chadwick Britain, 1800-1854. Cambridge.

Helsingin Sanomat (2005) 1.11.2005

Herranen T. (2001) Vettä ja elämää. Helsingin vesihuollon historia 1876-2001. Helsingin vesi, Helsinki.

Hietala M. (1987) Services and Urbanization at the Turn of the Century. The Diffusion of Innovations. Helsinki.

Hietala M. (2002) Helsinki : the innovative city : historical perspectives. Helsinki: Finnish Literature Society.

Hukka J. & Katko T. (2003) Water privatisation revisited – panacea or pancake? IRC Occasional Paper Series No. 33. Delft, the Netherlands. http://www.irc.nl/page/6003 or http://www.servicesforall.org/html/WaterPolicy/Summary_Revisited.html

Juuti P. (1993) Suomen palotoimen historia. Helsinki.

Juuti P. (2001) Kaupunki ja vesi. Doctoral dissertation. Acta Electronica Universitatis Tamperensis 141. Tampere 2001.

Juuti P. & Katko T (1998) Ernomane vesitehras. Tampereen kaupungin vesilaitos 1835-1998. With English summary. Marvelous water factory - Tampere City Water Works 1835-1989. Tampere.

Juuti P. & Katko T. (2004) From a few to all. Long-term development of water and environmental services in Finland. Pieksämäki.

Juuti P. & Katko T. (Eds.) (2005) Water, Time and European cities. History matters for the futures. EU. htpp://www.WaterTime.net, http://tampub.uta.fi/index.php?tiedot=80

Juuti P., Katko T. & Rajala R. (2005) For The quality of life - Evolution and lessons learnt of water and sanitation services in Porvoo, Finland, 1900- 2000. Natural Resources Forum 29 (2), 109-119. Blackwell Publishing, United Nations.

Juuti P., Rajala R. & Katko T. (2000) Ympäristön ja terveyden tähden. Hämeenlinnan kaupungin vesilaitos 1910-2000. Hämeenlinna.

Juuti P., Rajala R. & Katko T. (2003) Aqua Borgoensis. Porvoo.

Katko T. (1996) Vettä! - Suomen vesihuollon kehitys kaupungeissa ja maaseudulla. Tampere.

Katko T., Juhola P. & Kallioinen S. (1998) Declining water consumption in communities: sign of efficiency and a future challenge. Vatten 54 (4), 277-282.

Katko T.S., Luonsi A.A.O. & Juuti P.S. (2005) Water Pollution Control and Strategies in Finnish Pulp and Paper Industries in the 20th century. IJEP 23 (4), 368-387.

K.F.M. (1885a) in Duodecim (6-7), 66-73.

K.F.M. (1885b) in Duodecim (8-9), 92-131.

Koskinen M. (1995) Saastunut Näsijärvi terveydellisenä riskinä - Kulkutaudit, kuolema ja puhdasvesikysymys Tampereella 1908-1921. Master thesis, University of Tampere.

Lillja J.L.W. (1938) Helsingin kaupungin vesilaitos 1876-1936. Helsinki.

Nikula O. (1972) Turun kaupungin historia. Turku.

Nuoreva V. (1980) Suomen palotorjunnan historia. Jyväskylä.

Porter D. (1999) Health, Civilization And The State. A history of public health from ancient to modern times. London & New York.

Rajala R. & Katko T. (2004) Household water consumption and demand management in Finland. Urban Water Journal 1(1), 17-26.

Rasila V. (1984) Tampereen historia II. 1840-luvulta vuoteen 1905. Tampere.

Reid D. (1991) Paris Sewers and Sewermen. Realities and Representations. Harvard University Press.

Tampereen Sanomat (1867) 5.3.1867.

Tampereen Sanomat (1880) 10.3.1880.

Terveydenhoitolehti (1897a) No 1.

Terveydenhoitolehti (1897b) No 3.

Terveydenhoitolehti (1897c) No 6.

Terveydenhoitolehti (1897d) No 8-9.

Voionmaa V. (1929) Tampereen kaupungin historia II: Tampereen historia Venäjän vallan ensipuoliskon aikana. Tampere.

Ylppö A. (1922) Imeväisten ja pienten lasten huollon järjestämisestä sekä lastenhoidollisen ammattisivistyksen kokottamisesta. Duodecim, 183-190.

INTERNET:

http://www.fmi.fi/saa/tilastot_99.html#6, accessed 6.12.2005.

http://www.ymparisto.fi/default.asp?node=12760&lan=en, accessed 6.12.2005.

# 11
# COLONIAL MANAGEMENT OF A SCARCE RESOURCE: ISSUES IN WATER ALLOTMENT IN 19TH CENTURY GIBRALTAR

*Lawrence A. Sawchuk & Janet Padiak*

Figure 11.1 Location of Gibraltar.

## 11.1. GENERAL BACKGROUND

Gibraltar's unique topology and socio-political situation makes it an interesting case study in community water supply. In the 19th century, Gibraltar supported a population of around 20,000 civilians on a peninsula of under 5 square kilometres that had no source of standing surface water. Providing sufficient water for the colony's inhabitants was challenging, but the demands generated by the presence of a large military and colonial presence made water a political commodity. This chapter reviews the issues prevalent in water management during the 19th century, issues such as the collection and allotment of ground and stored water. Epidemics of yellow fever and cholera were exacerbated by the problems associated with poor sanitation. In the last two decades of the 19th century, saltwater condensers were introduced into the colony and large catchments were constructed to store winter rains in limestone caverns, providing a safer and more secure source of water.

## 11.2. GIBRALTAR AS A BRITISH COLONY

Gibraltar is situated on an isthmus at the entrance to the Mediterranean Sea, and is attached to Iberia by a sandy spit on the north side and surrounded by the sea on the west, south and east sides (Figure 11.2). Its surface area is limited (4.9 square kilometres) and dominated almost entirely by a large rocky projection rising, in some places vertically from the surrounding sea, to a height of 430 meters. (Gilbard 1881; Report of the 1881 Census of Gibraltar ) 'The Rock', as this outcrop is called, is composed of soft limestone breccia, riddled with natural crevasses and caverns and clad in shrubby vegetation such as geranium, aloe, myrtle, and cactus. The climate is salubrious, having mild temperatures all year, plentiful rains in the autumn and winter and an absence of precipitation in the summer months.

The Moors inhabited this peninsula along with much of Iberia until 1462, when the Spanish drove them out. In 1704, a joint Anglo-Dutch force fought Spain, and the British gained possession of this great military prize. The height of the Rock, with its view of the north coast of Africa, meant that whoever controlled Gibraltar could observe all traffic entering and leaving the Mediterranean. Gibraltar became an important naval station and fortress for the British and, until late in the 19th century, was the largest garrison outside the UK. The harbour was a crossroads of marine traffic in the western Mediterranean despite the absence of natural resources for supplying or re-victualing ships. Gibraltar had neither agricultural land to produce food, nor natural resources for refuelling, nor any source of surface water to supply ships in the harbour. Coal arrived from mines in Wales and elsewhere, to be transferred to ships of the British navy and merchant marine vessels of other nationalities. All food products were imported into the colony, many from across the land border from the Andalusian countryside and others from the north coast of North Africa. It was the provision of water, however, that greatly taxed the colony's ability to provide. The only source of water was precipitation that was trapped, either naturally by the sands, fissures and caverns in the Rock, or through artificial means, such as collection from rooftops into cisterns.

In addition to the soldiers, sailors, and British officials, Gibraltar was home to a civilian population. Following a siege by the Spanish from 1777 to 1783, most of the civilian population had fled, but in the 1790s some returned, joined by immigrants arriving from Mediterranean ports such as Genoa, Valetta, Leghorn and Tangiers. Whatever their origin, these new arrivals adopted the Spanish language and culture practised by the Gibraltarian people. (Kramer 1986) The British military understood that civilians were necessary in the colony. They were needed to provide the labour required by a large garrison and busy harbour but, in return for residing under the protection of the British crown and sharing in the colony's prosperity, civilians had to accept that their needs were secondary to those of the military. (Blake 1855; Preston 1946, 402–23) The civil population had no say in the governing of the colony; governors were military men appointed in London, answering only to the crown and the British parliament. The governors were, to all intents and purposes, omnipotent over civilians and there was never any doubt, right up to the end of WWII, that in the eyes of the colonizers Gibraltar was, first and foremost, a military fortress. (Sawchuk 1996, 863-867; Memorandum of A. Wolseley; Finlayson 1991)

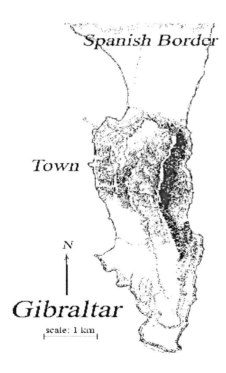

Figure 11.2 Map of Gibraltar in 1870.

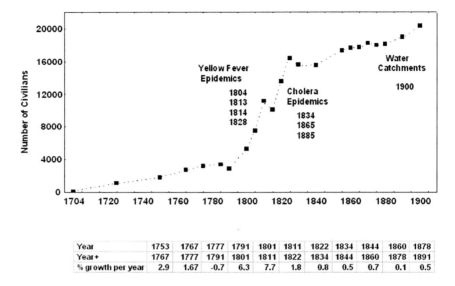

Figure 11.3 Population of Gibraltar, 1704 to 1900 (Data from Sawchuk 2001)

## 11.3. A SCARCE COMMODITY

Gibraltar's struggle with the provision of adequate clean water for its inhabitants began at the end of the 18th century with the solidification of the strategic importance of the garrison and the growth in the civilian population. Prior to this, the inhabitants depended on a fountain fed by an aqueduct originally constructed by the Moors by the 14th century. Reservoirs halfway up the Rock collected water trapped in spongy sands and crevasses and the aqueduct transported the water 1350 meters to a fountain in the lower part of the town. (Answers to Queries from the Army Medical Board 1831) The structure of the dams in the reservoirs and the progress of the aqueduct were complex, requiring constant attention by an overseer to keep the air flues clear and to prevent the earthen pipes from bursting. Nonetheless, except for renovation by the Spaniards prior to possession by the British and an occasional clearing of the underground channels, the aqueduct operated successfully for several centuries. (Answers to Queries from the Army Medical Board 1831)

The only other sources of water were the sands that formed the base of the town and those that formed the border with Spain. These spongy sands absorbed precipitation during the rainy season and, as they were slowly depleted, were replenished with water percolating through the crevasses and caverns in the limestone. Because these sands were close to sea level, the water drawn from bored wells was not potable, but was suitable for non-dietary uses. Much of the water drawn from the wells near the border with Spain was available to the public for a fee. (Extract of a dispatch to General William Houston 1831) In the town, of the

Figure 11.4 A *tinaja*, or water storage container.

dwellings that had their own well, most were inhabited by British officials, as this class was preferentially allotted the best houses in town. Ordinary Gibraltarians used the public wells along the border with Spain and carried the water into town. In the summer dry spell, as the water in the sands slowly sank, the water from these wells would become even more brackish and unusable; in years of low precipitation, they would dry up completely.

The water provided by the aqueduct and the few public wells soon proved to be insufficient for the growing garrison. In the years from 1784 to 1800, the combined civilian and military population of the peninsula increased fivefold (Figure 11.3). (Letter from J. O'Hara, Governor of Gibraltar, 18.12.1800) By 1814 the population had again grown, tripling to 16,800 individuals. (Howes 1994) Sufficient potable water for the inhabitants of Gibraltar was a formidable enough obstacle on its own, but the divisions within the colony, of military and civilian, of army and navy, exacerbated the problems. (Letter from J. O'Hara, Governor of Gibraltar, 18.12.1800; Sawchuk & Burke 1998) The military had needs that it considered superior to those of ordinary citizens; the navy needed large stores to supply ships and the army had a large strength of active men to supply, as well as horses that required large quantities of water daily. The two forces had preferential access to the water from the aqueduct, but there was no coordinated effort between the army and navy to allot or apportion water, and often the sudden appearance of the fleet could strip the colony of all its resources, both of food and of water. (Letter from J. O'Hara, Governor of Gibraltar, 14.12.1800) Gibraltarians were forced to utilise the poorest quality ground water, and some years this came at a price, for the British often taxed water drawn from wells. (Dispatch from George Don, Governor of Gibraltar 1831)

The other solution to the scarcity of water during the summer months was storage of the plentiful winter rains for use in the dry season. The military constructed cisterns to store rainwater from the roofs of large structures, such as barracks, the prison and hospitals. Gibraltarians lived in dwellings called patios, two or three-storied buildings constructed around an interior courtyard; each patio was occupied by several families. (Sawchuk, 1996) Patio courtyards were crammed with vessels such as barrels and *tinajas*, large earthenware pots, filled with precious water collected from roofs for use in the dry summer months (Figure 11.4). (Benady 1994) Most of the patios were owned by absentee landlords or were government property, and it was difficult for tenants to persuade the owners to erect private cisterns, particularly in the upper portion of the town where the poor lived. (Report on Barrack and Hospital Improvement 1836, 475) Patio inhabitants coped as best they could, filling whatever barrels and butts they could with water. When this supply ran out during the summer, they were forced to purchase water from itinerant vendors, who brought in supplies from Spain on donkeys, paying exorbitant prices for this service. (Sawchuk 2001) Such a shortage of water meant that many Gibraltarians, particularly the poor, often used the same water for several purposes in succession. (Report on the Sanitary Condition, 20)

The solution of rainwater storage to the problem of the supply of adequate sanitary water carried a health risk of which no one was aware at that time. Water stored in open containers, whether *tinajas* or cisterns, provided the standing water necessary for the reproduction of mosquitoes, and the warmth and humidity of Gibraltar's autumn season were suitable for the proliferation of the mosquito species that is the main vector for the yellow fever virus. (Sawchuk and Burke, 1998) Yellow fever is an endemic disease of African monkeys, and it is spread from monkeys to humans, and among humans, by bites from female mosquitoes. In Gibraltar, the introduction of the virus-carrying mosquito, *Aedes aegypti,* had probably accompanied marine cargo from Africa, possibly in pockets of water in barrels. The mosquitoes would have found the still, warm, stagnant waters of the *tinajas* and cisterns perfect for multiplying and the close proximity of Gibraltarians perfect for blood meals.

A severe case of yellow fever can result in liver infection, jaundice (hence the name) and haemorrhage. Those who experienced the disease as infants or children usually had only a mild case and were endowed with lifelong immunity. In 1804, yellow fever was a new disease to Gibraltar, and its introduction into a community with no previous exposure or immunity meant that it was a very lethal epidemic. It was reported that in 1804, of its military and civilian populations combined, 5733 persons died of yellow fever out of a total of 18,000, resulting in a mortality rate of 32%; civilians suffered greater losses than the military. (Sawchuk & Burke, 1998) The epidemic of 1804 was the first of five yellow fever epidemics that swept the colony, and it is likely that the method of storing rainwater directly contributed to these visitations. (Sawchuk & Burke, 1998)

At the time, there was no understanding of the role of stagnant water in the spread of yellow fever. The prevailing theory of disease transmission focused on 'miasma', or vaporous exhaustions, where diseases were spread through a noxious and offensive atmosphere.

(Sawchuk & Burke, 1998) The concept of disease caused by microorganisms or germs had not emerged, although there was the notion that some diseases, such as smallpox and syphilis, were 'contagious', or spread from person to person. Yellow fever was definitely believed to be a miasmatic disease, and the focus was on cleaning up piles of rotting organic debris that collected in heaps around the town of Gibraltar.

## 11.4. RESPONSE TO YELLOW FEVER EPIDEMICS

Despite the high death toll from yellow fever and other diseases common at the beginning of the 19$^{th}$ century, the population of Gibraltar continued to increase. In 1814, a second yellow fever epidemic broke out and, as the fever raged, Gibraltar was experiencing a change in governors. The new governor was George Don, and it is he who was responsible for the early development of Gibraltar's sanitary infrastructure over the seventeen years that he commanded the garrison. His introduction to the colony in the throes of a severe epidemic forced him to look closely at living conditions within the town. One of his first acts was to order the inclusion of a cistern as an integral part of all new dwelling construction. (Benady, 1994, 26) He also stipulated that cisterns must be covered, although he could not have been aware of the relationship of the yellow fever virus and the reproductive habitat of mosquitoes. (Benady, 1994, 26)

Don procured funds from the British government to plan and construct a sewage system for Gibraltar, a project that was completed in 1815. Unfortunately, the system was poorly designed; despite the natural incline of the town, the plan did not allow for sufficient flow to keep the waste matter moving out to the bay. (Report on Barrack and Hospital Improvement 1863) No system of sewer flushing had been planned, and attempts made to retrofit the system with pumps to flush with salt water failed. (Letter from Colonel C. Mann 1831) The entire system had the effect of bringing waste from the upper portion of the town to the lower portion of the town, and depositing it there, underneath the main thoroughfares. When, finally, hard winter rains dislodged the waste and forced it into the bay, the shortness of the mouths of the drains ensured that the effluvia was deposited above the low water mark, causing the seafront to be continually bathed with noxious organic matter.

By 1830, Gibraltar had endured three more yellow fever epidemics. The epidemic of 1828 was not as devastating as the 1804 epidemic, but over 1,500 people, military and civilian, died out of a population of about 17,000, resulting in a mortality rate of about 9 percentage. This time, it was the military that suffered greater losses than the civilians. Although the inhabitants did not know it, the epidemic of 1828 would be the last visit of yellow fever. At the time, the colony had endured four years of very low rainfall (less than 76 cm a year), making the water situation desperate, especially for the poor. The number of cisterns in private dwellings had increased to 250, although this was far short of the needs of the growing population. (Howes, 1994) Cisterns allowed Gibraltarians some autonomy in the provisioning of water, but the quality often left much to be desired. Because cisterns stored unfiltered water collected from roofs, and were located in the same places in which laundry was done and poultry was kept, a variety of unwanted debris would accumulate in the cisterns. Gibraltar was also the habitat of a colony of macaque monkeys, known as 'Barbary Apes', and they would often travel across the roofs of the upper portion of the town, fouling the surface as they went; this waste could also be collected in the runoff. (Letter from Colonel Smith-Dorrien 1920) Also, the cisterns were often contaminated because of their proximity to other functional areas of the patio, such as kitchens, shop activities and privies. It was, however, the best water available.

## 11.5. CHOLERA VISITATIONS

Between 1831 and 1834 a pandemic of Asiatic cholera spread through Europe, causing great fear and high mortality and, in 1832, Gibraltar was visited by the disease. Cholera is caused by a water-borne bacterium that thrives in untreated sewage and contaminated drinking water; affected individuals experience violent diarrhea that can cause severe dehydration and death. The connection of the spread of cholera with contaminated water was unknown at the time; like yellow fever, cholera was believed to be a miasmatic disease, acquired in the presence of malodorous vapours (see chapter 9 by Heikki S. Vuorinen). The dissemination of the disease was greatly facilitated by the crowded conditions, primitive sewage systems and the potential for cross contamination of drinking water with sewage common in urban Europe at the time. Although much had improved in Gibraltar as a result of the yellow fever epidemics, particularly in dwelling construction, sewage infrastructure and the cleanliness of the streets, the crowded town with a shortage of water and a large influx of travellers offered fertile conditions for the disease. In 1834, 242 civilians died, resulting in a mortality rate of 16.1 per 1000 inhabitants. The troops suffered greater relative losses: with 142 dead from a strength of 3,034, and a rate of 46.8 per 1000, mortality among the troops was three times that of the civilian inhabitants. (Army data from The Sickness and Mortality of Troops Serving in the Mediterranean, 1838, 129; civilian data from Howes 1994) Despite the high number of military deaths, there was no action on the part of the authorities in Gibraltar to improve conditions in the colony. Restrictions were, however, put on immigration to curb the rapid population growth.

The situation remained largely unchanged for the next three decades, and the only improvement in water supply was the growth in the number of private cisterns. (Report on Barrack and Hospital Improvement 1863 27) Cholera pandemics occurred again in 1848-49 and 1854, both times visiting the colony. In Gibraltar, the mortality from these epidemics was moderate and there was little impetus from the governor and his officials to make any further improvements in the sanitary situation, either for the military or for the civilians. It was, however, events in other spheres of the British Empire that caused an inquiry into the conditions in Gibraltar. First, there was a growing understanding of the role of sanitation in population health. Since the 1840s, reformers William Farr and Edwin Chadwick had been using mortality rates to support improvements in the sanitary conditions of the large cities in England. (Pickstone 1992) In 1854, John Snow, a London doctor, had shown that the source of cholera could be drinking water contaminated with sewage. (Bynum 1994, 79) For the military, it was the high death toll in the Crimean War that caused scrutiny of the living conditions of the soldiers.

Beginning in the 1860s, the army turned its attention to the health of the soldiers under its care. With the increasing manpower required by an expanding empire, and a decline in the number of recruits offering themselves for duty, the army was becoming aware that it must prevent unnecessary illness and death among its soldiers. In Gibraltar, many changes in the sanitary condition of the barracks resulted from this shift in focus, such as better washing facilities and better paving and drainage. In addition to concerns about the men, the gaze of the army medical establishment fell on the civilian population as a reservoir for disease that could spread to the troops. The high infant mortality among civilians of the colony was indicative of a high level of background (non-epidemic) mortality, and the occasional epidemics, not only of cholera but also smallpox, scarlet fever and measles, reminded the army medical men of the potential for high levels of disease among the soldiers. Some were beginning to see that a reduction of excess mortality among the civilians of the colony could benefit the troops. (Padiak 2004) At the same time, the military governors of Gibraltar were becoming increasingly concerned about the readiness of the fortress for a prolonged siege, and often this focused on the available water supply. (Letter from Edward Somerset, Acting Governor of Gibraltar 1878) These events prompted an inspection of the sanitary situation in Gibraltar in 1862 by a representative of the British government. (Report on Barrack and Hospital Improvement 1863) As a result, a report was issued that was scathingly critical of the situation for civilians in Gibraltar: it noted that the British had not laid one new water pipe in 150 years for the benefit of the populace. (Report on Barrack and Hospital Improvement 1863, 29) The people had been entirely responsible for the provision of their own water within the fortress, and had met the cost of any improvements themselves. They had, by this time, storage capacity for 122,744 cubic meters of water, providing between four to six litres of water per day per person, well below the recommended nine litres per day per person. (Report on Barrack and Hospital Improvement 1863, 30, 33) One third of households had no access to water, neither from cisterns nor wells, and were required to carry sanitary water from the public wells at the Spanish border or to buy potable water imported from Spain on the backs of donkeys. Altogether, an additional 45,000 cubic meters was imported annually, the cost of which fell unfairly on those in the poorer dwellings. The report noted that each person in Gibraltar paid as much for one day's supply of six litres of water imported from Spain as an entire household in England paid for one year's unlimited supply of centralized water. The military had somewhat better water reserves, either constructed or planned, but still, the most optimistic calculation was about seven litres of water per soldier per day. The navy had, by this time, constructed its own holding tanks to provide for the British fleet. (Report on Barrack and Hospital Improvement 1863, 31)

Figure 11.5 Wells on the sandy area near the Spanish border, around 1870.

## 11.6. THE SANITARY COMMISSION

In 1865, before any of the changes recommended by the report could be considered, there was another cholera pandemic. In Gibraltar, 400 people died as in 1832, resulting in a civilian mortality rate of 23.3 per 1000. The military suffered somewhat less, with 15.6 out of every 1000 troops dying of cholera. (Sawchuk et al. 2003, 202) Although the death toll was not especially high, this epidemic acted as a clarion call for the recommended improvements. (Sawchuk, 2001) One of the results of the 1865 cholera epidemic was the creation of the Sanitary Commission, a board made up of civilians appointed to oversee matters relating to public health, particularly in the allotment of sanitary water from wells. Water apportionment and use was such a contentious subject at the time that there is evidence that, while the board may have eased the availability of sanitary water for Gibraltarians, they may have created problems within the fortress. Certainly the military administration of the colony believed that the members of the board abused their power in the apportionment of well water. (Letter from Lord Napier of Magdala, Governor of Gibraltar 1882) Nonetheless, the sanitary commissioners were instrumental in developing Gibraltar's water distribution system late in the 19th century.

One of the ironies of the situation in Gibraltar is that the long inaction on the part of the colonial government in facilitating a centralized water system may have actually limited excessive mortality from cholera. Recent investigation into the 1865 epidemic suggests that Gibraltar's fragmented water supply prevented the problem of a common-source epidemic, in which contamination of one supply can cause disease in a large number of people. (Sawchuk

Figure 11.6 Water carrier filling barrels for distribution in the upper portion of the town, around 1890.

& Burke 2003) Using mortality and census records, it was shown that those inhabitants who depended on private wells or cisterns were much less likely to die of cholera than those who had no access to wells or cisterns and depended entirely on public wells or water imported from Spain. Cholera spreads fastest when a contaminated water supply is shared by a large number of individuals, and cisterns would have stored rainwater well before the start of the epidemic; contamination would have been on a household level only, not a community level. (Sawchuk & Burke 2003) As a fractured water supply had existed for many years in Gibraltar, this explanation may be applicable to the earlier cholera epidemics, in which mortality rates were also rather moderate.

In the years following the creation of the Sanitary Commission, the board oversaw the development of several new wells along the Spanish border (Figure 11.5). Although this water was brackish and unsuitable for dietetic uses, it was suitable for flushing sewers, cleaning streets and other general uses. (Letter from Carew Hunt 1882) The commission installed a pumping system that raised this salty water to a reservoir situated in the upper portion of the town, from where it supplied a small distribution system for the town. This formed the basis for one of the centralized water supply systems that exist in the colony today, delivering non-potable water for sanitary use.

## 11.7. THE PROVISION OF PURE WATER

Another result of the changes following the 1865 epidemic is the realisation that cistern water could not supply a sufficient quantity needed by the colony for human and animal consumption. (Report on the Sanitary Condition 1867, 33) Beginning in 1869, the colonial and military officials considered the purchase and installation of condensers; at that time condensers were capable of producing 122 cubic meters per day of pure water from seawater. (Letter to the Governor of Gibraltar from the Pure Water Company) The use of condensers was postponed in Gibraltar, partly because of the arguments between the War Office and the Colonial Office over who would pay for them. In the early 1880s, the possibility of being under siege loomed large in the considerations of Gibraltar's governor, and the condenser project was reconsidered and approved by the War Office. (Ayde 1895)

Pure water from the condensers was first available to the people of Gibraltar, both military and civilian, in March of 1885. (Report of Water Supply in Gibraltar 1885) The only system of water distribution in the town was for impure water, so this precious commodity was distributed by barrels in carts pulled by donkeys or by hand (Figure 11.6). Delivery was particularly problematic in the upper portion of the town, where the only access was through a maze of narrow alleys and steps. Despite the labour required, the people of Gibraltar appreciated that, for the first time in almost 100 years, they had access to pure water, as much as they could carry, throughout the year. (Letter From John Ayde, Governor of Gibraltar 1885) The days of importing questionable water from Spain were over.

There is some evidence that there was an immediate health benefit from the availability of good water. The condensers had arrived and were in operation only a few weeks when an outbreak of cholera appeared in Spain. In 1885, only 22 individuals died of the disease in Gibraltar, an improvement of 95% over the 1865 epidemic. Across the Spanish border in the town of La Linea, the mortality rate was high, with 191 deaths occurring out of a population of 12,000. (Sawchuk, 2001) It is likely, therefore, that Gibraltar's new supply of water prevented the spread of the disease. There is also some indication that there were long-term community health benefits from the provision of pathogen-free water. Analysis of deaths among infants due to weanling diarrhea indicates a reduction in fatal gastrointestinal infections; this reduction was particularly marked in the military community. (Sawchuk et al. 2002) Although there are many factors that could introduce harmful pathogens into the alimentary system of a child, impure water is most often the suspected source.

## 11.8. CATCHMENTS ARE CONSTRUCTED

At the end of the nineteenth century, an extraordinary water catchment plan was envisioned and initiated by engineers in Gibraltar. This project utilised the steep (35°) sandy slope on the east side of the Rock as a surface on which to collect rainwater for use as potable water (Figure 11.7). (Darton & Buckley 2001) In 1902, 140,000 square meters of the slope was clad

Figure 11.7 Water catchments after construction in 1902. The catchments were constructed on the steep slope on the east side of the rock, 100 meters above sea level.

in concrete and corrugated iron; precipitation was collected on these surfaces and guided into a channel at the base of the catchment, about 100 meters above sea level. This water was channelled into several large reservoirs. To reach the population in the town, a tunnel was bored through the Rock to the west side. The water was distributed to the dwellings in the town, particularly those in the upper town that had no other supply. The distribution system of this water was separate from the distribution of seawater, which was used for non-potable purposes such as sewer flushing and street cleaning. (Chairman's statement, Northumbrian Water Group, 2004)

The catchment provided potable water to Gibraltar for most of the 20$^{th}$ century, although increased demand after WW II required additional solutions to the problem of water supply. Strained relationships with Spain meant that piping water from Andalusian sources was not a consideration and Gibraltar began construction of desalting plants in the 1960s. (Darton and Buckley, 2001) In the next 30 years, the cost of desalination declined because of improvements in reverse osmosis technology, while the cost of maintaining the aging catchment increased and, in 1991, the decision was made to dismantle the catchment. This required removing the corrugated iron sheets and the concrete base upon which it sat, and returning the slope to its natural state. Soil nails and geotextiles are currently being used to provide a basis for the re-establishment of native vegetation. Private cisterns have also been dismantled and the areas converted to living space. Today, Gibraltar is totally dependent on seawater desalination for potable water and has a secondary system of seawater for non-potable, utilitarian uses such as toilet flushing and fire fighting. (Darton and Buckley, 2001)

## 11.9. CONCLUSIONS

Gibraltar's unique topology, limited land area and socio-political situation combine to make it a compelling case study in the evolution of community water supply. Part of this interest lies in the challenges inherent in supporting a high population density in a small area that has no source of standing surface water. Because of this, the development of community water in Gibraltar has focused on the collection and storage of precipitation and, in recent times, has included technological advances in desalination. Another reason for the study of community sanitation in Gibraltar is the illumination shed on the role of infectious diseases, such as yellow fever and cholera, as instigators of change: the high mortality accompanying major epidemics prompted the colony's administration to focus on ways of improving public health. Finally, study of the provisioning of a scarce commodity such as water exposes the tensions of colonialism, as the military preferentially allotted scarce resources to the garrison and naval yards over the domestic needs of native Gibraltarians. Ultimately, it was the needs of Gibraltar as a fortress, not as a home to 20,000 people, which brought pure and plentiful water to the civilians of the colony.

For much of the 19th century, the people of Gibraltar were left to their own devices and ingenuity to provide storage of potable water for use during the dry summer; their position as a civilian population inside a fortress placed them in a situation where military interests were always paramount. However, when the concerns of the military and colonial administrations overlapped with the needs of the people of Gibraltar, improvements could be executed on an extraordinary scale and with great speed. Unfortunately, this did not occur through the middle of the century, when disinterest and procrastination characterised the attentions of the military to the civilian situation. This changed after 1880, when concern for the health of the common soldier and understanding of the dangerous position in which the garrison would find itself in the event of a siege caused swift and sure action by the colonial government.

It was not until almost the beginning of the 20th century that the civilian population of Gibraltar had access to a reliable supply of potable water throughout the entire year. Improvements in technology have allowed the Gibraltarian people to utilise the unlimited amount of salt water in the sea surrounding the peninsula, rather than collecting and storing water or piping it across vast distances. Although Gibraltar is still a British colony, it is no longer a garrison, and the people of Gibraltar elect their government; AquaGib, which today provides water for the peninsula, is a joint venture between the Gibraltar government and a private firm. Issues in modern Gibraltar water management include minimising leakage, energy recovery and customer approval, rather than allotment and concern about the possible mortality implications of an unhealthy supply.

## 11.9. REFERENCES

PRIMARY SOURCES

Answers to Queries from the Army Medical Board: Query #9, answered by Governor Don, 10 January, 1831, Gibraltar Archives.

Dispatch from George Don, Governor of Gibraltar, to General Sir W. Houston, 23 September 1831, Gibraltar Archives.

Extract of a dispatch to General William Houston 7 September, 1831, Gibraltar Archives.

Letter from J. O'Hara, Governor of Gibraltar, 14 December 1800, Public Records Office of Great Britain file CO 91 41

Letter from J. O'Hara Governor of Gibraltar, 18 December 1800, Public Records Office of Great Britain file CO 91 41

Letter from Colonel C. Mann, to General Sir Alex Bryce, 5 November 1831, Gibraltar Archives.

Letter from Colonel Smith-Dorrien, Governor of Gibraltar, to Viscount Milner, 19th July 1920, Public Records Office of Great Britain file CO 091 474.

Letter from Edward Somerset, Acting Governor of Gibraltar, 25 May, 1878, Public Records Office of Great Britain file CO 537 30.

Letter from Lord Napier of Magdala, Governor of Gibraltar, to The Earl of Kimberley, Secretary of State, 16 October, 1882, Public Records Office of Great Britain file CO 91 360.

Letter from Carew Hunt to the Colonial Secretary of Gibraltar, 1 August, 1882, Public Records Office of Great Britain file CO 91 359.

Letter to the Governor of Gibraltar from the Pure Water Company, Public Records Office of Great Britain file CO 91 307.

Letter From John Ayde, Governor of Gibraltar, to A. Stanley, Secretary of State, 5 September, 1885, Public Records Office of Great Britain file CO 91 371.

Memorandum of A. Wolseley, 7 Aug 1887, Public Records Office of Great Britain file CO 883 4

Report on Barrack and Hospital Improvement Commission on Sanitary Condition and Improvement of Barracks and Hospitals at Mediterranean Stations, 1863, Parliamentary Papers XIII (1863).

Report on the Sanitary Condition of Gibraltar with reference to Epidemic cholera, 1867, Parliamentary Papers XXXVII (1867), 849, p. 20.

Report of the 1881 Census of Gibraltar, Public Records Office of Great Britain file CO 91 356.

Report of Water Supply in Gibraltar, 19 March, 1885, Public Records Office of Great Britain file CO 537 30.

The Sickness and Mortality of Troops Serving in the Mediterranean, Parliamentary Papers Parliamentary Papers XVI (1838), 129

LITERATURE

Ayde J. (1895) Recollections of a military Life. London: Smith Elder.

Benady S. (1994) *Civil Hospital and Epidemics in Gibraltar*. Gibraltar Books, Grendon.

Blake C. (1855) Letter to the Right Honorable Lord John Russell M.P. Woodfall and Kinner Blake, London.

Bynum W. F. (1994) *Science and the Practice of Medicine in the Nineteenth Century*. Cambridge University Press, Cambridge.

Darton E. G. & Buckley E. (2001) Thirteen years' experiences treating a seawater RO plant, *Desalination* 134, 55-62.

Finlayson T.J. (1991) *The fortress came first: the story of the civilian population of Gibraltar during the Second World War*. Gibraltar Books, Grendon.

Gilbard M. (1881) *A Popular History of Gibraltar, its institutions and its neighbourhood on both sides of the Strait*. Garrison Library Press, Gibraltar.

Howes H. W. (1994) *The Gibraltarian: the origin and development of the Population of Gibraltar from 1704*, 3rd ed. Gibraltar: Medsun.

Kramer J. (1986) *English and Spanish in Gibraltar*. Helmut Buske Verlag, Hamburg.

Padiak J. (2004) *Morbidity and the 19th century decline of mortality: An analysis of the military population of Gibraltar 1818 to 1899*. PhD thesis, University of Toronto.

Pickstone J.V. (1992) Death, dirt and fever epidemics: rewriting the history of British public health, 1780-1850, In Pickstone, J.V. (ed.). *Epidemics and ideas: Essays on the historical perception of pestilence*, Cambridge: Cambridge University Press.

Preston R. A. (1946) Gibraltar, Colony and Fortress, *Canadian History Review* 27, 402-23.

Sawchuk L.A. (1996) Rainfall, Patio Living, and Crisis Mortality in a Small-Scale Society: The Benefits of a Tradition of Scarcity? *Current Anthropology* 37, 863-867.

Sawchuk L.A. (2001) *Deadly Visitations in Dark Times*. Aquila Press, Gibraltar.

Sawchuk L.A. & Burke S.D.A. (1998) Gibraltar's 1804 Yellow Fever Scourge: The Search for Scapegoats, *Journal of History of Medicine and Allied Sciences* 53, 3-42.

Sawchuk L. A., Burke S. D. A. & Padiak J. (2002) A Matter of Privilege: Infant Mortality in the Garrison Town of Gibraltar, 1870 –1899. *Journal of Family History* 27, 399-429.

Sawchuk L.A. & Burke S.D. A. (2003) The ecology of a health crisis: Gibraltar and the 1865 cholera epidemic. In Herring, D.A. and Swedlund, A.C. (Eds.) *Human Biologists in the Archives*. Cambridge: Cambridge University Press, 178-215.

# 12

## COPING WITH DISEASE IN THE FRENCH EMPIRE: THE PROVISION OF WATERWORKS IN SAINT-LOUIS-DU-SENEGAL, 1860–1914

*Kalala J. Ngalamulume*

Figure 12.1 Location of Senegal.

## 12.1. GENERAL BACKGROUND[1]

The history of infrastructural building or improvement in colonial cities in Africa is still in its embryonic stage. In *Africa's Urban Past*, edited by David M. Anderson and Richard Rathbone, the chapters dealing with colonial cities focused mainly on colonial urban planning and on issues of land use and social control. (Anderson & Rathbone 2000) Little attention was given to the development of public works that made life in these cities comfortable. However, recently, at the trans-disciplinary conference on the history of water in Africa held at North-West University in South Africa, there were signs of interest among scholars in dams and power projects, and urban water. (Flows From the Past 2004) The same can be said of Senegalese cities, in general, and Saint-Louis, the first capital of French Senegal, in particular. This study contributes to the literature on urban Senegal. It focuses on the losing battle to provide fresh drinking water to Saint-Louis.

The process of growth of Saint-Louis in the context of international commerce and travel created the conditions favorable for the outbreak of infectious, bacterial, and parasitic diseases. The paucity and the unequal distribution of urban amenities, combined with compromised immune systems of the urban poor, determined the pattern of mortality and morbidity among city dwellers. Poor sanitation, poor housing, overcrowding, fecal contamination of the water supply and foodstuffs, and a lack of sewerage made the city especially vulnerable to water-borne diseases, and exposed family members and neighbors to the feces of the sufferers or immune carriers. (Figure 12.2)

The lack of water and sewage systems was part of the larger problem of waste removal in the city, including the household wastewater, garbage, animal waste, and even rain water on the city streets during the *hivernage* (or rainy) months, especially between July 15 and September 15. Plans for building the city's sanitary infrastructures were discussed but abandoned because they either lacked financial support or they were not viable. Some engineers favored the combined system, or *"tout à l'égout"*, which disposed of all the wastes and rain water at once; others argued in favor of a separate system, or *"le système séparé"*, which would evacuate only the kitchen and human wastes, leaving the rain water to be disposed of by other means. The combined system required building a large drainage system that would be used in full capacity only between July and September, but would have high maintenance costs in terms of the volume of water needed to keep it clean and avoid fermentation. The separate system had the advantage of eliminating the matters believed to present "the most immediate danger" to city health. House owners would be required to extend the sewer

---

[1] The materials for this study are drawn from my Ph.D. dissertation, entitled 'City Growth, Health Problems and the Colonial Government Response: Saint-Louis (Senegal), from mid-nineteenth century to 1914,' Michigan State University, East Lansing, MI, 1996. The archival materials were collected in 1994-5 during fieldwork research in Senegal and France assisted by grants from the Rockefeller Foundation and from the Joint Committee of African Studies of the Social Science Research Council and the American Council of Learned Societies with funds provided by the Ford, Mellon and Rockefeller Foundations, and in 2001 assisted by a Summer research grant from Bryn Mawr College, and in 2002 assisted by the Lindback Foundation Minority Grant. I thank them all for their support.

Figure 12.2 Water from Lampsar distributed in Saint-Louis. (Source: Faidherbe, 63)

service pipes to their houses themselves, while the administration would build public toilets in the slums. But neither plan was technically, politically, economically, or socially viable. City dwellers were left vulnerable to the "infected ponds" and "fermented substance-targets" present in and around their homes (ANS/H48-1) that were responsible for the diseases of the gastrointestinal system. Let us now turn to the problems of water supply in Saint-Louis.

## 12.2. WATER SUPPLY PROBLEMS

Saint-Louis residents enjoyed an abundance of water only four months per year during the *hivernage* season between June and November, with abundant rains particularly between mid-July and mid-September. City residents obtained fresh water from the Senegal River and the rain water collected in cisterns. Government cisterns met the needs of the Civilian Hospital, the Roignat garrisons, and the *Génie* workshops. But during the eight months of the dry season, between December and mid-July, the supply of water was a serious problem, since the river water turned brackish and became polluted (Sometimes, the river water remained salty until August. See *Moniteur du Sénéga* 1868, 218; see also Mémoire sur un avant-project 1869). Table 12.1 shows the transformation of the quality of the local water, as measured in a study conducted by Dr. Audibert, chief pharmacist at the military hospital in Saint-Louis.

Table 12.1 Changing Level of Salt in the Senegal River, Saint-Louis 1851-2 (per 100 gr) (Source: Berenger-Féraud 1873, 111-112.)

| June 1 | July 1 | August 1 | Sept 1 | Oct 1 | Nov 1 |
|---|---|---|---|---|---|
| 0.2530 | 0.0618 | 0.0640 | 0.0013 | 0.0009 | 0.0016 |
| Dec. 1 | Jan. 1 | Febr. 1 | March 1 | April 1 | May 1 |
| 0.4915 | 1.0906 | 2.0570 | 2.0325 | 2.0302 | 0.6250 |

The table shows that the Senegal River water around Saint-Louis became fresh in mid-July and could be used safely through December, which explains why Governor Valière fixed, in 1870, the date of January 4 for beginning the distribution of water from a government cistern to civil servants and the military. (Ministère de la Marine et des Colonies 1870)

During the period of water shortages, the well-to-do, estimated at 100 out of 14,900 residents in the 1850s and 1860s, utilized the water they had stored in the cisterns during the rainy season. The administration put in service ship water-tanks, *La Sénégambie* and *L'Akba*, which carried water from upstream at the Kassack *marigot*—a flood plain tributary of the Senegal River—to the city. Kassack was located near Makana village, sixteen kilometers east of Saint-Louis. The water was distributed as indicated in table 12.2. Animals owned by the administration and used for transportation received water from this source: 10 to 12 liters of water per day per mule and 12 to 14 liters per horse. (Bulletin Administratif 1892, 119-20)

The exclusive distribution of water to the officials and other administration employees underlined an important dimension of health inequality. The working class and the urban poor had to rely on unhygienic river water, that is, on ground water tapped by small wells that they dug outside the city in the dunes in Sor or in the Pointe de Barbarie south of the city.

The water drawn from these wells was poor in quantity and quality, and the wells rapidly filled with salty water. The poor continued to bathe and wash their clothes and utensils in the polluted river water; they were known for drinking the «bad water.» (Epidémie du Sénégal 1867) Yet, it was possible in Saint-Louis to buy water from private individuals, as seen in an advertisement published in the *Moniteur du Sénégal et Dépendances* in 1864:

NOTICE

*FOR SALE: Potable waters for 1 franc a cask, from the cistern owned by Gilbert. Contact Mr. Boudou."* (Moniteur du Sénégal et Dépendances 1864, 14)

Table 12.2 Distribution of water to government employees, Saint-Louis 1873. (Source: Berenger-Féraud 1873, 9)

| Civil servants and troops | Quantity of water/day |
|---|---|
| Head of administration & Comptroller<br>Chiefs of corps & services | 12 litres |
| Military officers & assimilated | 8 litres |
| Low-ranking officials | 5 litres |
| Low-ranking mil. officers, corporals, troops, sailors, other agents | 4 litres |
| Family members of officials | 4 litres/person |
| Administration employees | 2 litres |
| Total: | 35 litres |

Even when available for sale on the market, as in this case, the cost of water was still out of reach for the urban poor. Civilian prisoners also did not have access to potable water during the dry season. In 1868 Dr. Beaussier, *médecin* 1st class, observed that the water distributed to the prison population was brackish and contained a large amount of salty ocean water. He recommended that all the prisoners, military and civilian, be equally treated, and given water from the cisterns. (ANS/H2/AOF/9 1868)

By the early 1870s the price of one cask or barrel of water had increased from 12 to 15 francs. (Minutes of the *Conseil d'Administration* meeting 1878) But the poor could not afford to buy it. Some Muslims who held a marriage certificate issued by the *Cadi/Tasmir*, the head of the Muslim tribunal, showing that they had only one legitimate wife, were allowed to obtain the water distributed by the administration for 0 franc 13 cents a liter. (*Bulletin Administratif* 1885, 534, see also the decree of March 21) How many people qualified to buy the water is not known, but the majority of the poor were excluded for financial reasons.

During the dry season, water shortages and the contamination of water supplies were responsible for many cases of typhoid fever, diarrhea, dysentery, and gastrointestinal disorders. (ANS/H20 1868) City residents knew "from experience" that during the "season of the winds from the east", the diseases of the intestinal system as well as a variety of respiratory conditions from the bronchitis/pneumonia/influenza group increased. (ANS/H2/AOF/38)

## 12.3. EARLY EFFORTS TO BUILD WATERWORKS

Saint-Louis' expanding population led to more waste matter being discharged directly into the river, and to more exposure to infectious agents. The administration was well aware of the fact that water shortages constituted a bottleneck to urban growth. In the 1860s, Governor Faidherbe attempted to build artesian wells in the northern part of the city. The cost of building a well 250 m deep in order to bring water to the surface was estimated at 54,000 francs but the appropriate technology was not available in the colony. The construction of the well was suspended pending the arrival of the new machinery from France. (Rapport sur le forage) The Inspector of the *Service des Ponts et Chaussées* recognized that "the lack of water is one of the principal obstacles to the extension of our domination." (Rapport sur le forage)

Governor Pinet-Laprade decided to provide the city with a permanent source of clean drinking water by building waterworks. Indeed, in 1866, a year after he took office, the Governor erected a barrage on the Kassack flood plain at Lampsar in order to create a man-made reservoir of fresh water measuring 600 metres long, 70 metres wide and 2.5 metres deep. The estimated capacity was 8,500,000 cubic meters, which was sufficient to meet city residents' daily needs. (Mémoire complémentaire 1870) But the barrage was destroyed by floods. In the meantime, the working class and the urban poor were "at risk" due to the first cholera "epidemic".

## 12.4. THE 1868–1869 CHOLERA EPIDEMIC

The fourth cholera pandemic spread from India and China thanks to the intensification of international travel. Mecca, the pilgrimage center, rapidly became the central place for the transmission of cholera: 35,000 out of 90,000 pilgrims succumbed to cholera there in 1865. The pandemic had reached some port cities of North, and West Africa, including Saint-Louis and Bissau city in Portuguese Guinea—where it produced thousands of victims—and Central and Southern Africa (Belgian Congo, Tanganyika, Mozambique, and Madagascar).

One year later, a yellow fever epidemic struck Saint-Louis; the first four cases of cholera, followed by deaths, appeared in the city on November 24, 1868. Soon, the number of new cases reported at the military hospital rose to 164. The situation was very critical and the city was in a state of panic; city residents began to flee. Within days about 10,000 people, out of a population of 20,000, had left the city in panic. (Rulland 1869, 1-5. See also ANS/H27) The disease quickly spread throughout the city. It was "a situation of unseen gravity." Doctors had no effective drugs to combat cholera but strategies used against yellow fever, such as purging, or blood-letting, were also tried against cholera. In any case, the number of deaths increased rapidly. By the end of the epidemic, in December 1868, about 1,204 city residents, including 1,112 indigènes and 92 Europeans, out of a population of 20,000, had died in Saint-Louis as well as in various isolation camps. This was in just twenty-five days. From eyewitness accounts, the cholera epidemic had surpassed in intensity the yellow fever epidemic that had ravaged the city the previous year (Rulland, 1869. In his doctoral thesis in medicine, Gilbert-Pierre Eyoum mistakenly presents these mortality figures for Dakar instead of Saint-Louis; see Eyoum 1986, 22.) Missionaries' sources put forward the frightening number of 5,000 deaths. (Bulletin Général 6, 862-4; see also Sister Marie-Ange, 79) But there is no other independent source to corroborate this assertion. The data also indicate that cholera had concentrated its attacks among the urban poor, who had no reliable fresh water supplies, and who were more exposed to filth than the elite.

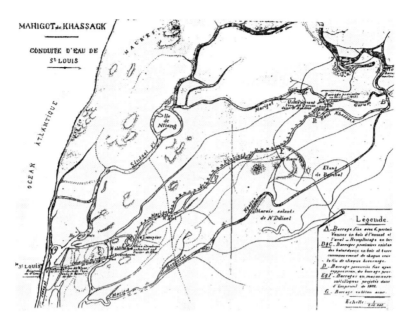

Figure 12.3 The water supply system. (Source: ANS/H48-1, Mission Sanitaire du Sénégal 1901)

Cholera reemerged first in Gambia in early May 1869. It then spread rapidly throughout various towns in Senegal, such as Gorée, Dakar, Podor and Dagana, at a time when the administration was at war with Amadou Sekhou in Cayor. The latter found in cholera an opportunity to recruit a following. The evidence suggests that there was a link between war and the spread of the cholera epidemic in the sense that war provoked the movement of people, caused famine, and exposed the refugees, whose immune system was already weakened by poor diet, to contaminated water and food. Soldiers were probably instrumental in the spread of cholera in the city.

The disease reappeared in Saint-Louis on May 25: two soldiers who became ill were admitted to the hospital and cured. But by early June the case rate drastically increased, and six to seven deaths were reported daily. News from Bakel, a town located in the Upper Senegal, was alarming: 800 people had died in the town, including the local physician. The beginning of July witnessed a stabilization of the number of cases at seven to eight deaths every day. The majority of the victims were poor city residents, who had little access to clean drinking water and safe food; until then, only three Europeans had died. However, on August 18, Governor Pinet-Laprade (forty-seven years old) succumbed to the disease, followed by the death of Doctor Maurel. (*Annales*, 23-5; see also *Bulletin de la Congrégation* 7, 131-2, 135-7) They were probably exposed to infection through the consumption of ice, as the governor was known for consuming "too much ice" that was made from river water. (Thiroux, 760) Panic and consternation spread throughout the city, prompting the flight of those who could afford it and religious devotion from those who could not escape. But the epidemic subsided, probably because many residents had deserted the city; in addition, the return of the rainy season contributed to the reduction of exposure to filth. In September, Valière, a Navy infantry colonel, was appointed Governor. The source material is far less abundant for the 1869 cholera epidemic mortality and morbidity than for the outbreaks of 1868 and 1893. Sources are silent about the last recorded cases and the ways in which the 1869 cholera epidemic ended. Very little evidence is available concerning the incidence of the disease in the city.

## 12.5. THE POLITICAL RESPONSE

The 1868–1869 cholera epidemic came as a warning call to the administration to find a permanent solution to the problem of a clean water supply. In early 1869 the director of the *Bureau du Génie* prepared a detailed project for the supply of water, which was approved by the Governor. Water would be drawn from Lampsar and taken to a pumping station at Makhana, and transported through a main aqueduct to a pumping station at Khor in Sor. From there the water would be taken to a high-level reservoir and a water tank in Saint-Louis where it would be stored before being distributed to city residents through smaller water pipes. Each resident would get at least 100 liters of water per day. The reservoir would be built either on the ground previously occupied by the marketplace, or near the church, or on the Pointe du Nord in the middle of the public garden. The cost of the project was estimated at 980,000 francs, and the annual operating expenses were estimated at 19,000 francs. The empty buildings of the old military post at Lampsar would be used to house the maintenance personnel, and to store the equipment and fuel for the operation of the pumps. (Mémoire sur un avant-projet 1869)

The project was approved by the *Conseil d'Administration* on June 30, 1869, as cholera struck the city for the second consecutive year. But Governor Pinet-Laprade's death due to cholera on August 18 put the scheme on hold. The issue of the establishment of a water supply system was again discussed by the *Conseil d'Administration* during a meeting held on July 12, 1870. The Interior Director, Trédos, indicated in his briefing that the execution of the project would require the mobilization of all the Navy infantry artillerymen, all the *Disciplinaire* personnel, and a section of the *Tirailleurs* battalion. He underlined the fact that the tasks they would perform, that is, digging and filling trenches, laying pipes, and sealing joints, would be detrimental to the health of the Europeans involved in the project. He also pointed out that the project would not cost less than 800,000 francs, that the administration could only mobilize 590,000 francs, and that, given the war between France and Germany, no funding would come from Paris. Governor Valière recognized that his administration had two options: either to postpone the construction of the waterworks or to execute the project at a cost of 685,000 francs. But at the end of the meeting, no decision was taken. (Minutes of the *Conseil d'Administration* meeting 1870) Nevertheless, the administration sent the Lampsar project to Paris for analysis and approval. In the meantime, floods during *hivernage* seriously damaged the dam built at the Kassack flood plain, and the reservoir was transformed into a stream. The damage revealed the vulnerability of the project and raised concern among the decision-makers. A new solution had to be found.

The members of the *Conseil des Travaux de la Marine*, an institution based in Paris, met on October 11, 1870 to discuss the viability of the Lampsar project. They recommended that the administration try to build another artesian well in Saint-Louis, to find ways to supply the Kassack reservoir with water intake pipe far upstream for a longer time period, to provide them with complete information concerning the quality of the water from the reservoir, and to locate the pumping station at Makhana. (Minutes of the *Conseil des Travaux de la Marine* meeting 1870) Two years passed without any tangible progress made concerning the construction of the project.

In 1872, the *Capitaine de génie*, Gouin, abandoned the Lampsar project, and designed a new water supply system focusing on the construction of eight vast reservoirs on Pointe Nord in Saint-Louis that could be covered and filled with river water during *hivernage* through pumps propelled by the northwest winds. Each reservoir would have the capacity of 4,000 cubic meters and would cost 75,000 francs. Gouin estimated the total cost of the project at 600,000 francs, and the annual operating expenses were estimated at 8,000 francs. According to his calculation, each resident would receive 10 liters of fresh water per day for 20 centimes. (Extrait d'un projet de réservoir 1872) But Gouin's project, despite the advantages it had in terms of labor and cost, did not find supporters among the decision-makers for reasons that remain obscure. A year later, Governor Valière sent a letter to the Minister of the Navy in Paris requesting that an engineer from the service of *Ponts et Chaussées* be sent to Saint-Louis for the establishment of a water supply system at Lampsar. He argued that the cost of maintaining and repairing the damage done to the dam by flood would amount to 8,000 to 10,000 francs per year. (Governor to minister 1873)

The next five years were spent in political squabbling over costs, technology, and funding. Important segments of the city's residents petitioned for a reliable water supply system. (Governor to ministry 1878) In early 1878, the Navy Ministry in Paris, under pressure from the *négociants* living in France, rejected the most recent Lampsar project presented by engineer M. Badois because the cost, estimated at 1,505,000 francs, was too prohibitive for the local administration as well as the municipality of Saint-Louis. But for "political as well as humanitarian reasons," the Governor decided to keep the Ministry's position secret until further notice. Instead, in order to avoid open hostility between the business community and city residents, and to give moral satisfaction to city-dwellers, he requested that the Ministry provide him with a hedge: the submission of the new Lampsar project and the Badois contract annexed to it to the *Conseil des Travaux de la Marine* for examination. Personally, he believed that there were signs that in the near future business interests would no longer oppose the execution of the project because they would be in the process of demanding the creation of a *Conseil Général* for Senegal with important budgetary prerogatives. He concluded his letter by emphasizing that the execution of the project would constitute "the most important achievement of any administration." The improvement of living conditions would put Senegal on an equal footing with the other colonies. (Governor to minister 1878)

As part of its internal policy, the administration had to make a complicated decision: to gain time by continuing to give the impression that it was committed to the execution of the Lampsar project. Lacking sufficient financial resources, the governor turned for funding to the municipality of Saint-Louis, which, since 1872, had the authority to borrow money. When the Municipal Council decided to take on debt, a group of 31 prominent taxpayers sent, on April 27, 1878, a letter of protest to Brière de l'Isle, former governor of Senegal and then Minister of Colonies, in which they argued that they were not consulted by the Municipal Council, as required by article 58 of the decree of August 10, 1872. But the governor had based his initiative on article 42 of the law of July 18, 1837, which stipulated that the Municipal Council had the obligation to consult the *habitants* only when the municipal income was below 100,000 francs. (*Habitants* to governor 1878a, 1878b; governor to *Habitants* 1878; minutes of the *Conseil des Travaux de la Marine* meeting 1870)

The governor also spent a great deal of time dealing with the logistics of the project, especially the search for a location for the high-level reservoir and the revision of the estimate. The powder-house (*poudrière*) was appropriated to become the water tank, and the initial estimate made by Badois was revised and adjusted. The new estimate of the Lampsar project was 1,675,500 francs. The estimate did not take into account the expenses related to the distribution pipes through the city, the construction of intermediary reservoirs along the distribution network, and the work to be done at the Gorun flood plain. The strategy adopted by Governor Valière finally paid off. The Ministry of Navy and Colonies in Paris, Vice-Admiral Jaureguiberry, who had served in Senegal as Governor between 1861 and 1863, agreed to back loans for a water supply system in Saint-Louis, and to put the project out to bids. (Minutes of the *Conseil d'Administration* meeting 1879) In December 1879, the contract was given to J. Le Blanc, engineer and supplier, who committed himself to providing Saint-Louis with a reliable water supply system within a reasonable time. The initial work then cost 1,100,000 francs; but additional funds were needed to complete the reservoir in Sor, the reservoir in Saint-Louis, the distribution systems for Sor and Saint-Louis, and miscellaneous expenses. Thus the total cost was estimated at 1,463,000 francs. (Cahier des charges 1879)

The news reached the administration in Saint-Louis on January 20, 1880, and the happy Governor hastened to communicate it to the population: the administration and the Municipal Council were authorized to borrow money for the execution of the Lampsar project.[2] City residents also learned that in order to accomplish the task with efficiency,

---

[2] The decree of May 29, 1880 authorized the administration to borrow 800,000 francs for the water supply system at an interest rate of 5%; the debt would be reimbursed in 12 years, using the ordinary resources of the colony, especially the taxes on alcohol and customs service. See Moniteur 1880b, 116.

J. Le Blanc delegated his authority to Mr. Colot, a local engineer. (*Moniteur* 1880, 15-16) The decision came at a time when the financial cost of operating and maintaining the ship for water distribution had sharply increased for reasons that are not yet clear; it went from 13,368.56 francs for 1,386,251 liters of water distributed in 1872 to 47,571.83 francs for 1,912,051 liters in 1880. (*Moniteur du Sénégal et Dépendances* 1873, 1; *Moniteur* 1881, 101) The resolution of the financial question was just one of the many problems with which the administration had to deal. By May 1881, initial excitement gave way to anxiety. The Lampsar water, measured by Richard, the Navy pharmacist, contained a significant quantity of vegetable debris and other impurities. It did not cook vegetables; beans boiled for two hours with the Lampsar water remained unchanged, while beans cooked with the water from the hospital cistern were well-done. In short, the Lampsar water was unfit for human consumption and for most aspects of domestic use. (*Moniteur* 1881, 133) Some members of the *Conseil Général* argued that the engineers of *Ponts et Chaussées* had known all along that the Lampsar water was not even close to safety standards, but, for selfish or other unknown reasons, they misled the administration to support a project bound to fail. (Minutes of the *Conseil Général* meeting 1882) These reactions underscored the competing agendas between the French, who controlled the colonial administration, and the métis, who dominated the *Conseil Général* as well as the Municipal Council, and who stood for the protection of local interests.

The administration continued to borrow money for Saint-Louis' water supply system, despite the succession of accusations and counter-accusations. In 1886 the project was completed. The water was pumped out of Makhana between December and June and out of Khor between July and early December. But when it arrived, Saint-Louis' residents realized that the water distribution network was incomplete. Certain sections of the city either were deprived of water for long hours during the day or received only a small quantity of water. Thus, 12,000 additional meters of smaller distribution pipes and more street taps and hydrants were needed. The total cost was estimated at 80,000 francs. The slums were provided with a few public fountains. (*Conseil Général* 1889, 23 and 25. See also *Bulletin Administratif*, 1889, 26-7) But the extension of the water distribution network through the city did not take place overnight. It was accomplished over the course of years and in several stages.

The water itself, between January and April, was pure and contained a small quantity of organic matter, salt, and clay that could be eliminated through a public sand-filtration system. But toward the end of the dry season (May-June), samples of the water from four public taps in the city and in Sor revealed that the water had become brackish, not drinkable, and loaded with vegetable matter called "*tambalayes*." (Journal Officiel du Sénégal et Dépendances 1889a, 120; 1889b, 291, 301; see also Rapport du chef de Service des Travaux Publiques 1889)

Table 12.3 The Cost of Bathing at the Military Hospital, Saint-Louis 1892 (in francs) (Source: ANS, Bulletin Administratif, 1892, 123-4.)

| Service | Cost | Beneficiaries |
|---|---|---|
| Partial bath | 1.50 | Officers |
| | 2.00 | Private individuals |
| Body | 2.00 | Officers |
| | 3.00 | Private individuals |
| Shower | 1.50 | Officers |
| | 2.00 | Private individuals |

The population growth that characterized the 1890s increased the demand for water, resulting in a deficit of approximately 800 tons. The shortage, mostly toward the end of *hivernage* (early November), obliged the administration in 1893 to borrow 700,000 francs in order to improve the water supply system. (Journal Officiel du Sénégal et Dépendances 1893, 10) They tapped the small arm of the Senegal River to supply the military hospital and the Roignat garrison and the big arm of the river to meet the needs of the *Subsistances* Service; and officials continued to use the "Akba" ship to bring water from upstream near Richard Toll to the city. (D'Anfreville de la Salle, 31) By then physicians became alarmed by the trends in incidence of all types of bowel infections, labeled as "dysentery," among city residents as a result of access to polluted river water. Cases of "dysentery" were the fourth largest cause of hospital admissions in 1889, the third in 1891, and the second in 1893 and 1895. Dysentery was followed by typhoid fever, which dominated the category of "fevers." (ANS/H16, Tableau synoptique) The working class and the urban poor paid the heaviest toll to water-borne diseases.

The quantities of water were insufficient to cover government employees' daily needs. Bathing in particular was a crucial problem. The bath was available at the military hospital between 2 and 4 p.m. every day on a fee basis for those who carried an authorization from the head of the Administration Service and a medical certificate. (See table 12.3)

When cholera struck Saint-Louis in June-October 1893, there was a serious water shortage in the city. How did the administration handle the new crisis?

## 12.6. THE 1893 CHOLERA EPIDEMIC

Although cholera cases had been observed in Saint-Louis for some time, the Sanitary Commission officially declared the re-emergence of the disease in the city in July 1893. The number of deaths had reached an average of twenty per day on July 8, and fifty-six deaths were reported on July 21 alone. By the end of July, Saint-Louis had suffered 753 deaths among the urban poor, out of a population of 20,173. (ANS/H16, Dr. Carpot, 238) Only four

Europeans had died from cholera. In early August the case rate started to decline with less than ten cases per day. The last case was reported on October 11. By then cholera had killed 897 poor city residents, according to official records. But missionaries reported between 1,500 and 2,000 deaths. (*Bulletin de la Congrégation*, XVII, 205)

The re-emergence of water-borne diseases, especially cholera, remained one of the main preoccupations of the administration. The quality of the river water taken in November 1895 in front of the military hospital and the Roignat garrison revealed that Saint-Louis' water continued to contain organic matter and pathogens responsible for diarrheal diseases, dysentery, and typhoid fever. In order to reduce the high incidence of the diseases of the digestive system, the Chief Medical Officer recommended the division of each cistern into two compartments: the first would contain the water for human consumption from the rain collected during *hivernage* and the "Akba" water supplied during the dry season; and the second compartment would contain river water that could be used for cleaning, bathing, washing, and, if boiled, cooking and tea-making. (ANS/H15/2; ANS/H15/3; ANS/H15/5) But the *Commandant en Chef des Troupes* made a different suggestion to the governor general: to equip each military settlement with two separate cisterns, one cistern containing the water to be used for human consumption, and the other containing the water for other domestic uses (cleaning, bathing, cooking, etc.). He recommended that, following the wishes expressed by General Borgnis-Desbordes during his visit to Senegal in 1892, the water be filtered (using Chamberland filter type) as it was a routine practice in the metropolitan army. He reminded the governor general that the issue of improving the quality of water through filtration had been brought to the attention of the Ministry of Colonies in 1894, and the Undersecretary of State for Colonies in early 1895, but it had not been resolved because of the heavy operating expenses related to the use of filters. He concluded that if the administration was determined to preserve the colonial society and the very existence of the European soldiers, the officials needed to obtain special loans in order to buy the much-needed equipment. (ANS/H15/6) A temporary solution was found in 1896 when the authorities decided to stop taking water from the Senegal River around Saint-Louis, which was polluted and contained pathogens, such as *E. coli*. Instead, the city would be supplied with the water from the rain stored in the cisterns during *hivernage*, and with the river water from upstream distributed during the dry season. (ANS/2G/1,7)

Thus, almost ten years after the establishment of the water supply system in the city, the demand for clean water was far from being met for natural, financial, and technological reasons. The water from the Kassack flood plain, taken from a pumping station at Makhana since December, was polluted by March: it was loaded with vegetable debris and animal waste, contained a high level of salt, and had a bad smell. It was not even adequate for sprinklers. By May, the flood plain was dry, forcing the pumping engines to shut down for many hours every day. The water technology used was not well developed in those years. In 1899 the level of salt in the water often damaged the pipe transporting the water from Makhana, thus reducing the volume of water transported from 2,200 to 1,800 cubic meters per day. (*Conseil Général* 1899, 26) Alternative projects were tried at Richard-Toll and N'dial and abandoned.

In order to find a permanent solution to the problem of deteriorating water quality, the *Conseil Général* approved, in May 1899, an administration plan aimed at improving the water supply system, and authorized a 700,000 franc loan for the building of a new dam at Kilère (150,000 francs), the repair of the main distribution pipe near Makhana (80,000 francs), the installation of sand filters at Khor (80,000 francs), the extension of the distribution pipe network in the city, especially in the Pointe du Nord *quartiers*, as well as in the slums of Guet-Ndar and Sor (80,000 francs), and miscellaneous tasks (55,000 francs). (*Conseil Général* 1899, 30) In a report sent to the *Conseil Général*, the Head of the *Travaux Publics* Service, Malenfant, suggested to improve the quality of the Makhana water by purification before its distribution. Such an operation required either the construction of covered reservoirs measuring 40,000 cubic meters and costing 1,800,000 francs, or recourse to a less expensive operation of "*ozonization*" used at Lille and Osborne, costing 350,000 francs. (*Conseil Général* 1899, 33-4) But the implementation of the new plan was delayed by a terrible yellow fever epidemic that struck the city in 1900. (Ngalamulume, completed book manuscript; for more on yellow fever epidemics in Saint-Louis, see Ngalamulume 2004, 183-202)

The evidence suggests that by 1908 the renovation of the water supply system did not result in an improvement of the water quality. One study found the water 'highly suspect', not drinkable, and recommended an additional chemical and bacteriological analysis, which revealed that the water was mediocre and unfit for human consumption even after decantation, filtering, or boiling. (ANS/H15/54) The quantity of water was also insufficient, even after the building in 1913 of a third pumping station at Boudoun, located upstream of the Kassack marigot. In October of the same year the administration once again created a commission charged with the study of the issue of the water supply in the city, but World War I intervened. (ANS/H11/Senegal; ANS/H8/Senegal; Journal Officiel 1913)

## 12.7. CONCLUSIONS

The evidence presented here suggests that the transformation of Saint-Louis from a trading station into a colonial port city posed serious problems of city planning and city improvement, and exposed its residents to infections by sea routes and by land. Cholera epidemics revealed the dimensions of inequality in Saint-Louis. The overwhelming majority of the victims of the epidemics were the working class and urban poor residents who had no access to fresh water supplies during the dry season, had a poor diet, lived in unhygienic conditions in overcrowded slums, and had little formal education to even prepare them to read the posters and understand the ordinances. The installation of the technological sanitary systems, such as water and sewer systems, was crucial to the creation of a new way of life in which the 'diseases of civilization,' such as cholera, would be eliminated first from the rich quarters, and later from the slums. In Saint-Louis efforts made to provide a supply of water were plagued by financial constraints, technological challenges, and competing agendas of the colonial administration and the Municipal Council, dominated by the *Métis* and local interests. Moreover, the water distribution network also revealed the inequalities between rich and poor *quartiers*. Several improvement plans failed to materialize due to financial and technological difficulties, and, sometimes, for lack of political support. But World War I changed the agenda.

## 12.8. REFERENCES

PRIMARY SOURCES:

ANS/H2/AOF/9, médecin 1$^{st}$ class, Beaussier, to médecin en chef, Dr. Julien-Henri Rulland, Jan. 29, 1868.

ANS/H2/AOF/38, médecin en chef to governor, n.d.

ANS/H8/Senegal, Dr. L. Huot to governor, October 119, 913;

ANS/H11/Senegal, governor's decision, no. 1568 of October 10, 1913

ANS/H15/2, Général de Brigade Boilève to governor general, Nov. 15, 1895.

ANS/H15/3, Médecin-major Henry Reboul to Commandant en Chef des Troupes de l'A.O.F., Nov. 12, 1895

ANS/H15/5, Médecin en Chef to governor general, no. 294 of Nov. 24, 1894

ANS/H15/6, Commandant en Chef des Troupes to governor general, no. 551 of Dec. 13, 1895.

ANS/H15/54, Dr. Bourret, Director of the Bacteriological Laboratory, to the Chief Medical Officer, September 2, 1910.

ANS/H16, Tableau synoptique des maladies et mouvement des malades, 1889-1913.

ANS/H16, Dr. Carpot, "Nosologie ...," 238

ANS/H20, Séries Sénégal, minutes of the HPSC meeting, Jan. 10, 1868.

ANS/H27 no piece nb. ANFSOM.

ANS/2G/1,7, Rapports des chefs d'administration, 1896: Médecin en chef to governor general, July 15, 1896.

ANS/H48-1, Mission Sanitaire du Sénégal (Février-Mars 1901): Rapport Technique, by engineer Jacquerez.

Annales. ACSE.

Berenger-Féraud (1873)Etudes sur la Sénégambie. In Moniteur du Sénégal et Dépendances, 14.1.1873

Bulletin Administratif, 1885. ANS.

Bulletin Administratif, 1892.

Bulletin Administratif, 1889. ANS.

Bulletin de la Congrégation 7 (1869-70).

Bulletin de la Congrégation, XVII. ANS.

Bulletin Général 6 (June 1868- March 1869). ACSE.

Cahier des charges relatif à l'entreprise des Travaux à exécuter pour l'etablissement d'une conduite d'eau potable à Saint-Louis, Paris, Dec. 10, 1879. Série géogr Sénégal XII/19, ANFSOM.

Conseil Général (1899) Récueil des Rapports, minutes of the meeting in May 1899. ANS.

Epidémie du Sénégal (1867) In Univers, Nov. 23, 1867. Chevilly-Larue, Boîte no. 157 B. Sénégambie. Sénégal. Gambie. Affaires diverses, 1864-1876, piece 033, ACSE.

Extrait d'un projet de réservoir à la pointe du nord de l'île de Saint-Louis (1872) by the Capitaine de génie, Gouin, s.d.. Séries géogr. Sénégal XII/17c/14, ANFSOM.

Governor to Habitants (1878) 3 May 1878. Série géogr. Sénégal XII d. 18 a, ANFSOM.

Governor to minister (1873) Jan. 9. Séries géogr. Sénégal XII/17b, ANFSOM.

Governor to minister (1878) April 22, 1878. Série géogr. Sénégal XII/18a, ANFSOM.

Governor to ministry (1878) April 18, 1878. Série géogr. Sénégal XII d. 18, ANFSOM.

Habitants to governor (1878a) 27 Apr. 1878. Série géogr. Sénégal XII d. 18 a, ANFSOM.

Habitants to governor (1878b) 3 May 1878. Série géogr. Sénégal XII d. 18 a, ANFSOM.

Journal Officiel du Sénégal et Dépendances, March 28, 1889. ANS.

Journal Officiel du Sénégal et Dépendances, August 8, 1889. ANS.

Journal Officiel du Sénégal et Dépendances, July 1, 1893, 10. ANS.

Journal Officiel, October 16, 1913. ANS.

Mémoire complémentaire sur le projet de conduite d'eau du marigot de Lampsar à Saint-Louis,» by the director of Ponts et Chaussées, May 12, 1870. Séries Sénégal XII 17 b, ANFSOM.

Mémoire sur un avant-project de conduite d'eau destinée à l'alimentation de Saint-Louis et de ses faubourgs, April 6, 1869 by the chief of Bureau du Génie. Séries géographique Sénégal XII d. 17 b, ANFSOM.

Ministère de la Marine et des Colonies. Sénégal et Dépendances, Bulletin Administratif des Actes du Gouvernement, 1870 (Saint-Louis: Imprimerie du gouvernement, MDCCCLXXI).

Minutes of the Conseil d'Administration meeting, July 12, 1870. Séries géogr. Sénégal XII/17b ANFSOM.

Minutes of the Conseil d'Administration meeting, Febr. 4-5,1878. Série géographique Sénégal d. 18 a, ANFSOM.

Minutes of the Conseil d'Administration meeting, Jan. 2, 1879. Série géogr. Sénégal XII/18c, ANFSOM.

Minutes of the Conseil des Travaux de la Marine meeting, Oct. 11, 1870. Séries géogr. Sénégal XII/17b ANFSOM.

Minutes of the Conseil des Travaux de la Marine meeting (1878) 4 June 1878. Série géogr. Sénégal XII d. 18 a. ANFSOM

Minutes of the Conseil Général meeting, Febr. 6, 1882. ANS.

Moniteur (1880a) 3.2.1880, 15-16. ANS.

Moniteur (1880b) 22.6.1880. ANS

Moniteur (1881) 14.6.1881. ANS.

Moniteur du Sénégal (1868) 2.8.1868.

Moniteur du Sénégal et Dépendances (1864) 26.1.1864. ANS.

Moniteur du Sénégal et Dépendances (1873) 14.1.1873, ANS.

Rapport sur le forage du puits artésien de Saint-Louis, by the Inspecteur du Service des Ponts et Chaussées. Séries géogr. Sénégal XII 17 b, ANFSOM.

Rapport du chef de Service des Travaux Publiques, July 30, 1889. Série géogr. Sénégal XII/21, ANFSOM.

Rulland H. (1869) «Rapport sur l'épidémie de choléra de Saint-Louis (Sénégal) en 1868,» January 3, 1869, Interim Chief Medical Officer to Governor: Séries géographique Sénégal XI/30

LITERATURE:

Anderson D. and Rathbone R. (eds.) (2000) Africa's Urban Past. James Currey, Oxford.

D'Anfreville de la Salle, Dr. Notre Vieux Sénégal.

Eyoum (1986) Le choléra en Afrique Noire francophone, doctoral thesis in medicine, Université P. et M. Curie, Paris VI, 1986, 22.Faidherbe (s.d.) Le Senegal.

Flows From the Past (2004). In: A Trans-Disciplinary Conference on the History of Water in Africa, The North-West University, Vanderbijlpark, South Africa, 8-10 December 2004.

Ngalamulume K.J. (1996) City Growth, Health Problems and the Colonial Government Response: Saint-Louis (Senegal), from mid-nineteenth century to 1914. PhD Thesis, Michigan State University, East Lansing, MI.

Ngalamulume K.J. (2004) Keeping the City Totally Clean: Yellow Fever and the Politics of Prevention in Colonial Saint-Louis-Du-Senegal, 1850-1914. Journal of African History, 45 (2004), 183-202.

Ngalamulume K.J. (in preparation) The Invention of Tropicality: Epidemics and Public Health in the Reconfiguration of Power Relations in Colonial Saint-Louis-du-Senegal, 1867-1920.

Thiroux A. Le n'diank, choléra du Sénégal.

# 13

## WATER SUPPLY IN THE CAPE SETTLEMENT FROM THE MID-17TH TO THE MID-19TH CENTURIES

*Petri S. Juuti, Harri R.J. Mäki & Kevin Wall*

Figure 13.1 Location of South Africa and Cape Town.

## 13.1. GENERAL BACKGROUND

The immediate catchment area of Cape Town is too small to give rise to large rivers, so, until major schemes of the 20th century brought water from distant catchments, the city depended on water from small streams flowing off Table Mountain. Water supply problems during the period reviewed were the direct result of the smallness of this catchment and its gradual urbanisation.

This article presents a brief history of the development of water supply to and wastewater disposal from the settlement at the Cape of Good Hope, from the settlement's foundation in the mid-17th century as a way station en route to the East Indies, until the mid-19th-century and the advent of the first effective municipal government.

## 13.2. EARLY TIMES

In 1503 the ship of the Portuguese explorer da Saldanha made the first recorded landfall on the shores of Table Bay. The crew came ashore and filled their casks with sweet, fresh water. This water was far more important to the survival of the sailors than the cow and ten sheep they bartered from the Hottentots - a crew could sail longer without meat than they could without water. The water of the Fresh River was thus the first product of the land to be exported. (Wall 1983, 265)

In subsequent years, ships of many nations called to refresh their men, replenish water supplies and barter for livestock, and Table Bay rapidly became the recognised way station on the route to the Indies. The leading trading company in the 17th century was that of the Netherlands, and various circumstances led to the Dutch East India Company establishing a permanent refreshment station in 1652. The instructions to the officer in command were to ensure a regular supply of water, meat and vegetables, which would save the lives of many who would otherwise die of scurvy and other illnesses. (Wall 1983, 267)

Although rainfall was good, the difficulty always was that it was confined to the winter months. On February 9 1654, the commander wrote: *"During the night the Almighty gave a fine rain, the first in seven weeks, so that the crops on the lands, which were quite withered, will now be considerably revived".* (Burman 1962)

To combat the summer water shortages, some means of storing water was needed. Accordingly, in 1663, the second commander of the settlement, Zacharias Wagenaar, constructed a masonry reservoir near the Fort, with a sluice at its seaward end where water was drawn off. Excavations in the 1970s for an office block and shopping centre revealed the remains of this pioneer water engineering work, and they are now on public display. (Wall 1983, 267)

From the earliest days the settlers received complaints of water pollution. The poor quality of the water sometimes supplied to shipping, and sickness in the settlement, were blamed on the infiltration of mineralised excess irrigation water from the Company's gardens and on the washing of clothing upstream of the points from which drinking supplies were drawn.

Figure 13.2 Greenmarket Square, 1762 drawing. Fountain on the right-hand side -- in the middle background, buckets being carried. (The building in the centre of the drawing, the Burger Watchhouse, is now a national monument.)

Steps were taken to prevent the infiltration, and a "placaat", or proclamation, was issued forbidding washing in the river itself. Thus did South Africa enact its first environmental pollution control measure. (Wall 1983, 267)

## 13.3. 18$^{TH}$ CENTURY

Throughout the 18th century, the main water resource was the several streams that ran off Table Mountain (some of which dried up in summer) and springs. This water was brought into the town by canals or "grachts". Fountains were also built in the course of time. Some homeowners dug wells - the well that served the masonry Castle can still be seen today. The construction of the castle to replace the mud-brick Fort, which was the first major public work of the settlers, commenced in 1666.

Because the pipe network was minimal, water had to be carried in buckets from the fountains and canals. In the absence of pipes to their houses, the more affluent 18th-century citizens relied on water being brought to them by their slaves or by water sellers. It was also the slave's duty to empty the sanitary buckets. There were many complaints that slaves did not do this hygienically – they were supposed to empty these buckets on wasteland or into the sea, instead they avoided walking so far. *"One of the most maddening habits of the slaves was that of emptying sanitary tubs into the canals, and frequently even before other people's doors, instead of into the sea".* (Fehr 1955, 13) It is easy to criticise such habits, but the job was not a pleasant one. Those who don't have to do unpleasant jobs may not be sympathetic to the need to change the way in which they are performed. (Juuti & Wallenius 2005, 136)

Children, once they were old enough to carry a full bucket (from the ages of five or six), were the water carriers of the poor families. (Knox 1992, 147) Also in other countries, children and women were those who carried water. (Katko 1997, 35)

The open canals served also as the only source of water for the purpose of fighting fires. Fires were frequent so long as thatch remained the normal roof covering, and, with a southeast wind, the result of careless smoking of a pipe might be the demolition of a considerable section of the town. (Hattersley 1973, 14) Consequently pipe smoking in the streets was strictly forbidden in the year 1704. (Fehr 1955, 17)

Wood is an easy material with which to make water pipes, but wooden water pipes in the course of time leak, and eventually rot. Wooden pipes were replaced by lead pipes in Cape Town in 1799. (Grant 1991, 29) Another reference states that lead pipes were "in general use by 1750". (Burman 1969, 98) Yet another reference states that lead was used for selected pipes (for example, to the jetty from which ships drew their supply) from early in the 18th century onwards. (Laidler 1952, 99)

Until early in the 19th century, the town itself grew steadily but slowly. The immigrant population, and their descendants, established farms and small settlements in the interior of the country. By the late 1770s, the "trekkers" had come into contact with the vanguard of the Xhosas on the banks of the Fish River in what is today the Eastern Cape Province. The British took the Cape in 1795, gave it back to the Dutch (not giving it to the Company this time, but to the Batavian Republic) in 1803, and took it again, this time for keeps, in 1806.

## 13.4. 19TH CENTURY

When the Batavian Republic took back the Cape, the Burgher Senate (the committee of prominent citizens) recommended the laying of new pipes. But:

> It was left to the British to take the first steps towards ensuring an adequate permanent supply for Cape Town [...] In 1811, a "waterhouse" [or reservoir] was completed, holding 250 000 gallons [1100 kilolitres]. At the same time cast-iron pipes were laid in Orange, Long and Strand Streets. [...] As time went on, more and more fountains were erected, and by 1834 there were 36 free-flowing fountains in the city. This waste [of water flowing continuously] left the inhabitants short of water, so [starting] that year, the fountains were [replaced] by hand-pumps, operated by "swaying" a lever back and forth. [...] The last of these pumps–the "Hurling Swaai" pump–now a historical monument" can still be seen. (Burman 1969, 98)

Figure 13.3 Drawing of the Hurling-Swaai pump in Prince Street. Source: Laidler 1952, 168.

Pumps were a centre of social life for Cape Town's poor, as hours were spent collecting this vital commodity. A mass of municipal regulations dealt with the management of the sources of water, regulating the size of the vessels which might be used, the hours the pumps were opened, and the use made of them. [...] There was no sanitation. In the absence of any expedient for draining off the filth, it might be left to putrify in a yard or allowed to meander across the street. Inspectors in 1840 reported in several cases quantities of rotting fish and human excrement in the very sleeping apartment. Such examples may have been extreme. But it was in general difficult for people to keep clean [...] (Worden et al. 1998, 120)

The Municipality of Cape Town, established on 3 March 1840 according to Ordinance No. 1 of that year, inherited the responsibility for the water supply. From this date revenue was raised through a rate on immovable property, thus laying the foundation of the municipal rating system. A Water Superintendent was appointed. In 1854 the first Town Engineer of Cape Town was appointed. (Morris 1970, 4)

In an effort to improve the position, the [municipal] Council bought up a number of private springs and water rights. But in 1849 the Superintendent reported that the scarcity of water was so great that he had stopped street services, to keep a reserve for fire-fighting. [...] The Council now decided to build a reservoir with a capacity of 2,500,000 gallons (11,000 kilolitres). This reservoir (No. 1) was completed in 1852, and proved highly successful. Greatly encouraged, the City Fathers in 1856 embarked on a larger reservoir (No. 2). The latter had a capacity of 12,000,000 gallons (55,000 kilolitres). (Burman 1969, 99. Both reservoirs were taken out of service many years ago.)

Figure 13.4 Photograph of the Prince Street pump from 1875. (Source: Cape Archives.)

Although these two reservoirs provided a welcome addition to Cape Town's water supply, they did not solve the water problem. In 1869 a filter-bed was built in the gorge of Platteklip Stream [the largest of the streams flowing down Table Mountain on the city side]. [...] As the city grew, so did the water shortage. [...] It was decided to build a storage reservoir big enough to cope with the problem, and in 1877 work commenced on the Molteno reservoir. [...]When finished in 1881 it was 40 feet deep, and could hold 41,000,000 gallons [nearly 200,000 kilolitres].(Burman 1969, 69, 100)

Following Molteno reservoir's breaching in 1882, it was reconstructed, and since 1886 it has served the city faithfully.

In 1882, the Cape Colony Parliament enacted legislation that laid down the powers and the duties of municipalities and the procedure to be followed in constituting future municipalities. They were 'given control of water supplies, slaughtering, washhouses and sewage disposal'. (Morris 1970, 4)

Although the Molteno reservoir now gave the town sufficient storage space, it was useless having the storage if there was an insufficient flow of water to fill the reservoir. Attention turned to tapping the waters of the Back Table of Table Mountain, that is, the very much larger catchment (compared to that draining towards the city) that drains southwards, i.e. away from the city. But how this was done, is another story, for another time!

Figure 13.5 First big reservoir, later called No. 1 Reservoir. (Photo: Petri Juuti)

## 13.5. CONCLUSIONS

The small catchment area in which the growing town was located provided sufficient water for more than 200 years, although a difficulty that had to be overcome from the start was how to store water through the long dry summer months. This chapter has outlined the developing need to not only provide engineered storage facilities for the growing town, but also to prevent pollution of the water supplied and distributed, and to dispose of the wastewater hygienically.

Figure 13.6 Second big reservoir, later called No. 2 Reservoir. (Photo: Kevin Wall)

## 13.6. REFERENCES

Burman J. (1962) Dams and Drains – the story of Cape Town's water supply. The Cape Argus, 16.9.1962.

Burman J. (1969) The Cape of Good Intent. Cape Town.

Fehr W. (1955) The Town House, Its Place in the History of Cape Town. Cape Town.

Grant D. (1991) The Politics of Water Supply: The History of Cape Town's Water Supply 1840-1920. M.A. Thesis, University of Cape Town.

Hattersley A. F. (1973) An Illustrated Social History of South Africa. Cape Town.

Juuti P. S. and Wallenius K. J. (2005) Brief History of Wells and Toilets. Pieksämäki.

Katko T.S. (1997) Water! Evolution of Water Supply and Sanitation in Finland from the mid-1800s to 2000. Tampere.

Knox C. (1992) Victorian life at the Cape 1870-1900; illustrations by Cora Coetzee. Vlaeberg.

Laidler P.W. (1952) A Tavern of the Ocean. Cape Town 1952.

Morris S.S. (1970) Water and waste management in the Cape Town metropolitan area. Proceedings, Conference of the Institute of Water Pollution Control (Cape Town, March 1970).

Wall K.C. (1983) A new history of South Africa from a water resources viewpoint. The civil engineer in South Africa, May 1983, 265-271.

Worden N., van Heyningen E. and Bickford-Smith V. (1998) Cape Town - The Making of a City: An Illustrated Social History. Cape Town.

# 14

# DEVELOPMENT OF THE SUPPLY AND ACQUISITION OF WATER IN SOUTH AFRICAN TOWNS IN 1850-1920

*Harri R.J. Mäki*

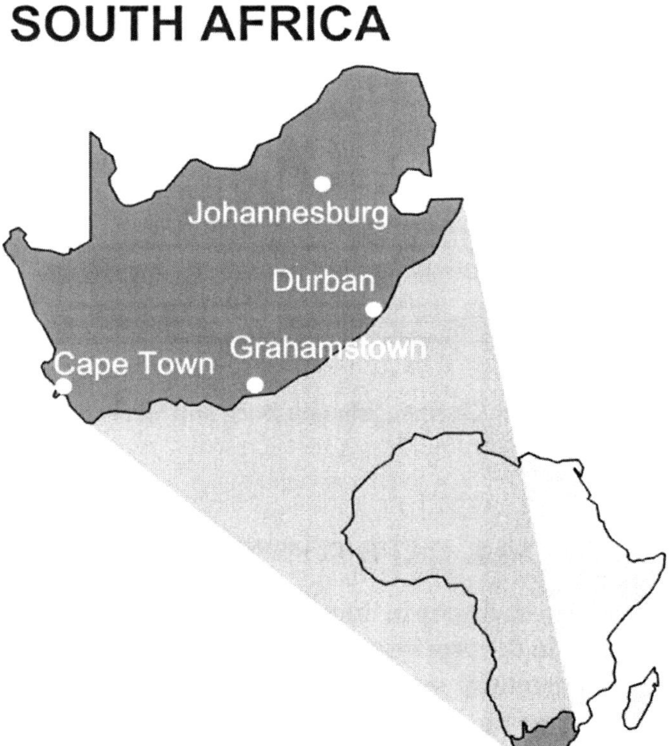

Figure 14.1 Key places mentioned in the chapter.

## 14.1. GENERAL BACKGROUND

The main objective of this chapter is to analyse solutions to water-related problems and the creation of water management in four South African towns in 1850-1920. The role of health issues in the development of the water supply is also discussed. During these years water and sewerage were generally identified as major urban problems in towns. The towns examined are Cape Town, Grahamstown, Durban and Johannesburg. Figure 14.2 below shows the population growth of these towns.

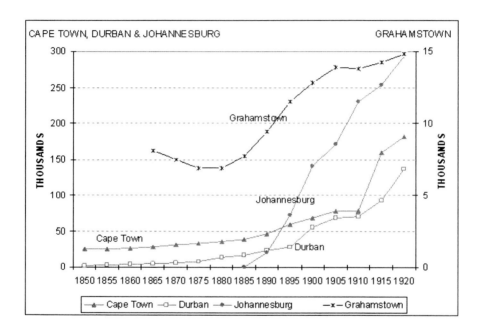

Figure 14.2 Population of Cape Town, Durban, Johannesburg and Grahamstown in 1850-1920.

## 14.2. CAPE TOWN

Cape Town is the oldest European-style city in South Africa. Early stages of the development of its water supply are presented in the previous chapter 13 by Juuti, Mäki & Wall. In sanitary matters, Cape Town was in a lamentable state in the late 1850s.

In 1894, the famous journalist R.W. Murray recorded that

> Forty years ago[...] There was a broad open drain, about six feet wide, running from the Government Gardens through Adderley Street to the Bay. This drain was the main sewer of the city into which all the dirt and offal were emptied [...] (Murray 1894, 2)

A lack of water affected two areas of Cape Town's life. Firstly it was essential for trade and commerce, and the functioning of shipping and industry. Secondly it was a central aspect of sanitation reform. While these were the motives for improving the water supply, the inherent reason for water being a source of conflict was that it represented an enormous source of expenditure. (Grant 1991, 198)

In 1862, Cape Town bought a part of the Waterhof Estate and in 1865 "Kotze's Spring" was acquired. In 1868 all the watermills along the Platteklip watercourse were bought up. Water pipes could be spread over the whole town through these new purchases. (Picard 1969, 85)

The first water restrictions were in order already in the late 1800s and in 1872 the water supply was not equal to demand according to the City Water Works Committee. Average daily consumption was about 6 l/person/day, with a population of some 30,000. The task of reporting on the water supply was given to John G. Gamble, the Hydraulic Engineer to the Colony of the Cape of Good Hope. (Hodson 1980, 349)

Gamble researched Table Mountain and in his report pressed for the construction of a tunnel through the Twelve Apostles range. The purpose was to tap the largest single catchment area of the Backwater Stream, running into the sea at Hout Bay. Besides the tunnel, also a pipeline from the tunnel down to a reservoir was envisaged. (Hodson 1980, 349)

In 1875, there was a series of articles in the Cape Monthly Magazine that spoke of the need for street improvement and an increased water supply. This started a campaign between 1875 and 1882 to improve the sanitation and infrastructure of the city; this campaign was also a part of the promotion of Englishness in the town. (Bickford-Smith 1995a, 45)

In the 1880s daily suspensions of the water supply were not unheard of. In 1902, Cape Town had approximately 64,000 inhabitants and in summer there was a severe water shortage. Population growth had been considerably strong in the preceding years; in 1891 the population had been only 48,000. (Grant 1991, 114) Partly because of these reasons, the city water engineer presented a plan to get salt water from Camps Bay in 1902 in order to use it for watering streets. Salt water was used for this purpose for the first time in 1861. (Barlow 1914) The water engineer presented this plan several times during the following years but it was never realized, mostly because city councillors thought the plan too expensive. (Memorandums of Cape Town water engineer)

In 1882, the Cape Parliament enacted legislation that gave municipalities control of, for instance, water supplies and sewage disposal. The town council of Cape Town was allowed to abandon a distribution of a free supply of 114 to 227 litres (25 to 50 gallons) of water per day to every dwelling house. Instead, the council had now the power to make property owners supply 454 litres (100 gallons) of water daily to each house, and the owners had to pay for the water. (Wall 1998, 1; Bickford-Smith 1995a, 58)

A sanitary engineer was appointed in 1887. The initiation of plans for a comprehensive drainage scheme in 1888, however, only occurred due to pressure from reformers outside the city council. They succeeded in getting parliament to appoint a select committee to investigate sanitation. The committee recommended a comprehensive drainage scheme. (Bickford-Smith 1995a, 59)

Figure 14.3 Aerial view of reservoirs on Table Mountain. On the left side are the Woodhead and Hely-Hutchinson reservoirs and on the right side the Victoria and Alexandra reservoirs of the Wynberg municipality. (A.G. 6530, Cape Archives Depot)

The municipality delayed comments on the emergency works Gamble had suggested for 15 years, until 1887. Some of the delay was occasioned through selling the plans and the development rights to a private enterprise, the Table Mountain Water Supply Company, which in the end profited only by the resale to the municipality of the original plans and rights. (Hodson 1980, 349)

The delay was critical, as water consumption continued its rapid increase. When the diversion became operative in 1891, it was evident that the scheme was overcommitted and not viable without major storage. In 1892 a start was made to the building of a reservoir on the Lower Plateau of Table Mountain. This Woodhead Reservoir was completed in 1897. (Figure 14.3) (Hodson 1980, 349) As part of this scheme, Mocke Reservoir was finished in 1896. The bordering Hely-Hutchinson Reservoir was started in 1898 and completed in 1904. It was considered to be an engineering triumph.

In 1889, the Cape Town District Waterworks Company was established for the building and maintenance of reservoirs and related installations for the supply of water to some of the smaller municipalities. The Company set about the purchasing of properties that contained strong springs of water. For instance, the springs at Rondebosch-Newlands were bought and a reservoir built at Newlands. (Slinger 1968, 13)

Waterborne sanitation was introduced to Cape Town only in 1895. W.T. Olive was appointed as a Town Engineer, primarily for the purpose of supervising this process. The pipe installation was laid in tunnels hewn through rock and a sea outfall was led to Green Point. (Slinger 1968, 13)

New water reservoirs and dams in Table Mountain and the pipe track to carry water from there to town were ready by 1902. At the same time, most stone and brick houses also had water pipes laid on and the pumps had practically disappeared. (Hattersley 1973, 179; Worden et al. 1998, 226-27; Simkins and van Heyningen 1989, 99-100) Why did it take nearly 50 years to solve these problems when they were clearly seen already in the middle of the 19$^{th}$ century? (For more about the sanitary situation in the early 1890s, see: A Sanitary Crusade 1891, 8-9)

One answer to these questions could be the identification of the water question amongst the working classes as a matter that was put forward by house owners and the middle class. Tenants were afraid that the securing of a better water supply would mean higher rents and so poorer inhabitants of the city opposed these reforms. (Grant 1991, 198-205)

A bubonic plague hit Cape Town in 1901. At that time, Cape Town was an old, slum-ridden town where whites were living surrounded by native servants. The outcome of the plague was the call for an African location by the Medical Officer of Health, Dr. Jasper Anderson, where they could be housed in controlled and sanitary conditions without flooding into the already overcrowded central area. The Ndabeni location was constructed rapidly and soon there were 7000 Africans living there. (Maylam 1990, 61; Slinger 1968, 14)

This plague led to changes in African locations also in other towns of South Africa, for instance, in Port Elizabeth all the old locations were demolished and a new one, New Brighton, was built six kilometres from the city centre. (Christopher 1987, 197; Swanson 1977, 400-407) When the plague hit Johannesburg, local authorities removed people from the inner city "coolie location" to the new township of Klipspruit. (Maylam 1995, 24-25)

In the beginning of the 20th century, the Municipality of Cape Town again became concerned about its water supply and requested a commission to enquire into the matter. The Government appointed The Cape Peninsula Commission and instructed it to cover, besides the water issue, the whole question of local government in the Peninsula. After more than a year's labour, the majority report recommended one municipality in the Peninsula. Most of the municipalities, however, desired to be left alone. Cape Town's wish was to be constituted as a kind of water board for the Peninsula. The report was not acted upon. (Wall 1998, 2)

As a result, the City Engineer of Cape Town, J. Cook, submitted in 1904 a report on various water supply schemes, amongst which were Steenbras, Palmiet, Zachariashoek, Berg River Hoek, Wahmbers, Wemmershoek, Muizenberg, Twenty-four Rivers and Franschhoek. The Cape Town Council favoured the Franschhoek reservoir. First the scheme got the support of the southern municipalities, but when it was delayed in the parliament for a next session, the support faded away. (Wall 1998, 5)

In 1905 Cape Town was again in trouble with its water supply and several Table Mountain schemes were put forward. A dam was proposed in the Waai Vlei; another reservoir was suggested in Disa Gorge. A third scheme was the heightening of the Wynberg municipality reservoirs or the building of another reservoir in Orange Kloof. These schemes had, however, their limitations because the most of the water in Table Mountain was already being utilized. (Wall 1998, 5)

The passing of the Southern Suburbs of Cape Town Water Supply Act in 1907 finally killed the Franschhoek scheme. The act authorized the municipalities of Claremont, Mowbray, Rondebosch and Woodstock to continue with a scheme on the Wemmershoek. (Wall 1998, 5) The Kloof Nek Reservoir was built in 1908. During the same year a system of metering was introduced in order to prevent the wasting of water.

In the study period the last big change was the unification of suburban municipalities with Cape Town in 1913. The main reason for this was water, or more correctly, the scarce availability of it. Surrounding municipalities were totally dependent on Cape Town for their water supply, which was pointed out already in 1902. (Bickford-Smith et al. 1999, 46-47) Still it took 11 years to reach the only viable solution.

In 1915 the City Council approved a sewerage scheme for the Southern Suburbs. Work was begun during the next year and it was completed in 1917. Also in 1915 the Board of Engineers decided that it would be expedient to go outside the Peninsula for water supplies. After investigating both the Wemmershoek Valley and the Steenbras Valley, they recommended that Steenbras be developed. There was a referendum among ratepayers between Steenbras and Wemmershoek. Steenbras won the poll and the Council adopted the Steenbras Scheme in 1917. (Wall 1998, 6) The contracts for the dam and the pipeline were signed in 1918, but the project was not completed until 1921, because of difficulties with the supply of equipment.

## 14.3. GRAHAMSTOWN

Grahamstown was established in 1812 as a military camp. Situated inland, its problems with water management differed from those in Cape Town. Unlike most other inland towns, in addition to being a market and administrative centre, it was also a major educational and ecclesiastical centre.

Iron pipes were installed in the town centre in 1844–45 to replace the open furrows in the streets. Furrows carried water from the small dams built on numerous watercourses passing through the town. These furrows were becoming unhygienic so there was an urgent need to do something. This installation was coupled with the building of dams and reservoirs to meet an increased demand for water. From the beginning there were problems; pipes were not dug deep enough to endure wagon traffic in the streets, dams were badly constructed, the pipes were vandalized, etc. To help with these matters, a "Superintendent of Water" was appointed in April 1850. He was expected to concern himself with all matters connected with the supply of water. (Hunt 1961, 189-96)

Figure 14.4 View of Douglas reservoir near Grahamstown. (A.G. 13435, Cape Archives Depot)

The biggest problem, however, was the inadequacy of water storage capacity. This situation was aggravated by a drought in the summer of 1858-59. In April 1859 a City Engineer was appointed to supervise all the public works of the town. One of his first tasks was to supervise the construction of the new reservoir, which was opened on 25$^{th}$ January 1861 by Governor Sir George Grey and named after him the "Grey Reservoir". The reservoir was an important milestone in the development of Grahamstown's water supply. Yet this solution to the problem of water storage only emphasized the inadequacy of the piping. (Hunt 1961, 196-98)

In 1887 water from springs and hills was conserved in five reservoirs holding about 50 million gallons. (Figure 14.4) Water was distributed in 1000 pipelines to the 1500 houses and stores. The sanitation of the town was fairly good. The old cesspool system was replaced by the pail system; the sewage was removed nightly and covered up with earth outside of the town. (Souvenir of Grahamstown 1887, 35, 39) At the end of the 19$^{th}$ century Grahamstown was considered to be a very healthy town, with a fine climate, pure air, well looked-after sanitary arrangements, and a good water supply. (Atherstone 1886, 159-62; A Sanitary Crusade 1891, 32)

The 20th century started with the building of Milner and Jameson dams in 1897 and 1904 in the Niuwejaars river in order to get better water sources. (Management of the Water Resources of the Republic of South Africa ) The next important step was the building of a Water filtration plant in 1914. (Invitation to the opening ceremony of the water filtration plant in Grahamstown, 1.9.1914)

## 14.4. DURBAN

Durban was established in 1824 as Port Natal on the eastern coast of South Africa, where there was plenty of water; the problem was the quality of it. In 1854 Bishop Colenso complained that the water was the greater devil in Durban, because wells were not sunk deep enough to keep organic material from polluting them. The only remedy was to drink rainwater or very clean water from the Umgeni River four miles away. (Hattersley 1956, 96-97)

In 1856 the erection of an embankment at the head of the *vlei* (Afrikaans: valley, vale) above the Umgeni brickfields improved the sanitary condition of the town. (Ellis 2002, 38) The first wells in Durban were probably private, but the earliest public well dates in all likelihood from the year 1864. A typical dug well, its wall was made of alternate double rows of brick and single rows of slate. It was situated in Berea Road near the Old Dutch Road intersection. The well was found in 1968 during excavations. (Bjorvig 1994, 321-322)

As a cholera epidemic was advancing and a smallpox epidemic was ravaging Cape Town in 1871, the Government of Natal charged Durban community with a total want of sanitary precautions. The town had, however, already established a committee for this. There was, however, a big difference between the governor's point of waste accumulation and refuse creating conditions for diseases and Durban politician's consideration about the Indian immigrants. The main result was the tight regulation of building standards in the town. (Swanson 1983 407-409; Popke 2003, 252-255)

In 1877 the number of public pumps and wells had increased to 18. The unsatisfactory quality and sparse quantity of the water from these pumps had been the source of discussion by the councillors already for years.[1] Boring operations had been carried out during 1876–77 and finally in 1878 a shaft was sunk on the Flat near the Botanic Gardens. From this well 227,000 litres (50,000 gallons) of water were obtained per day in July 1879. Storage tanks were erected on the Flat and the water led in pipes into the town. This "Currie's Fountain" was the main source of water until the Umbilo Waterworks were completed 1887. A storage reservoir with a capacity of 227,000 litres (50,000 gallons) was erected in the Botanic gardens in 1884. (Henderson 1904, 225-26)

---

[1] There is some conflict in the sources, as Bjorvig says that the Berea Road well was the first public well and Henderson says that these discussions started in 1861. So where were those public pumps installed if there were no public wells until 1864?

Figure 14.5 Clear water reservoir on the Umlaas River. (Photo by Mr. Stanley Fletcher in Henderson 1904.)

In September 1883, the Borough Engineer submitted reports and estimates on the schemes to supply water from the Umlaas, the Umhlatuzan and the Umbilo Rivers. In December the Council decided after careful consideration of this report that under existing conditions the Umbilo River was the most suitable source. The "Durban Corporation Waterworks Law of 1884" authorized the Council to construct waterworks, acquire lands, levy a water rate, and frame bye-laws in connection with the water supply. (Henderson 1904, 226-27)

These Pinetown Waterworks, as they were usually described, were formally opened on 21st July 1887. They were constructed at the bend of the Umbilo River some nine miles from Durban. Water was conveyed from there to an open service reservoir on the Berea. (Twentieth-Century Impressions of Natal 1906, 439)

The Borough Engineer submitted a storm water drainage scheme in 1889. This scheme provided channelling for the principal roads in the central town and Berea. (Bjorvig 1994, 320) In the years 1888-90 serious droughts were experienced and this, combined with the demand for water for manufacturing purposes, made it necessary to extend the water supply. As a result, the Umlaas temporary pumping plant was opened in 1891. It was situated some 14 miles from Durban. (Figure 14.5) (Twentieth-Century Impressions of Natal 1906, 439-440)

Figure 14.6 The Camperdown Dam. (Photo by Mr. Stanley Fletcher in Henderson 1904.)

In 1890 Robert Boyle commented that the water supply was fairly good and that there was a scheme to provide an unlimited supply of water to the town. (A Sanitary Crusade 1891, 28) This was realized in 1894 when the Umlaas gravitation scheme was built. In it, the distance between the water intake and the filter beds was shortened and the delivery of water augmented by the construction of tunnels and conduits. It was still necessary in 1898 for the Medical Officer to recommend that the people of Durban boil their drinking water. (Bjorvig 1994, 324)

An effective sewerage system became operational in 1896. Lavatories were provided throughout the town. There was also an outfall for waterborne household sewage, which was discharged into the sea during the first few hours of the ebb tide. (Bjorvig 1994, 327-328)

A population increase between 1897 and 1900 forced the town council to give permission to build a temporary dam at Camperdown in 1900 (Figure 14.6). In 1905 it was decided that this would be changed into a permanent dam. Work was started the next year and in 1908 the retaining wall was laid. (Bjorvig 1994, 325-326)

## 14.6. JOHANNESBURG

Johannesburg is the youngest of the case towns. It sprung up suddenly in the South African Republic after gold was found in the area in 1886. Its multinational citizenry and location in the watershed between the Vaal and Limpopo rivers 60 kilometres from the nearest major river created problems for supplying water.

First, water was needed to satisfy the mining industry's needs. In mining fields water was needed, of course, for household purposes but also in the mining process. Water was not a problem when there were only a few people. Gold diggers took their own water from shallow wells, whilst convicts appeared in the capacity of sanitary gangs. (Hattersley 1973, 238)

In 1887 a Sanitary Board was elected to govern the town. The provision of sanitary services was one of its duties. It was not easy to discharge this duty. It would have been ridiculous to install proper drainage and sewage disposal when it was assumed that the urbanization was only temporary. Even when it began to appear that Johannesburg was becoming a permanent town, the scarcity of water made any proper scheme of sanitation impossible. Therefore the only practicable means of providing a sanitary system was the collection of refuse in buckets. (Maud 1938, 23)

At first there was enough water but as the population grew rapidly, the first indications of a water shortage could be seen already in early 1887. The government of the South African Republic tried to solve problems by granting a private concession to provide Johannesburg with water in December 1887. The result was the establishment of the Johannesburg Waterworks and Exploration Company in March 1888. (Tempelhoff 2003, 28-30) This is an important point of difference between Johannesburg and Cape Town. While Cape Town kept the water supply in its own hands, in Johannesburg it was given to a private company.

In 1888 the Sanitary Board imposed charges for the nightly collection of sanitary pails and for the daily collection of rubbish and slop-water. It was not allowed to run dish- or bath water into the streets. In every house there was a tank in the back yard, where such "slop-water" was to be collected. These tanks were emptied twice a week when a large wagon with several tanks collected the slop-water. The use of this service was compulsory and non-payment was punished by imprisonment. (Leyds 1964, 32)

In 1889 Barney Barnato, one of the mining magnates and an extremely rich man, acquired the Johannesburg Waterwork and Exploration Company. The company then built a dam and pumped water to a reservoir on Saratoga Avenue. (Figure 14.7) This was the first proper gravity reservoir in Johannesburg. The company claimed that during the rainy season there was available a daily supply of nearly 3.4 million litres (750,000 gallons). (Leyds 1964, 53)

The success of the company and the growing demand created competition. A concession was granted in 1891 to the owners of Wonderfontein farm and in 1892 the Braamfontein Water Company was founded. (Cartwright 1965, 186) Another water business venture was the Vierfontein Syndicate. As long as local sources provided enough water, there was no need to look somewhere else. This changed when rapid urban and industrial growth began. It

Figure 14.7 Above Waterworks – December 1890. (Drawing by Harry Clayton in Leyds 1964.)

was then that the sources further afield (like the Vaal River) were contemplated. (See for instance Tempelhoff 2000, 88-117)

Robert Boyle's comment about Johannesburg in 1890 was:

> The sanitary condition of Johannesburg, though not yet all that could be desired, has been greatly improved within the last year or so [...] The pail-closet system is used. There is a splendid opening here for an improved dry-earth closet, as the system at present employed is very unsatisfactory. The water supply is not good, and care must be exercised in using it, especially for drinking purposes. (A Sanitary Crusade 1891, 20-21)

Indirect taxation and the policy of concessions kept the cost of living in Johannesburg unnecessarily high. In 1892 there was a widely signed petition demanding a better water supply. (Maud 1938, 39) In January 1893 the Johannesburg sanitary committee formed a special water committee. From the deliberations of this committee, it became clear that the search for water must go beyond the borders of Johannesburg. Mining magnate Sammy Marks also made recommendations. He thought that coal could be used to provide power for the scheme to pump water to the Witwatersrand from the Vaal River. (Mendelsohn 1991, 51) (See also figure 1.1. Sammy Marks Fountain.)

In the spring of 1895 Johannesburg experienced very severe drought. The rains were late and the situation was serious enough for the Waterworks Company to introduce rationing. Water had to be supplied by mule carts to the higher parts of the town. (Neame 1960, 51) In the same year, there was a poll amongst the white inhabitants of Johannesburg about the Wonderfontein proposal. If the proposal were to be accepted in the poll, the sanitary board was to be allowed to take over the concession. The poll, however, went against the proposal and the following negotiations between the Sanitary Board and the owners of the concession came to nothing. (Maud 1938, 127-128)

In 1897 James Bryce's description of Johannesburg included following:

> The streets and roads alternate between mud for the two wet months and dust in the rest of the year; and in the dry months not only the streets but the air is full of dust, for there is usually wind blowing. But for this dust, and for the want of proper drainage and a proper water-supply, the place would be healthy, for the air is dry and bracing. But there had been up to the end of 1895 a good deal of typhoid fever, and a great deal of pneumonia, often rapidly fatal. In the latter part of 1896 the mortality was as high as 58 per thousand. (Bryce 1897, 316)

Only in September 1897 did Johannesburg get the Municipal Bill, which made it possible for the Sanitary Board to make the institutional arrangements for providing reliable services, for instance a good water and sanitary infrastructure. (Tempelhoff 2003, 29)

The town council saw an improved water supply as one of its immediate objectives. The takeover of the Waterworks Company by the Johannesburg Consolidated Investment Company in 1895 had brought an improvement, but not enough. On 1 December 1897 the council decided to negotiate with the company about taking over the water supply itself. When the council intimated that it might open its own waterworks, the J.C.I. directors agreed to discuss their taking over the water supply. The murder of the J.C.I. chairman, however, caused the negotiations to be broken off. (Appelgryn 1984, 124)

In 1898 the Town Council succeeded in preventing the Pretoria Government from selling out the monopoly of sewage disposal in the Town. By doing this there was some chance of solving the problem with sewage scientifically. (Maud 1938, 45, 58) In the same year, the Waterworks Company secured a stable supply of water from the farm Zuurbekom, which lies about 17 miles to the southwest of Johannesburg. A pumping station was completed in 1899. The water supply from this source was continuous and didn't need purification. (Leyds 1964, 54)

In 1898 there was again drought in Johannesburg with a real danger that mines would be forced to close down. Sammy Marks again offered to bring water from the Vaal River. He already got the backing of the government but unfortunately he couldn't get the backing of the Wernher Beit mining house. Because of their stalling, Marks had to withdraw his scheme. (Mendelsohn 1991, 83-4) Plans about bringing water from Vaal River to Johannesburg were realized only in 1916, when the building of the Vaal Barrage was started.

By the turn of the century, the sanitary situation was getting even worse and the need for water was growing all the time. There were only a few wells in the town at that time.

The Municipality found that if it were to provide the town with essential services such as sewerage and refuse removal, its labour force would require easier access to the town than was possible from the municipal location at Klipspruit. Accordingly permission was granted for the accommodation of bona fide African employees in compounds dotted about the town, and in servants' quarters in the backyards of white residential properties. (Parnell & Pirie 1991, 130-131)

The first full-time Medical Officer of Health, Charles Porter, was appointed in 1901. During his tenure of office, he became an influential person throughout South Africa in campaigning for public health and town-planning legislation. (Parnell 1993, 476-78)

In 1901 the town council of Johannesburg appointed its committees. Water and sanitation issues belonged to the engineer's department, which went under the supervision of the works committee. Clearance of the insanitary area, the development of electricity and transport undertakings, and the initiation of a sewerage scheme would, however, have put too much strain on this committee. Therefore, in 1902, a new committee was constituted, and in 1904 this was divided into two committees, the tramway and lighting committee and the parks and estates committee. Water wandered for some time from one committee to another, until 1910, when it found a companion in the fire brigade. (Maud 1938, 231-231)

An important decision for the water supply of Johannesburg was the establishment of the Rand Water Board in 1903. This board was born as a result of a joint appeal of the Johannesburg town council and the Chamber of Mines to the Transvaal Government for a better solution to the water supply problem in the area. The government appointed a commission to examine the matter and the Rand Water Board was established according to its recommendations. (Maud 1938, 57; Tempelhoff 2003, 58-66, 77, 607)

In 1905 the Rand Water Board finally assumed ownership of earlier concessions after a fight in an arbitration court about the price. The Board got control over water sources that could produce some 2.5 million gallons a day in April 1905 (Maud 1938, 130)

The first waterborne sewerage system was also introduced in 1903. (Shorten 1970, 232) Between 1903 and 1906 the town council spent three and a half million pounds (approximately 331 million euros) on sanitation and sewerage, road construction and storm water drainage, water supply, electric tramways, and lighting. (Maud 1938, 70)

By 1908 the Rand Water Board's sources of supply to Johannesburg were as follows: "There were two unused wells at Milner Park. There was a well at the Staib street depot of the board and three boreholes at New Doornfontein. The rest of the water was drawn from Zwartkopjes and neighbouring properties in the Klip river valley, as well as at Zuurbekom." (Tempelhoff 2003, 94-95)

Table 14.1 Key events in the four towns..

| | Cape Town | Grahamstown | Durban | Johannesburg |
|---|---|---|---|---|
| 1850 | | Appointment of the "Superintendent of Water" | | |
| 1852 | No. 1 Reservoir completed | | | |
| 1854 | First Town Engineer | | Borough of Durban proclaimed, first Mayor | |
| 1859 | Cape Town Water Supply Bill | First City Engineer | | |
| 1860 | No. 2 Reservoir completed | | | |
| 1861 | Parliamentary select committee to investigate water supply; Sea water used for street watering | Grey Reservoir Graham's Town municipal board act | Municipal corporations Law | |
| 1864 | Cape of Good Hope Municipalities Act | | First public well | |
| 1867 | Cape Town municipal council act; First Mayor | | | |
| 1869 | Intake and small filter beds constructed | | | |
| 1872 | Special committee appointed to report on the water supply | | Municipal corporations Law | |
| 1878 | | | "Currie's Fountain" | |
| 1881 | Gamble's report on Cape Town water supply | | | First general municipal law in the Transvaal |
| 1882 | Table Mountain Water Supply Bill | | First Borough Engineer appointed | |
| 1883 | Public Health Act of Cape Colony; First Medical Officer of Health | | Barnes' report on water supply | |
| 1884 | | | Botanic Gardens storage reservoir; Durban Corporation Waterworks Law | |
| 1886 | Molteno Reservoir completed | | | Birth of the Johannesburg Diggers committee |
| 1887 | Cape Town and Wynberg Water Supply Bills; First Sanitary Engineer | | Umbilo Waterworks opened | Sanitary Board elected |
| 1888 | | | | Johannesburg Waterworks and Exploration Company |
| 1889 | Cape Town District Waterworks Company | | Storm water drainage scheme | Saratoga Avenue Reservoir |
| 1890 | | | Fletcher's report on water supply | |
| 1891 | Tunnel constructed through the Apostles range and a pipe-line laid to the Molteno Reservoir | | Umlaas temporary pumping plant Fletcher's report on sewerage | Wonderfontein Concession |

|  | Cape Town | Grahamstown | Durban | Johannesburg |
|---|---|---|---|---|
| 1892 | | | | Braamfontein Water Company |
| 1894 | | | Umlaas gravitation scheme | |
| 1895 | Waterborne sanitation | | | Water poll |
| 1896 | Mocke Reservoir and the Sea Point Service Reservoir constructed | Grahamstown Water Supply Bill | Storm water drainage scheme waterborne sanitation | Municipal councils wet |
| 1897 | Public health Act of Cape Colony; Woodhead Reservoir completed | Milner Reservoir built in Slaai Kraal | | Johannesburg Municipal Bill Stadsraad constituted First Mayor |
| 1899 | | | | Zuurbekom pumping station |
| 1901 | Bubonic plague | | Public Health Act of Natal Camperdown temporary dam | Johannesburg council proclamation; First full-time Medical Officer of Health |
| 1902 | Waterworks committee appointed to consider and report upon the water supply | Grahamstown Municipality Act | Togt Labour Act | |
| 1903 | Joint committee of the Cape town and Suburban Municipalities appointed | | Clear Water Reservoir completed | Municipal Corporations Ordinance Rand Water Board |
| 1904 | Hely-Hutchinson Reservoir completed; French Hoek Water Supply Scheme Bill approved (withdrawn later) | Jameson Reservoir built in Slaai Kraal | Umgeni water supply scheme | Municipal corporations ordinance |
| 1905 | | | Flood; Umbilo Water Scheme abandoned | |
| 1906 | Cape Peninsula Water Supply Bill | | | Johannesburg Municipal Ordinance |
| 1907 | Southern Suburbs of Cape Town Water Supply Act | | | |
| 1908 | Kloof Nek Reservoir completed | | Camperdown permanent dam | |
| 1910 | | | | Fire-Brigade and Water Committee |
| 1912 | Cape Municipalities Act | | Coedmore Filters built | Local Government Ordinance of Transvaal |
| 1913 | Unification of sub-urban municipalities | | | |
| 1914 | | Water Filtration Plant | Service reservoir at Congella | |
| 1915 | | | Service reservoir at Stella | |
| 1916 | | | | Starting of Vaal Barrage |
| 1917 | Sewerage scheme for the Southern Suburbs completed; adoption of Steenbras scheme | | Flood | |
| 1919 | | | Water Engineer's Department created | |

## 14.7. EXAMPLES FROM OTHER TOWNS

For comparison, some facts about the situation in some other towns in South Africa:

Port Elizabeth started its first extensive water scheme only in 1903. (Simkins 1989, 83) Robert Boyle's description of the town a decade earlier included the following:

> From the natural advantages of its position it ought to be a well drained and thoroughly healthy town; but, like Cape Town, it is at present certainly not the one, nor can it truthfully be said to be the other, the smells arising from the defective drainage, so far as there is any attempt at drainage, being very bad indeed. It is now proposed, however, to have all this changed […] (A Sanitary Crusade 1891, 32)

The population in 1855 was 5,000, and in 1921 54,000. (Christopher 1987, 196)

About situation in Kimberley Boyle commented:

> Result being that what was formerly simply a pest-hole, where human beings died off like flies, may now be ranked as, owing to the improvements that have been made in the sanitary arrangements, including the introduction of a splendid water supply from the Vaal river at a cost of £40,000 (approx. 42 million euros), one of the healthy towns of the Colony. The pail-closet system is at present used; but it is proposed, at some future time, to adopt a water-drainage system. (A Sanitary Crusade 1891, 11-12)

In Kimberley a private company, the Griqualand West Water Company, controlled the water supply. It was not taken over by the municipality until 1921. (Roberts 1978, 25) The population in 1880 was 13,600, and in 1921 it was 39,700. (South Africa: Historical Demographical Data of the Urban Centers)

Pretoria had, according to Boyle, an abundant and good water supply and it was using a pail-closet system. (A Sanitary Crusade 1891, 23) The population in 1875 was 1,500, and in 1921 it was 74,100. (South Africa: Historical Demographical Data of the Urban Centers)

Bloemfontein began to improve its sewerage only after the Spanish Flu epidemic in 1918. The municipality was horrified about the conditions poor whites were living in and in 1920 it took over the sanitary services that were still in the hands of private contractors. The result was that in 1924 the proud Mayor could announce that all of *white* Bloemfontein was now connected to the waterborne sewerage system. (Phillips 1987, 229-230) This example goes beyond my examination period but is still, I think, quite good for showing which were the priorities amongst the municipal decision makers.

The population in 1868 was 1,200, and in 1921 it was 39,000. (South Africa: Historical Demographical Data of the Urban Centers)

## 14.8. SEGREGATION POLICIES AND HEALTH

The level of urban segregation varied between these towns. Cape Town and Grahamstown were less rigidly segregated than the others. In Durban Africans were early on forced into barracks under so-called togt system and it had the most developed form of administrative control over its African population. (On the togt system, see Popke 2003, 256-258) In Johannesburg blacks were housed in the town by their employers or in locations built outside the town. (Lester 1996, 54-55. See also Maylam 1990)

In the middle of the 19$^{th}$ century the miasmatic theory of the origin of diseases still had strong support. According to this theory, diseases were born in wet and dirty soil when the organic material rots. (For more on miasmatic theory, see: Porter 1999, 82) This theory could very well exist with the segregationist policies. An ideological link existed already in the early 19$^{th}$ century between blackness, dirtiness and disease. (Deacon 1996, 289-91) The blacks were thought to be naturally more susceptible to diseases than the whites; the reluctance to make any environmental improvements in black compounds or locations was partly resulting from this view. (Packard 1987) In the early 20$^{th}$ century black slums were considered to be nurseries of infection by white health officials. Many diseases, for instance, plague, tuberculosis and smallpox, were said to have originated within the slum districts. (See, for instance, Packard 1989 and Parnell 2003)

In 1899 in Cape Town the Medical Officer of Health recommended the appointment of female sanitary inspectors and ordered a campaign against tuberculosis. He estimated that tuberculosis caused one out of every nine deaths among whites and one out of every seven among non-whites. Infant mortality was reported to be 155/1000 amongst the whites and 261/1000 amongst the coloureds. (Slinger 1968, 14; Shorten 1963, 272-73)

As Maynard Swanson has shown, with the advent of industrialization and urbanization as well as the development of public health consciousness, "fear of epidemic cholera, smallpox and plague both roused and rationalized efforts to segregate Indians and Africans in municipal locations", especially in Natal and Transvaal. In the towns of the Cape Colony, as well as in Durban and Johannesburg, "the accident of epidemic plague became a dramatic and compelling opportunity for those who were promoting segregationist solutions to social problems". The equation of blacks with infectious disease and the perception of urban relations in terms of "the imagery of infection and epidemic disease" provided a rationale for rural separation and the removal of African housing to the edges of the towns. (Swanson 1977)

Outbreaks of diseases, like smallpox in 1882 or bubonic plague in 1901, overcame the government's reluctance to pay for the building of separate hospitals or locations. Improved state revenues enabled the government give attention to segregating institutions like prisons. (Bickford-Smith 1995b, 73-74; on segregation in other institutions of Cape Town, see Deacon 1996)

A public health department was set up in 1897 in the Cape Colony. In Transvaal public health was organized only after the South African War. By 1910 most towns had municipal medical officers and hospitals for diseases. (Marks and Andersson 1988, 260)

## 14.9. CONCLUSIONS

These four towns represent different kinds of geographical locations, their ethnographic structures were different and at first they were part of different political units. If, for instance, average annual rainfall is considered, they fall into different categories, in descending order: Durban, Johannesburg, Grahamstown and Cape Town. (Lester et al. 2000, 23, fig. 2.5)

In three of these towns, population growth was remarkable. In 1865 Cape Town had bit less than 30,000 inhabitants and in 1920 it had nearly 200,000 (Worden 1998, 177; Bickford-Smith et al. 1999, 71). Durban's population in 1860 was approximately 5,000 and in 1921 it was nearly 150,000 (Kuper et al. 1958, 53). Johannesburg had 15,000 inhabitants in 1889 and nearly 300,000 in 1921 (South Africa: Historical Demographical Data of the Urban Centers). Grahamstown was a small town, which didn't experience a rapid population growth in contrast to the other towns; it had 6,900 inhabitants in 1875 and 14,900 in 1921. (South Africa: Historical Demographical Data of the Urban Centers)

This chapter shows that there were important differences between these towns in how they developed their water supply. Some of these differences were because of the different geographical environments, some because of the different governance cultures. This difference in governance is especially seen in the case of Johannesburg, which differs most from the others.

As can be seen, the path to an effective water supply was different in each of the examined towns. In Cape Town water had to be fetched from farther and farther away. First from the town area, then from Table Mountain; finally they had to rely on rivers and new reservoirs and dams, which were situated quite a long distance from the town centre. This development continued in Cape Town after the end of the study period; today it is taking water from nearly every source that was discussed in the early 20th century.

In Durban there was enough water in nearby rivers. It just had to be utilized. Between 1887 and 1904 all three of the main water sources there, namely the Umbilo, Umlaas and Umgeni rivers, were taken into use. You could say that the situation was the same in Grahamstown: there was enough water around, and they just had to build effective reservoirs and dams to collect it. Of course, in the case of Grahamstown, you have to take into account that the population growth was not at the same level as in the other towns, meaning that the need for water was not as urgent as in the other towns.

In Johannesburg we have a totally different case. The town was born in a watershed, where there was not much water. Already in the 1890s some far-sighted people saw that the only solution was to get water from the Vaal River. This, however, was such a major undertaking that it was realized only in 1916, when the building of the Vaal Barrage was started. For approximately 30 years the water supply of Johannesburg was on quite shaky ground.

An interesting difference between Johannesburg and the other towns is that from the beginning the water supply in Johannesburg has been in private hands, whereas Cape Town, Grahamstown and Durban have kept it under the municipality's control. Of course, one reason for this was the habit of the South African Republic to give up concessions and get money from them. An interesting point is that this continued after the South African War and the British takeover in 1901. The establishment of the Rand Water Board formalized the situation where a private company was controlling the water supplies and was providing municipalities and mines of the area with water. The city of Johannesburg, however, continued to control the distribution of water to its citizens until a few years ago, when even this was privatised. More about the early private operators can be found in Chapter 18.

In the Durban area the development has followed the pattern of Johannesburg: currently Umgeni Water (established in 1974) is providing municipalities with the water. In Grahamstown and in Cape Town the municipalities have so far kept the whole water supply in their own hands.

## 14.10. REFERENCES

Primary Sources

Cape Archives Depot

Town Clerk, Cape Town (3/CT)

Memorandums of the Cape Town water engineer to the Electric and Water Works and Fire Brigade Committee, 3.8.1902, 26.9.1902, 3.11.1902, 31.10.1903, 6.5.1905, 30.5.1905, 13.6.1905, 6.7.1905, 17.9.1906, 2.11.1906, 6.11.1906, 15.11.1906, 3.6.1908, and 19.12.1911, 3/CT 4/1/1/90, E37/1.

Barlow, Chas. R. (1914) Report on Cape Town's Water Supply, October 1914, 3/CT 4/1/4/298, F134/4.

Town Clerk, King William's Town (3/KWT)

Invitation to the opening ceremony of the water filtration plant in Grahamstown, 1.9.1914, 3/KWT 4/1/167, M7/17.

Literature

Appelgryn M.S. (1984) Johannesburg: Origins and Early Management 1886-1899, Pretoria.

Atherstone W.G. (1886) 'Graham's Town and the Eastern Districts'. In Noble, J. (ed.) Official Handbook: History, Productions, and Resources of the Cape of Good Hope, 159-62.

Bickford-Smith V. (1995a) Ethnic Pride and Racial Prejudice in Victorian Cape Town: Group Identity and Social Practice, 1875-1902. Cambridge.

Bickford-Smith V. (1995b) South African Urban History, Racial Segregation and the Unique Case of Cape Town? Journal of Southern African Studies 21(1), 63-78.

Bickford-Smith V., van Heyningen E. and Worden N. (1999) Cape Town in the Twentieth Century: An Illustrated Social History, Cape Town.

Bjorvig A.C. (1994) Durban 1824-1910: the formation of a settler elite and its role in the development of a colonial city. Ph.D. Thesis, University of Natal.

Bryce J. (1897) Impressions of South Africa, New York.

Cartwright A.P. (1965) The corner house: the early history of Johannesburg, Cape Town.

Christopher A.J. (1987) Apartheid Planning in South Africa: The Case of Port Elizabeth. The Geographical Journal 153(2), 195-204.

Deacon H. (1996) Racial Segregation and Medical Discourse in Nineteenth-Century Cape Town. Journal of Southern African Studies 22(2), 287-308.

Ellis B. (2002) White Settler Impact on the Environment of Durban, 1845-1870. In Dovers, S., R. Edgecombe and B. Guest (eds.) South Africa's Environmental History: Cases & Comparisons, Cape Town, 34-47.

Grant D. (1991) The Politics of Water Supply: the History of Cape Town's Water Supply 1840-1920, M.A. Thesis University of Cape Town.

Hattersley A.F. (1956) More Annals of Natal, London.

Hattersley A.F. (1973) An Illustrated Social History of South Africa, Cape Town.

Henderson W.P.M. (1904) Durban: Fifty Years' Municipal History, Durban.

Hodson D. (1980) The Woodhead Reservoir, Cape Town. The Civil Engineer in South Africa, November 1980, 349-51.

Hunt K.S. (1961) The Development of Municipal Government in the Eastern Province of the Cape of Good Hope, with special reference to Grahamstown, 1827-1862. Archives Year Book for South African History 24, Pretoria, 137-289.

Kuper L., Watts H. & R. Davies (1958). Durban: a Study in Racial Ecology, London.

Lester A. (1996) From Colonization to Democracy: A New Historical Geography of South Africa, London.

Lester A., Nel E. & Binns T. (2000) South Africa, Past, Present and Future: Gold at the End of the Rainbow, London.

Leyds G.A. (1964) A History of Johannesburg, Johannesburg.

Management of the Water Resources of the Republic of South Africa (1986) Pretoria, appendix 2.

Marks S. & Andersson N. (1988) 'Typhus and Social Control: South Africa 1917-1950'. In MacLeod, R. and M. Lewis (eds.) Disease, Medicine, and Empire: Perspectives on Western Medicine and the Experience of European Expansion, London, 257-83.

Maud J.P.R. (1938) City Government: The Johannesburg Experiment, Oxford.

Maylam P. (1990) The Rise and Decline of Urban Apartheid in South Africa. African Affairs 89(354), 57-84.

Maylam P. (1995) Explaining the Apartheid City: 10 Years of South African Urban Historiography. Journal of Southern African Studies 21(1), 19-38.

Mendelsohn R. (1991) Sammy Marks: The Uncrowned King of the Transvaal. Cape Town.

Murray R.W. (1894) South African Reminiscences, Cape Town.

Neame L.E. (1960) City Built on Gold, Johannesburg.

Packard R.M. (1987) Tuberculosis and the Development of Industrial Health Policies on the Witwatersrand, 1902-1930. Journal of Southern African Studies 13(2), 187-209.

Packard R.M. (1989) White Plague, Black Labour, Pietermaritzburg.

Parnell S. (1993) 'Creating Racial Privilege: The Origins of South African Public Health and Town Planning Legislation', Journal of Southern African Studies 19(3), 471-88.

Parnell S. (2003) Race, Power and Urban Control: Johannesburg's Inner City Slum-yards, 1910-1923. Journal of Southern African Studies 29(3) 615-37.

Parnell S.M. & Pirie G.H.(1991) Johannesburg. In Lemon, A. (ed.) Homes Apart: South Africa's Segregated Cities, Cape Town, 129-45.

Phillips H. (1987) The Local State and Public Health Reform in South Africa: Bloemfontein and the Consequences of the Spanish Flu Epidemic of 1918. Journal of Southern African Studies 13(2), 210-33.

Picard H.W.J. (1969) Grand Parade: The birth of Greater Cape Town 1850-1913. Cape Town.

Popke E.J. (2003) Managing Colonial Alterity: Narratives of Race, Space and Labor in Durban, 1870-1920. Journal of Historical Geography 29(2), 248-267.

Porter D. (1999) Health, Civilization and the State: A History of Public Health from Ancient to Modern Times, London.

Roberts B. (1978) Civic Century: The First One Hundred Years of the Kimberley Municipality, Kimberley.

A Sanitary Crusade through South Africa (1891) London.

Shorten J.R. (1963) Cape Town. [Cape Town].

Shorten J.R. (1970) The Johannesburg Saga. Johannesburg.

Simkins C. and van Heyningen E. (1989) Fertility, Mortality, and Migration in the Cape Colony, 1891-1904. The International Journal of African Historical Studies, 22(1) 79-111.

Slinger E.W. (1968) Cape Town's 100 Years of progress: A Century of Local Government. Cape Town.

South Africa: Historical Demographical Data of the Urban Centers. Http://www.library.uu.nl/wesp/populstat/Africa/safricat.htm, visited 2.10.2004.

Souvenir of Grahamstown: A Health and Holiday Resort (1887) Grahamstown.

Swanson M. (1977) The Sanitation Syndrome: Bubonic Plague and Urban Native Policy in the Cape Colony, 1900-1909. Journal of African History 17(3), 387-410.

Swanson M.W. (1983) The Asiatic Menace: Creating Segregation in Durban 1870-1900. International Journal of African Historical Studies 16(3), 401-421.

Tempelhoff J.W.N. (2000) On Laburn's 'mystery' query – A prehistory of the Vaal River as water source of the Witwatersrand (1887-99). Historia 45(1), 88-117.

Tempelhoff J.W.N. (2003) The Substance of Ubiquity: Rand Water 1903-2003, Vanderbijlpark.

Twentieth-Century Impressions of Natal: Its people, commerce, industries, and resources (1906) Natal.

Wall K. (1998) Water, civil engineers and multipurpose metropolitan government for the old Cape Peninsula municipalities. SAICE Journal 40(3), 1-8.

Worden N., van Heyningen E. and V. Bickford-Smith (1998) Cape Town: The Making of a City: An Illustrated Social History, Cape Town.

# 15
## WATER, LIFELINE OF THE CITY OF GHAYL BA WAZIR, YEMEN

*Ingrid Hehmeyer*

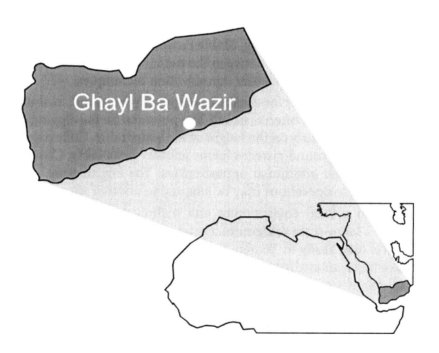

Figure 15.1 Location of Ghayl Ba Wazir on the coastal plain of southern Yemen.

## 15.1. GENERAL BACKGROUND

Since 1982, the Royal Ontario Museum (Toronto, Canada) has sponsored an archaeological programme in what is now the Republic of Yemen. Vastly different material was examined through the investigations of several different sites. A common element has been the question of what resources were available to sustain the cultural record. By far the most important aspect is the availability of water and human-environmental relationships in this arid region of southern Arabia.

The treatise presented here is a site study of Ghayl Ba Wazir, which gives a unique example of water engineering principles and management strategies, defined by local characteristics. But the solutions found to cope with environmental constraints match the overall theme of the Canadian Archaeological Mission's research programme in Yemen.

In arid lands, the pre-eminent factor limiting development is water. Aridity implies "an imbalance between the demand for water and its supply, the supply being too scarce to meet the demand" (Hillel 1992, 108). This imbalance is caused by an insufficiency of direct rainfall to sustain agriculture, which generally calls for a minimum amount of 250 to 300 mm per year (Kopp 1981, 48). In addition, aridity is characterized by high variability in precipitation, and while in some years there may be a sufficient amount of rain to sustain a crop, in most years this will not be the case. Therefore, in arid environments permanent settlements cannot be supported exclusively by rain-fed farming. Stable farming populations, however, can and do exist. But this requires closing the gap between the demand for water and its supply through irrigation and the implementation of water conservation techniques.

Settlements also need clean water for domestic purposes year round. In some cases, springs exist fortuitously, though they are often seasonal. The potential for the digging of wells depends on various factors, but particularly on the height of the water table. Collection of surface run-off and its storage in underground cisterns forms another possibility. Characteristics of the local hydrogeology can offer additional opportunities. The engineering principles applied successfully to one of these operations may be adapted to another situation.

Human sensitivity to a local environment and technical creativity can turn the most marginal of sites into a large-scale settlement. Outstanding examples abound in Arabia. The following describes a case study in Yemen, in the southern part of the Arabian peninsula, where humans have taken an environmentally hostile situation and turned it into a source of economic success.

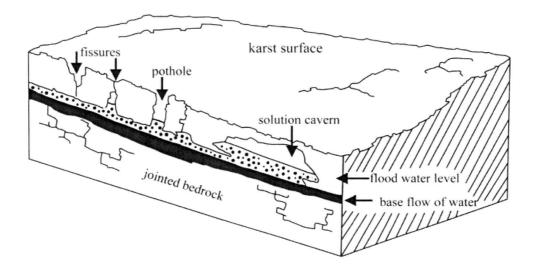

Figure 15.2 Typical water movement in a karst landscape (after Chorley et al. 1985, fig. 8.6 & fig. 8.8 B).

## 15.2. THE WATER TUNNELS OF GHAYL BA WAZIR

Ghayl Ba Wazir is an urban centre just inland from the southern coast of Yemen (see Figure 15.1). The term *ghayl* as part of the town's name underlines the fact that the settlement's existence is intimately connected with underground water flow: in colloquial Yemeni use in an urban context, *ghayl* means "an artificial often partly subterranean water-channel" (Serjeant et al. 1983, 19). The underground water gave rise to a prosperous settlement.

Ghayl Ba Wazir lies in a barren gypsum shield, or karst. Surface weathering of the exposed rock has led to the forming of a distinctively pitted, hard crust. The calcareous rock beneath is fractured by numerous joints. It is also moderately soft and water-soluble. Over time, water seeping along sub-surface fissures dissolved the rock, and an underground network developed (Bloom 1978, 136-148). Where joints were enlarged through the solution process, and particularly at their intersection, underground stream-flow resulted in the erosion of large caverns holding considerable volumes of water (Figure 15.2). When the roofs of these solution caverns collapsed, abrupt openings appeared at the surface, with almost vertical walls, or even overhanging cliffs (Bloom 1978, 149-151, and fig. 7-10(b); Chorley et al. 1985, 187 and fig. 8.8.B). The technical term for them is "collapse sink" (Figure 15.3). In Arabic they

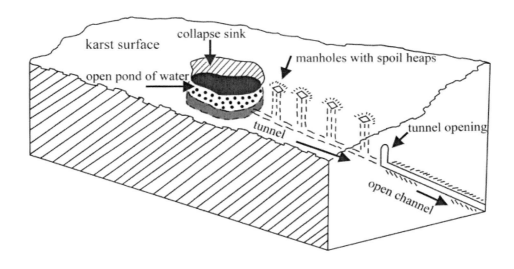

Figure 15.3 A water tunnel drawn from a collapse sink.

are known as *hawma*, pl. *huwam* (Serjeant 1964, 58). The abruptness of their opening is well carried by their local names: *Hawmat al-'Arus* implies "the Pit of the Bride", and another—*Hawmat al-Kabsh*—"the Pit of the Ram", from the characters who, according to local legends, fell unsuspectingly down these holes in the dark. See Serjeant 1964, 58 for the story of Hawmat al-'Arus.

The reservoirs that punctuate the otherwise totally dry, rocky landscape behind the town were a potential source of water in a bleak and arid environment, an invaluable resource. Because the water originates from the underground aquifer, it is sediment-free. Therefore, besides irrigation agriculture, the water could be used for drinking and domestic purposes. The limitation to exploiting the water in the immediate vicinity of the reservoirs was the bare rock and the lack of soil for agricultural use, as well as the removed distance of the settlement. The challenge, then, was to find a way to lead the water from the natural open pools both to the town and to the market gardens on its outskirts. The technical solution employed in Ghayl Ba Wazir was to tunnel through the bedrock, creating underground channels that allowed the water to be directed where it was needed, using gravity flow (Figure 15.3). Locally, the term used for these water tunnels is *ma'yan*, pl. *ma'ayin*. After running below ground for some considerable distance, depending on the terrain, the tunnels ultimately lead into open surface channels.

Figure 15.4 A line of rock-cut shaft openings in the landscape of Ghayl Ba Wazir.

The engineering of the tunnels through the bedrock involved cutting vertical shafts down from the surface to provide access for tunnelling below and to facilitate removal of the excavated debris, which was dumped in the immediate vicinity of the shaft openings. The landscape of Ghayl Ba Wazir is dotted with thousands of these precisely rock-cut manholes (Figure 15.4). The shafts were generally sunk at regular intervals of between 5 m and 10 m along the length of the tunnels, though examples exist with shafts only 3 m apart from centre to centre. While the spacing of the vertical shafts may seem unnecessarily close, it can be argued that given the huge investment in capital to dig a tunnel, the most efficient strategy would have been to excavate it as quickly as possible in order to get a fast return on investment. Numerous clean-out shafts to accommodate a large team of diggers were required as a result, to avoid congestion in the tunnel from so many men being involved.

The rectangular openings of the shafts vary in size from c. 0.5 m x 0.7 m, to 0.7 m x 1.05 m. The depth of the bedrock that needed to be cut for the purpose of engineering the tunnels could be considerable. Shafts more than 13 m deep have been measured. Toeholds were normally provided on two opposite faces of the shafts, to allow climbing up and down. Their traces consist either of small projections left protruding from the shaft faces, or small cavities cut back into them. The finishing chisel marks are well preserved on the straight sides of the shafts (Figure 15.5). Their curving line can be interpreted as the result of an artisan's arm rotating as the hammered chisel or sharply pointed pick struck the rock. For technical variations on the principle of a tunnel with vertical shafts set at regular intervals

Figure 15.5 Curving tool marks on a shaft face.

along its length, see Hehmeyer et al. 2002, 87-88. The article provides a detailed report on the technicalities of the water systems of Ghayl Ba Wazir.

A proper gradient was essential for the functioning of the tunnel system. This does not mean that the channel bed had to be perfectly even. Water flowing under gravity can easily overcome the occasional undulation in the channel floor, so long as the flow is constant and sufficiently strong. Nonetheless, it is clear from our measurements taken in Ghayl Ba Wazir that the supervising engineer worked with precision in the long run.

Regrettably, there are few figures to cite for comparative purposes because in so many cases the tunnels have fallen into disuse. Debris accumulated in the derelict underground channels makes it impossible to record a precise measurement. Recent water development schemes in the district have resulted in a rapidly dropping water table, which has caused many of the collapse sinks and the tunnels fed by them to run dry. (For details, see below: "Problems with a dropping water table".) However, in some instances, diesel pumps were successfully installed to tap the groundwater from below the former water level. The pumped water is then fed back into a still existing tunnel, taking advantage of its engineered gradient to deliver water to the same destination as before. This practice has preserved the original construction and allows the gradient to be measured, which in one instance was as low as 0.03 % (a drop of 0.47 m in 1.5 km).

At first glance, with regard to the engineering principles involved, the water tunnels of Ghayl Ba Wazir are reminiscent of the famous *qanat* systems of Iran. (For these, see Lambton 1978. A discussion concerning the engineering principles applied in the water tunnels of Ghayl Ba Wazir in comparison with the *qanat* systems of Iran is given in Hehmeyer et al. 2002, 94-95.) A *qanat* is a horizontal gallery tunnelled back into an aquifer, to lead water under gravity flow to its destination. Typically, a *qanat* is found in the alluvial fan outwash that forms below a high range, which means that a *qanat* is dug through porous sediments. The water tunnels of Ghayl Ba Wazir, on the other hand, were cut through bedrock to tap an open reservoir, the *hawma*. But in both cases, vertical shafts provided access and facilitated removal of the spoil during excavation.

While there is the possibility of diffusion of a technical principle, a strong argument can be made for the indigenous development in Ghayl Ba Wazir, based on observation of the local hydrogeological conditions and experimentation. Seepage along faults and joints was not uncommon. There is firm evidence for this in the exposed gully faces where large gypsum crystals can be seen. These were formed through the evaporation of water slowly moving through bedrock joints. Even today, condensation is visible around some of the fissures in the early morning, an indication of underground water nearby. It is logical that in the past someone may have dug back into the rock in order to increase the flow. Technically speaking, this would relate the water tunnels of Ghayl Ba Wazir to the so-called spring flow tunnels, a principle which was well-established in Palestine in the first millennium BC (Ron 1989, 227, 229). While rock-cutting is a typically Yemeni way of engineering water systems (for other examples, see Hehmeyer et al. 2002, 95), in the case of Ghayl Ba Wazir it was the specific hydrogeological and geomorphological characteristics of the terrain that dictated the design and execution. This means that even in a general Yemeni context the water tunnels of Ghayl Ba Wazir are a special case, an original local response to a particular natural condition.

## 15.3. THE AGE OF THE WATER TUNNELS

The town of Ghayl Ba Wazir is traditionally held to have been founded by a certain Shaykh 'Abd al-Rahim bin 'Umar Ba Wazir who came to the area in AD 1306-07 (Ibn Shaykhan 1999, 19, 22). An inscription over the entrance of the congregational mosque states that the building was established by him in the year AD 1317-18. According to local legend, 'Abd al-Rahim bin 'Umar discovered the subterranean water in the 14$^{th}$ century by firing an arrow in the air. From the spot where it landed, a spring immediately flowed forth. The story is clearly apocryphal, and similar stories are ubiquitous in the Middle East. But it reflects the reality of the terrain, and the fact that in this particular landscape the sudden appearance of springs was not to be unexpected. Local legend also states that 'Abd al-Rahim bin 'Umar had the first tunnel dug to supply water to the mosque. It is notable that the site of the mosque and its adjacent cemetery are located in the lowest part of the town now in existence. While the settlement grew subsequent to the mosque's foundation, the ground surface rose over time as a result of the accumulating urban waste. Only the mosque and the cemetery remained preserved at the original level. Upon 'Abd al-Rahim bin 'Umar's death in 1346, he was buried in the mosque where his tomb can be visited today.

An organic sample taken from sediments at the bottom of one of the collapse sinks gives an (uncalibrated) radiocarbon date of 600 BP (600 radiocarbon years "before present". The collapse sink from which the sample was taken is Hawmat al-Harth. For its location, see Hehmeyer et al. 2002, 85, fig. 2). Calibration results in calendar dates that fall within the 14$^{th}$ century, which can be judged to be more than a coincidence. The justification for this claim is that the aforementioned underground water movement in such a karst landscape erodes the soft bedrock and leads to the development of caverns filled with water. After the collapse of the roof of the cavern, an open reservoir is formed. Before the roof caved in, the underground water would have been clean, filtered by movement through the rock. Upon becoming a collapse sink, the now open pool of water would be subject to accumulation and sedimentation of wind-borne material or aquatic organisms. The calibrated date establishes that the particular natural reservoir from which the sediment sample was taken must have been exposed at least since the 14$^{th}$ century. Clearly, in an arid environment such as this, the local people would have been motivated to find a way to exploit water from the sudden formation of a collapse sink.

## 15.4. WATER USES

Generally speaking, in Ghayl Ba Wazir the water from the open reservoirs was directed first towards the settlement, so that people could make use of the clean water for drinking and domestic purposes. Based on accounts of local people, before the modern era and the introduction of deep-well pumped water, the town's inhabitants preferred traditional well water for drinking, because the water flowing from the karst had a high calcium content. The exception to this rule was a tunnel drawing water from the gravel beds located beyond the gypsum shield, which had sweet water. The allocation from this stream, for the specific collection of drinking water, was made at prescribed hours of the night to ensure minimal contamination by animals or humans.

Provisions were made to accommodate different urban needs. Until today in the town, wherever the water flows below ground, access is afforded via a staircase, for women to fetch water. There are also public stations designated for laundry use and facilities for bathing, furnished through an enclosure straddling the stream. In addition, religious institutions require clean water. It is one of the principles of Islam that a believer has to be ritually pure before prayer. Formal ablution facilities are usually provided as part of the mosque compound. In Ghayl Ba Wazir, specific branches of the tunnels are directed towards the mosques, where they supply water to the ablution facilities (Figure 15.6). The story of Abd al-Rahim bin 'Umar digging the first tunnel to the mosque underlines the significance of the religious requirement for water in an Islamic city.

Figure 15.6 A channel supplying water to the ablution facilities of a mosque.

Following ritual or domestic use, the water is delivered to vegetable gardens and orchards on the outskirts of the city. This principle of water allocation, first to the settlement and then to the field systems at the other end, implies conservation of the scarce resource by recycling grey water for irrigation. Fruit, vegetables and fresh fodder are cultivated in well-kept market gardens (Figure 15.7). In addition, cash crops such as tobacco and henna are grown. Ghayl Ba Wazir is famous for the good quality of both of these crops, which thrive in the calcium-rich soil.

## 15.5. WATER ALLOCATION

Whether for agricultural purposes, domestic, or religious use, the water flow to these different places needed to be allocated systematically and equitably to everybody involved. The issue of central importance was how to divide day and night into units of time without the use of clocks. Traditionally in the Middle East, the day is not partitioned into clock-measured hours of equal length, but rather the daylight period is apportioned astronomically, i.e. based on the sun's daily passage in the sky, into twelve hours which begin at sunrise and end at

Figure 15.7 Irrigated market garden on the outskirts of Ghayl Ba Wazir.

sunset. Since at any given locality the length of day varies in the course of the year, the length of the daylight hour also changes: during the winter months one hour is shorter than during the summer. Astronomical subdivision of the day thus results in so-called seasonal or unequal hours, with each a duration of one twelfth of the time between sunrise and sunset (King 1990, 250; see also King 1996, 172).

The length of these daylight hours was traditionally measured by using the sundial principle in its most general sense: time during the day was determined by observing the shadow cast by an object placed in the sunshine. In Ghayl Ba Wazir, until the appearance of mechanical timekeeping pieces in the 1960s, partition of the day into units of time for water allocation was based on measurement of the farmer's shadow length at a given time of day, with a prescribed shadow increase (or decrease) indicating a given unit of time. Obviously, shadow lengths vary throughout the seasons, depending on the sun's changing altitude in the sky. Therefore, the changes in shadow length during the daytime were determined and recorded in charts for any calendar date, throughout the astronomical year, for the location of Ghayl Ba Wazir. For details, see Hehmeyer 2005.

During the night, water allocation was timed based on a star calendar. It identifies a total of twenty-eight conspicuous stars or constellations, the so-called lunar mansions, of which thirteen to fourteen are visible during any given night of the year, according to the season. The passing of time is indicated by the succession of these constellations. Each farmer—depending on the size of his land—would have one or more designated mansions. The farmer's turn for irrigation would start as soon as his mansion had reached its highest position in the sky, and it would last until the next one in succession had proceeded to the same position.

The schedule for a unit of land irrigated by a specific tunnel was based on an odd number of divisions, so that an individual farmer's time slot for irrigation was not repeated in a fixed way, but changed between day and night. In other words, if a farmer had received water in the course of the previous irrigation cycle during the day, his next time slot would be during the night. Since the duration of the day and night vary throughout the year, it was felt that this measure ensured an equitable share for each farmer in the long run.

Whether during the day or night, timing water allocation in the Ghayl Ba Wazir area was the responsibility of the *muqaddim*, the supervisor of a specific water tunnel. The *muqaddim* was a farmer (and land owner) in the unit of land irrigated by the respective tunnel. He was elected by his fellow farmers for his astronomical knowledge and experience, as well as for his trustworthy and responsible character. He also monitored the required maintenance of the underground water systems. We were fortunate to be able to find a former *muqaddim* who remembered operating the traditional system for measuring time that had been in use until the 1960s. Muqaddim Salim from al-Suda', a village which is located a few kilometres to the northeast of Ghayl Ba Wazir (Hehmeyer et al. 2002), was willing to share his knowledge with us. Together with a number of farmers who were equally familiar with the procedures, he provided priceless information that allowed us to understand the principles involved in water allocation and management of the tunnels. Interviews were conducted between December 2003 and January 2004.

The timekeeping practices for water allocation in Ghayl Ba Wazir go back at least several centuries. The earliest written record is a manuscript that lists the changes in shadow length throughout the year. The text indicates a 16th-century origin for the measuring system. For details, see Hehmeyer 2005.

## 15.6. PROBLEMS WITH A DROPPING WATER TABLE

Karst landscapes are distinctive in their relief and drainage, due to the solubility of the rock through the movement of natural water. Until today, rises and drops of the water table can suddenly occur because of the ongoing processes of solution and collapse in the fractured bedrock, which may result in changes to the inter-connected underground network. Periodic heavy rainfalls—both locally and in the catchment area—lead to an immediate rise of the flow rate in the tunnels, with the inherent danger of the increased volume of water seeking its own independent passage and causing damage to the system. A sudden drop of the water table, however, has the opposite, more dramatic consequence. For example, two of the smaller collapse sinks—*Hawmat al-'Arus* and *Hawmat al-Kabsh*—have been totally dry for years. Yet we can deduce from a written report that *Hawmat al-'Arus* contained water in 1920, with its level cited as 10 m below ground surface (Little 1925, 55).[1]

In the past, technological strategies were developed to guarantee a continuous water flow in the tunnels. In response to a drop of the water table in one of the collapse sinks and the drying-up of a tunnel fed by it, the tunnel had to be deepened along its entire length below the floor level at its original intake. If measures of this kind were still not successful, other water sources had to be found and underground diversions needed to be dug to direct water into the existing tunnel. These measures are evident in the archaeological record (see Hehmeyer et al. 2002, 91-92). Either scenario meant an enormous effort for the community. While the exact dates of the events could not be determined, stories abound in Ghayl Ba Wazir that corroborate such emergency action.

There have, then, been water flow problems in the Ghayl Ba Wazir area for quite some time. As far as one can judge from both the technical remains and the reported regulatory mechanisms, these challenges were resolved through community effort, based on social consensus. But while the problems of the past were manageable, drastic changes of recent origin imposed from outside the community have left the local population in a state of helplessness.

Ghayl Ba Wazir is located some 40 km to the northeast of al-Mukalla, capital of the province of Hadramawt. Until Yemeni unification in 1990, Hadramawt was part of the People's Democratic Republic of [South] Yemen. Prior to 1967 it was the so-called Protectorate, occupied by the British, but hardly developed. Following independence in 1967, major political and economic changes occurred, including nationalization of land as a prerequisite for reform and modernization of the agricultural sector. A detailed report of the measures introduced is given in Pritzkat 1999. It was in the post-1967 era that the traditional practices in agriculture were increasingly replaced by so-called modern ones. As an example, in Ghayl Ba Wazir knowledge of the star calendar and timekeeping based on shadow lengths were skills that became no longer needed, because of the increasing availability of watches. In addition, watches were judged to be more accurate. But the modernization and nationalization process ignored the fact that traditional practices were part of a complex system of regulatory means based on community consensus. Its abandonment has resulted in animosities within the community that are felt until today. What has, however, caused the most severe problem is the fact that basic principles of sustainability were neglected.

---

[1] Little does not name the collapse sinks—for which he erroneously uses the term mi'yan—but from his description of three of them that he visited, it is apparent that he is referring to Hawmat al-Harth, Hawmat al-Sirkal, and Hawmat al-'Arus. The same level of water, much higher than it is today, was reported for all three. For their location, see Hehmeyer et al. 2002, 85, fig. 2.

Figure 15.8 Water pumped from a collapse sink is fed into the existing underground channel network.

In order to supply al-Mukalla with drinking water, the government authorities decided in the 1980s to tap the water resources of Ghayl Ba Wazir. In 1986, the Mukalla Water Project started to pump from five wells. After Yemeni unification in 1990, al-Mukalla became one of the fastest-growing cities of Yemen. (For details, see Pritzkat 1999.) As of 1998, deep-well pumping in the area of Ghayl Ba Wazir was expanded dramatically to meet the increasing needs of the provincial capital (estimated population in 2003: approx. 300,000).

The decision to use the water resources of Ghayl Ba Wazir was made without any local participation in the formulation of the pumping policy. The current over-exploitation of the aquifer has been catastrophic for water management in the area. Each year, the water in the collapse sinks decreases considerably. While the drinking and domestic water of the town of Ghayl Ba Wazir itself is now provided by the Mukalla Water Project, for the local farmers the dropping water table spells the end of the traditional water extraction system. They have no choice but to resort to the use of diesel pumps. In one of the collapse sinks, while there is still open water (observed in January 2004), albeit at a much lower level than in the past, water is pumped directly from the reservoir. Near the now totally dry collapse sinks, deep wells have been drilled. Paradoxically, while the extraction method has changed, the distribution system remains the same, for the water is fed back into the existing tunnel network (Figure 15.8).

The local farmers realize that the use of diesel pumps will not be a long-term solution, since the water table drops at a dramatic rate. Therefore a local initiative has been started, which aims at improving the situation through making use of the hydrogeological characteristics of the karst. Run-off from a nearby wadi—an ephemeral flood course fed by seasonal rains in the Yemeni highlands—will be trapped by a barrage and directed into naturally eroded fissures in the karst (Figure 15.2). Because the underground network is inter-connected, it is hoped that the water will seep through the fissured bedrock and cause the water table to rise in the entire system.

## 15.7. CONCLUSIONS

Until recently, the community of Ghayl Ba Wazir flourished through adherence to centuries-old practices that were based on respect for natural resources, and consideration for the needs of the community's members. Interestingly, in the historical sources there is no evidence of disputes over water shares. The fact that 16$^{th}$-century regulations were applied successfully until the late 1960s indicates that the community accepted them as an equitable solution to a potentially volatile situation. Today's Ghayl Ba Wazir faces a fundamental conflict between the interests of a traditional local community that is concerned about the sustainability of water use, and the needs of a fast-growing urban society that calls for more and more water.

## ACKNOWLEDGEMENTS

The Canadian Archaeological Mission of the Royal Ontario Museum, Toronto, operates in Yemen under a licence from the General Organization for Antiquities and Museums, San'a'.

## 15.8. REFERENCES

Bloom A.L. (1978) Geomorphology. Englewood Cliffs, NJ.

Chorley R.J., Schumm S.A. & Sugden D.E. (1985) Geomorphology. London and New York.

Hehmeyer I. (2005) Diurnal time measurement for water allocation in southern Yemen. Proceedings of the Seminar for Arabian Studies 35: 87-96.

Hehmeyer I., Keall E.J. & Rahimi D. (2002) Ghayl Ba Wazir: applied qanat technology in the fissured karst landscape of southern Yemen. Proceedings of the Seminar for Arabian Studies 32: 83-97.

Hillel D. (1992) Out of the Earth. Civilization and the Life of the Soil. Berkeley and Los Angeles.

Ibn Shaykhan S.M. (1999) Nafahat wa-'abir min tarikh Ghayl Ba Wazir [Scents and fragrance from the history of Ghayl Ba Wazir]. Aden and San'a'. [In Arabic]

King D.A. (1990) Science in the service of religion: the case of Islam. Impact of science on society 159: 245-262.

King D.A. (1996) Astronomy and Islamic society: Qibla, gnomonics and timekeeping. In Rashed R. (ed.), Encyclopedia of the History of Arabic Science. London and New York. i: 128-184.

Kopp H. (1981) Agrargeographie der Arabischen Republik Jemen. (Erlanger Geographische Arbeiten, Sonderband 11). Erlangen.

Lambton A.K.S. (1978) Kanat. The Encyclopaedia of Islam. (New Edition). iv: 528-533. Leiden.

Little O.H. (1925) The Geography and Geology of Makalla (South Arabia). (Survey of Egypt, Geological Survey). Cairo.

Pritzkat T. (1999) "Marktwirtschaft" statt Sozialismus in Südarabien: Die Reform der Bodenordnung im Südjemen am Beispiel Mukalla. Jemen-Report 30 (2): 25-31.

Ron Z.Y.D. (1989) Qanats and spring flow tunnels in the Holy Land. In Beaumont P., Bonine M. & McLachlan K. (eds), Qanat, Kariz and Khattara: Traditional Water Systems in the Middle East and North Africa. London. 210-236.

Serjeant R.B. (1964) Some irrigation systems in Hadramawt. Bulletin of the School of Oriental and African Studies 27: 33-76.

Serjeant R.B., Costa P. & Lewcock R. (1983) The Ghayls of San'a'. In Serjeant R.B. & Lewcock R. (eds), San'a': An Arabian Islamic City. London. 19-31.

WATER, LIFELINE OF THE CITY OF GHAYL BA WAZIR, YEMEN

# 16
## HISTORY AND PRESENT CONDITION OF URBAN WATER SUPPLY SYSTEM OF TASHKENT CITY, UZBEKISTAN

*Dilshod R. Bazarov, Jusipbek S. Kazbekov & Shavkat A. Rakhmatullaev*

Figure 16.1 Location of Tashkent city, Uzbekistan.

## 16.1. GENERAL BACKGROUND

The water supply system of Tashkent city, the capital of Uzbekistan, has a long history of development and has been modified to various extents under different political and economic systems. Drinking and domestic water use, watering of home yards and gardens, sewage are major constituents of the urban water supply system. The water supply has contributed greatly to the creation of a microclimate and small unique oases in urban surroundings—recreation-parks, "*khauz*", and others (*khauz* in the Uzbek language means a pool or basin where water was stored temporarily for later use).

This chapter discusses the development and evolution of the water supply system of Tashkent City: how water has been managed, delivered, and allocated as well as quality control systems during Tsarist Rule (Russian Empire), the Soviet Communist Era, and the Post-Independence Period. The authors outline three main points related to the advantages and disadvantages that have greatly impacted the transformation of the urban water supply system. The first point is population dynamics, the second point is political changes that dictate management infrastructure, and the last aspect is mechanisms of operation and maintenance (O&M), i.e. market-oriented services versus centralized government control. In addition to an historic overview of the development of the water supply system, current water quality and quantity standards of the water supply system and sanitation as well as future perspectives are discussed.

## 16.2. WATER SUPPLY DEVELOPMENT UNDER TSARIST RULE (RUSSIAN EMPIRE)

The semi-arid, dry climate of Uzbekistan has forced local people to bring water from distant sources in order to exist. The first stages of the water supply system mainly comprised a conveyance network of small earthen ditches that transported and delivered water to its end water users.

In historical prospective, the water supply system of Tashkent city has been developed as the city was transforming into an urban metropolis. It should be noticed that during the early period, the water supply system has been used mainly for drinking and population water supply purposes. In the middle of the 19$^{th}$ century, there were a number of large canals and "*aryk*" (from the Uzbek language, meaning earthen ditch) –water canals that have been flowing through city territory, namely Bozsu with its diversion canals Ankhor and Kechkuruk, and Burdjar that branches out from Ankhor at the city limits (Kadyrov 2002, 68–82).

Brick walls surrounded city territory; within the city limits there were four large residential areas called "*dakha*" (Kukcha, Sebzar, Shaikhantakhur and Beshagoch). The large canals and aryks, which were mentioned above, flowed through these residential areas—"*dakha*". There were numerous smaller scale intake ditches that served as the water supply of the local population of adjacent streets and neighborhoods ("*makhallya*" in the Uzbek language means public union of community members and local population) (Kadyrov 2002, 70–72) In fact, large single makhallya or several makhallya owned aryk, from which water had been distributed among residents and different civic institutions. Each makhallya had its own "khauz", a basin or pool where water was stored for later use during the winter–spring period. *Guzar* (meaning in Uzbek inter-makhallya trade and a social centre) basins with larger volume capacities supplied water to several makhallya during scarce water periods (Temporary Regulations, 1878.)

The management, operation and maintenance of the municipal water supply system have been carried out according to *khanate* regulations and laws. Both in villages and cities, the water supply system was managed by local initiative administrative groups, including *aryk-aksakals*, *mirabs* and *tuganchi*. Before the Tsarist Russian period, people elected on a democratic basis such responsible personnel, but after the Russian conquest local authorities appointed them. Aryk-aksakal was responsible for the management of large aryks and headwater intake structures that diverted water from rivers. In the case of long canal-aryk, several aryk-aksakals were responsible.

On the other hand, mirabs mainly administered the distribution and allocation of water resources among farmers (dekhans), whereas in cities it was done by makhallya aryks. Tuganchi, who "constructed dams or dykes" was responsible for the operation and maintenance of ditches and canals with practical knowledge and experience for constructing simple nomadic water intake structures from rivers and canals by using local materials – wood, stone, soil, brush wood, etc. During that period, concrete and metal structures were not used widely in Central Asia. Nevertheless, more than 1.5 million hectares of lands had been irrigated within the modern territory of Uzbekistan. Before the 19th century, the city wall of Tashkent City was round, with the total length of the wall constituting approximately 12 km with 12 gates. The area of the city was about 11 km$^2$ (1100 ha), with a population of 106 thousand people (in the year 1866) (Kadyrov 1998, 140.)

After the conquest of Central Asia by the Russians and the establishment of the Turkestan general-governorship administrative system, Tashkent became both the military and administrative centre of all the Turkestan territory (1867). Tashkent began to sprawl rapidly in a southeast direction; on the left bank of Ankhor a new city or "Russian part" was established according to a special general plan with its city centre, known in Soviet times as the "Square of the revolution", presently Amir Timur square. Thus, all streets and avenues in the new part of Tashkent had flowing aryks along both sides of the roads with perennial trees – plane trees, oaks, acacias, Lombardy poplars, etc. In order to answer the question how the whole water supply system was maintained and managed, we provide a fragment of the document, to be exact, from "Temporary rules on water use in Turkestan territory" approved by the first general-governor von Kaufman in 1876 that was implemented as an experiment in Tashkent city and the Kuramin district (territory that was irrigated from the river Chirchik up to its discharge with the river Syrdarya). A special section in this document states: "Costs of the water supply: 29". The costs of the water supply of Tashkent City and the Kuramin district are calculated according to the following sources (Table 16.1):

a) 8993.33 is the amount that was provided from the Official Government according to the "City regulation of the year 1870" after city council approval.

b) All other costs of the sum 15,966.67 rubles refer to the funds of the district, in particular tax collection that was used for the maintenance of local administration.

c) Maintenance and repair of dams, ditches, bridges and other engineering facilities on main irrigation canals refer to: in cities—to municipal incomes and local raw material supply, whereas in districts—to raw material supply; in particular each single community should have provided the required materials.

Table 16.1 depicts the cost category of O & M of the water supply system of Tashkent City (1876)—Source Kadyrov 2002.

| Category | Asian Part of City (Rubles) | Russian Part of City (Rubles) | Total (Rubles) |
|---|---|---|---|
| 1. Wages of irrigator, his assistant, and land surveyor and equipment | 2022.22 | 1011.11 | 3033.33 |
| 2. Wage of Aryk-aksakal | 1600 | 400 | 2000 |
| 3. Wages of 8 Mirab | 2160 | 960 | 3120 |
| 4. Two Tuganchi | 400 | 200 | 600 |
| 5. Two Mirab on Zakh-aryk | 160 | 80 | 240 |
| TOTAL | 6342.22 | 2651.11 | 8993.33 |

In late 1870's 5 aryk-aksakals, 28 mirabs and 2 tuganchi served the aryk-water supply system of Tashkent City. One aryk-aksakal with 8 mirabs and 2 tuganchi served the needs of the new part of the city and only 1/3 of direct costs were covered from the treasury (municipal income) for the needs of the water supply in Tashkent city. 2/3 of costs were funded from funds of the district and the raw materials supply from the city and district population. It should be noticed that "Temporary rules" were implemented within a short period and did not consider the experience, knowledge and traditions of the local people. In 1882 these "Temporary rules" were cancelled and substituted in 1886 with "Instruction on rights and duties of irrigation officials, district heads, aryk-aksakals and mirabs on water supply management in Turkestan territory". Before the "Temporary rules", aryk-aksakals and mirabs were elected democratically at the meetings of dekhkans or their representatives who used their service. "Instruction" was partially rehabilitated in this rule keeping the right of the military governor to recommend to the district head which aryk-aksakals to assign and dismiss.

## 16.3. CONCLUSIONS: WATER SUPPLY EVOLUTION DURING SOVIET COMMUNIST REGIME AND PRESENT

In 1966, Tashkent witnessed a catastrophic earthquake and the major part of the water supply system was destroyed or severely damaged. Since the earthquake, Tashkent city has changed significantly; it was reconstructed and almost 100% of the water supply system was reconstructed with the construction of a sewage system too. Presently the city population exceeds 2.5 million people. The territory of the city now covers an area over 300 km² (11 km² in 1866 and 34 km² in 1917, with 280,000 inhabitants), 30,000 ha (about 30%) of which has been occupied by parks, avenues, gardens, and flower gardens (Khabirov & Gutnikova 2000, 247–255).

In the early 1940s, tap water was introduced widely to the public. One of the tap facilities was built on Chigatai Street (presently Farobiy Street) in the guzar of makhalla Chakichman. Water was sold for coupons, which were distributed in advance in the form of a checkbook, with a 2-kopeck coupon per bucket of water. Water from aryks was used for domestic purposes (washing of clothes) and other purposes – watering of flower gardens and trees. The situation remained during World War II. In the late 1940s, three more such tap facilities started to operate in different parts of the city and by the end of the 1950-1960s almost all neighborhoods had such tap water access. During that time water from aryks lost its value. Water in the aryks began to disappear and became muddy and had an unpleasant smell. Groundwater and surface resources comprise the centralized water supply system of Tashkent City. The main water intake facilities Kadyrya and "Bozsu" are being operated for withdrawing water from the Bozsu canal (volume is more than 700,000 m³/day). The extraction of groundwater is carried out through the Kibrai water supply facility, the south water supply station and also by using several small water intakes (8 areas with 1 to 9 wells) with a total yield of 726,000 m³/day. Thus, presently the total possible water intake is more than 2,326 million m³/day (Khabirov & Gutnikova 2000, 250–252)

According to official municipal statistics, the actual use of water per capita increased from 1960 to 2000 (Figure 16.2). The water norm per capita has been established as 250 (litres/day) in 1965, 1970 and 1980, and 330 (liters/day) in 2000 by a decree of the Tashkent City Khokim (governor). However, water users have overused water resources and regulations are observed neither by local authorities, industries nor by the local population. There is no available published systemized statistical data. Some numerical data (population, total flow that is being conveyed through water pipelines per year) are taken from the atlas of Tashkent City that was published in 1983 and the presented data have been interpolated. The population aspect could have played a major role in the increase of per capita water use. For example, in 1965, 1975 and 1981 the population reached 1.22, 1.83 and 1.87 million people respectively.

Figure 16.3 depicts water conveyance efficiencies of water supply system performance over time. The pipes have deteriorated to various extents and their performances decreased from 0.85 in 1965 to 0.75 in 2000. On the other hand, water losses increased accordingly. Thus

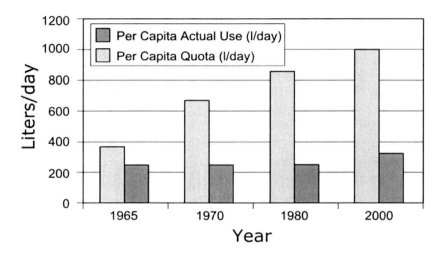

Figure 16.2 Actual Per Capita Use in Comparison with Quota in Taskent City, Uzbekistan, 1965–2000.

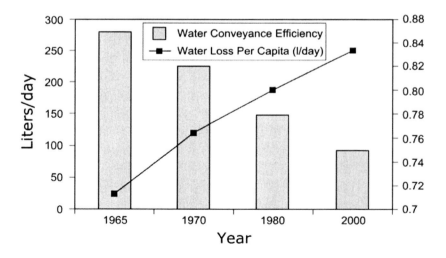

Figure 16.3 Water Conveyance Efficiency vs. Water Loss Per Capita in Tashkent Water Supply System, Uzbekistan 1965–2000.

preventive measures can include the rehabilitation of water supply systems by replacing the existing network of pipes, and other facilities. In addition, water prices should more realistically reflect actual prices. Local authorities must monitor and press charges to local industries and businesses for discharging pollution into the public water supply system.

Construction sites are spread all over the Tashkent area and use water from the urban water supply system. The concern is that they use water in the category of drinking water, wasting high quality water resources instead of using water from aryks or other irrigation canals that flow through Tashkent.

Tashkent City has sophisticated water treatment facilities, which is part of a single urban water supply system. Water is treated through mechanical, biological, and chemical water treatment procedures, and is finally disinfected against disease-causing microorganisms. After all this process, water quality must meet the Governmental Standard GOST "Drinking Water"—on water purity, chemical concentration (content of soluble salt), and odour. Hypothetically, water should not contain such substances as nitrates, heavy metals, or phenols. However, source water quality from the canal Bozsu and groundwater resources are not of satisfactory quality and need supplemental treatment such as ozone treatment, radiation by ultra-violet rays, and membrane technologies. Unfortunately, the Bozsu canal and groundwater are within a zone of pollution and contamination. Consequently, it is necessary to adopt new technologies on a large-scale that require large investments.

The major part of treatment plants and facilities are in poor condition and served for many decades without considerable renovation. Local authorities cannot fund or purchase new equipment for updating their old facilities. A second major issue is the attitude of different water users toward the rational use of water. Many practice wasteful technologies in the production process or in domestic water use. There is water-pricing policy but it is not operational and feasible.

Finally three recommendations are presented:

1. The paramount recommendation the authors argue is water policy development. This policy will create favourable conditions and a regulatory basis for replacing old management, maintenance, and operation mechanisms.

2. The financial aspect is the most problematic. All of post-Soviet people from childhood used to think that only the government using the state budget solves such large-scale and complex tasks. In a market-oriented economy, the problem is solved with the involvement and active participation of all stakeholders. Certainly, without the support and assistance of governmental authorities the considered problem cannot be solved, but local city communities should contribute their own efforts and funds in maintaining and operating the water supply system. One option can be the creation of a special fund that will serve the water supply development of Tashkent City in all its aspects. The city khokimiyat can act as a founder of the foundation that will select other founders for the establishment of the fund. The first significant investment of the fund should be financing a large-scale inventory of the existing conditions of the water supply system of Tashkent.

3. The institutional aspect of the problem is also a complex task. Making the right decision on transferring these tasks to someone and specifying implementation conditions is half way to achieving success in the planned measures.

## 16.4. REFERENCES

Kadyrov A. (1998) Stories from Irrigation of Uzbekistan. [original in Uzbek] Tashkent, "Khalk Merosi", p. 140

Kadyrov A. (2002) History and Present Condition of Water Supply in Tashkent City. (in Russian). Issues of Potable Water Supply and Ecology. Tashkent, "Universtitet", pp. 68-82.

Khabirov R.S. & Gutnikova R. I. (2000) Issues on Potable Water Supply in Uzbekistan. [original in Russian] Water resources, Aral Sea and Environment. Tashkent, "Universitet", pp. 247-255.

Salikhojaev Z. (2000) Live Water. [original in Russian]. Tashkent, "Shark".

Temporary Regulations on Irrigation in Turkestan Territory (Krai) approved by General-governor K. P. Kaufman in 1878.

# 17
# PHILADELPHIA WATER INFRASTRUCTURE 1700-1910

*Arthur Holst*

Figure 17.1 Location of Philadelphia, USA.

## 17.1. GENERAL BACKGROUND

This chapter takes a look at the water supply system in Philadelphia, Pennsylvania, USA, from its beginnings in 1700 when the City's population was 4,400 up to 1910 when it served 1.5 million people. It highlights technical innovations adopted by the City of Philadelphia and improvements made to its distribution system during this period.

Before the late 1700s, Philadelphia residents drew their water supply from wells and reservoirs, like everyone else had been doing for centuries. These water sources were unprotected from pollution and contamination and they were often a source and carrier of diseases. In his will, Benjamin Franklin allotted a substantial amount of money to be used in the building of a dam in the Wissahickon Creek in order to supply clean water to the city. During the 1790s, yellow fever rampaged through Philadelphia and it was responsible for the deaths of very large numbers of the population. Residents were concerned about the inadequate supply of water. They believed that the yellow fever epidemics were caused by contaminated water and that a better, cleaner supply was needed to fight the disease and to also help firefighters to do their work more efficiently.

The City Council began to look for new sources of water. Benjamin Henry Latrobe, an engineer and architect, was hired by the city to find an adequate source that would supply the city with continuous drinking water. Latrobe reported that the best solution would be to build a basin on the Schuylkill River at Chestnut Street and use a steam engine to lift the water and force it up to Centre Square. A second steam engine would lift the water up and into a reservoir which was built nine meters above ground and was able to hold 60.57 cubic meters of water. From here, the water would be distributed through wooden pipes all through the city. Despite strong opposition, his plan was approved (Birkinbine 1859, 9-10).

## 17.2. CENTRE SQUARE WORKS

A contract for building the steam engines, capable of providing the city with 3,800 cubic meters of water per day, was made with Nicholas I. Roosevelt, a talented inventor and machine builder from New York. There was great skepticism at the time that the idea of transportation of water through steam engines could actually work, since there were only three steam engines with this kind of power at the time, and none of them were used to supply water. Steam engines were still a very new technology at this time, as evidenced by Oliver Evan's high pressure steam engine model pictured in figure 17.2. The basin that was built was 61 meters long and 26 meters wide and the bottom extended ca. one meter below the low water mark. The pumps of the steam engine at Schuylkill and Chestnut Street sucked water from a well which was 12 meters deep and three meters in diameter. The engines raised the water into a brick tunnel that was approx. two meters in diameter and 960 meters long, leading to Centre Square.

Figure 17.2 Model of Oliver Evans high pressure steam engine. (Source: Philadelphia Water Department Historical Collection)

Water was distributed through wooden pipes throughout the city, underneath the streets, for the first time in January 1801. Tree trunks served as the material for these pipes, which were cut into 3.65 meter sections with a hole drilled through its center. The trunks were connected to each other by straps and iron couplings. One of the many problems with these pipes was that when they intersected, they formed sharp right angles, which greatly reduced water pressure, making it harder to effectively distribute water.

The engine at Schuylkill and Chestnut Street was considered big for its time. It was one meter in diameter with a 1.8 meter stroke. The pump was 44.5 centimeters in diameter and had a 1.8 meters stroke. This engine was able to pump 5,580 cubic meters of water per day. The one at Centre Square, within the pump-house pictured in figure 17.3, was less than a meter in diameter with a 1.8 meter stroke. Its pump was 0.45 meters in diameter and also had a 1.8 meter stroke. It was able to pump 3,640 cubic meters of water per day. These engines were expensive to operate due to the large amount of coal that they needed in order to operate continuously. At the top of the engine house were two tanks built from wood, located ca. 15 meters above the brick tunnel. Both of the tanks were 13.7 meters deep. The bigger one was 4.2 meters in diameter and contained 65 cubic meters of water while the smaller one was only 2 meters in diameter. The pumps needed to run continuously in order to keep these tanks full since water was used up at the same rate that it took to fill them.

The engines were a great innovation in the distribution of water, as depicted in the cross section of the Centre Square and the Schuylkill pump-houses in figure 17.4. and 17.5. Nevertheless, these new technologies had many defects. They mostly had problems with the parts made from wood, like the flywheels, shafts and arms, lever beams, cold-water pumps and cisterns. The boilers were also built of wood. They were 2.7 meters high, 2.7 meters wide, 4.6 meters long and had leaking problems. The fire box within the boiler was made of wrought iron.

The first pumps were built without air chambers, and in 1810, one was added to the pump in Centre Square. The watering committee experimented with a cast iron boiler, which proved more successful than the boilers made of wood. Because of this, two cast iron boilers were installed at Centre Square. These would also prove to be problematic from time to time with their plates and flues cracking. People who could afford to pay for the construction of a connecting pipe and a yearly fee could have an indoor water supply. Since a more constant supply was needed, construction began for a new water works in 1812 on the Schuylkill River at Fairmount.

Figure 17.3 Painting of the Centre Square pump station. (Source: Philadelphia Water Department Historical Collection)

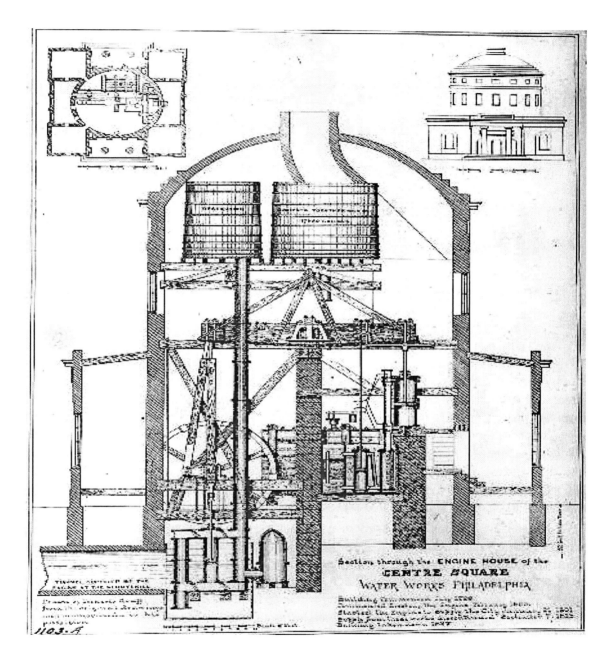

Figure 17.4 Cross-section of the engine house at the Centre Square pump station. (Source: Philadelphia Water Department Historical Collection)

Table 17.1 Philadelphia Population Growth.

| Year | Population | Year | Population |
|------|------------|------|------------|
| 1700 | 4,400 | 1810 | 53,722 |
| 1710 | 6500 | 1820 | 63,802 |
| 1720 | 10,000 | 1830 | 80,462 |
| 1730 | 11,500 | 1840 | 93,665 |
| 1740 | 12,000 | 1850 | 121,376 |
| 1750 | 12,700 | 1860 | 565,529 |
| 1760 | 18,750 | 1870 | 674,022 |
| 1770 | 28,800 | 1880 | 847,170 |
| 1780 | 30,000 | 1890 | 1,046,964 |
| 1790 | 33,000 | 1900 | 1,293,697 |
| 1800 | 41,220 | 1910 | 1,549,008 |

## 17.3. FAIRMOUNT STEAM WORKS

A high-pressure engine, built by Oliver Evans and seen in figure 17.4, was installed in the Water Works and was made operational in 1817. This engine was capable of pumping 11,631.27 cubic meters of water per day. The engine did 24.75 revolutions per minute carrying 88 Kg of steam. The steam cylinder was 50 centimeters in diameter and had a 1.5 meter stroke. Its pump was 50 centimeters in diameter with a 1.21 meter stroke. The engine had 4 cylinder boilers, which were 76 centimeters in diameter and 7.31 meters long. They held a pressure of 99.79 Kg of steam. It cost $84.50 ($1162.21 in 2003 dollars using CPI (Consumer Product Index data series) per day to raise 8,706.45 cubic meters of water. The reservoir contained 12,363.63 cubic meters of water and was distributed though wooden pipes to the distributing chest located at Centre Square, 2,900 meters away.

In 1815, cast-iron pipes were used in constructing the Fairmount Water Works and it wasn't until one year later that cast-iron water mains were added. Cast-iron pipes are made by pouring liquid iron into a cast to form any desired shape. This greatly decreased the loss of water pressure that was caused by the sharp right angles that were formed when wooden pipes intersected, thereby allowing water to flow more efficiently. By 1815, the city had laid about 73,640 meters of wooden pipes, some of which were unearthed in 1901 during construction efforts as shown in figures 17.6. and 17.7. The rapid growth in the population, as evidenced by the chart in table 17.1., continued and the cost of supplying water to the city increased. This growth forced the City Council to search for cheaper ways to supply water. A dam was built at Fairmount Water Works in 1821 in order to use natural water power from the Schuylkill River to supply the city with water.

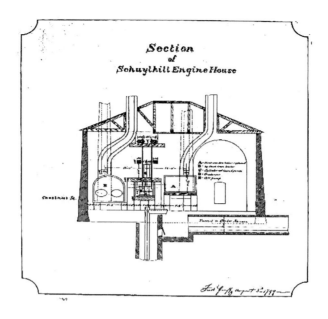

Figure 17.5 Cross-section of the engine house at the Schuylkill pump station. (Source: Philadelphia Water Department Historical Collection)

Figure 17.6 Wooden pipes unearthed in Center City in 1901. (Source: Philadelphia Water Department Historical Collection)

## 17.4. WATER POWER

The city entered into a contract with Ariel Cooley, from Massachusetts, to build the dam for $150,000 ($2,427,176.47 in 2003 dollars using CPI; Consumer Price Index data series). The work began in 1819 and was finished in 1821. The river is about 274 meters wide and nine meters deep at high water. Cooley placed cribs made of logs 15.24 meters high and 5.18 meters wide in sections made of rock. The cribs were filled with rocks and earth to prevent leakage. In the muddy section of the river, the mound was built 4.5 meters higher than the other part and it served as an over-fall. The base in this section is at least 45.72 meters and the top is paved with building stone in order to prevent damage by water and ice. A stone pier is located between the mound and the over-fall. It is 8.53 meters high and 7.01 meters wide, which supports the mound and protects it from damage. The dam is built diagonally running upstream and towards the end, the rest of the dam forms a right angle towards the western shore joining the pier of the guard-lock and creating a large over-fall which is 366.98 meters long. The dam was reconstructed well after its inception in 1872, as shown in the plan and section of the dam in figure 17.8. The whole dam is 488 meters long.

To make use of the water-power, water wheels were created. The wheels were placed below the high water level. Even though one would think that the wheels would stop when there is high water, this only happened during spring tides when the wheels would stop about 64 hours per month. The pumps were powered by the rotating of the water wheels. These double forcing pumps were individually connected to a main, made from iron, that has a diameter of 41 centimeters, and which discharges into a reservoir. The shortest main is 86.5 meters long. A stop-cock was located at the end of the mains to stop the water when needed. After a later expansion of the Fairmount Water Works, there were a total of nine water wheels that were in use. Eight of the wheels ranged from 4.5 to 5.4 meters in diameter and were able to pump from 290 to over 300 cubic meters of water per hour depending on their diameter, with the bigger ones raising more water. The ninth wheel was a Jonval turbine and was seven feet in diameter. Its buckets were 0.33 meters wide and 0.25 meters deep. It powered the pump through a pair of bevels and a pair of spur wheels. The pump had a diameter of 41 centimeters and a 1.8 meter stroke. It did about 12 double strokes per minute and was capable of raising 330 cubic meters of water per hour. Water was stored in four reservoirs built at different times in response to the demand for water. They were capable of holding approx. 102,000 cubic meters of water.

There were three water mains distributing water from these reservoirs with diameters of 0.76, 0.55 and 0.50 meters. Another reservoir was supplied with water from the Fairmount Water Works and was located between 22nd street and Corinthian Avenue, and Poplar and Parrish streets, a location that is higher than the reservoirs at Fairmount. It was capable of holding up to 77,000 cubic meters of water. This reservoir was filled with water through a stand pipe that was 15 meters high and had a diameter of 1.2 meters, and water was distributed by one main 0.76 meters in diameter. The water pumps were arranged so that

Figure 17.7 Photo of the measuring of one of the unearthed pipes. (Source: Philadelphia Water Department Historical Collection)

they could work one or all at a time in order to fill either the Fairmount reservoirs or the one at Corinthian Avenue.

The distribution of water into the city had been largely done through wooden pipes, which had proven to be very unreliable due to their perishability and inability to handle high pressures for long periods of time. The Watering Committee laid down a small number of cast-iron pipes to test their performance. They were very impressed with the outcome and in 1820 authorized the replacement of wooden pipes with cast-iron ones. To this day, we can still find wooden pipes that were left in the ground, unused, during the installation of the metal pipes, as seen in figure 17.6 and 7.

By 1875 the population had risen to 817,000 and with more water being supplied, water consumption also rose to 0.30 cubic meters from 0.01-0.02 cubic meters per day per capita. Notably, Philadelphia's City Council wanted to provide all of its constituents with water, which was a difficult task given the size of the city. So much water was used, that open pools of water, springs, brooks, etc. overflowed the city streets due to inadequate sewers that were not built to handle a population this large.

Sewers were built as early as 1762, which worked well at the time, since the population was low. Water can break down organic wastes by oxidation in a natural process that occurs after water has flowed for a few kilometers. Typhoid and cholera were killing hundreds of Philadelphians in the 1890s and the city finally decided to take action. Sewers were

# PHILADELPHIA WATER INFRASTRUCTURE 1700-1910

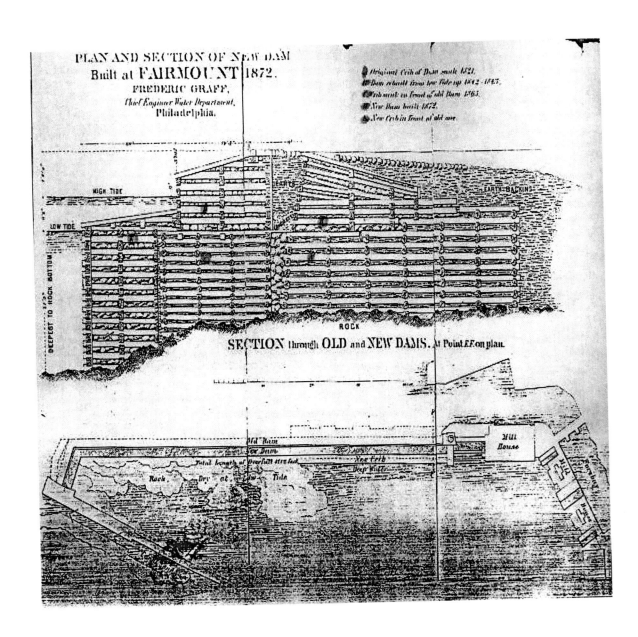

Figure 17.8 Plan and cross-section of new dam built at Fairmount in 1872. (Source: Philadelphia Water Department Historical Collection)

constructed all across the city, such as the Mill Creek Sewer built in 1883 (Figure 17.9). In 1899, the Belmont Filtration Plant was built. Its underground layers of sand were used to filter out impurities, but this simple system was not enough, since the watercourse was not long enough to filter out all wastes produced by polluters. Another filtration plant was built, the Torresdale Filtration Plant, on the Delaware River, which was much larger and cleaner at the time. The Plant was completed in 1908. In addition, large iron pipes, such as those seen in figure 17.10., had effectively replaced the wooden pipes which had been laid throughout the city. Soon after, all city water was being transported and chlorinated more efficiently and on a regular basis, providing much cleaner water and ending many diseases that had been spread at the time.

## 17.5. CONCLUSIONS: CONTINUING DEVELOPMENT

In year 2005, the Philadelphia Water Department has continued its tradition of working to deliver safe and clean water to the citizens of the city. Now, there are more than 1,600 miles of combined sewers, 1,200 miles of sanitary and storm overflow sewers, 150 miles of intercepting sewers, 169 regulating chambers, roughly 85,000 manholes, and around 75,000 storm water intakes. As it always has been, development and improvement of the water system remains an integral component of Philadelphia's drive to economic prosperity.

This commitment to improve has existed in Philadelphia since the construction of the first pump-houses and water works. In fact, the Schuylkill pump station of 1799, depicted by a cross-section in figure 17.5, was viewed as an engineering marvel for its time and it was in fact one of the most popular tourist sites at the end of the 18th century. Just as the push for clean water guaranteed an end to the yellow fever epidemics of the 18th and 19th centuries, the Philadelphia Water Department continues this legacy today. It works to guarantee that every citizen receives clean and safe drinking water in his or her home, as well as providing the basis for the city's public health system and economic growth. In addition, there is no longer only a concern for public health, but also for environmental health, as the Philadelphia Water Department addresses the problems caused by storm water runoff.

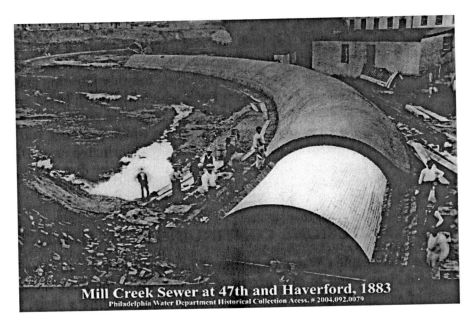

Figure 17.9 Photo of construction of the Mill Creek Sewer in 1883. (Source: Philadelphia Water Department Historical Collection)

Figure 17.10 Photo of iron pipe installation in 1895. (Source: Philadelphia Water Department Historical Collection)

## 17.5. REFERENCES

Hardy III C. (2004)The Watering of Philadelphia. Pennsylvania Heritage, p. 26-30.

Birkinbine H. P.M. (1860) History of the Works and Annual Report of the Chief Engineer of the Water Department of the City of Philadelphia. Philadelphia: C. E. Chichester, Printer. February 9, 1860, p. 5-25.

Weigley R. F. (1982) Philadelphia: A 300-Year History. W. W. Norton & Company.

# 18

## PRIVATISATION OF WATER SERVICES IN HISTORICAL CONTEXT, MID-1800S TO 2004

*Petri S. Juuti, Tapio S. Katko & Jarmo J. Hukka*

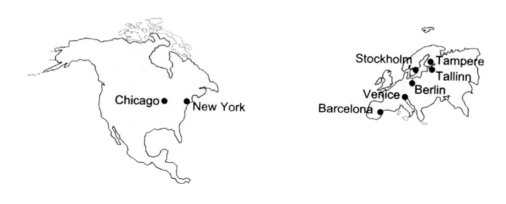

Figure 18.1 Some keyplaces mentioned in the chapter.

## 18.1. GENERAL BACKGROUND

The start of the industrial revolution and the related growth of cities gradually created the need for centralized water and sanitation. In many respects, England was the forerunner of modern water supply and sanitation systems, but the innovations soon spread to other parts of Europe and the major cities of the world. As the cities of Europe grew, sanitary and environmental problems overwhelmed city governments to a greater degree than before, and modern technology was often seen as the solution (Hällström 2002, 17; Juuti 2001). Melosi (2000) shows how European water technology was transferred to, and eventually developed, in America.

The role and development of municipal organizations was another important feature of this development (Hällström 2002, 18). The establishment of modern water systems was largely based on private initiatives. Yet, the evidently unsatisfactory quality of private company supplies led to a re-evaluation of the organizational means (Hassan 1998, 18).

From 1861 to 1881 the share of municipal water supply in larger provincial towns in England grew from 40 to 80 per cent, and reached some 90 per cent in 1901. The growth of the urban infrastructure was the most dynamic element of the British economy from the 1870s to the 1930s. If housing is ignored, the investments in public health, local transport, water, electricity and gas were by the early 1900s as much as one quarter of all capital formation in Britain (Millward 2000). In North America a considerable number of urban water supply systems were built in 1830–1880 while it proved more difficult to fund sewerage systems (Ling 2003). Most US citizens drew their drinking water from private wells or other water sources until the last quarter of the 19th century. According to some researchers – e.g. Joel A. Tarr, Stuart Calishoff and Nelson Blake – the needs of businesses and industries, real estate owners, fire fighting companies and health authorities hastened the birth of water works, making public works necessary. New York and Chicago, among others, started water acquisition and distribution with the help of private enterprises (Keating 1989).

In Rhenish Prussia, the rising income of the middle-class voter and demand by industrial users, rather than public health crises, created demand for improved water supply (Brown 1988). Brown further points out that historians credit the sanitation revolution with the decline in mortality, while the spur that the sanitary reform gave to municipal intervention in the local economy through the regulation of housing and land markets as well as the provision of services such as water and sewerage, is less well known.

Hassan (cited by Brown 1988) argues that the demands of industries, such as cloth finishing and dyeworks, persuaded cities to take an active role in water provision rather than addressing the concern for public health. On the other hand, Gaspari & Woolf (1985) show that in 122 cities in the U.S., sewage systems reduced mortality significantly, while water filtration systems had no impact. More recent impact studies from developing countries show certain variation depending on conditions. Yet, the overall trend is that improved water supply results in somewhat reduced mortality and the impacts are bigger when sanitation is introduced. Yet, the best results will be gained if health education is also introduced. In Finland the first water supply and sewerage systems of urban centres in the 1870s to 1890s were in most cases constructed simultaneously although often under separate organizations. There was demand mainly for fire-fighting water (Hietala 2002; Juuti 1993 & 2001), but drinking water supply and sanitation, and in some cases industrial needs, also played a role. Thus, it is obvious that the impacts of improved water supply and sanitation depend on local conditions, as does demand.

Historically, Barraqué (2003) recognizes three main time-related paradigms in public water supply and sanitation: quantitative and civil engineering, qualitative and chemical/sanitary engineering, and the most recent one – environmental engineering and integrated management. Infrastructure and the built environment of today are the results of decisions and efforts made decades and even centuries ago (Kaijser 2001). Besides, decisions concerning building and rebuilding these systems and structures will shape the material world of future generations. Already for some time historians like Melosi (2000) have been interested in the concept of path dependence – how decisions made bind our alternative development paths. These decisions may be of binding, limiting or postponing nature (Kaivo-oja et al. 2004).

The objective of this chapter is to discuss several private proposals and granted concessions for establishing waterworks in Europe and the USA. It also describes the cultural and historical background of water resource and service management. The chapter refers to decisions, arguments and policies that largely resemble those presented over one hundred years later, in the 1990s, when privatisation was again promoted. However, the objective of the chapter is not to compare the economic or other types of performance of various undertakings but to particularly question the ethics and rationale of selling and reselling the concessions or ownership of public water undertakings.

## 18.2. EXPERIENCES FROM THE 1800s

It is important to understand the current differences and cultures in water resources and services management and their historical background. Barraqué (2003) has formulated a rough typology of water resources management and institutional cultures in Europe. This typology is based on Germanic vs. Roman legal origin and, on the other hand, centralised vs. subsidiary (decentralized) tradition (Table 18.1). The only three states covered by river-basin institutions are the ones that have historically been centralised monarchies: Spain, England/Wales, and France. Yet, they have evolved differently. Besides, in some countries river basin authorities, like those in the Nordic counties, have been formed on a voluntary basis.

In England and Wales, water resources policy has been centralized in the post-war period, particularly after the introduction of River Basin Authorities in 1963. Water supply and sewerage systems became centralized in 1974 with the establishment of ten Regional Water Authorities. The more recent extreme example of the privatisation of ten regional water

Table 18.1 A rough typology of water institutional cultures in Europe. (Barraqué 2003, modified by the authors)

| (Predominantly) | ROMAN ORIGIN | GERMANIC ORIGIN |
|---|---|---|
| CENTRALISED | Spain | England |
| SUBSIDIARY | Portugal, Italy | Netherlands, Germany, Nordic countries |

and sewerage undertakings during Prime Minister Thatcher's regime (1979–1990) sets England and Wales clearly apart from other European countries. Spain, Portugal and Italy have systems built on Roman law, while those of England, the Netherlands and Germany are based on Germanic law. In Spain, Portugal and Italy, the political history of the 20th century explains also largely the ways and emphasis of water resources management. Germany has a long tradition of local drainage associations, while river basin management has not been institutionalised except for the famous Ruhrgenossenschaft. Due to the strong subsidiarity, water policy is in the hands of 16 Länder (states) rather than with the Bund (federation). In the Netherlands, historical development has led to water-user associations, while wastewater management is largely based on water boards by 2000 (Uijterlinde et al. 2003).

The Nordic countries are perhaps the ones with the strongest subsidiary tradition and do thus fall in the same category with the Netherlands and Germany. According to Barraqué (2003), it is difficult to place France in any of these categories. On the one hand, France is clearly a follower of Roman law and the centralized tradition. Yet, the six water basin authorities have become largely subsidiary institutions. As for water services, the role of municipalities has declined over time. Several Central European, as well as the Baltic countries were subject to the highly centralized Soviet tradition of state water management after WWII. It will be interesting to see to what extent they will "go back" to the municipal tradition, or whether they will choose the private company tradition for the short or long term. Although the typology described above applies mainly to water resources management, it also explains the differences in subsidiarity tradition, and thus the role of local governments. This difference is crucial when we take a closer look at the evolution and strategic decisions concerning the management options for water and sewerage services.

**Early private operators**

Although public-private cooperation and related issues are not defined as the only strategic decisions to be considered in this study, they often seem to be the only ones of strategic importance that have been written about. There are examples of cities' involvement through some level of water services provision already from the Middle Ages. It may have occurred in the form of public wells or, like in the case of Tallinn, the services of water carriers that were paid by the town council in the 1330s. As for waste disposal, the so-called Lübeck Act was adopted in Tallinn in 1257. This Act stated how toilets and pigsties were to be located in relation to streets and neighbouring bounds (Kaljundi 1997).

In the mid-1800s most Western nations, if not all, started to develop urban water and sewerage services through privately owned companies or private operators. Yet, in most countries the utilities were fairly soon taken over by municipalities. Only in France have private operators, such as Veolia Water (earlier Vivendi and the Compagnie Générale des Eaux) survived and expanded since 1853. This is largely due to the fact that in 2000 France still had some 36,000 municipalities. It is very difficult to imagine that all municipalities would have their own water and sewerage undertaking. There are also several other policy instruments that have favoured, and still favour, the use of private operators. In some places, such as Barcelona in Spain and Venice in Italy, private companies have maintained concessions for over a century. One of the basic tenets of water and sewerage services (WSS) is that the WSS infrastructure is a natural monopoly – a concept introduced by John Stuart Mill (1806–1873) in 1848 (Sharkley 1982, 14). Accordingly, it is feasible to construct only one such system per service area.

## Establishing local governments

The first modern water systems were built on the basis of builder-owner or concession models in many European countries, and particularly in North America. In most cases, however, municipalities soon took over these water and sewerage systems. For example, in the early 20th century, 93 per cent of the systems in German urban centres were municipal, as were all the urban WSS systems in Sweden and Finland (Wuolle 1912).

In the middle of the 1800s, a clear distinction developed between the public/general and private spheres of society. The private sphere was considered to consist of "private social groupings" – individuals, families and local communities. Local level services were largely managed by private entrepreneurs because there was hardly any legislation on local governments. The state could have an impact on these matters only through legislation, such as the laws enacted in the 1860s and 1870s in many parts of Europe (Kilander 1991, cited by Nygård 2004a, 164; Nelson & Rogers 1994, 27).

During the 19th century, the previously private systems came under public ownership and public provision because of the inefficiency, costs and corruption connected to them. In the late 19th century, the emphasis was on municipalization. Democratically elected city councils bought existing utilities and transport systems and set up new ones of their own. This resulted in more effective control, higher employment, and greater benefits to the local people. Councils also gained the right to borrow money to invest in the development of their own systems (Hall 2003, 7).

Nelson & Rogers (1994) pointed out the background and birth of the First Public Health Law in Sweden. It came into force in 1874. Initially it was clearly influenced by the British Public Health Act of 1848. The committee drafting the 1874 Act considered the promotion of preventive health care of utmost importance. Along with the Act, for instance, public health boards became compulsory in each town. The Swedish Act also served as a model for the Health Decree of 1879 in Finland (Nygård 2004b). Yet, in historical context it is good to remember that it was characteristic of the whole of Europe that the working classes had no representation in municipal government. For example, it was not until 1903 that the first representative of the working class became a member of the Stockholm city council (Hietala 1987, 55–56).

**British and German experiences**

In Britain, the continuous period of private water undertakings started in London in 1681 following the Great Fire of 1666, as the city administration granted the use of the first arch of London Bridge for water supply purposes to Peter Morris for 500 years (London's Water Supply 1953, 9: 1–15). The private sector started operating other water works at the beginning of the 19th century, and parliamentary regulation was introduced gradually. As in many other countries, the responsibility for most water works, and practically all sewerage systems, belonged to local governments by the end of the 19th century or, at the latest, in the beginning of the 20th century.

During the 20th century financing by the central government increased continuously. Modern water supply and sewerage systems were born in England, although the birth of sophisticated WSS systems dates back to antiquity or even earlier. The novel idea that emerged in England in the 1840s was that these services were the responsibility of the Government (Juuti 2001).

From England these "modern" systems spread to Germany and other parts of continental Europe starting in the 1840s. The first systems in England were privately owned. On the continent, however, the public sector had a more central role from the beginning, except for France. The British were still more advanced in water management than their continental counterparts around 1870, and English companies became involved in the establishment of water and sewerage works in many cities in continental Europe. This came about in three ways (van Craenenbroeck 1998):

(i) an English enterprise provided financing or became an owner;
(ii) an equipment and appliances manufacturer established the water works to guarantee a market for itself; and
(iii) the development of the works was started by hiring English experts as managers.

Figure 18.2 Lord Avebury in *Figaro* cartoon 1881. (Source: http://www.sil.si.edu/digitalcollections/)

In England, the lobbying for state-run WSS services began in the 1880s. For instance, Joseph Chamberlain (1836–1914), a member of the Liberal Party, campaigned strongly for the state to take over responsibility. He argued in 1884, for instance, that:

"It is difficult, if not impossible to combine the citizens' rights and interests and the private enterprise's interests, because the private enterprise aims at its natural and justified objective, the biggest possible profit."

Castro et al. (2003) point out that in England, particularly in London, the poor quality of private water services prompted complaints since the 1850s, but it took half a century to put the water companies in public hands. In London it happened in 1902.

The arguments raised by Lord Avebury (John Lubbock, 1834–1919; Figure 18.2) in 1906 in England, in favour of privatisation, seem interestingly similar to those presented by its promoters some 90 years later. These are contrasting points – efficiency vs. market failure. Bakker (2003) pointed out that the experience from the laissez-faire management of the water supply as well as other utility services in the 19th century in the UK and the notion of "market failure" made public ownership of water supply infrastructure justified.

In Germany, the idea of municipal enterprises was already so well established by 1900 that there were no protests against the transfer of electricity, gas or waterworks to municipal ownership. Many politicians, in fact, thought that if services were owned and managed by municipalities, the general public could benefit from the profits. As Hietala (1987, 152) points out, "the services were not only useful but also lucrative for the cities". Thus, from the municipal focus the two countries are now in the early 21st century quite far apart.

## "Eternal" concessions

In France private companies have participated in the operational management of water works for over a hundred years (Kraemer 1998b, 335). The best-known enterprise, the Compagnie Générale des Eaux (later Vivendi and Veolia Water), was founded by an imperial decree already in 1853 (Goubert 1989, 175). The Antwerp water system in Belgium also dates back to that period. An English engineering company, Easton & Andersson, started work on it on the basis of a concession contract in 1879. The concession ended in 1930, although the city had tried to take over the works earlier (van Craenenbroeck 1998).

According to Barraqué (2005), in France local authorities have historically built close relationships with private companies – to protect themselves from centralization. Barraqué argues that very special relations exist between local and central governments in France. For instance, lending at cheap rates through local councils was forbidden for long. French local governments have been mature politically, but not financially. This, again, has to do with the large number of local councils there mentioned earlier. In Barcelona, Spain, the concession has run for 125 years, until early 2002.

Societé General des Eaux de Barcelonne (Agbar) was created in 1867 as a Belgian Company. Some years later it was bought by Societé General des Eaux et de l'Eclairage (Lyonnaise des Eaux) which, again, was bought by Catalan Banks (Sociedad General de Aguas de Barcelona) in the 1920s. Except for the period of the Spanish Civil War in 1936–1939, Agbar has been a private company as it still was in 2005. Thus, the entire network and all pumping stations, reservoirs and treatment plants are owned by it. Agbar grew with the city and connected villages (3.2 millions inhabitants). The City Council is, by Spanish law, ultimately responsible for the service and fixes tariffs and service conditions as a regulator, not as an owner. The Agbar company has various types of contracts also in other parts of Spain and abroad: private, delegated-managed, mixed companies, concessions, etc. (Molina 2003).

In Venice, Italy, Generale des Eaux held the concession for water supply for about a century, from the end of the 19th century to 1973 (Lobina 2003). In Portugal, the Companhia das Águas do Porto of the second largest Portuguese city was municipalized in 1927. It was a long process, which started in 1920 with the creation of a committee to analyse the conditions of water supply in Oporto. In Lisbon, the process was very similar, but in the end municipalization did not occur. The high cost of the compensation for dismantling the monopoly before the expiry of the contract was one of the reasons. Another one was the different situation in Oporto compared to Lisbon with regard to the water supply. The situation in Oporto was worse as concerns both the quantity and the quality of the water provided. The French company invested very little in the waterworks, and its response to the difficulties generated by the war was lowering the quality of the service provided. In addition, the Companhia das Águas de Lisboa had a large number of mainly Portuguese stakeholders (da Silva 2002, 29).

A clear example of the limited competition in connection with re-tendering for long-term concessions is provided by the case of Aguas de Valencia in Spain. In 1902, the city of

Valencia awarded a water concession to a private company, AVSA. The contract specified that the monopoly would last for 99 years. Thus, in the late 1990s, for the first time since 1902, the city of Valencia began to draw up tender documents. At this point AVSA, part of the SAUR-Bouygues multinational group, on advice of the international accounting firms Price Waterhouse and Arthur Andersen, announced that if it lost the tender, it would demand a compensation of 54 million euros for investments it had made in the system (Expansion 2001).

The invitations to tender contained a clause stating that the winner would have to pay 54 million euros to AVSA. Not surprisingly, there was not a single competing bid. AVSA, now in a joint venture with the council, will enjoy the concession for a further 50 years. In 2050, the city of Valencia will have had 150 years of private water monopoly without a single competitive bid (Alfonso 2001, 6).

In Barcelona, Aguas de Barcelona has enjoyed an unbroken indefinite concession for 136 years, with no prospect of a competitive tender in the near future for the same reasons as in Valencia (Hall & Lobina 2004).

## The experience of the USA

In the USA, the biggest cities usually owned the water works and were responsible for their operation at the end of the 19$^{th}$ century and in the beginning of the 20th century. Only a few of them had the engineers and expertise required for design and construction. Many cities hired consultants to make designs and supervise construction. Some engineers of the consulting companies later became city engineers and chiefs of public works while others designed and supervised construction works in different parts of the country (Keating 1989).

This is an early example of public-private cooperation – although not in the same sense as somewhat misleadingly claimed by several current donors. In the western parts of the country, in California in particular, individuals and private companies took care of the provision of water services to a large extent. This situation lasted considerably longer on the west coast than in other parts of the country, until cities assumed the responsibility for the provision of public water services. Yet, public water systems in the west were established primarily for irrigation – not for domestic water use (Keating 1989).

According to Melosi (2000, 74, 119–120), the water works was the first important public utility, and the first municipal service in the US, that demonstrated a city's commitment to growth. In 1830, some 20 per cent of the 45 water works were publicly owned and 80 per cent privately owned. The share of public water works increased gradually, being about 50 per cent in 1880 and 70 per cent in 1924. Melosi (1999) points out that US cities wanted to take the responsibility for public services for several reasons. They wanted to look after their own interests (so-called home rule), and to direct the cash flow from the services to the city instead of giving money to the private sector. In the case of water and solid waste management services, cities considered that the safeguarding of hygienic and healthy conditions was their responsibility, and they did not have confidence in the private sector's ability to manage these affairs.

The cities also considered the feedback from the customers regarding the quality of the services. Further, according to Melosi, the nature of the political system in the States explains the tendency to emphasize services owned by the public sector. Private sector involvement has been greater in solid waste management than in water and sewerage management.

## Experiences from the Nordic countries

In Tampere, the growing and later the leading industrial centre of Finland, the industrialist William von Nottbeck (1816–1890; Figure 18.3) offered to build a water pipe after a major city fire at the request of the municipal authorities in 1865. He proposed that a wooden pipe be constructed from Mältinranta at the head of Tammerkoski Rapids to the Central Square at a cost of 7,500 silver roubles (105,000 euros in 2004). In his second proposal, a network covering the whole town would have cost 28,000 roubles (400,000 euros). He was then asked to submit his conditions for running the water supply. These conditions (Box 18.1), consisting of ten paragraphs, actually meant that the industrialist would make a tidy amount of money and the town would take all the risks (Juuti 2001; Katko et al. 2002).

Although the implementation of the plan would have been a considerable financial risk for the town, revenue from the planned water pipe would have been only a tiny fraction of the enormously rich aristocrat's income. His dividend income alone was in six figures at that time. The town decided, however, not to accept his offer and started developing the waterworks as part of the municipal administration (Katko et al. 2002; Juuti 2001; Juuti & Katko 1998).

A rather similar proposal was made also in Sundsvall, Sweden, in 1874. The industrialist J.W. Bergström from Stockholm made an offer to build a water pipe for 250,000 rikstaler (5 million euros). The town, however, approached J.G. Richter from Gothenburg and asked him to submit a plan for both a water pipe and a sewer. In Linköping, another Swedish town, a private water system was constructed in the 1870s based on a 30-year concession. There may have been a few other similar arrangements in Swedish municipalities, but the works have for the most part been under municipal administration (Isgård 1998).

In 1866 in Helsinki, the capital of Finland, a proposal for the establishment of water works was made, originally at the request of the Senate. Yet, at that time municipal legislation made it too difficult to establish a municipal water and sewage works. Instead, tenders were requested for private concessions. Subsequently, the entrepreneur W.A. Abegg made two separate proposals to implement the approved plan, and after lengthy negotiations the town signed a concession with Abegg in 1871. He was also given a special permit to distribute water against payment. The concession was given for 75 years, but Abegg withdrew from the project and sold the concession further to the Neptun Company from Berlin in the summer of 1872. Under the direction of the engineer Robert Huber (1844–1905), the new company started constructing the water works, but because of the Europe-wide recession,

Figure 18.3 William von Nottbeck (1816-1890)

Box 18.1 von Nottbeck's terms for establishing a private water works for the city of Tampere in 1866.

> 1. Arrangements must be made to allow piping water initially from Mältinranta (upstream) or Mustalahti through the City to Kauppatori (marketplace) so that pipes of practical size can be used.
>
> 2. Permission to lay water mains in the middle of streets must be granted so that the owners of the houses along the streets can be supplied the necessary water at a reasonable price.
>
> 3. The conveying of water is to be considered the property of myself or my heirs and I shall therefore be entitled to use the water for my own needs. The City Board agrees not to allow the building of any other water supply system within the City without my permission as long as I maintain the water supply system in proper condition and gradually expand it to other parts of the City.
>
> 4. Should I decide to sell the facility, I agree to offer it first to the City Board at the offered price.
>
> 5. Should an act of God render the water supply inoperable for a period of time, I agree to repair it, but will not pay compensation for any resulting damages.
>
> 6. The City Board agrees to impose a large enough fine to prevent acts of vandalism against the system or any part of it.
>
> 7. The City Treasury shall pay me annually 200 Finnish marks for the use of the water system by the City Hall, the upper elementary school and the hospital, even if said institutions do not use it.
>
> 8. As the water system contributes to public safety and, among other things, benefits the City dwellers in the form of lower insurance premiums on their properties, I consider cheap a charge of 40 pennies for each one sixteenth (market fee) charged by the City Board, which entitles people to use water free during fairs and regular market days.
>
> 9. I agree to set as reasonable a price as possible on the water consumed in the marketplace and from fireplugs installed in certain locations.
>
> 10. In the future, should the income from the one sixteenths fall below 2,500 marks, the City Board agrees to make up the difference.

the project could not be completed within the agreed time (Herranen 2001, 18; Norrmén 1979, 7; Turpeinen 1995, 223)

In a sarcastic article dated 14.5.1875, *Uusi Suometar*, the local newspaper, put the whole Abegg adventure in its right place:

> Before we start to evaluate this contract, we must tell our readers, that Mr. Abegg has nothing to do with the whole matter anymore. When he had signed the contract, he left the country and sold his right to build the waterworks to a German company called Neptun for 150,000 marks [authors' note: 500,000 Euros]. That was indeed a good deal at the expense of the stupid Finns. As we innocently wondered about Mr. Abegg's generosity, who wanted to build a waterworks for us, Mr. Abegg explained to his fellow citizens in Berlin the real nature of the contract. As a result of his convincing story he was paid 150,000 marks for that piece of paper. We think that this minor deal will encourage Mr. Abegg also in future to let us enjoy his company here and to provide similar piped water services for other Finnish towns, too.

In spite of these, at least from the public's point of view, questionable experiences, politicians seem to have short memories. Abegg was active and successful also in other sectors in Helsinki: he was one of the owners of the Helsinki city tram monopoly. In this quite similar case, Abegg and his partner got the horse-driven tram monopoly for 30 years in 1885 – just thirteen years after he had sold the water concession. Yet, this time they did not sell the rights immediately further but tried indeed to start city traffic services by themselves. Since they did not get it started, the city of Helsinki cancelled the franchise in 1888. (http://www.nettilinja.fi/~ahellman/ratikat/raitiot.htm)

## 18.3. REINTRODUCTION OF PRIVATISATION SINCE THE 1980S

**Britain**

Until 1974, most water supply and all wastewater services were developed by local governments, with increasing support in the form of central government subsidies. During that time, there was no enforcement system to safeguard drinking water quality. In 1974, the UK Parliament decided to transfer the provision and production responsibilities for water and sewage services in England and Wales from local authorities to regional water authorities (RWAs). The boundaries of these authorities were set in accordance with the watershed areas. The RWAs were owned and managed by boards nominated jointly by national and local governments (Summerton 1998; Gustafsson 2001).

Twenty statutory private water supply companies, serving some 25 per cent of the population, were allowed to exist alongside the RWAs (Castro et al. 2003). Sewerage systems were still operated by local governments, but as agents of the RWAs. During the oil crises of 1974 and 1979, the British government used the water sector as a macroeconomic regulatory instrument. In order to control public sector borrowing requirements, and to keep water charges low for political reasons, the central government cut its financing to

RWAs heavily. As a result, investments at the beginning of the 1980s were only one third of those at the beginning of the 1970s. This meant that the government overlooked the long-term environmental protection requirements while also disregarding the fact that Britain was about to join the European Union. It had actually promised to increase sector investments, which at the time of reorganization were already barely enough to maintain the systems. As the WSS systems were very old, there was a desperate need for extra finance (Summerton 1998).

One of the key problems was that the central government did not give RWAs permission to borrow enough funds (Semple 1993). Thus, the main reason for the privatisation of water and sewerage, which happened with many other infrastructure services as well, was political and ideological. Okun, a "grand-old-man" in the water sector (1992, cited by Kubo 1994, 36), too, pointed out that an important factor leading to the privatisation was that the Thatcher Government limited the RWAs´ ability to borrow money for capital projects. Originally, the government owned the works completely. The companies were privatised by floating the majority of the shares in December 1989. The rest of the shares were sold through the stock exchange. French water companies also became owners of English and Welsh water companies. At the time of the privatisation, the decision was considered feasible by many, and as Bakker (2003, 9) notes, "few would have predicted that, within a *decade, managers of more than one water company would be proposing a return to public ownership of assets*".

The private companies reacted to the 1999 reduction of prices, and of expected profits, by cutting down investment programmes, reducing staff, and searching for alternative management models. At this stage, at least two of the ten water and sewerage companies presented plans for partial or total mutualisation and becoming not-for-profit operators (Castro et al. 2003).

This was first proposed by the Kelda Group in Yorkshire, but OFWAT, the regulator, rejected the application mainly on technical grounds. Interestingly enough, Yorkshire Water promoted its mutualisation proposal as follows: "Much of the debate over the water industry in recent years has its origins in public discontent with the concept of privatisation of this particular industry: the importance of clean water and efficient sewage disposal to the health and well-being of the community contribute to an intuitive feeling that these assets are more appropriately held in community ownership" (Yorkshire Water 2000, cited by Bakker 2003, 10).

This is quite close to what Bernard Wuolle (Figure 18.4) stated in his presentation on the roles of municipalities and public services at the first Finnish Municipal Days in Helsinki in September 1912: "Would it not, therefore, be better to reform legislation instead of encouraging a return to private enterprise even in sectors which are more naturally served by municipal utilities?" (Hukka & Katko 2005).

Another quite similar proposal for mutualisation put forward by Welsh Water (Glas Cymru) was, however, approved by OFWAT in July 2001 (OFWAT 2001, cited by Castro et al. 2003). This Welsh case represents the first serious departure from the model institutionalised in

# PRIVATISATION OF WATER SERVICES IN HISTORICAL CONTEXT ...

Figure 18.4 Bernard Wuolle, one of the leading Finnish infrastructure experts in the early 1900s, since 1922 a professor in Industrial economics. (Source: The Student Union of Helsinki University of Technology - TKY)

1989 with the full privatisation of the water industry. Privatisation and regionalisation of water services in England and Wales are quite obviously linked to the relative long-term declining role of local authorities.

Millward (2000) mentions that in the early 20th century a more regional focus for utilities became necessary – electricity grids and water basin development being the examples. This shift to a regional focus was one of the first steps in undermining the influence of local governments in the provision of utilities. One could, though, ask why technical infrastructure services were considered a whole instead of distinguishing between water resources management and water services as almost everywhere else.

While Okun (1977) considered the regional and river basin approach of England and Wales highly positive in terms of water and sewerage services, he considered that the benefits of this approach were lost through privatisation (Okun 1992, cited by Kubo 1994, 36).

## France

France has a particularly large number of small municipalities and a long tradition of private companies and municipal water and sewerage works competing for the production of WSS services. In 1990, there were about 37,000 municipalities and 14,000 independent water works (Morange 1993). Around 2000 there were some 16,000 water utilities and some 16,800 sewage utilities (Barraqué 2002), thus about a half compared to the number of municipalities.

In France, the prevalence of private companies in the water and sewerage sector can be explained largely by the great number of municipalities. In general, the municipality or a federation of municipalities has responsibility for the provision of the services. Municipalities also have the right to determine service charges independently, regardless of who the service producer is. The law allows municipalities to select their service production model (Haarmeyer 1994).

Water and sewerage works can be managed by any of the following options (Barraqué 2002, Morange 1993, Moss 2002):
(i) Ordinary municipal works,
(ii) Economically independent municipal works,
(iii) Concession,
(iv) Lease contract (affermage),
(v) Management or service contract,
(vi) Fixed management contract (gérance),
(vii) Service contract based on results (régie intéressée),
(viii) Public corporation under private status (Société d'Economie Mixte).

A municipality or a federation of municipalities can produce the services or contract them out to private companies. Contracting out is very common – 45 million French citizens, or three quarters of the population, are supplied water on the basis of management or lease contracts (Beecher 1997).

Concessions were widely used in France in the 1800s but are quite rare nowadays. On the other hand, French companies have increased the use of concession-type contracts in other countries (Barraqué 2002, Kraemer 1998a). With regard to sewerage services, some 55 per cent of the population is served by public utilities (Barraqué 2002, Moss 2002).

The relative share of the population served by private operators has also grown. The share has almost doubled during the last 40 years. In any case, with as many small municipalities in France as noted above, it is practically impossible to have WSS utilities run by individual municipalities. In principle, French municipalities have the option of returning to municipal management, though many would question how realistic it is. According to Moss (2002), French government statistics show that the concessionaire has been changed in some 15 per cent of the cases, and in one per cent operations have reverted to direct municipal management. Desmars (2003) reports that more than 500 contracts expire annually. In some 20 per cent of the cases, local councils study the possibility of reverting to public management, but only in one per cent is a new "regie" established.

## The Netherlands

The history of water and sewerage works in the Netherlands can be divided into three stages:
(i) 1854–1920, when private companies operated the majority of the works;
(ii) 1921–1975, when the majority of the works were under municipal administration;

(iii) After 1976, when the works have been owned by municipalities and provincial governments, but have been managed quite autonomously according to commercial principles (Blokland et al. 1999).

The number of water and sewerage companies peaked in 1938, after which it has constantly decreased. In 1957 a new water law was enacted, which required the establishment of larger companies. At the end of the 1990s there were about 30 companies. Some water and sewerage companies were merged with public energy utilities. Even if an energy utility were to be sold to a foreign company, water would remain public. The intention was to change these rules, but in September 1999 the Dutch government decided that water services would remain in public hands (Petrella 2001, 14).

**Germany**

In Germany, municipal enterprises have proved very successful in the late 20th century and the early 21st century, and it has been acknowledged that they have been managed according to sound business principles. About 6,000–7,000 small, mostly municipal organizations supply the bulk of the water. Germany's Federal Environment Agency (FEA) has expressed its concerns about liberalization of the water industry. It concluded that the current structure has guaranteed drinking water quality and has protected the environment.

The report stated (Reina 2001):
> These accomplishments would be jeopardized if regional influence was to be reduced and the consumer's connection to 'his' waterworks was weakened by opening the market, which would further increase the rising costs of companies.

In the 1990s, to some extent encouraged by the privatization of WSS services in England and Wales, and particularly on the basis of the so-called "Washington consensus", international donors started to support concession and operational contracts – in practice the interests of multinational companies.

## 18.4. DISCUSSION WITH RECENT EXAMPLES

Based on the literature, we can argue that this quite dramatic policy change was not based on evidence and experiences but rather on ideology as in the UK (e.g., Hukka & Katko 2000, Stiglitz 2002). Based on German experiences, Schramm (2004) points out how "the new debate on privatization of the water infrastructure is led without historical consciousness". As for bilateral donors, it would be logical for them to first look at their own development experiences instead of promoting models that they have hardly any experience of in their own countries. No wonder that the naive belief in the benefits of private concessions or long-term contracts in developing countries proved erroneous as noted, for instance, by a representative of a leading financial institution in 2003:

We and others largely overestimated what the private sector could and would do in difficult markets (Anon 2003).

Around the mid-1990s it was believed that private investments would increase substantially while the opposite has happened, as is plain to see now. Thus, it seems that after a short period of experimentation, the real constraints related to such operations have become apparent. Comparative analyses of the EU-funded PRINWASS research project – covering some 15 countries in Latin America, Africa and Europe – clearly support the need to recognize the importance of diversity and adaptation to local conditions of the institutional arrangements and policy options for WSS services (Vargas & Seppälä 2003).

Based on the US experiences, Melosi (2004) noted that cities are particularly wary of multinational companies that have no local ties but are the driving force of privatisation. Handing over control of water supplies and water delivery to the private sector is not the same as making a public good into a private one, but is rather a "fundamental erosion of local authority well beyond more traditional tensions between city and region, city and state, and the city and the federal government". See also the chapter by L. Phoenix.

Jacobsen and Tarr (1996, cited by Melosi 2004) stated that: "Although it is widely believed that today's movement toward privatization represents a major shift from public to private supply of infrastructure, history provides examples of many shifts in both directions". Urban environmental history also draws attention to the 'software' dimensions of environmental problems: "certain patterns of wasteful and inefficient resource use and pollution have developed as the result of social and cultural adaptations to historically new technologies" (Schott 2004).

Public-private partnerships are receiving increasing attention in international debates. Partnerships have their goals, and the parties should have their own specific objectives and goals. The Association of Finnish Local and Regional Authorities (1997, p. 9) defined the four key elements of partnerships:
1. The private sector has its own, for-profit-oriented objectives.
2. The public sector has its own, development-oriented goals.
3. Both parties participate in the implementation of the project at their own risk.
4. The project would not be implemented in the first place without the partnership, or effective and efficient implementation would not be possible without the partnership.

According to Rees (1984), the only proper way to assess the performance of a public organisation is to determine how well it meets set goals and objectives. This fact has generally been forgotten, when the public sector organisations have been criticised for being ineffective and inefficient. Martin (1994, 23) pointed out that the debate on privatisation has focussed on which ownership or management model is better, the private or the public one.

Indeed, instead of seeing public-private partnership (PPP) in a very narrow sense, taken merely to apply to private operators, concessions, or the like, we should remember the most commonly used type of private involvement – consultants, contractors, manufacturers, etc. selling their goods, equipment and services to public utilities based on continuous competition. Outsourcing of non-core operations to private contractors through competitive bidding is one simple example, and is really just an extension of the procurement policy to cover both delivery of services and supply of materials. In such arrangements the private sector is obtaining some 60 to 80 percent of the operating expenditure and nearly 100 percent of the expenditure on capital investment projects (Hukka & Katko 2004). Such a tradition has a long history in many countries.

Thus, a combined approach, considering both the material infrastructure as well as the related manifestations in law, administration and urban culture, is to be considered. In any case, our historical and long-term experiences show that there are, on the one hand, obvious traditions of WSS services related largely to the role of local government and subsidiarity, and, on the other hand, external pressures that seem to completely ignore local conditions.

As for transparency, the study by Transparency International shows (www.icgg.org/corruption.cpi_2004_data.html) that countries which have selected public ownership and the municipal approach for their water services seem to have the least corruption. In some cases dramatic changes have taken place in terms of ownership. Instead of any historical or other evidence of their applicability, let alone superiority, they rather seem to reflect sudden ideological changes in the societies – in the East and the West.

## 18.5. CONCLUSIONS

In historical perspective, water management, both of services and resources, is substantially a local issue. In the development of water services provision and production, the local government has played an important role in many countries, though not everywhere. This local focus is also in harmony with the subsidiarity principle.

It seems that through path dependence each of the developed water management cultures tends to continue along a "naturally taken path" while other options are not necessarily always considered. Among the most binding constraints seem to be "eternal concessions" that are still valid e.g. in some European cities. Such a "historical burden" may seriously undermine the use of available options and good governance principles such as openness and transparency. Although such approaches may provide continuity, they can be highly questionable from the viewpoint of the basic aims of modern WSS services.

Recently, the concept of public-private partnership (PPP) has been widely introduced to international water policy discussions. Unfortunately, PPP has been understood in a very narrow sense, taken merely to apply to private operators, concessions, or the like, while at least in practice ignoring the most commonly used type of private involvement – consultants, contractors, manufacturers, etc. selling their goods, equipment and services to public utilities based on continuous competition. By this definition, the concept of public-private cooperation includes all possible forms. Besides, current EU legislation on public procurement requires that such services be subjected to competition in projects exceeding certain cost limits – in the case of both public and private operators. In fact, many public utilities have bought such services, equipment and goods from the private sector since the establishment of modern systems, and even before that.

Data from one country, Finland, shows that over the last 150 years probably more than a half of the annual cash flow from utilities has gone to the private sector as payment for such services and goods. A clearly larger share has been used for investments than annual operating costs. The situation is likely highly similar in other countries with market economies, at least in principle.

As concerns private concessionaires, they played a remarkable role in the mid-1800s in establishing the systems almost everywhere. In several cases the concessions were bought back by the city before the contract expired. The full privatisation in England and Wales in 1989 was a dramatic change although a result of longer-term purposeful policy. In the 1990s, only a few concessions and a few additional operation contracts were awarded, and this seems to be the most recent trend in PPPs. Yet, the argument that the private sector would make additional investments in the sector is largely false, or at least exaggerated in the historical as well as the more recent context.

As for institutional arrangements, private concessions have sometimes become very extended, which might be the case also with the recent full privatisation of utilities. Fundamental strategic changes have been decided upon often without any evidence of their potential superiority. It is possible that in the early phases of establishing the systems, options and alternative ways were discussed and considered relatively more than later on when the established systems were expanded.

It is obvious that institutional changes are needed, but they should not be done for the mere sake of change – like the idea of reinventing private concessions or operators in a completely ahistorical context: not recognizing the earlier models, let alone the experiences gained.

Since lessons once learned are not necessarily always, if ever, remembered, it is very important to look back to history when considering key strategic choices for present and future water and sanitation services. Historically private operators have tended to use "hit and run tactics" and have left the real problem-solving largely to municipal administrators. Therefore, what is really needed in water services is public sector reform and visionary management. Certainly selling further concessions without even trying to establish and start the systems is unethical and unjustified.

The responsibility of organizing water and sewerage services of necessity cannot be left to private operators, since they tend to, and are obliged to, think only of the interests of their shareholders, not those of the public. The failures of private concessions in many cases – the inability and unwillingness to construct large and complex infrastructures and massive financing needs prevented the success of private water supply systems. Only in cases like France were the private concessions quite successful but not because of their superiority but because the local governments were not allowed to have access to cheap public funding, as was obviously the case in most other countries that time.

Unfortunately, in policy-making and, particularly, in political decision-making, the time frame is just too short for proper analysis. It is for this reason that also researchers should take institutional and management issues more seriously instead of concentrating on issues that may seem to bring "easier" scientific merits but are of less societal importance.

Proper policies should be based on sound analysis of experiences and evidence of the viability and sustainability of the selected options and strategies. In the 1870s, as in the 1990s, the adopted strategies were weakly, if at all, justified. Whatever strategies cities and municipalities select, they certainly should make proper feasibility studies to compare and analyse *various options* of ownership, operational management and financing before making major strategic selections. Besides, they should not tie their hands with options that will be difficult to get rid of.

## 18.6. REFERENCES

Alfonso J. (2001) Aguas de Valencia gana el concurso de suministro local. Cinco Dias, 3 October 2001 (Original in Italian).

Anon (2003) Return to resources for the World Bank. Water21. June, pp. 13-14, 16.

Association of Finnish Local and Regional Authorities. (1997) Public-private partnership. Helsinki.

Bakker K. (2003) An Uncooperative Commodity: Privatizing Water in England and Wales. Oxford University Press.

Barraqué B. (2002) Personal communication. LATS, Paris. 25 April, 2002.

Barraqué B. (2003) Past and future sustainability of public water policies in Europe. *Natural Resources Forum* 27(2), 200-211.

Barraqué B. (2005) Personal communication. 16 Jan, 2005.

Blokland M., Braadbaart O. & Schwartz K. (Eds.) (1999) Private business, public owners. Government shareholdings in water companies.The Ministry of Housing, Spatial Planning and the Environment.

Brown J.C. (1988). Coping with crisis? The diffusion of waterworks in late nineteenth-century German towns.*The Journal of Economic History* 48(2), 307-318.

Castro J.E., Kaika M. & Swyngedouw E. (2003) London: structural continuities and institutional change in water management. Special issue:Water for the city: policy issues and the challenge of sustainability. *European Planning Studies* 11(3), 283-298.

van Craenenbroeck W. (1998) Antwerpen op zoek naar drinkwater. 1860 – 1930.

Desmars M. (2003) Different types of management for water supply in France. University of Greenwich, 1st WaterTime Workshop. April 11, 2003.

Expansion (2001) Informe De Pricewaterhousecoopers Valencia Pagara A Avsa 54 Millones Si Rescata La Concesion. 17.1.2001. (Original in Italian)

Gaspari K.C. & Woolf A.G. (1985) Income, public works, and mortality in early twentieth –century American cities.T*he Journal of Economic History* 45(2), 355-361.

Goubert J-P. (1989) The conquest of water.The advent of health in the industrial age. Polity Press.

Gustafsson J-E. (2001) Vägen till privatisering, vattenförvaltning i England och Wales (Way to privatisation. Water management in England and Wales). Forskningrapport TRITA-AMI 3082. KTH. 80 p.

Haarmeyer D. (1994) Privatising infrastructure: options for municipal systems. Journal AWWA. Vol. 86, No. 3. pp. 42-55.

Hall D. (2003) Public services work. PSIRU. Sept 2003.

Hall D. and Lobina E. (2004) Private and public interests in water and energy. *Natural Resources Forum* 28(4), 268–277.

Hassan J. (1998) A history of water in modern England and Wales. Manchester University Press.

Herranen T. (2001). Vettä ja elämää. Helsingin vesihuollon historia 1876-2001 (Original in Finnish). Helsinki Water.

Hietala M. (1987) Services and urbanization at the turn of the century. The diffusion of innovations. Studia Historica 23, Helsinki.

Hietala M. (2002) Fears of fires. Impact of fires on towns in Finland at the beginning of the 19th century. In: Platt H. & Schott D. (Eds.). Cities and catastrophes. Coping with emergency in European history. Peter Lang, 141-161.

Hukka J. & Katko T. (2000) Privatization of water services – puzzling experiences, yet little discussion. *European Water Management* 3(3), 43-44.

Hukka J.J. & Katko T.S. (2003) Water privatisation revisited – panacea or pancake? IRC Occasional Paper Series No, 33. Delft, the Netherlands. Available: http://www.irc.nl/page/6003

Hukka J. & Katko T. (2004) "Liberalisation of water sector"- a way to market economy or to monopoly market? *Water & Wastewater International.* 19(9), 23-25.

Hukka T. & Katko T. (2005) (under final edition). Municipal or private ownership of water works? In: Cooper R., Hatcho N. & Janski L. (Eds.) History of Water: Lessons to learn. UNU Press, Japan.

Hällstrom J. (2002) Constructing a pipe-bound city. A history of water supply, sewerage and excreta removal in Norrkoping and Linköping, Sweden, 1860-1910. Doctoral dissertation. Dept. of Water and Environmental Studies, Linköping University.

Isgård E. (1998) I vattumannens tecken. Svensk va-teknik från trärör till kväverening. Olsson & Finfors. (Original in Swedish)

Juuti P. (1993) Suomen palotoimen historia. Helsinki. (Original in Finnish)

Juuti P. (2001) Kaupunki ja vesi. Doctoral dissertation. Acta Electronica Universitatis Tamperensis; 141. University of Tampere. (In Finnish, Summary in English).

Juuti P. & Katko T. S. (1998) Ernomane vesitehras – Tampereen kaupungin vesilaitos 1835–1998 (In Finnish, Summary in English and Swedish).

Juuti P. & Katko T. (2004) From a Few to All: long-term development of water and environmental services in Finland. KehräMedia Inc. http://granum.uta.fi/english/kirjanTiedot.php?tuote_id=9527

Kaivo-oja J.Y., Katko T.S. & Seppälä O.T. (2004) Seeking for Convergence between History and Futures Research. *Futures, Journal of policy, planning & futures studies* 36, 527-547.

Kaijser A. (2001) Redirecting infrasystems towards sustainability. What can we learn from history?

Kaljundi J. (1997) 660 years of water business in Tallinn, 1337-1997. Tallinna Vesi.

Katko T., Juuti P. & Hukka J. (2002) An early attempt to privatise – any lessons learnt? Research and technical note. *Water International* 27(2), 294-297.

Keating A.D. (1989) Public-private partnerships in public works: A bibliographic essay. pp. 78-108. In: APWA. 1989. Public-private partnerhips: privatization in historical perspective. *Essays in Public Works History.* No. 16.

Kraemer A. (1998a) Privatisation in the water industry. *Public Works Management & Policy* 3(2), 104-123.

Kraemer A. (1998b) Public and Private Water Management in Europe. In: Correia F. N. (ed.). Selected Issues in Water Resources Management in Europe. Vol. 2. A. A. Balkema, 319–352.

Kubo T. (1994) Reorganisation of water management in England and Wales 1945-1991. Japanese Sewage Works Association.

Ling T. (2003) Some lessons from the historical experience of industrialized countries in the development of modern water systems. National Audit Office.

Lobina E. (2003) Personal communication. University of Greenwich. 11 Sept, 2003.

London's Water Supply 1903-1953. A review of the Work for the Metropolitan Water Board. Staples Press, London.

Martin B. 1994. (Original in Finnish) Kenen etu? Yksityistäminen ja julkisen sektorin uudistaminen. Who benefits? Privatisation and public sector reform. Vastapaino. Tampere.

Melosi M.V. (1999) Personal communication. University of Houston. 28 Jan, 1999.

Melosi M.V. (2000) The Sanitary City. Urban Infrastructure in America from the Colonial Times to the Present. The John Hopkins University Press. 578 p.

Melosi M. (2004) Full circle: Public goods versus privatization of water supplies in the United States. International Summer Academy on Technology Studies – Urban Infrastructure in Transition. Graz, Austria. pp. 211-226. (http://www.ifz.tugraz.at/index_en.php/article/articlev-iew/658/1/30/)

Millward R. (2000) The political economy of urban utilities. pp. 315-349. In: Daunton M. (Ed.) The Cambridge urban history of Britain. Volume III, 1840-1950.

Molina J. (2003) Personal communication. Agbar. 16 Oct, 2003.

Morange H. (1993) Ranskan kunnallistekniset palvelut. Kuntien ja yritysten yhteistyö. The Association of Finnish Local and Regional Authorities. Helsinki. (Original in Finnish)

Moss J. (2002) Personal communication. 11 April, 2002.

Nelson M. C. & Rogers J. (1994) Cleaning up the cities: application of the first comprehensive public health law in Sweden. Scandinavian Journal of History, Vol. 19, 17-39.

Norrmén G.W. (1979). Oy Huber Ab 1879-1979. (Original in Finnish)

Nygård H. (2004a) Bara ett ringa obehag? Avfall och renhållning i de finländska städernas profylaktiska strategier, ca 1830-1930. Doctoral dissertation. Åbo Akademi University Press. (Original in Swedish)

Nygård H. (2004b) Personal communication. Pännäinen, Finland. 21 Nov, 2004.

Okun D. (1977) Regionalisation of water management. A revolution in England and Wales. Applied science publishers.

Petrella R. (2001) The water manifesto. Arguments for a world water contract.

Rees R. (1984) Public enterprise economics. London. Weidenfeld & Nicholson.

Reina P. (2001) Germany warns on water liberalisation. Water21. Magazine of the International Water Association. Feb 2001. p. 6.

Schott D. (2004) Urban environmental history:What lessons are there to be learnt? Boreal Env. Res. 9: 519–528.

Semple A. (1994) Privatised water services in England and Wales. pp. 35-38. In: Hukka J. J., Juhola P., Katko T. & Morange H. (eds.) 1994. Sound institutional strategies for water supply and sanitation services. Proceedings of the UETP-EEE short modular course. TUT, Finland. 8-9 Dec, 1993. TUT, IWEE. B 59. 112 p.

Seppälä O. (2004) Visionary management in water services: reform and development of institutional frameworks. TUT. Publ. 457.

Sharkey W.W. (1982) The theory of natural monopoly. Cambridge University Press.

da Silva A.F. (2002) Public and private management in water supply systems, Portugal (18501930). Universidade Nova de Lisboa, Faculdade de Economia (UNL).

Stiglitz J.E. (2002) Globalisation and its Discontents.W.W. Norton & Company, New York.

Summerton N. (1998)The British way in water.Water Policy. Vol. 1, no. 1., 45-65.

Turpeinen O. (1995) Kunnallistekniikkaa Suomessa keskiajalta 1990-luvulle. Finnish Municipal Engineering Association. (Original in Finnish)

Uijterlinde R.W., Janssen A.P.A.M & Figueres C.M. (Eds.) (2003) Success factors in self-financing local water management. A contribution to the Third World Water Forum in Japan 2003.

*Uusi Suometar*, 14.5.1875.

Vargas M.C. & Seppälä O.T. (2003) Cross-comparative Report on Water Sector Trends regarding Policy, Institutional and Regulatory Issues. Reflections and Findings on Five Selected Countries (D19). PRINWASS project: Barriers and Conditions for the Involvement of Private Capital and Enterprise in Water Supply and Sanitation in Latin America and Africa: Seeking Economic, Social, and Environmental Sustainability.A European Commission Fifth Framework Programme Research Project, INCO2 Research for Development.

Wuolle B. (1912) *Kuntain teknilliset liikeyritykset*. (Municipalities' technical enterprises). First Municipal Days, Helsinki. 16-17 Sept, 1912. Minutes of the meeting. Helsinki. pp. 88-103. The archive of the Association of Finnish Local and Regional Authorities. (In Finnish).

INTERNET:

www.icgg.org/corruption.cpi_2004_data.html, accessed 6.12.2005.

http://www.nettilinja.fi/~ahellman/ratikat/raitiot.htm, accessed 6.12.2005.

# 19
# CONCLUSIONS

*Petri S. Juuti, Tapio S. Katko & Heikki S. Vuorinen*

Figure 19.1 Water vendor from Dar es Salaam, Tanzania in 2005, operating when public service is not available. (Photo: Osmo Seppälä)

Sanitation in towns around Europe was one of the great achievements of the 19th century. First sanitation was based on the miasmatic theory of the origin of diseases, but during the century the role of water in the transmission of several important diseases – cholera, dysentery, typhoid fever and diarrhoeas - was realized. The final proof came when the microbes causing these diseases were discovered.

Chemical and microbiological examination of water was added to the evaluation of the quality of water based on the senses. During the 19th century, filtering of the entire water supply of a town was introduced and the systematic chlorinating of drinking water was started in the early 20th century. The discovery of microbes and the introduction of efficient ways of treating large amounts of water paved the way to an era in which the public health problems caused by polluted water seemed to belong to the history of mankind.

A "new" disease – cholera – started worldwide epidemics in the early 19th century. It was especially this disease that served as a justification for the sanitation movement around the world. The example of Gibraltar in the 19th century shows that a community with a fractured water supply might even be less vulnerable in the face of waterborne epidemics like cholera before the era of bacteriology and modern water treatment methods were introduced.

The water closet was seen as a victory for public health care, but where did the human excreta go through sewer pipes? It went directly to bodies of water. When it comes to the health of humans and the state of the environment, the WC was a major drawback. Nutrients were getting out of use in agriculture and instead discharged into nature's metabolism in the water bodies. This development is described in later chapters of this book.

The example of Gibraltar shows how the military had needs that were considered superior to those of ordinary citizens; the navy needed large stores to supply ships and the army had a large strength of active men to supply, as well as horses that required large quantities of water daily. When considered together with the evidence from the Roman Empire (Chapter 6 by Ilkka Syvänne), one may conclude that in an imperial power needing a strong military machine, water supply and sanitation for military needs was a primary concern of the authorities.

The example of Saint-Louis-du-Senegal reveals that the emerging colonial rule created the conditions favourable for the outbreak of infectious, bacterial, and parasitic diseases. The transformation of Saint-Louis from a trading station into a colonial port city posed serious problems of city planning and improvement, and exposed its residents to infections by sea routes and by land. Cholera epidemics also revealed the dimensions of inequality in Saint-Louis. The overwhelming majority of the victims of the epidemics were in the working class and urban poor residents, who had no access to fresh water during the dry season, had a poor diet, lived in unhygienic conditions in overcrowded slums, and had little formal education to even read the posters and understand the ordinances. The experience of Saint-Louis is also a reminder of how unplanned city and town areas are still a great challenge in many parts of the developing economies.

In arid lands, the pre-eminent limiting factor to development is water. In such environments, permanent settlements cannot be supported exclusively by rain-fed farming. Stable farming populations, however, can and do exist. But this requires closing the gap between the demand for water and its supply through irrigation and the implementation of water conservation techniques.

The examples of Gibraltar and Ghayl Ba Wazir in Yemen show that constructing a sustainable water supply system is not an easy task in an arid environment. Until recently, the community of Ghayl Ba Wazir flourished through adherence to centuries-old practices that were based on respect for natural resources and consideration for the needs of the community's members. The fact that 16th-century regulations were applied successfully until the late 1960s indicates that the community accepted them as an equitable solution to a potentially volatile situation. Today's Ghayl Ba Wazir faces a fundamental conflict between the interests of a traditional local community that is concerned about the sustainability of water use, and the needs of a fast-growing urban society that calls for more and more water.

The example of Tashkent in Uzbekistan shows that there are three main factors that have greatly impacted the transformation of the urban water supply system: (i) population dynamics, (ii) political changes that dictate management infrastructure, and (iii) mechanisms of operation and maintenance. In this semiarid area, the main problem is that water resources have been overused and regulations are observed neither by local authorities, industries nor by the local population. Besides, pipes have deteriorated to a large extent. Furthermore, the water-pricing policy is not operational, nor otherwise feasible.

Although rainfall was adequate in southern Africa, the difficulty was always that it was confined to the winter months. To combat the summer water shortages, some means of storing water was needed. An interesting difference is that water supply in Johannesburg has from the beginning been in private hands, whereas Cape Town, Grahamstown and Durban kept it under the municipality's control. The establishment of the Rand Water Board in 1903 formalized the situation where a private company was controlling the water supplies and was providing municipalities and mines of the area of Johannesburg with water. The city of Johannesburg, however, continued to control the distribution of water to its citizens until a few years ago, when even this was privatised. Industrialization was one of the key drivers for establishing water and sewerage services for the case cities. Although the political development has been totally different in the case cities, the main developments in water and sewerage services have many surprisingly similar episodes.

The establishment of modern water systems was largely based on private initiatives, while the role and development of municipal organizations soon became an important feature of this development. The growth of the urban infrastructure was the most dynamic element of the British economy from the 1870s to the 1930s.

In the mid-1800s, most Western nations, if not all, started to develop urban water and sewerage services through privately owned companies or private operators. Yet, in most countries the utilities were fairly soon taken over by municipalities. Only in France have private operators survived and expanded since the mid-1800s. This is largely due to the fact that in 2000 France still had some 36,000 municipalities. In such conditions, it is very difficult to imagine that all municipalities would have their own water and sewerage undertaking.

In the 1980s, full privatisation was brought back in England and Wales followed soon thereafter by promoting transnational private operators in Latin America (see Chapter 28 by Esteban Castro and Leo Heller). In any case, the ideologies of excessive private financing of water and sewerage services did not come true and by 2005 several private concessions had been withdrawn. Yet, in the majority of cases the private sector has been for long involved through selling their goods, services and equipment to public utilities based on competition.

The experience of Cape Town and Philadelphia show how different materials can be used in water supply. Wood is an easy material to make water pipes, while wooden water pipes in the course of time may leak, and eventually become rotten, unless staying continuously under water. Wooden pipes were replaced by lead pipes in Cape Town in 1799 and in Philadelphia by cast-iron pipes in the early 19$^{th}$ century.

In today's Global Village, proper wells and latrines would mean a dramatic improvement for a great number of people – from one to two and a half billion – who are lacking these services. Some 10,000 or even more people in the world die every day due to diseases like dysentery, various diarrhoeal diseases, and cholera, caused by the lack of safe water and adequate sanitation. These diseases are the biggest catastrophe in the world. Yet, since most of those who die are children and old people, whose death is considered "natural" or/and people who are more or less marginalized in their societies (e.g. refugees, poor) or outside areas that are important for the global economy, mortality due to these waterborne diseases is too often considered unavoidable.

Figure 19.2 Water services directly from the street in Dar es Salaam, Tanzania in 2005. (Photo: Osmo Seppälä)

# PART III

# MODERN URBAN INFRASTRUCTURE

In the world there is nothing more submissive and weak than water. Yet for attacking that which is hard and strong nothing can surpass it.
- Lao-Tzu

How dear to this heart are the scenes of my childhood,
When fond recollection presents them to view.
. . . .
The old oaken bucket, the iron-bound bucket,
The moss-covered bucket, which hung in the well.
- Samuel Woodworth, The Old Oaken Bucket

a) Drawing water from a traditional unprotected and open pit in Tanzania in the late 1970s. b) Sewerline construction in the mid-1950s in Vaasa, Finland. (Photo: Vaasa Water)

# 20 INTRODUCTION

*Petri S. Juuti, Tapio S. Katko & Heikki S. Vuorinen*

Figure 20.1 Famous well from Broad Street, London. Original pump is believed to have been situated outside the nearby "Sir John Snow" public House. Dr. John Snow (1813-1858) removed pum hande in time of cholera epidemic, 1848-49. (Photo: Tapio Katko Drawing: Source: The Broad Street Pump, Safe & Sound, Penguin, 1971 in English MP. Victorian Values - The Life and Times of Dr. Edwin Lankester, 1990.)

# INTRODUCTION

In the period of the late 19th and early 20th centuries, the so-called bacteriological revolution took place.It was discovered that the main health problems were caused by micro-organisms. Several diseases, such as cholera, typhoid fever and dysenteries, were found to be caused by bacteria spread via polluted water. In the early 20th century these health problems associated with water pollution seem to have been resolved in the industrialized countries when chlorination and other water treatment techniques were developed and widely taken into use.

Today there is a global shortage of potable water. While microbiological problems related to water are largely focused on the developing world, new types of biological health hazards transmitted by water are also emerging in the post-modern Western world. Anxiety about chemical and radiological environmental hazards and their impacts on human health mounted in the 1960s.

In the early 20th century, drinking water standards were developed nationally in various industrialized countries. After World War II and especially in the 1950s, the requirements for safe and potable water became urgent with the increase in travel, particularly by air. Safeguarding public health in the international community was then allocated to the World Health Organization (WHO), which took an active role in developing international standards for drinking water.

Several of the problems and challenges faced during the third urbanisation phase were to a large extent related to the increasing global population growth. Some 300 years ago, global population started to grow very fast, and with accelerating speed. Since 1700, global population numbers have grown from about 610 million to nearly ten-fold by 2000. Industrialisation on its part required urbanisation – or vice versa. This meant increasing rural migration to urban and suburban areas. Gradually agriculture also experienced the effects of modernisation through the introduction of various types of machinery. This made it possible to have excess labour to work in factories. Thus, a structural change of the whole society took place, resulting in urbanisation. By 2000 A.D., in almost every country, over one half of the population lived in urban areas. Urbanisation as such created new demands for various types of services, such as traffic, distribution of products, energy supply, water supply and sanitation as well as solid waste management.

Industrialization based on fossil energy, advances in medicine and public health, global transport and communication systems are some of the reasons behind this development. Certainly the water and sewerage services had a very important role.

The 1900s was a period of extensive population growth – on average the global population quadrupled but at the same time the urban population grew up to 13-fold (Figure 20.2). Based on the rapidly developing chemical industry, manufacturers of medicine developed a multitude of efficient chemical compounds such as contraceptive pills and antibiotics. Inventions in fuel-based motors and the mass production of cars, aeroplanes and other vehicles started to increase connections and mobility. During the century industrial production increased by 40-fold, while the use of energy tens of folds.

In the first years of the third millennium, people are still migrating from the countryside to cities and from one country or region to another. Most of them are searching for better living conditions and standards of living, although they do not always find what they had expected.

The case studies in this section of the book start with a chapter by Nyangeri & Ombongi on the history of water supply and sanitation in Kenya. The evolution of these services and the related institutional and organisational set-up is divided into four periods: (i) British East Africa Protectorate (1895–1920); (ii) Kenya Colony and Protectorate (1920–1963); (iii) Independent Kenya (1963–1980); and (iv) Independent Kenya (1980–2002) with the latest institutional changes.

The next case takes us to the southern neighbour, Tanzania. The chapter by Jan-Olof Drangert follows the history of water supply during the last hundred years in a rural area some sixty kilometres from Lake Victoria. From Tanzania we move further to Nigeria. The chapter by Olukoju provides a history of water supply since 1915, with an emphasis

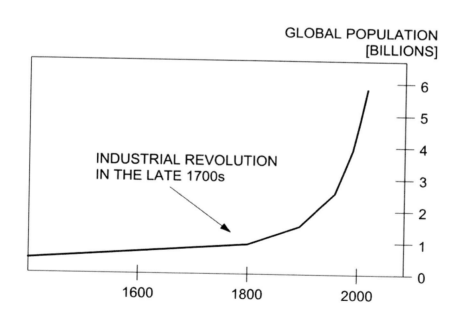

Figure 20.2 Global population growth and the effect of industrialisation, from 1500 to 2000.

on current problems of water supply in the city of Lagos, Nigeria's former capital and the leading commercial, industrial and maritime centre of West Africa. The chapter highlights the activities of water tanker operators and the producers and retailers of "sachet" or pure water, citizens' self-help efforts - such as the sinking of boreholes and wells - and the role of government in water supply and regulation of the quality of non-state supplies, especially "pure water".

In 2005 billions of people, particularly in developing economies, lived largely in agrarian-like conditions described in Chapter 24. This chapter, by Juuti, Katko, Mäki & Toivio, describes altogether six rural cases in the Finnish and South African countryside - with various types of small-scale and expanding water supply systems in the rural environment at various times.

In Chapter 25 by Juuti and Pal we move to Europe and urban history. In this chapter the history of water supply and sanitation in two industrial towns, Miskolc in Hungary and Tampere in Finland, are compared from the 19th century onwards.

From countryside and rural systems, we move back to the urban environment. The city of Riga, Latvia in the Baltic region was founded in 1201, and during the medieval period developed into an important trading centre within the Hanseatic League. In the second half of the 19th century, Riga transformed into a modern industrial centre of great importance within the Russian Empire. Latvia, having been part of the former Soviet Union for over four decades of the 20th century, regained her independence in 1991 and joined the European Union in 2004. During the transition period from the centrally planned system to a market economy, attitudes towards environmental issues, such as drinking water and wastewater treatment, have changed. In 2005, Riga was the largest city in the Baltic States. A short history of the water supply and sanitation of Riga is presented in the chapter by Springe & Juhna.

From Europe we sail to the American continent, to Mexico. The chapter by García analyses water history in an indigenous region located in central-western Mexico. The scarcity of water in the area is due to peculiar geological and topographical conditions and dates back to remote past. The traditional sources of water supply have been rainwater and small, intermittent springs and waterholes and water has central place in the indigenous religion.

Next we travel southwards to Brazil and Argentina. The central hypothesis of the chapter by Castro and Heller is that understanding the historical trajectory of water and sanitation services requires an analysis of how the development of these services is interwoven with, and often even determined by, wider processes of socio-economic, political and cultural development. The chapter by Madaleno discusses the issue of geopolitics in Chile. The author argues that the Aymara Indians survived in remote and harsh environments of Northern Chile, whereas in a few years modern water legislation and the regulation of natural resources have given way to water exhaustion and ecological depredation.

From South America we move to the Far East, to Japan. Tokyo, until 1868 called Edo, was a very small city in the beginning of the 17th century, but experienced unprecedented rapid growth after 1603. As there was not enough drinking water in Tokyo, great efforts were made to establish the Edo water system (EWS). This EWS worked well until the 19th century, when cholera prevailed and the sanitary conditions became serious. Yet, in the 20th century Tokyo dismantled the EWS and completely modernized it regardless of the obvious advantages of the old system. This development is described in the chapter by Otaki.

The chapter by Dixon describes the role of waterways and their connections to health, and as sources of traditional food resources for the Maori people of New Zealand. The waterway locations chosen for this study are the coastal mudflats of Tauranga and the inland Waikato River of the North Island of New Zealand.

Chapter 32 by Vuorinen describes the identification of numerous new waterborne health hazards in the 20th century. After the Second World War the safeguarding of public health in the international community was allocated to the World Health Organization (WHO), which took an active role in developing international standards for drinking water.

# INTRODUCTION

# 21
# HISTORY OF WATER SUPPLY AND SANITATION IN KENYA, 1895 – 2002

*Ezekiel Nyangeri Nyanchaga & Kenneth S. Ombongi*

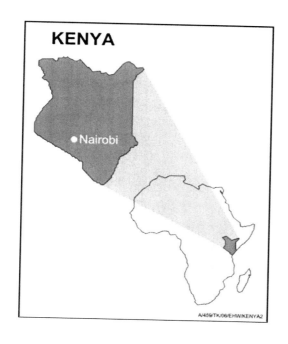

Figure 21.1 Location of Nairobi, Kenya.

## 21.1. GENERAL BACKGROUND

With an urgent agenda to find effective and sustainable solutions for problems facing the water sector in the present and future, there has been an increasing demand for historical investigations. It is interesting that today we are confronted by the same problems, perils, and opportunities that faced the water sector in yester years. Today past decisions limit the potential for future developments.

The process of water development is intricately intertwined with the history of Kenya and its people. This chapter is a historical outline of the utilisation and management of water and sanitation services. It focuses on key actors and policy shifts in a study that covers wide-ranging issues including the trajectories of water technologies, governance, operations and maintenance. The information used to compile this chapter was gleaned from documentary evidence from the Kenya National Archives (KNA), Nairobi, past issues contained in newspapers, international organisations reports and books from various libraries in and around Nairobi including the British Institute of Eastern Africa, Nairobi.

Modern Kenya can be traced back to the late 1880s when the region came under the administration of the Imperial British East African (IBEA) Company. After the advent of British rule, the country underwent a series of rapid changes that transformed it from a scarcely delimited entity of autonomous ethnic groups to a colonial territory, the British East Africa Protectorate (BEAP). In 1895, IBEA handed over the territory to the British government following the company's financial woes that constantly undermined its activities starting in 1888 when it was granted the charter (Huxley 1935).

## 21.2. KENYA AS A BRITISH EAST AFRICA PROTECTORATE (1895 – 1920)

The development of water supply in Kenya followed patterns of colonialism including urban development, the evolution of the colonial state and the exigencies of the colonial administrators. From the beginning, the general water supply administration was undertaken by the Hydraulic Branch (HB) of the Public Works Department (PWD) under the Director of Public Works (DPW). The general remit of the DPW, with regard to water, was the administration of the Water Law of the Colony and undertaking a hydro-graphic survey. The HB started its activities in the coastal town of Mombasa. The branch's services expanded correspondingly with the emergence of urban centres and the establishment of colonial posts in the hinterland. In 1902 and 1903, it opened its offices in the colonial capital, Nairobi, and in Kisumu on Lake Victoria respectively. By 1910, HB had offices in the Rift Valley towns of Naivasha and Eldoret and in the Mount Kenya region in Nyeri.

Of the several factors that provided impetus for the development of water supplies in Kenya, the Uganda railway is regarded as the pioneer. Before the start of the building of the Uganda Railway, there was no single piped water supply in Kenya. Natives derived their water supply from rivers and springs and transported it home in various types of receptacles ranging from clay pots to buckets made from hides. In dry areas people moved from place to place in search of water. Generally, no attempts were made to bring water to the people; rather it was the person who went after the water (Mats 2004).

After the IBEA was bought out by the British Government and Kenya was declared the BEAP in 1895, the next move was to find a cheaper and faster means of transportation and to improve accessibility to the mainland. Consequently, the construction of the Kenya-Uganda railway was mooted and the actual construction started in Mombasa in 1896 in order to reach Nairobi in 1899 and later to reach Port Victoria, present day Kisumu City, in 1901. (Marsh & Kingsnorth 1965) The first piped water supplies were developed and managed by the Railway to serve towns such as Mombasa, Nairobi, Nakuru, Kisumu, Eldoret and Kitale. In Nairobi, for example, the initial water supply was developed from the Kikuyu springs in 1906. It was later sold to the Nairobi Municipal Corporation in 1922 (Colony and Protectorate ... 1913-1923).

In Nakuru, the railway put in a dam on the Njoro River by 1901 to supply water to the Nakuru station and when the Public Works Department (PWD) developed water supply from the Mereroni River in 1913/15, the railway managed it for the PWD. As the railway grew stronger and expanded its operations to Eldoret in 1924, the need for water became inevitable and hence the railway developed a water supply to the station and the town in 1928 from the Sosiani River (Doria. Undated.).

Records at the Kenya Railways water supplies register shows that almost 70 percent of all the railway water supplies were developed in the first 20 years of its operations. The mode of delivery included gravitation or pumping by diesel engine or hydram from water sources including springs, boreholes and rivers. The railway water supply system was untreated. As the Public Works Department developed new township water supplies, some railway water supplies were abandoned as the railway connected to the new supplies.

**Private sector**

During 1898, Mombasa had a small water supply. It was during this time that attempts were made to licence a private water provider following the application from a Mr Smith. The existing mode of water supply prior to this requisition was by 'condensing'. The British East Africa Protectorate had a condensing plant at Mombasa for the whites while the rest of the population in the town got water from privately owned wells. The condensing plant was brought from Britain in 1895 due to a water shortage that occurred during the construction of the Uganda Railway. (British East African ... 1898; Whitehouse 1951) Subsequent reports on the question of water supply for Mombasa from 1899 show that the private developer did not succeed.

By 1914, private organizations and individuals had ventured into public water supplies. On 23rd September 1914, the Government entered into an agreement with Captain James Archibald Morrison of the Upper Nairobi Township and Estate Company Limited for the supply of water to the up-market areas of Nairobi such as Muthaiga. Consequently, the Muthaiga Water Supply Company was formed. The private Company had sought to construct its own source from the Ruaraka River but was denied. However, permission was granted to tap water from the existing Kikuyu Springs Railway water supply to Nairobi. The company operated until around 1923 when the Muthaiga water supply was taken over by the Nairobi Municipal Corporation (Colony and Protectorate ... 1913-1923).

The large-scale European farmers played a key role in private water services mostly for their own farms. For example, Lord Delamere, the earliest settler in Kenya, planned a water supply from the Rongai River to his Soysambu farm in 1914. This, however, was not put in place immediately owing to the effects of the war even though it had been surveyed. The pipeline began After World War I. Later, in the 1950s, it was the white farmers who led the colonial government into the development of the first rural water supplies that opened up vast agricultural land in mainly European-settled areas (East African Standard 1959).

The Kilifi administrative station was started in 1918 but experienced difficulties of water supply as it was not well endowed with surface water. In the 1920s, a private operator was on contract to supply the town with 720,000 gallons per year (3,273m$^3$/year). The supply depended on well water drawn manually till 1929, when pumping was introduced. One of the wells was sunk by an individual, Mr Sheikh Ali bin Salim. The water thereof was used by the public, including Government employees in the station. When pumping was introduced, a contract was awarded to an individual, Mr. Lillywhite, to pump and hence supply the township with water. (Colony and Protectorate ... 1929-1949) By the terms of Mr. Lillywhite's contract, he received £150 ($300) every year for supplying up to 720,000 gallons (3,273m$^3$) and extra remuneration if the amount exceeded that figure.

**Water Legislation**

The first water legislation was contained in Section 3 of the Crown Lands Ordinance of 1902 and only covered the issuance of water permits. In succession to this Ordinance, the Water Rules of 1903 and later 1909 were formulated. Mr. Lewis in his report on irrigation, drainage, water legislation, etc, of 1926, described both sets of rules as ultra vires on account of their not having been based on legislative authority covering them (Sikes 1926a).

The Crown Lands Ordinance of 1902 was repealed and re-enacted as the Crown Lands Ordinance of 1915. In this ordinance, two brief sections on water legislation were included in sections 75 and 145, chapter 140 of the Laws. Section 75 denied the person buying, leasing or occupying Crown Land the express right to the spring, river, lake or stream on such Land except in the case of abstraction for domestic use. Section 145 prevented the damming of a spring, river, lake or stream on Crown Land acquired in the above-mentioned ways. The section also provided for the formulation of rules on the issuance of water permits in case one desired to construct a dam. It was under this provision that the Crown Lands Water Permit Rules of 1919 were enacted, giving the Director of the Public

Works Department the power to consent or refuse to permit the abstraction of water from a spring, river, lake or stream.

The above legislation, in lieu of water governance, proved very inadequate and efforts were made towards comprehensive water legislation. The first was a solo effort by the then Director of the Public Works Department Sir McGregor Ross. In 1916, the British East Africa Protectorate requested Sir Ross to prepare draft water legislation for the colony. He faced a number of challenges, including lack of town planning; however, he produced and presented a document that later came to be known as the Draft Water Ordinance of 1916. (British East Africa … 1916) It was not drafted into a bill probably due to World War I.

## 21.3. KENYA COLONY AND PROTECTORATE (1920 – 1963)

In 1920 BEAP was changed into the Colony and Protectorate of Kenya. (Huxley 1935) Over the years HB's activities, administratively and operationally, were affected by the prevailing situation during the colonial period. For example, in the 1930s the operations of the Public works department (PWD) dwindled following the Great Depression. The department was unable to undertake most of its programs and frequent reorganization occurred. Some offices of the HB merged and others were abolished. (Kenya National Archives 1995) However, one of the most remarkable changes in the administration of water supply in the colony came during 1956, when a commission of inquiry under Sir Herbert Manzoni, a city engineer with the Birmingham Corporation, was set up to look into the re-organisation of the PWD. Manzoni recommended the transfer of the HB to the Department of Agriculture.

In the early 1960s, the 'variegated' nature of the water administration in Kenya continued just like in the decades before. At this time, three sections were involved in water supplies provision, namely: the Ministry of Works (MoW) which was to maintain the township water supplies in certain urban centres whose water supplies had not been transferred to local authorities. Second, the Water Development Department would be responsible for developing new water supplies for urban centres together with rural water supplies. It also was charged with all the technical work that was formerly conducted by the hydraulic branch under the MoW. Finally, some of the water supplies would be transferred to Local Authorities where such authorities were deemed to posses the capability to manage them. The District Agricultural Committees (DAC), with the assistance of the divisional water engineers, were responsible for proposing the water supplies to be developed within their respective areas. This arrangement later became cumbersome as too many water projects were proposed and sifting them down to a few that could be developed took a long period of time. To solve this problem, the Provincial Agricultural Water Committees (PAWCs) were formed in place of the DACs. The Central Water Board (CWB) was the paramount body that ratified all proposed water projects earmarked for development (Water Development Department 1965).

The Mombasa Pipeline Board played an important part in the field of water supply through the Act Cap 373 that came into effect in 1957. The Board was charged with the responsibility of administering the bulk supply of water to Mombasa and its environs. Similarly, it was in charge of constructing bulk delivery pipelines to other towns around Mombasa. For the day-to-day operations of its installations, the water department acted as the operating agent.

In the late 1950s and early 1960s, a number of piping schemes were installed in the "non-scheduled areas", the principal ones being Kabare, Zaina, Lolkeringet and Kibichori in the Bungoma District. All these schemes were designed and installed by the African Land Development (ALDEV) organization and handed over to the Local Authority to run. Financing for the major schemes was normally loans, but in the case of Zaina there was also an element of grants. The loans for rural water development were administered by the Agricultural Finance Corporation, which managed them on behalf of the Treasury (Water Development Department 1965).

The role played by the Ministry of Agriculture in water development cannot be overlooked. To increase the carrying capacity of the land, water was required; however, the portfolio for water provision lay with the Ministry of Works, which was limited in terms of its capacity to handle all water projects, including those in support of agriculture. This state of affairs led to the proliferation of three units within the Ministry of Agriculture, Animal Husbandry and Natural Resources, namely: African Land Development (ALDEV) and its technical staff for the design of water supply; the Soil Conservation Service Unit (SCSU); and the Dam Construction Unit (DCU). ALDEV provided the finance and at times the technical assistance through its technical unit to mainly rural water supplies in African areas. When the water ordinance was enacted in 1951, the Minister for Agriculture was given the overall mandate over water development policy. This meant that although the overall mandate over the water policy was in the hands of the Minister for Agriculture, the implementing agency was the hydraulic branch in the Ministry of Works. This caused a serious conflict. The existence of the ALDEV, SCSU, DCU as well as the hydraulic branch of the MOW was viewed as a duplication of resources. After the Manzoni report of 1957, the government viewed it as its policy to transfer the HB from MOW to the Ministry of Agriculture, Animal Husbandry and Natural Resources. The new policy was necessitated by the need to ensure harmonisation of the various water administration responsibilities shared by various government departments besides enhancing co-ordination and a clear chain of command (Colony and Protectorate ... 1957b).

Rural water supplies in mostly agricultural areas were funded under a vote called rural water grants. The funds provided under this vote were for rural schemes with an agricultural emphasis. Funds provided by the Treasury for the purpose of rural water grants were included in Ministry of Agriculture and Animal Husbandry vote and the authority supervising the expenditure was the ALDEV through its Executive Engineer. Immediately prior to the setting up of the Water Development Department, policy with regard to Rural Water Grants was controlled by the Central Agricultural Board in consultation with Regional Agricultural Boards and District Agricultural Committees. The organization initiating a request for a scheme was the District Agricultural Committee to whom proposals made by the local people came in the first instance.

Dam construction work was primarily a water augmentation exercise. Due to scarcity of water for either irrigation, watering the stock or for general human consumption, dams were constructed to curb the water shortage problem particularly during the dry season. The water development division advised on the dam construction sites.

The construction of dams can be looked at in two lights: firstly, the dams constructed by the public works as a means of augmenting water supplies to mainly townships -- a good example are the Ruiru and Machakos dams; secondly, the dams constructed by the Dam Construction Unit (DCU) working under the Soil and Water Conservation Service Unit (SCSU) of the Ministry of Agriculture, Animal Husbandry and Natural Resources (Soil conservation Engineer 1953).

The dams constructed under the DCU were mainly aimed at both controlling floodwater as well as conserving the water for agricultural purposes especially in the semi-arid areas. Most of the dams were constructed in mainly European-held areas, as most of the European farms required a whole dam for the water requirements of one farm.

The total number of hydraulic schemes accomplished by 1953 was 706 dams, 90 waterholes and 12 other works that included rams, drainage, minor irrigation etc. Thus the total number of hydraulic structures of various kinds dealt with by the SCU, including dams and waterholes built by the dam units, was 808. The dam construction activities went at a feverish pace in the 1950s and a lot was accomplished in the period 1950–1953 (Ministry of Agriculture ... 1953).

Initially when the Public Works Department had not been fully developed and the numbers of staff were few, the health officers under the Ministry of Health always supported water development. The provincial and district medical officers of health were always involved in the approval of town plans and eventually the development thereof or therein. The building of schools, hospitals, etc had to be approved by health officers on health grounds and hence the matter of water provision as central to public health could not be ignored. The water to be supplied to a township was certified by a medical officer of health before being developed. Besides the direct role played by the persons in the health department, the presence of water borne diseases accelerated the need for clean drinking water. In 1959, the UNICEF Executive Director recommended an allocation of $60,000 to the Kenya colony and protectorate for a pilot project in Environmental Sanitation during the two years 1960/61. The environmental Sanitation Programme for Kenya was a ten-year development plan divided into two five-year phases. UNICEF supplied equipment for drilling and the construction of wells, latrine boring, concrete making and survey equipment, utility vehicles for supervision and trucks for the transportation of supplies.[1]

The United Nations under UNICEF recognised the role clean water played in the prevention of communicable water-borne diseases as well as in improving the health and economy of Africans. The "Environmental Sanitation Programme" assisted by WHO/UNICEF was started in August 1960. The main objective was to develop water supplies for the smaller rural communities. In addition to promoting awareness in the community of the benefits of adequate and safe water supplies, this integrated programme was concerned with improved methods of waste disposal in schools, health centres, markets and public meeting places (Wignot 1974).

In accordance with the policy determined by the agreement of the Kenya government/WHO/UNICEF for Environmental Sanitation, the programme had the following objectives (Taylor & Silveira 1962): (a) To supply safe and adequate quantities of water for rural populations; (b) To provide excreta disposal systems for rural populations; and (c) The intensification of sanitary education and the systematization of compost preparation and use.

Nairobi, because of its relative significance as a colonial capital and rapid population growth, witnessed a fast growth of private water providers. For example, by 1957 Nairobi had 17 gazetted private water undertakers including in areas such as Spring Valley Estate, Roselyn estate, Ridge Ways Estate, Mwitu Estate, Kibigare Estate, Ruaraka Mango Farm, Gadern Estate, Roysambu estate, Karen Estate, Kitisuru Estate, Kuwinda Estate, Kisembe Estate, Mirema Estate, Tigoni Estate, Kirawa Water Co. Ltd, Kikuyu Estate, and Green Hills Estate near Limuru (Chairman's Report 1957).

**Water policy and legislation**

Until the 1940s, although water development continued mainly in townships, there was no written national plan on how to mobilize the country's water resources. The water development was in most cases ad hoc and usually undertaken at a township level and basis. Each township had its own water development plan cut out.

The first formal water development plan known as the "Dixey scheme" was put forward in 1943 and covered the water-scarce areas of the Northern Frontier District, namely Moyale, Marsabit, Wajir, Garissa, Lodwar, Isiolo, Samburu and Mandera. The report covered various aspects including the existing water supplies, the necessity for improvement, topography and geology and, most importantly, recommendations for improvement (Dixey 1950).

The second formal water development plan known as "the Swynnerton Plan" was established in 1954 under the Ministry of Agriculture by the Deputy Director, Mr. Swynnerton. It was for agricultural development mainly in African areas. But agricultural

---

[1] The Recommendations of the Executive Director for an Allocation, Kenya, Basic Maternal and Child Welfare Services, Environmental Sanitation, 13th August1959. File no BY/29/2, ref: E/ICEF/R.790.

development was synonymous with water development and the provision of water formed an integral part of the Swynnerton plan. Most parts of the Nandi District and the Coast Province benefited well from this development plan (Bolony and Protectorate ... 1954).

Mr. Sikes proceeded with the work left behind by his predecessor and being pretty knowledgeable in the matters of water legislation, he revised the previous document and formulated another draft ordinance of 1921 and that of 1922. It was at this time that he also wrote and published a book entitled Modern Water Legislation. He established that, although there were sections of Crown Land Ordinances No. 21 of 1902 and No. 2 of 1915, relating to water, no legislation providing for the adequate control of water in the interest of the community had been enacted in Kenya by May 31, 1921. The result of this lack of legislation and inadequate legislation caused waste of water resources of the colony, the construction of ill-considered projects, interference in privileges and rights, and an increasing tendency towards litigation. Disputes regarding water matters were on the increase and those charged with the responsibility of guarding the water resources in the country had a very difficult task. (British East Africa ... 1923; Sikes 1922).

The draft water legislation was, however, again never passed into a bill owing to the economic slump of 1922 occasioned by World War I and the changeover of currency from the Rupee to the Shilling. This economic recession led to the trimming of the Public Works Department staff by 50 per cent, including the post of Deputy Director of Public Works and the Hydraulic Engineer due to financial difficulties. This state of affairs in 1922 caused action to be suspended, as it became obvious that the ordinance, if enacted, could not be administered without an addition to the existing staff. The reduction of staff entirely precluded the administration of a coded law on a subject of much complexity (British East Africa ... 1923).

The issue of water legislation was held in abeyance until 1926 when a specialist on matters of irrigation, drainage, water legislation, etc from the Union Republic of South Africa, Mr. Lewis, was invited to make an investigation and give recommendations on the same. At the time of the completion of his investigations, a conference of East African Legislatures was held at Tukuyu in 1926 to discuss and see how to harmonize water law in the region (Sikes 1928).

The conference together with Mr. Lewis's report provided the required impetuous for the enactment of the water legislation. One of the recommendations in Mr. Lewis's report was that an ordinance would have to be formulated to suit the unique situation of Kenya and that a select committee be composed to oversee the process of enactment of the water legislation (Lewis 1926).

Pursuant to Lewis's recommendation, the Water Legislation Committee was formed in 1927 with the Director of Public Works, Mr. Sikes, as the Chairman. The committee took up office and by 1928 a draft bill to make provision for the employment and conservation of waters and regulate water supply, irrigation and drainage was put in the official gazette on 7$^{th}$ July, 1928 (Sikes 1928).

The process of formulating the water legislation in Kenya at that time was marred by a number of problems. Firstly the towns had not been planned; this was one of the problems encountered by Sir Ross during the preparation of the Water Ordinance of 1916. Secondly, the economic slack of 1921/22 led to trimming of the work force a delay in the formulation and implementation of the water legislation occurred. Thirdly the white settlers saw the legislation as a threat to them mainly because it was going to remove the right to possess water from the public and to vest the powers in one person, "The Governor", and hence deny them the right to water which they so desperately needed to run their expansive farmlands. (Ross 1916; Sikes 1926b; Dunn 1928)

Before the first comprehensive water legislation was acceded to, further rigorous discussions on the draft bill were held, culminating in the enactment of the Water Ordinance of 1929. Many amendments were made to the Water Ordinance of 1929 before a new water ordinance entitled the Water Ordinance of 1951, which was again re-enacted in 1952 and renamed the Water Act Cap 372 of Laws of Kenya. (Bruce 1929; Provincial Administration ... 1966-68)

## 21.4. INDEPENDENT KENYA (1963 – 1980)

In 1963, the country gained its independence from the British. Many actors who had previously been involved in water governance proceeded with their activities in water provision, while other changes came in as a result of independence and reshaped the landscape of water development. The many changes that were alongside the administration of water by various ministries and bodies. The first such major change was made by Sir Herbet Manzoni regarding the organisation of the Public Works Department. Manzoni's recommendation was implemented in 1964, during which the Water Development Department was formed under the Ministry of Agriculture, Animal Husbandry and Natural Resources (Colony and Protectorate ... 1957a). Thus, from July 1964 to 1974, the policy on rural water expenditure was made within the Ministry of Agriculture (Water Development ... 1965).

Until 1964, the Hydraulic Branch of the Ministry of Works (MoW) was responsible for water and sewerage development in urban areas. Rural water development was under ALDEV of the Ministry of Agriculture (MoA). The two organizations were amalgamated under the Ministry of Natural Resources (MoNR) in 1964 and transferred to the MoA in 1968 when the water development division was established. However, the responsibility for provincial setups of the division was divided between the Director of Water Development (DWD) and the Provincial Director of Agriculture. The distribution of authority and responsibility was very vaguely defined. The seeds of weakness in the management of water supplies were therefore sowed (WHO 1973).

Before ALDEV was merged with the Hydraulic Branch in 1964, it had carried out the investigation and design of several new major piping schemes, principally in Inoi,

Thegenge, and Kahuti in the Central Province on behalf of the Local Authorities. In late 1965, it was viewed that if the development of schemes worth, say, more than £5,000 ($10,000) was to go ahead, then the only alternative was to take the schemes out of the control of the Local Authorities, and install and run them as Central Government schemes (Water Development ... 1965).

It was suggested that the appropriate authority to manage and administer loans for piping schemes on behalf of the Central Government would be the proposed National Water Authority. The question of the collection of revenue with which to repay the loans remained a bone of contention. It was seen that a top-level policy would be made available by the Central Government on whether it would wish to recover: (i) capital costs plus interest and operating and maintenance expenses, or (ii) interest on capital plus operating and maintenance expenses (capital repayment being made out of general revenue), or alternatively (iii) operating and maintenance expenses only (Water Development ... 1965).

To shorten the communication channels, it was proposed that the new Provincial Water Advisory Committees take over from District Advisory Committees submit their proposals directly to the Director of Water Development for investigation. The composition of these Provincial Water Advisory Committees was made up of members such as Agricultural Officers and Forest Officers, as well as other members both official and unofficial, and was given technical advice by the local senior representative of the Water Department. This setup was the one in operation up to 1974, when the Ministry of Water Development was formed to take charge of all matters concerning water development (Water Development ... 1965).

In 1972, the water development division was upgraded to a department and the director became directly responsible for the provincial organizations. The water department was given the overall responsibility for water development in the country. The degree of water service for the urban population in 1972 was judged to be by world standards relatively high with regard to access to safe water supply (WHO 1972a).

Another important player at this time was the Ministry of Local Government (MoLG), which was in charge of water supplies in eight municipalities. The DWD was responsible for assisting the Ministry of Local Government in meeting all technical requirements in these urban water supplies. The rural water supplies developed then, by the WHO/UNICEF programme, were operated by County Councils, which were also under the MoLG (WHO 1973).

In 1970 a credit agreement was signed between the Governments of Kenya and Sweden for financing Rural Water Supply development. Part of the agreement was that the Kenya Government should prepare a Water Master Plan encompassing all sectors and WHO was selected as the executive agency for the study. The study was limited to community water supplies, sewerage and water pollution (WHO 1971).

The WHO co-ordinated study came up with interesting findings that informed the process of water management in the country. For example, it became clear that external funds were available to cover the major portion of the funds required for development; however, the need for local funds to cover recurrent expenditure was rapidly increasing. There was a shortage of manpower, primarily senior staff. The organizations involved in development work grew more rapidly than the Government administration systems could absorb. Thus, most problems related to water supplies grew primarily due to finance, manpower and administrative procedures. In this respect, the following were major problems that affected the rural water supplies (WHO 1971):

- Lack of coordination between Government agencies involved in the community water supply sector and administration in the water development division;
- Lack of an organization with authority on water development;
- Lack of Long-Term planning in the community water supply sector and poor design criteria;
- Lack of data on the state of water development and availability of water sources in the country;
- Shortage of qualified Kenyan staff;
- Lack of technical staff within the Ministry of Local Government and staff for the development of Lands and Settlement water schemes;
- Shortage of development funds for urban water supplies operated by WDD and general recurrent funds;
- Low efficiency in revenue collection from the rural water supplies developed by WDD;
- Lack of inter-linkage between the development plans and the selection procedure for water schemes to be developed; and
- Poor purchasing procedures.

The WHO Sectoral Report (WHO 1973) on the administration and organizational structure for water supply and sewerage development made a number of recommendations, some of which were implemented and appreciably changed the management of water supplies in the country; among these were:

- Establishment of a ministry in charge of water supplies.
- Making the Nairobi water and sewerage department a parastatal independent of the Nairobi City Council, but reporting to the Ministry of Water when created and its operations carried out on a commercial basis.
- Establishment of a marketing section by the DWD to improve the economy of urban water supplies under the water department.
- Clear indication of drastic cuts, which appeared arbitrary to the Ministry of Finance, on annual recurrent estimates, which included operation and maintenance prepared by the water department.
- Taking over private water supplies within the city of Nairobi by the water and sewerage department within Nairobi City Council as the quality of water supplied was dubious.
- Introduction of commercial accounting since the accounting systems used by the seven Local Authorities under the MOLG were inadequate.
- Handing over the urban water supplies under the DWD to LA until they can prove that they have the technical and financial resources to properly operate and maintain the supplies.
- Taking over of the County Council schemes by the DWD where the County Councils have insufficient resources to operate and maintain them.

## Water policy and legislation

Immediately after independence, Kenya produced the "Sessional Paper No. 10 on African Socialism and its application to Planning in Kenya" (Republic of Kenya 1965). The paper directed Government policy towards priority concerns for Africans, which at that time were identified as the eradication of poverty, illiteracy and diseases. These policies meant that major basic services including water were going to be provided free of charge or subsidized by the Government. Water was seen as a social service that the Government had to provide to its citizens, either free of charge or subsidized (Aseto & Okello 1997). A further interpretation of the paper was that the commercialization or even privatization of some basic services, water included, would be minimized or done away with. Therefore, effort was directed towards establishing water supply facilities so as to reach as many people as possible in order to reduce the incidence of disease. Little or no effort was directed towards the sustainability of the facilities being put up.

Aseto and Okello (1997) noted further that as a result of this policy document, the Government increased its involvement in direct economic activities by nationalizing, as well as establishing, many new state enterprises in a variety of sectors and locations. They further noted that no steps were taken to encourage the growth or strengthening of the private sector commensurately. This policy therefore dictated the water tariff revisions in the country. It can therefore be seen that between 1974 and 1979, full cost recovery as a policy had been dropped from the official government policy in the sector. The only consistency in the official policy was the financing of operation and maintenance costs. It is a well-known fact that costs of production change over time. The implication of the above government policies was the need for continuous reviews of water tariffs. But this was never espoused, and remains a problem even to date. The tariffs set between 1970 and 1981 took the subsidy implication of the Sessional paper into account.

## Rural water supplies

The first rural water supplies project (Rongai pipeline) was started in 1948 with the aim of opening up agriculture in the mainly European-settled areas. Several rural pipelines were envisaged; these included Rongai, Vissoi, Olabanaita, Westacre, Elburgon, Enarosura, Kinja and Kinangop ring main. All of these rural pipelines were in place and operating by the end of 1959. Although the above pipelines were the major ones initiated in the mainly European-settled areas (with the exception of Enarosura) also other schemes ranging in number between 50 and 60 were developed in African areas (East African Standard 1959), where the tendency was to put in a greater number of schemes of smaller dimension. Such schemes included those designed to protect water catchments by conveying piped water supplies to grazing areas sufficiently remote from important catchments to afford them protection.

The rural piping schemes were executed and operated in different ways, depending on whether they were in the old "scheduled" areas or the "non-scheduled" areas. In the "scheduled" areas, eight pipelines (Vissoi, Enarosura, Kinja, Westacre, Kinangop, Rongai,

Elburgon and Olobanaita) were financed by the Central Colonial Government and installed by the Hydraulic Branch of the Ministry of Public Works. These water projects were either in the former Naivasha Country Council or Nakuru County Council areas (i.e. in the white settler farms) (Water Development ... 1965).

Out of the eight schemes, five were controlled by Associations of Operators and three were operated by the Water Development Department as normal gazetted water supplies. However, the loan repayment position with regard to these three supplies was not the responsibility of the Water Development Department. For the other five pipelines, a loan repayment to Government was satisfactory.

According to the Water Development Department (1965), most of the rural water supplies in the old scheduled areas ran into trouble as the European settlers sold their land to African farmers and left. Thus the association of operators broke down and these schemes ran into difficulties, namely in: (a) operation and maintenance; (b) management (repayment of loans, etc); (c) sale of water under prevailing circumstances of settlement and changing ownership; and (d) physical collection of revenue.

**Ministry of Health**

From 1960 to 1972, UNICEFs provided assistance to the programme in the form of mechanical water pumps and diesel engines to power them, hydrams, hand pumps, piping and related materials such as asbestos-cement roofing sheets for rain catchment at schools. The WHO provided engineers and health inspectors as technical advisers to the programme (Wignot 1974).

Initially, UNICEF aid was being matched by contributions from the local communities in the ratio of 60 to 40 percent respectively. The latter contributed labour and materials as well as money. At the beginning of the fiscal year 1970, the central government began contributions to the programme. UNICEF assistance decreased thereafter and ceased at the end of the fiscal year 1972 with the central government and the local communities now sharing the costs (Wignot 1974).

At the time of UNICEF's withdrawal, some 561 rural water supply schemes throughout the country had been completed or were being designed and initiated. They were designed to provide a piped water supply to an estimated population of 664,000. The work was carried out in nine phases over a thirteen-year period. Authorities agreed that this had the desired effect of showing the benefits of a permanent, safe water supply in the rural areas. Numerous communities in the country started organizing committees to develop their own water supplies (Wignot 1974). They were trained to follow a standard design system, including hydraulic calculations, costs and submission methods. For reasons of standardization, all equipment used in the programme was generally of one make.

In 1974, UNICEF carried out a follow-up study on the projects it had implemented covering 197 out of 561 schemes in eleven districts and six provinces. Some of the problems encountered by the Demonstration Rural Water Supplies Programme projects are as outlined below (Wignot 1974):

- Operation and Maintenance: Many of the schemes were found not to be operational during the evaluation, owing to improper attention by the caretaker due to a lack of funds to buy spares parts, employ a permanent pump attendant and buy fuel and lubricating oil.
- Lack of transport: Transport of parts or material to make repairs was often lacking.
- Lack of standard procedure for operating and maintaining a supply: Supplies using mechanical pumps and engines were affected as the Ministry of Health had no training school for pumps and engines operator.
- Failure of water source: Along the coastal hinterlands, several dry years resulted in the drying up of wells and springs and streams and the abandonment of water supplies dependent on these sources.
- Sabotage or theft of materials: Sabotage was thought to be by those who owned hand-carts, donkey-carts, etc. and engaged in bringing water to the village from rivers, springs or other sources. The Ministry of Health supply was considered to be a direct threat to their livelihood and these people were thought to be responsible for the pipe breakage and destruction of equipment that rendered the supply inoperative. In other communities, thieves stole the brass and copper fittings from pumps, engines and hydrams for their scrap value, putting the supply out of commission.
- Poor design and construction: Some schemes had failed to deliver water due to design errors, the wrong type or size of equipment and piping being installed.
- Revenue collection: Schemes failed where the local people were not willing to support the projects by paying for a water supply, hence it could not function properly.
- Distance between schemes: This made it difficult for a maintenance man or unit to service the schemes regularly, or for an operator to pump water regularly, as in some divisions where one operator was responsible for a group of schemes.
- Human factor: It was evident from field visits that in those areas where the health officers were generally interested in and concerned with the development of good water supplies, the supplies tended to be well operated and maintained than where the health officer was not so interested.
- Breakage of hand pumps: Some types of hand pumps such as Craelius, Oasis and Simac, could not withstand hard use by the public and were very expensive to repair when they broke down. Hand pumps were installed and left to run as long as possible and when one broke down it was abandoned, or removed and never replaced, or kept in repair only as long as funds were available.

## Water legislation

The Act (Cap 372) was amended in 1962 and 1972. Under the Water Act cap 372, many organisations were involved in the urban water supply and development and management. Some were concerned with policies, others with implementation, operations and maintenance and others had multiple roles, which led to conflicts in the sector.

The WHO report (1972b) No 6 concentrated on a holistic study of the Water Act (Cap 372), and this was considered by the consultants that the change of an act required time and serious thought. The following were the principles recommended:

- A comprehensive water act should be introduced in due course, but it should not be drafted in haste;
- Meanwhile much can be done under the existing law to improve matters, particularly in the fields of sewerage disposal, abstraction and pollution control, so that amendments to the existing law could be limited to those necessary to meet changing circumstances; and
- There should be a move as quickly as possible towards a single enforcement agency.

Therefore, the main problems in the sector were blamed on institutional weaknesses resulting from a lack of a clearly defined framework apportioning roles and responsibilities. In particular, the role of MWRMD as a primary service provider and principal regulator undermined the performance of the sector.

**Ministry of Water Development**

In the early part of the 1970s, the Government recognized the crucial role played by the water sector in the general economic growth of the country. Under the new Ministry, a new plan of action was established aimed at improving efficiency and extending the water services to as many citizens as possible. This increased access to water tremendously in the rural areas.

A full-fledged Ministry in Charge of Water Development affairs was created in November 1974 a year after the recommendation by WHO. One of the Ministry's first decisions was to take over the management of not only Government-operated water schemes but also the self-help and County Council-operated schemes. The Government's decision to create such a Ministry was due to the awareness that water supply and environmental sanitation are the biggest contributors to an acceptable health situation. In effect, therefore, the government was in agreement with WHO that water was not receiving the attention it deserved under the MoA. Indeed, within its first decade of creation, major development programmes to provide improved water supplies to the people in rural areas and the improvement and extension of services in the urban areas were undertaken.

## 21.5. INDEPENDENT KENYA (1980 – 2002)

**Ministry of Water Development**

By late 1970s, development agencies had realized that the ministry could not continue with business as usual. The ministry agreed with their observation and as a result a number of studies supported by the Swedish International Development Agency (SIDA) were commissioned. The water use study of 1983 argued that the MOWD should be divested of its operation and maintenance responsibilities. Similarly the operation and maintenance study of 1983 made strong arguments in favour of decentralization. In effect, therefore, these reports called for reforms revolving around the initiation of changes in the management of schemes with a view to: enabling the MOWD gain effective control over its schemes; decentralizing management, operation and maintenance to appropriate levels for rapid and effective response to scheme-specific happenings; increasing the level of consumer participation and responsibility in the management; increasing the level of equity in the social distribution of scheme waters; and generating resources needed for operation and maintenance from the consumer.

These reports, however, warned that without these reforms the water sector would increasingly find it difficult to operate and maintain the schemes; and generate the resources required for the much needed expansion of its investments to reach the majority of the population without access.

In order to ensure the success of this new, ambitious programme, the Government took over several water supplies previously managed by the local communities, local authorities and other public and private institutions. For instance, in 1980 during the water decade, there was the ministry's slogan of "Water for All by the year 2000". This undertaking was, however, short-lived as the Government soon realized that resources for sustaining the services were not forthcoming as the economy of the country was on the decline.

On June 24th 1988, through Legal Notice No. 270, the President ordered that the National Water Conservation and Pipeline Corporation (NWCPC) be established, under the State Corporations Act. The NWCPC was to operate those water supplies placed under its care on a commercial basis. The main objectives were: to commercialise water sector operations, to achieve financial autonomy in water operations, to improve the performance and efficiency of water schemes and to reduce dependence on public funding of independent water schemes.

The first National Water Master Plan (NWMP) was developed in 1979 through Swedish assistance and remained the main guiding plan for water development until 1992, when Japan International Corporation Agency in conjunction with Kenya Government formulated the second NWMP. This was still the guiding document for water development in 2005. Up to the early 1990s, the implementation of rural water supplies was based on a supply-driven approach (SDA) to development strategy with high construction targets that left little opportunity for community involvement. The SDA strategy was found to be non-viable and thus a demand-driven approach (DDA) strategy was introduced. (Republic of Kenya 1979; Republic of Kenya 1992; Skytta et al. 2001)

Table 21.1 presents the number of water supply systems in Kenya by the year 2000 managed by various agencies. These systems supply water services to a total of about 18.6 million people, which meant average service coverage of 59 per cent of the estimated total population of 31.6 million people (Aide M'emoire 2000).

Table 21.1 Number of water supply systems in Kenya by service provider and population served.

| Service provider/producer | Number | Population served (millions) |
|---|---|---|
| Ministry of Water Resources Management and Development | 628 | 6.1 |
| National Water Conservation and Pipeline Corporation | 48 | 3.7 |
| Communities | 356 | |
| Non-governmental organizations | 266 | }4.9 |
| Local authorities | 243 | |
| Total | 1 549 | 18.6 |

The 2002-2008 National Development plan (Republic of Kenya 2002a) indicated that about 75 percent and 50 percent of the country's urban and rural population respectively have access to safe drinking water. This has been achieved through the provision of some 330 gazetted water sources countrywide, accounting for 80 percent of the served population, while the rest (20 percent) of the population is served by non-gazetted schemes. There are over 1,800 water supplies that are operational, out of which about 1,000 are public operated schemes. Non Governmental Organizations, self-help groups and communities run the rest. (Figure 21.2) Similarly, there exists about 1,782 small dams and 669 water pans in the country, of which 1,183 are operational and 1,168 silted but operational; 100 have dried up or have been abandoned. In addition, there exists about 9,000 boreholes the majority of which require rehabilitation or replacement.

The 2002-2008 National Development plan (Republic of Kenya 2002a) indicated that out of the 142-gazetted urban areas in Kenya, only 30% have sewerage systems. According to Nippon Koei (1998), the majority of sewers constructed in the 1950s, 60s and 70s have never been inspected and are in poor structural condition. Development has exceeded the available hydraulic capacity of older sewers, resulting in frequent blockages, overflows and surface flooding. This situation has posed serious environmental and health problems

Figure 21.2 Water well with hand pump in Lokubae, Turkana District. (Photo: Ezekiel Nyangeri Nyanchaga)

with main sewer systems in most urban centres suffering from constant breakage or leakage, and inadequate capacity to handle their full sewerage load. A number of factories and enterprises are known to discharge effluents through mainstream rivers and valley depressions, causing high pollution levels. Thus, effluent pollution makes rivers, streams and dam water unsafe for domestic and livestock consumption.

## Kenya Railway & Donor Communities

Although the railway developed most of its water supplies in the nascent years of colonialism, this did not deter it from extending its distribution system over the years and as late as the 1980s the railway was still providing water to both its employees and the public. It has also participated in the development of major water pipelines like Kilimanjaro (Nol-Turesch Pipeline), the Ngong - Magadi Pipeline in 1915 as well as the Mzima springs Pipeline to Mombasa. However, the railway handed over most of its water supplies in the 1980s to the Ministry of Water Development with the exception of a few, such as the Nanyuki Water Supply (Kenya Railways ... Undated). This was especially occasioned by the development of petrol/diesel driven engines, which led to the steam engines being phased out; thus the Kenya Railway found itself using little of its water system, hence the handing over.

The Swedish government through SIDA offered assistance to mainly the Eastern, Coast and Western provinces. SIDA has since 1970 continuously supported the water sector, mainly gearing its efforts towards the implementation of large piped rural water supplies. In the 1980s the cooperation focused on operation and maintenance, training of staff and the rehabilitation of existing supplies while simpler technologies were tried for the few new water supplies. SIDA, through the Kenya-Sweden Water and Sanitation Program, supported water, sanitation, health and hygiene interventions in Kwale District from 1983 to 1997. During 1983 and 1984, SIDA, through UNDP/World Bank/MoWD, supported hand-pump testing in Kwale that successfully led to the development of the AfriDev hand pump that is currently produced locally. This project aimed to develop a Village Level Operation and Maintenance (VLOM) hand pump by installing and testing 12 different kinds of hand pumps. In the 1990s, the emphasis was on community-managed water supplies and support for Ministry of Water Resources in the efforts to develop a National Water Policy (BG Associates 2001b).

The Minor Urban Water Supply Programme (MUWSP) started in 1974 and assisted by the Government of Norway (NORAD) concentrated not only on the provision of potable water, but also on the disposal of wastewater in minor urban centres located in medium to low potential areas. A total of 42 water supply projects and 8 sewerage projects together with five other minor cost categories of water laboratories, district stores, workshops and offices, sewerage training, water resource survey and a central unallocated stores fund were undertaken. The activities of NORAD stopped in Kenya in 1991 before the end of phase (1988-92) (Hifab ... 1988).

The Finnish support for water development in Western Kenya started in February 1981 as the Kenya-Finland Rural Water Supply Development Program (later called Kenya-Finland Western Water Supply Program, KFWWSP, also called "Kefinco"). It began with studies and planning and proceeded through four implementation phases up to 1997. After that, the Finnish support continued with a new project called Community Water Supply Management Program (CWSMP) from 1997 to the end of June 2003.

During the earlier phases of Finnish support up to 1993, the supply-driven approach (SDA) development strategy was applied with high construction targets that left little opportunity for community involvement. The SDA was found to be non-viable, and thus the mid-term review for Phase III recommended a demand-driven approach (DDA) which would lead to a choice of technology and/or level of service reflecting the wishes and long-term organisational, management and financial strengths of the community. Thus, the fourth phase adopted a demand-driven approach (DDA). Towards the end of the KFWWSP, there was increasing concern over the sustainability of the results, as the responsible ministry and the consumer community did not have the needed capacity to develop and maintain water resources and facilities (Nyanchaga & Owano 2001).

The community water supply management project (CWSMP) was to support the Government policy direction, and thus adopted the demand-responsive approach (DRA) that emphasised the communities' own initiative and commitment to contribute, own, improve and replicate facilities put in place.

The German Government development agency has been supporting urban water supplies right from the early seventies. It was the first to take seriously the need to improve the management of water supplies in urban areas. However, after the programme moved from MOWD to MoLG, it stopped fully involving the Ministry in charge of water affairs in its plans; this implied that the issues they advocated were not incorporated in the policies for urban water supplies by the department that was in charge of the water policy.

Some of the urban water supplies in the country that were built through the support of KFW include Eldoret, Kitale, Malindi, Kericho, Nyeri etc. Unlike the Germany Agency for Technical Corporation (GTZ); sustainability was not an issue in its loans/grants until the mid-1990s.

The Japanese International Development Agency (JICA) did not want to tie conditions to its support to Kenya generally. Therefore, some of their projects like Taveta-Lumi and the Greater Nakuru Eastern Division did not have a component of management training. Sustainability was assumed provided by either the Ministry of Water Development or NWCPC. Specifically, JICA assumed that with its annual training in the management of a small number of local technical staff in Japan, the benefits would be passed over to the water supplies. Unfortunately this was never the case.

It was only after JICA supported the preparation of the National Water Master Plan of 1992 that Japan realized that the issue of sustainability was serious and required addressing. Indeed their current supported water project in Meru is piloting a management arrangement, which puts commercial aspects first. The success of this Japanese-sponsored project may lead to another mode of management of urban water supplies that may be adopted for ministry-managed projects in the post-reform period.

The Government of the Kingdom of Belgium (BoG) implemented a pilot phase of the Water Users Support Programme (WUASP) from January 1995 to June 1999 under a bilateral programme with the Kenyan Government targeting three semi-arid districts (that is, Kajiado, Machakos and Makueni). The main phase of WUASP was formulated in 2001 as a follow-up of a pilot phase expected to address water schemes in the three districts (Muchena & Dik 2003)

The Americans through USAID and the Canadians through (CIDA) also contributed to the development of water supplies in various rural areas in Kenya. The Dutch joined in the development of rural water supplies in the Nyanza province. Each of the agencies from different countries was given a special entry point to water development, thus each of them contributed in a special way in different areas of the republic.

## Private Sector

Throughout the history of water development in Kenya, the private sector has been involved on a small scale and in an intermittent way. However, today the role of the private sector is becoming more common as improved water provision is increasingly becoming an economic good rather than a social one. The public institutions, which after independence held water more as a social good, have been unable to render effective service, hence paving the way for the private sector to inject commercial values into water supply in the country.

To carry out the commercialization of the services, the Ministry of Local Government with the support of the GTZ established the Urban Water and Sanitation Management (UWASAM) project aimed at assisting local authority in self-sustainability for their water and sanitation services through commercialization and privatization. The initiative was carried out in phases, that is, a pilot phase in July 1987 — December 1993 (phases I and II) in three municipalities. In the January 1994 — December 1996 period, or phase III, the financial management guidelines developed during the pilot phase were implemented in nine participating municipalities. It was recognized that if the financial viability was to be attained, financial autonomy from the urban councils would be required. Therefore, the water and sanitation departments were established in those nine municipalities. Although the creation of the departments created some improvements in the service provision, it soon became apparent that certain problems were inherent within the local government structure and even the creation of separate departments within that structure could not solve the problems, which included: Lack of introduction of tariff costs due to politicians quest to gain short term popularity; delays in the approval of tariffs and budgets due to long bureaucratic approval procedures; diversion of water revenue to unrelated expenditures at the expense of the water services; and difficulty in the recruitment and retention of competent staff at all levels (Urban Water ... 2001)

In April 1996, the Ministry of Local Government accepted, in principle, the need to introduce a commercial approach to the water and sanitation operations in the local authorities. Various options were considered. The most preferable was the adoption of public sector ownership and private operations and management models that were accepted for implementation on a pilot basis for four municipalities (Eldoret, Nakuru, Kitale and Nyeri). The water and sanitation companies, fully owned by the municipalities, were established under the Companies Act, Chapter 486. A corporate management team ran these companies, comprising a managing director, a commercial manager, and a technical manager, who all were accountable to the board of directors. By the end of 2002, it was only the Nyeri and Eldoret companies that were in operation whereas the Department of Water Development took up Kitale and Nakuru operations due to the poor operations of the companies. Legal changes under the Water Act of 2002 paved the way for Kericho, Kisumu and Nyahururu starting operations as commercialized water and sewerage companies.

By 2000, there were two Water Management Contracts (Malindi and Tala) in the country. In Malindi, the current management contract has been developed from a service contract, that is, an "Improvement in Billing and Revenue Collection" contract, that was funded from the Second Mombasa and Coastal Water Supply Project (financed by the World Bank) and which ran from 1995 to December 1997. In Tala, in July 1999, the Kangundo County Council resolved to privatize the provision of water services in Tala Town by entering into a 30-year contract with Romane Agencies. The company operates all the assets for the entire water system and took over an accumulated electricity bill of $12,000 (BG Associates 2002).

Runda Water Supply Company Ltd, established in 1975 as a Housing Development Estate and appointed a water undertaker the same year, is the only big private water supplier that by 2001 was supplying water to over 500 households of Runda Estate in the suburbs of Nairobi City (BG Associates 2002).

**Water policy and legislation**

A very important policy directive, which had an effect on the water sector, was the Presidential directive of April, 1981 that revised the then existing rural tariffs on account of (a) abolishing temporary metered rural tariffs, and (b) unified official rural tariffs throughout the country in place of the then existing geographically different tariffs. This directive started off the direct involvement of politicians in dictating the policy in the water sector. Indeed the bureaucrats then got scared of revising even the urban tariffs such that the legal Notice No. 94 of 1975 (Republic of Kenya 1975) for urban water schemes but set in 1977 remained unchanged for fifteen years. Thus, the sustainability of urban water supplies was affected directly. The overall long-term objective of achieving self-financing for the water sector through the policy of full cost recovery for urban schemes was thus defeated. In the same breadth the policy of recovery of the direct operation and maintenance costs for rural water supplies was also jeopardized. The Ministry of Water

Development, not only standardized the tariff throughout the country, but also almost abolished metered connections in rural areas.

Therefore, the policy of the ability to pay in different parts of the country that had been expounded in the development plan was not strictly adhered to (Legal notice No. 194, December 1981; Republic of Kenya 1981). Whereas the fifth development plan (1983-1988) reverted to the theme of the third development plan, as far as the policy on water was concerned, the legal notice no. 194 of 1981 was not withdrawn. It is interesting to note that this legal notice was not revised till 1992, in spite of even the publication of the Governments Sessional paper No.1 (Republic of Kenya 1986) that introduced cost recovery in the provision of social services.

In 1983, the District Focus for Rural Development policy was promulgated by the government whose intention was to decentralize planning and administration for multi-sectoral development to the district level. This had an initial benefit as decision-making cycles were reduced from the national to the district level, but capacity problems in financial monitoring created unforeseen bottlenecks, with endless exercises of national financial transaction reconciliation based on both headquarter- and district-based documents. In 1986, the Government launched Sessional Paper No.1 of 1986 on Economic Management for Renewed Growth. Under this new policy, the Government was to address strategies for the provision of basic services, which would accelerate economic growth and reduce inflation (Republic of Kenya 1986).

In 1992, the Ministry of Water Development released two important documents that continued to guide the sector up to the end of the decade. The first was the Delineation Study: The Water Sector in Kenya and the second, the National Water Master Plan (Japan International ... 1992). The main outcome of the delineation report was a defined and improved delineation of the roles, functions and responsibilities of the principal actors in the sector, with special focus on those roles, functions and responsibilities which best suited the MOWD and NWCPC. On the other hand, the National Water Master Plan set out long-term plans for the much-needed reforms in the management and development of the water sector. One of the most important recommendations to come from the two reports was that the Ministry should develop a water policy.

Subsequently, between 1995 and 1999, the ministry was involved in a policy development process for the sector. This was published as Sessional Paper No. 1 of 1999 under the title "National Policy on Water Resources Management and Development" (Republic of Kenya 1999). This document is the blueprint that has since then guided legal, administrative and investment reforms in the water sector. The document also proposed the necessary framework and provided a mechanism for mobilising resources to safeguard and develop the country's water sector.

The delay in publishing the document implied that the policy did not capture the emerging trends in the water sector that were fast gaining pace, such as commercialization. The policy spelt out the need to decentralize decision-making in the water sector, but failed to clearly explain where the services would be decentralized. Though the policy stated that urban water supplies would be handed over to Local Authorities with adequate capacity, no specific moves were made to build the capacity in the Local Authorities. Therefore, as proposed in the document, a new legal framework to back up the new policy directions had to be developed.

Prior to the launching of the National Water Policy, the sector had been steered through strategies that often resulted in unsustainable water utilities, and ultimately, poor service. The policy paper was specifically meant to address the following issues:

- Water resource management; water supply and sewerage development; institutional framework; and financing of the water sector.
- Propose institutional reforms that separate water resources management from water services provision.
- Outline separation of policy, WSS regulatory board and implementation functions within the sector.

Following the National Water Policy that heralded a new era for the water sector, the ministry prepared the Country Strategy on Water and Sanitation Services that aims to further develop the policy aspirations and define an implementation framework. The main problems facing water services delivery have tended to revolve around a lack of clarity with regard to the institutional framework, the unsustainability of services and inadequate financing. The strategy was therefore, developed with a focus on the specific roles of the various actors clearly defined in an institutional framework that underscores the separation of service delivery from regulation. Two key institutions to be established as part of this new framework were the Water Supply and Sanitation Services Boards and the WSS Regulatory Board, which underpins this separation and ensures fair play among various actors. The new institutional framework also underscores the need to place water and sanitation services under single utilities in view of the close linkages in operations, maintenance and commercial aspects. This is in contrast to past practice where the two utilities were managed separately. The strategy emphasizes an increased role for the Private Sector in service delivery taking into account the need to ensure that the commercial principles that drive PSP do not undermine Government's aspirations as defined in the Poverty Reduction Strategy Paper by limiting the access of the poor. The strategy emphasizes more than ever before the role of communities in service provision by refining a framework that will enhance the communities' role, and access to finance as well as technical support (Ministry of the Environment ... 2001).

Water legislation: Finally, the water policy noted that the Water Act, Cap 372, needed to be revised. This Act held until 2002 when a new act known as Water Act 2002 was enacted to accommodate the increasingly complex administration of the water sector. The Water Act 2002 was enacted with new institutions specified in the new decentralized setting. The legal language used in the act generally refers to policy-making, monitoring and coordination as the responsibilities of the Minister. Indeed within the Act, the Minister's statutory and discretionary powers are mentioned frequently touching on all areas of water resources management and water services. However, the decentralization leading to accountability and efficiency that is the cornerstone of the Act calls for a clear separation of functions within the sector (Republic of Kenya 2002b). Overall, the Act provides a sound basis for implementing water sector reforms, although there are some areas where it will need to be augmented if the policy reforms are to be effective. The three most important issues are institutional coordination, decentralization, and financing water resources management. The Act still has its weaknesses, but a bigger problem than inadequate legislation and regulations is poor enforcement and compliance. If the Act ultimately provides the enabling environment for a simple and straightforward methodology for developing regulatory and institutional procedures, then a number of the potential legal issues and hurdles will be resolved. The Act creates two statutory bodies, that is, the Water Resources Management Authority that is responsible for catchment management, water apportionment, pollution control and the enforcement of laws relating to water resources management; and the Water Services Regulatory Board (WSRB), which are the licensing body and economic regulator. Under the Act, this regulatory board is an autonomous body with responsibility for regulating water and sewerage services in the country.

The problem with the Act is that all the critical interventions in areas susceptible to monopoly power abuse appear to be totally emasculated. For instance, section 47(g) of the Act only give the regulator the power "to develop guidelines for the fixing of tariffs for the provision of water services" whereas section 73(1) permits the WSBs to regulate themselves: "a licensee shall make regulations for or with respect to conditions for the provision of water services and the tariffs applicable". It should be made clear that water and sewerage services that have public interest issues in respect to matters of safety, cost, access and availability cannot be permitted to regulate themselves.

Tariffs or any other regulation of the WSBs are subject to approval by the Regulatory Board, but there are no specific legal tests the Board can apply to accept or reject a set of tariffs or regulations. Even if the WSRB were to develop guidelines for tariff setting, the guidelines would not have the force of law and could easily be contested or rejected by licensees (WSBs) or their agents — the water service providers.

The upshot is that the Act fully exposes consumers to the rent-seeking activities of monopolists. At the very least, the Act ought to be amended such that the WSRB can apply specific criteria to make judgments on the level and reasonability of tariffs charged by Water Service Boards. The Water Services Regulatory Board has no independent source of funding. The Act only provides for the retention of revenues from license fees with the approval of the Minister and the Treasury. In this respect, decisions of the Board that displease the Minister or the Treasury can result in denial of funding.

## 21.6. HISTORY OF SANITATION IN KENYA

**Sanitation in Kenya, 1895-1920**

Traditionally, people used 'natural sanitation', open space, plantations or bush for human waste disposal. For some, the alternative was to burry in shallow holes referred to as the "Cat Method". The use of the river was also reported in some communities. Mombasa is known to have existed in the $2^{nd}$ century A.D. and to have been most prosperous in the fifteenth century as a commercial town. It was at the same level as Italy by then (Edward 1985). Although no sanitation records regarding that period exist, it is assumed that sanitation works were well organized to warrant such affluence. However, the earliest and most widely used modes of conservancy were the bucket latrine and the pit latrine. Bucket latrines consisted of single and double sanitary pails. The single bucket system was introduced by the colonial administration around 1900/04, at a time when the railway townships came up. Initially single bucket latrines were more prominent than double bucket latrines. At the same time, some crude forms of pit latrines were evident. Williams records that pit latrines were in use in the Naivasha railway station by 1907 to serve the police lines (Williams 1907).

The development of sanitation in Kenya also followed closely the development of the railway and the proliferation of townships along with stations. The building of the Uganda railway produced townships like Nairobi, Nakuru, Naivasha and Kisumu then referred to as railway stations. It is along these townships that the earliest sanitation issues in Kenya are recorded. In Nakuru, the latrines for the lower class of railway servants, the police, and the inhabitants of Indian bazaars were deep open trenches. The single bucket latrine was later upgraded to the double bucket system. Disease outbreaks like plague, typhoid, cholera, and dysentery had a pronounced effect on sanitation in the colony as early as 1902 in Nairobi.

Night soil and refuse collection were the main activities of sanitation. The night soil was either buried in trenches and covered with soil as it was done in Nairobi or just transported away from the residential areas and poured out on the open ground (Kisumu). Pit latrines were common in the reserves in areas where the geo-hydrological conditions allowed. In Kisumu, the refuse was dumped in a hole left after quarrying (Williams 1907).

Williams (1907), a consultant, designed sanitation works for Nairobi, which involved a comprehensive drainage system and water supply and was intended to run for three years from 1907 to 1909. By 1913, very few of Mr. Williams' recommendations on sanitation in Nairobi had been executed. Mr. Simpson recommended the implementation of a drainage and sewerage scheme in the provision of toilets with flushing arrangements for both Asians and Africans (Simpson 1914).

## Key players in administration of sanitation

The proliferation of towns pre-empted the need for a proper sanitary administration and management. Various departments played different roles in the installation and administration of sanitation works, the ministry of health under the public health department and the railway authority being the principal participants in the beginning. The district/provincial administration and the public works department played a secondary role (Samez Consultant 2004).

In Nairobi, for example, the control of conservancy work was vested in a municipal committee, which was formed at the railway station in 1901 while night soil removal in Naivasha was under the management of the collector. Meanwhile, in Nakuru a board managed the conservancy of all the railway houses. At the same time, in Kisumu the municipal committee constituted by an ordinance of February 13 1904 undertook the conservancy and refuse collection (Williams 1907).

From 1900 to 1913, sanitation lacked institutionalized control to deal effectively with unsanitary conditions and disease epidemics played a major role in accelerating the need for proper sanitation. In Nairobi a plague epidemic attack in 1902 and 1905 impelled commencement of coordinated sanitation works. In this respect, Dr. Moffat made a report on the disposal of night soil, refuse, water supply and drainage (Williams 1907). In 1913, cerebral-spinal fever in the East Africa protectorate caused thousands of deaths significantly affecting labour supply. This prompted a review of sanitation in the protectorate. This situation meant that a strong sanitary administration was required especially in Mombasa, and Nairobi. However, the whole colony in general was also in dire need of a proper sanitary dispensation. The lack of proper sanitation delayed the implementation of proposed projects. (Simpson 1914)

With the realisation that, by 1913, the East Africa protectorate had no sanitary administration and control and thus lacked an adequate sanitary organization or power to deal effectively with diseases and insanitary conditions, Simpson conducted a study on sanitation, proposed a board of sanitary staff that would consider and advise on all questions of town planning sewerage, drainage, water supply, the making and repealing of bylaws, and matters of policy or administration involving expenditure (Simpson 1914).

Simpson proposed further that the division of Medical Department be split into two branches, one solely concerned with sanitary administration and the prevention of disease and the other with the treatment of disease and administration of hospitals and other medical matters. He recommended two branches, one central and the other local.

The legislation on sanitation and the regulations for the preservation of public health were contained in a series of rules and ordinances published in the official gazette from time to time (Simpson 1914). The earliest ordinance relevant to sanitation was the East Africa Township ordinance enacted in 1903. This ordinance not only gave the governor the mandate to make and repeal rules that provided for the control of sanitation in various townships but also the power to approve sanitary rules made by other departments (A Supplement of ... 1903). Under the same ordinance, the setting and cancellation of fees for conservancy was made. Some of the rules made under the East Africa Township Ordinance include the Township rules 1904/05, Township sanitary rules of 1917, the Mombasa township building rules, and the Coast province eating houses rules (Acting Governor 1917).

The Township fee and conservancy rules of 1908 repealed all the rules made under the East African Townships Ordinance 1903 and gave provision for the adjustment of conservancy fees in 1919 and 1922. By 1914 the public health ordinance had not been drafted (Simpson 1914).

**Sanitation in Kenya, 1920 – 1963**

From the beginning the administration of sanitation was not vested with any particular ministry, department or organisation. It was rather fragmented into the Ministry of Health, Provincial administration and Public works departments. Through the Director of Medical Services and the Sanitary Inspector, Health Inspector, and Medical Officers of health in most districts, the public health department controlled the overall sanitation. This included coordinating, inspecting, planning, and overseeing the execution of sanitation works. However, in some instances native control was also encountered. A good example is the Digo district, which by 1928 had appointed native overseers responsible for the proper usage of latrines and the general cleanliness of the small villages (Medical Officer ... 1928).

Provincial and District administration oversaw the enforcement of sanitation rules and mobilized participation in most districts. The District Commissioner, Nyeri in 1938 (Provincial Comissioner 1941) and Kakamega (Plat 1936) in 1936, for instance, were in charge of conservancy. The drainage authority was the Public Works Department, which carried out surveys and prepared sewerage work plans.

Anderson, the director of medical services in 1950, noted that Kenya required a more definite indication of policy on the part of the government in areas like housing for all races in the urban areas and trading centres and conservancy measures with a greater degree of coordination between the departments. The medical division and its sanitation department had achieved a lot though the government had not responded accordingly. He suggested setting up of an advisory committee. The committee was to be responsible for (a) Implementation of recommendations of Vassey Report for the provision of African housing; (b); an overhaul of the existing statutory position with regard to building and overcrowding; (c) The introduction of a suitable conservancy system in urban areas and the smaller townships; and (d) The provision of staff to supervise conservancy and cleansing departments (Anderson 1950).

By 1950, a sewage section of the public works department was established. This department was responsible for the preparation of detailed designs for government buildings and institutions, carrying out preliminary surveys and reports for local authorities and townships and the preparation of detailed designs for work accepted for construction for local authorities and townships (Rhodes 1950).

In 1956, the Ministry of Local Government investigated the sewerage problems in new townships for preparation of development plans. The result was that there were very many townships which needed urgent attention. This led to a consultation of the director of medical services and the secretary for African affairs, the local government, health and housing. They concluded that the sewerage problem could be remedied by a colony-wide approach rather than dealing with townships. Despite the immediate and urgent need to address sanitation issues in these townships, the financial situation could not allow for all of them to be dealt with at once. Henceforth, the schemes were prioritized according to the significance of individual towns (Jones 1956). Priorities were allotted for the provision of township sewerage as follows: priority A (urgent townships), B under investigation and C was covering proposed townships. The bottlenecks for the provision of sewerage were funds, services of the Drainage Engineer, provision of an adequate water supply and township development plans (Crichton 1956a).

## Institutional Conflicts

As stated earlier, there was no particular institution mandated with the overall sanitary administration but there was a rather fragmented approach. This state of affairs created conflicts as exemplified by recorded incidences presented below:

- In Nanyuki, the ministry of health and housing, the ministry of works, the ministry of local government and the treasury did not coordinate over the mode of sanitation to adopt, in 1962: whether to continue with a bucket system or to embrace the more modern but expensive water-borne sanitation (Hall 1962).
- A major problem affecting the smooth development of sanitation in Nyeri was a disagreement of principles between the Medical authorities and the Council Engineer over sewer construction in 1960 (Davies 1934). In 1967, the ministry of health and the ministry of housing could not agree over the construction of temporary houses in Karatina (Lloyd 1968). In 1980, another difference arose between the Ministry of Works and the Nyeri Urban District Council over the construction of the proposed municipal sewerage system (District Puplic ... 1980).

## Legislation on sanitation

The first comprehensive statutes touching on sanitation were enacted in the 1920s (Sikes 1923). The more relevant of these pieces of legislation was the Public Health Ordinance (Cap 130), which was crafted in 1921 (Sikes 1922). The Public Health Ordinance gave the Ministry of Health the power to administer public sanitation. This ordinance provided for the setting of the Public Health (Port Health) Regulations of 1923, which were amended in 1937. Under the amended Ordinance, the Public Health (Drainage and Sanitation) rules of 1950 were completed[2]. Later on in the 1950s as more and more townships gained the capacity to run Municipal councils, the Local government Act was enacted and it became apparent that some responsibility as regards water supply and sanitation would have to be passed on to the Local Authorities[3].

Several rules regarding sewerage and drainage were made and others amended[3]. The rules were meant to apply to all townships, trading centres and factories throughout the colony. They addressed the following issues: (a) Specifications and detailed requirements for the construction of subsoil, storm, and surface water drainage; (b) General matters affecting sewers and drains, protection of public sewers, connection with public sewers, drainage works in public places and through private land, and plan execution and testing of drainage works; (c) Specifications and detailed requirements for the construction of foul water drains and the erection of fittings etc; and (d) Special provisions relating to latrine accommodation. The Chiefs Act enacted during the colonial period, also had a certain significance for sanitation (BG Associates 2002).

With the upcoming of Nairobi and the formulation of a Municipal council, there was born another ordinance called the Municipal Corporations Ordinance of 1922. This ordinance empowered the Municipal council to make bylaws that controlled the management provision of water and sewerage services. The ordinance empowered the municipal council to take over conservancy of Nairobi Township. The main areas concerning sanitation covered under this law included dangerous and offensive trade, conservancy, refuse receptacles and the prevention of insanitary premises, overcrowding and the provision of latrines (Town Clerk 1923).

Later on as more municipalities emerged, the need for an ordinance to govern the affairs of the local authorities was inevitable and henceforth the Local Government (municipalities) Ordinance of 1928 was enacted. Under this law, the Townships' refuse receptacles and refuse removal together with Township conservancy services and conservancy fees of 1949 were established (Louis 1950).

There were townships which did not have municipal councils and thus were catered for under the Townships ordinance of 1930 (Marchant 1932), which provided for the establishment of rules on refuse receptacles, conservancy fees, as well as slaughterhouse sales and conveyance of meat for such townships. An example of a township which established rules under this ordinance is Kisii (By Command of ... 1948a).

There were also well-established trading centers that had not been given the status of townships and in such cases the trading centres ordinance of 1932 applied. This ordinance allowed for the Trading Centres Dustbin Rules as well as the Trading Centers Conservancy Rules of 1948 (By Command of ... 1948b).

---

[2] The Amendment of the Public Health Ordinance (Cap 145, Revised Edition, 1935) and The Public Health Amendment Ordinance, 1937 into the Public Health [Drainage and Sanitation] Rules, 1950. Kenya National Archives, BY/29/44

[3] Adapted form the public health ordinance (Cap. 145, revised edition, 1935) and the public health amendment ordinance, 1937. Rules. The public health (drainage & sanitation) rules 1950. File no: Kenya National archives BY/29/44, 1950-1958.

Meanwhile the industrial development continued to grow and the need for better sanitation for industrial establishments as well as the populations such set-ups attracted became manifest. Thus it became necessary to formulate the factories ordinance of 1947, which provided for control of sanitary matters within factory settings and in the estates established for factory workers (A Draft of the Factories … 1947).

Later it was realized that some trading centers had acquired the status of County Councils and thus the Local Government (County Councils) ordinance of 1952 was enacted to provide for sanitation in the county councils (Tope 1960).

Although the greatest need for sanitary management was felt in the Trading centres, townships and factory estates, there were some African areas especially around major townships that required proper sanitation. This situation prompted the enactment of the African district council ordinance of 1950 (Africa District … 1950), which was an offshoot of the Native Councils Ordinance enacted in the early years to provide for the formation of the Local Native Councils and the African District Councils. The African district council Ordinance of 1950, on the other hand, recognized that Africans could play a role in the betterment of sanitation in African areas. A good example is the Digo District (District Comissioner 1929) and the Nandi District councils. For the Nandi District Council there was even the formulation of bylaws on the cleanliness of villages adopted in 1953 and bylaws on the construction and maintenance of latrines, adopted in 1957 (President … 1953).

## Influence of World War II and the emergency period on sanitation

World War II and the emergency period in Kenya in the 1950s had both a negative and positive impact on sanitation. For instance, a high influx of population in Nairobi during and after World War II put pressure on sanitary facilities, prompting the need for improvement. However, World War II also severely affected the supply of sanitary materials, especially buckets and labour, thus leading to the deterioration of sanitation. The acute scarcity of materials required for bucket-making in Kenya hampered the changeover from a single-bucket system to a double-bucket system. The importation of steel was prohibited and local materials had to be used for the manufacture of buckets (Ling 1947).

The effects of World War II caused inflation, a shortage of materials for construction of sanitary apparatus and labour. The cost of one bucket increased fourfold. Locally made buckets were unsatisfactory since they had a short life and thus were only purchased as a supplementary means during the war as the export of steel during this time was prohibited. (Ling 1947)

Major sanitary issues during the emergency period reached a peak in 1954 when the demolition of markets and shopping centers significantly affected sanitation. The impact of the emergency was felt to varying degrees in different places. In Nanyuki, the emergency had resulted in a marked deterioration of sanitation. The beforehand clean police stations turned out to be shameful sites with poor sanitation and without a shed or tent for dead

Mau Mau bodies. The number of personnel at the police station tremendously increased while the sanitary facilities remained the same used at the beginning of the emergency. Nuisances increased at the stations from the overcrowding of the detention camps, inadequate latrines and drainage arrangements (Jones 1954).

The emergency due to political disturbance in Fort Hall had an adverse effect on sanitation. This was so because it was always difficult to convince the local African that the teachings of a health officer were aimed at improving his health and his home conditions. This fact was more noticeable in the Kikuyu tribe who were eager to educate themselves in academic subjects but were very difficult to convince to depart from their normal way of life, even if to their own advantage. It was expected to be worst in the Fort Hall district, which was described as the "worst of the three Kikuyu Reserves". However, rather than being an impediment to sanitary progress, the emergency led to a sudden boost and health work was by March 1953 at a higher level than any time before. Progress was achieved without additional pressure from the office since the beginning of the emergency. This was attributed to the fear of refusing to cooperate with those trying to improve existing conditions or the local people's appreciation of the advantages of cooperation (Health Inspector ... 1953).

Under the village construction programme introduced in 1954, the village supervisors assumed a more important role and it was possible to observe stricter supervision. Africans were concentrated in the detention camps. As a security measure, most native markets remained closed for harbouring subversive Mau Mau[4] elements. Houses and shops were demolished, consequently affecting the established sanitation works. The few shops that were open disregarded sanitation. A shortage of labour for latrine construction arose as most of the able-bodied men were serving in the Kikuyu Guard. This led to the installation of more unsanitary communal latrines (Tord 1954).

**Latrine campaigns**

The indigenous population was unwilling to espouse sanitary measures imposed by the colonial government. In various reserve areas the majority did not appreciate the need for latrines or other sanitary work (Medical Officer ... 1927). Latrine campaign or propaganda was the term use to describe public health education to convince the indigenous population to embrace the exotic sanitary practice i.e. the use of pit latrines or bucket latrine system. The campaign also involved transmission of general information on sanitation (Colony and Protectorate ... 1928 to 1938).

---

[4] The major organized group fighting for freedom in Kenya.

The period between 1929 and 1939 was one of the most vibrant periods in public health education. Latrine campaigns were carried out through out the colony. By 1929 low quality but incessant campaigns and health exhibitions saw the sanitation situation remarkably improved (Colony and Protectorate ... 1928 to 1938). Although the campaigns were largely effective, other areas remained adamant in the use of traditional methods, e.g., in Machakos (Karanja 1949). Karanja deemed this a sanitation problem since people indiscriminately disposed off the refuse and most people who came from the rural areas did not use the latrines; instead they used every bush within the township. Majority regarded latrine campaigns as rather officious interference with the liberty of the subject (Medical Officer ... 1934). Another incidence happened where, despite the appointment of native overseers to look after and educate the natives on matters of sanitation in the six large coastal locations of the Digo district, the pit latrine was a big failure. People did not adopt the campaign at all. This appeared as a deliberate barrier to stop the campaign (Karanja 1949).

The latrine campaign was a major strategy that led to assumption of sanitary work in general and pit/bucket latrines in particular. However, the degree of effectiveness of the campaign, determined by the time span, varied by region and tribe. For instance, the latrine campaign in Muranga, Kerugoya/Embu was more positively responded to (Donovan 1935) than in coast province districts and Machakos. In Digo district by 1935, North Kavirondo, the Nandi reserves and the Kakamega majority took an unnecessarily long time to appreciate the campaign (Medical Officer ... 1927).

**Trends in the use of pit and bucket latrines**

The pit latrine and bucket latrine developed hand in hand, with pit latrines being prominent in native reserve areas and bucket latrines common in townships/urban areas and in labour camps. Initially, pit latrines were considered of inferior quality. However, as time went by the adoption depended upon the suitability of the area concerned.

Double bucket latrines consisted of two buckets; one bucket for replacing the other with contents in the process of disposal. This mode was considered more convenient and effective and was widely embraced in the 1930s. The use of bucket latrines dominated conservancy works in the 1920s and 1930s despite the development of townships, labour camps, schools, etc. The use of pit latrines in the native reserves and quarters went hand in hand. In most instances, bucket latrines were converted to waterborne sanitation. A wholesale conversion of all buckets in the government buildings was proposed with a conviction to eradicate the sanitary inconveniences and particularly amoebic dysentery (Pedraza 1936).

The nuisance created by the bucket system prompted the idea of waterborne sanitation (Sikes 1928). The Provincial Medical Officer, Nyanza (Provincial Medical Officer ... 1952), enumerated the limitations of bucket-type latrines and pit latrines. He observed that the bucket system failed because of the following reasons: They were quite burdensome if dwellings were scattered over a large area; also heavy rainfall, bad roads, and steep hills rendered night movement difficult and the use of motor vehicles impossible. There was an acute shortage of conservancy staff, especially sweepers; and very many latrines were needed to serve a small population; when overused it was unhygienic and least desirable.

According to Blomfield (1947), bucket latrines were not only expensive but the system was becoming obsolete after a few years. Russell (1957) in Mwea/Tebere observed that the bucket system needed constant supervision throughout the collection and disposal process. On the other hand, the pit latrine system was ineffective because of the following reasons (Provincial Medical Officer ... 1952). Pit latrines were favourable fly bleeding zones, hence they were a source of fly-borne diseases. If subsoil water was near the surface, the latrines became offensive and were dangerous, as they could collapse. They caused water pollution if built near aqueduct mains and pipes in the event that the soil became fouled. They required constant emptying or closure, the frequency of which was determined by the number of users and depth. This brought a problem in finding alternative sites since more than half the plot was built over. In addition, pit latrines were made as temporary measures, since the conservancy system was awaited.

However, not all townships adopted waterborne sanitation as it was complex to implement and some of the major townships such as Kisumu, Kisii (Davies 1946), Kakamega (Colchester 1947), Nakuru, Naivasha, and Laikipia (Medial officer of Health 1940) and Machakos (Provincial Commissioner ... 1947) did not adopt it at that time.

The years 1950 to 1953 were the peak of investigations into sanitation works. The Public Works Department carried out an intensive survey in the Central and North Rift Valley, Nyanza, as well as in the Central South (Machakos), Central North (Meru) and Coast areas. The inspection entailed personal visits to schools, hospital, townships, police posts etc. where public sanitation was the main concern. They investigated sewerage systems, proposed drainage systems under construction, septic tanks and pit latrines and came up with recommendations on the cornerstone of the sanitation work in the colony[5].

By 1954, various parts of the colony had adopted different modes of sanitation; almost all the native reserves used pit latrines (Jones 1954). Most townships had both single and double pit latrines and bucket type latrines. Waterborne sanitation dominated European quarters in the major townships (Member for Health ... 1954).

Construction of township sewage treatment works started around 1956 after the formulation of the priority program. Between 1956 and 1962 less than 10 sewage treatment works were constructed. This marked the period when waterborne sanitation was seriously considered, mainly in major townships. Pit latrines, septic tanks and cesspools remained

---

[5] General summary for the file No BY/29/44, 1950 to 1958. Kenya National Archives.

the main mode of sanitation in the less important towns and in the rural areas (Crichton 1956b).

**Challenges to Progress of Sanitation**

In Kitale, Mombasa and Kisumu, for instance, conservancy staff was very inadequate for the rapidly growing township. The state of affairs was entirely due to a lack of funds as there was no financial provision made for the corresponding increase in conservancy requirements. In Kisumu, for instance, considerable funds to put the drains in order were needed. In Mombasa the problem of allowances was apparent; the Medical Officer of Health expressed the importance of increasing the monthly allowance for the administration staff (Medical Officer … 1926).

Construction of pit latrines in some districts such as Digo in 1938 was a laborious exercise as the ground quickly became waterlogged during heavy rain and the usual pit latrine collapsed (Wright 1938). Other places like Voi had underlying rock and it was impossible to dig a pit latrine. Public pit latrines at Voi were blasted in solid rock underneath the surface soil and lasted for about six months (Wundanyi 1958).

Lack of township plans or improper planning, the poor siting of Nairobi in the low lands and a lack of any complete and accurate maps hampered the initial sanitation works, necessitating extensive and costly surveys and causing a permanent drainage problem. Most townships had no plans and others had poorly made plans; this led to a delay in executing sanitation works or produced overly poor systems, as was the case of Tala market in Machakos (Todd 1949).

Africans were unwilling to learn and adopt the sanitation measures advised. For instance, in the Digo district by 1935, North Kavirondo, Nandi reserves and Kakamega, the majority did not appreciate the need for latrines or cooking meat despite rigorous campaigns (Medical Officer of Health 1934). Sanitation works were hampered by a lack of or inadequate labour due to cultural orientations whereby it was against the African customs to clean or sweep particularly in Kisumu.

Poor management, maintenance and repair of sanitary works such as septic tanks at Kabaa mission and Voi native hospital discharged a considerable quantity of solids due to the bad design of the tank outlet and the choking of radial armed soakaways (Squires 1950).

Due to the presence of the white ant, which ate away the bearer timber on which cement tops rested, eventually the latrines[6] collapsed. Such a latrine was a danger to the users' life.

Wilful damages or theft of the parts of the installed machinery at different points was witnessed during the implementation of the environmental sanitation scheme in 1962. Deliberate damaging of sewers through filling the manholes with stones by people grazing cattle along the sewer line in 1981 was a major problem. This called for more supervision of the sewers[7].

## Sanitation in Kenya, 1963 – 1980

This period was marked by the entrenchment of the Environmental Sanitation Programme and significant financial and technical assistance from bilateral and multilateral sources, such as SIDA, UNICEF and WHO, the Federal Republic of Germany and Norway through NORAD grants, UNDP, IBRD, and SWECO (WHO 1975). The principal objectives of environmental sanitation included the provision or assisting in the provision of proper excreta disposal methods, demonstration of the value of such improvements and helping the population understand the part such improvements play in disease transmission and prevention with the help of properly conducted health education campaigns. In addition, the programme sought to improve the training of health personnel by the provision of equipment and material to the medical training center for sanitation studies and intended to create enthusiasm amongst the local district councils, local communities and health workers for a greater effort to improve sanitation, among other objectives (The Goverment of Kenya ... 1964).

Excreta disposal and composting patterns varied from district to district; in all cases it was proposed to provide borehole latrines. Excreta disposal called for making concrete slabs, etc and assistance was needed to provide concrete mixtures, reinforcing mesh and latrine boring equipment. Sanitary inspectors introduced the making of compost experimentally, which was proposed to be systematized, and the compost was put to use in the first instance in health centre garden plots. Health education was carried out since the need for public understanding of health issues was fundamental. A great deal of manual work was achieved through the voluntary labour of the communities and families concerned, which involved directly the community development movement. Thus education was a necessary part of the campaign. This was proposed to be realized through health care centre staff, the use of posters, information and other display materials (United Nations ... 1959).

Environmental sanitation was the turning point of sanitation throughout the country. The construction of latrines, both pit and waterborne, building demonstration houses, health centres, slaughter slabs, successful health education campaigns, the provision of drainage and sewage disposal systems and the provision of water supplies verified the success of the programme (Gichaaga 1967).

In December 1970, WHO entered into an agreement with the government of Kenya to assist in developing a national programme for community water supply, sewerage and water

---

[6] Choo is a Swahili word meaning latrine, the meaning associated with this text is pit latrine.

[7] Berger writing on behalf of the director for water development department to the town clerk on 26th January 1981 to inform him of the deliberate damage of sewers. File No. By/29/58, Ref WD/2/6/34-SM.

pollution control financed partially as a part of a development credit agreement between the Kenya and Sweden governments through SIDA and partially through a Swedish grant. Phase I of this project commenced and reports were issued by WHO to assist the Kenya government in the implementation of a water supply and sewerage programme in the country (WHO 1975).

Further assistance in the development of a sewerage programme continued under phase II of the project by the provision of a WHO Sanitary Engineer who had been instrumental in the establishment of a sewerage division within the water department of the then new Ministry of Water Development. As a result of consultation between the Kenya government, SIDA and WHO, it was agreed that an evaluation and updating of the sectoral study should be carried out by the WHO mission (WHO 1975).

Between 1958 and 1973, little or no attention was given to sewerage development in the country outside of Nairobi. A Sewerage Division was established in the Water Department in 1973. There were 37 urban sewerage projects at various stages under the technical control of the Sewerage Division. A water pollution control section was established in 1972. The section made considerable progress in establishing systems for the preservation of quality of the country's water resources. Between 1971 and mid-1974, 64 rural water projects serving about 350,000 people were commissioned (WHO 1975).

A water pollution and control section was established within the water resources branch of the water department in 1972 and by 1975 was responsible for the preservation of the quality of the country's water resources. The water quality monitoring network was started and finances for additional laboratory equipment at the water department was provided. Prior to 1973, the standards of water and sewerage projects varied considerably because there were no national design criteria. The sectoral study presented design criteria for water and sewerage.

The selection of water supply and sewerage projects was carried out on an unplanned basis, thence adoption of recommendations of the "sectoral study" set out a more rational procedure for the setting of priorities in the selection of water and sewerage projects.

## Sanitation in Kenya, 1980 – 2002

One of the objectives of the government in developing health services as stated in the development plan was the promotion of general good health through the prevention and control of diseases. The health strategy for 1981 emphasized preventive and promotive programmes aimed at assisting the population to take action to improve good health and reduce the incidence and spread of diseases. These included[8]: the practice of general good hygiene and the observance of sanitary habits; water protection; control of pests and disease vectors; provision and maintenance of healthy living.

By 1981, the responsibility of individual members of the public in respect of sanitary practices rested under various sections of the Public Health Act. However, the law did not deal with the promotion of positive preventive programmes but infringement and penal aspects. A national sanitation council was established in 1980 (Nganga 1980) to influence the community to adopt a positive attitude towards community health, formulate the national sanitation code for observance by all Kenyans and advise and guide the local authorities. In addition it was supposed to prepare and implement a national environmental sanitation action programme and expand and improve community-based environmental health programmes[8]. Its main part was carried out through latrine campaigns.

The National Sanitation council seems to have fizzled out without achieving its mandate. It is not clear what milestones were achieved by the Council. However, currently the main responsibility for providing adequate sanitation lies with the Ministry of Health as mandated by Law. There are also other actors including government Ministries, NGOs and agencies such as UNICEF, AMREF, World Bank WSP, SIDA etc. The Chief Public Health Officer manages sanitation services at MoH under the Department of Preventive and Promotive Health services, which is divided into Environmental Sanitation, Food Control, Water Quality Monitoring, Approval of Buildings, Law Enforcement for the Food, Drug and Public Health Act, Nuisance Control, Port Health, Malaria and Vector Control and Rural Health development.

The current MoH organization has its origin in the Kenya Health Policy Framework of 1994 and the National Health Sector Strategic Plan 1999–2004. These two policy documents laid down the development and management of the health services in the country on a long-term basis. One of the plan's key policy objectives is to expand safe

Table 21.2 The common sanitation technologies in Kenya

| Sanitation facilities | Population served | |
| --- | --- | --- |
| | Number (millions) | Per cent |
| Conventional sewerage | 1.4 | 6.0 |
| Septic tank | 0.9 | 4.0 |
| Aqua Privy | * | 0.1 |
| Bucket Latrine * | * | 0.1 |
| Pit Latrine | 11.3 | 49.4 |
| No sanitation | 9.2 | 40.4 |

*Less than 0.1 million.

---

[8] Nganga, the PS ministry of health draft of 'the establishment of national sanitation council' to all ministers on 19th December 1980. File no. By/29/58.

Figure 21.3 VIP latrine in Lokubae, Turkana District. (Photo: Ezekiel Nyangeri Nyanchaga)

water supply and sanitation coverage. The intended reforms in improvement of sanitation are to be achieved by (BG Associates 2001a): development of a national sanitation policy; capacity building for rural communities; enforcement of the provision of refuse disposal systems in urban areas; enforcement of sanitation laws once created; and encouragement of the private sector and NGO participation in the provision of water and sanitation services.

The National Sanitation Policy is being developed by MoH in conjunction with UNICEF, WHO and the World Bank. The new policy is expected to lay down a strategy and guidelines to be followed in the management of the environmental sanitation sector.

## Trends in sanitation technology since 1980

The greater part of the Kenyan population lives in the rural areas and in the absence of piped water supplies, disposal of human waste (excreta) has been primarily by means of pit latrines. A survey conducted in 1983 and reported in the sanitation field manual for Kenya (Republic of Kenya 1983) showed the number of Kenyans and the percentage served by the most common sanitation technologies to be as shown in Table 21.2.

By 1983 the Kenyan situation proved that although the aqua privy and bucket systems had been common in the early 1900s through to the 1930s, they had become inappropriate, were only used in minimal numbers and were obsolete as well as injurious to health. At the same time, the other technologies introduced or advocated for use in Kenya include the small bore sewerage, VIP latrine (Figure 21.3), alternating twin pit latrine, pour flush latrine, and the composting latrine.

The composting latrine, which had been introduced in various areas, had been met with social stigma in various societies and was being discarded as an alternative. The VIP, twin pit, pour flush latrines and septic tanks were the preferred alternatives. Although the aforementioned technologies were preferred, their planning and implementation was influenced by economic factors including available resources and finance at both the institutional and individual levels.

The report entitled "The Kenya Water Sector" by Hukka, Katko and Seppala of 1992, which also dealt in less detail with sanitation, put the sanitation coverage for the rural areas at 20% and that of urban areas at 40%. The report recognised that the government expenditure in both water and sanitation in the period between 1980 and 1989 was about K£430 million, within the correct magnitude. The services were provided for about 3.5 million people but the population growth was about 7.5 million.

Scrutiny of the study on the establishment of a public sewerage system database in Kenya shows that there are various undertakers, including the Ministry of Water Resources Management and Development, public companies, local authorities and the National Water Conservation and Pipeline Corporation, which are in charge of the provision of water and sewerage facilities in Kenya (Samez Consultant 2004).

The pit latrine has been used as a successful excreta disposal facility in Kenya with about 73% percent of the population having access to this facility (BG Associates 2001a).

There were indications that although financial ability was a major constraint in the provision of sanitary facilities, many poor people could afford pit latrines. A major revelation was that of increased demand for sanitation facilities due to improvements in education, health and hygiene awareness.

For the period covering 1980 to 2002, altogether 31 sewerage treatment plants were constructed.

Advancement in sanitation technology has been slow indeed. The pit latrines introduced at the turn of the century remain the most widely used mode of excreta disposal in Kenya. The progression towards waterborne sanitation and sewerage has been slow and limited to the cities and urban centres.

There is no specific legislation on sanitation and the Public Health Act remains the most relevant statute as far as sanitation is concerned. There are, however, other statutes fragmented along other ministries that touch on matters appertaining to sanitation, such as the Local Authorities Act Cap 265, and the Chiefs Act. For example, the Chiefs act has been used to enforce adherence to the Public Health Act when there is an emergency. Section (10) and (12) of the Act have occasionally been used by the government to force rural households to dig and construct pit latrines during cholera or typhoid outbreaks.

Figure 21.4 Oxidation ditch in Kiambu STW (commissioned in 1979) (Photo: Ezekiel Nyangeri Nyanchaga)

Figure 21.5 Phase I Dandora STW Facultative pond - commissioned in 1982 (35km East of the Nairobi City Centre) (Photo: Ezekiel Nyangeri Nyanchaga).

Figure 21.6 New Changamwe treatment works aeration tanks. (Photo: Ezekiel Nyangeri Nyanchaga)

Figure 21.7 Kangemi sewerage works in Nyeri (High rate-percolating trickling filter- commissioned in 1988) (Photo: Ezekiel Nyangeri Nyanchaga)

Chiefs orders were used in Tharaka in 1994 after an outbreak of cholera and many households were forced to construct pit latrines.

## 21.7. CONCLUSIONS

The key argument in this chapter is that the water and sanitation development in Kenya over the years has been multi-sectoral. However, the state has played a lead role in policy formulation and implementation. Initially, water use and supply policy initiatives were ad hoc but later on clear development plans were put in place. The dominance of the state in development led water to be regarded as a social good as opposed to an economic commodity. This inhibited but not completely excluded the private sector participation in services. The political and administrative portfolio under which the water and sanitation development existed has constantly changed over the years. This was mainly because of the differences in the relative significance the various regimes attached to water and/or sanitation; also the changing water and sanitation needs and requirements as well as local and international concerns about services led to this state of affairs. However, throughout the colonial period, just like it would be during the post-colonial era, water use, supply and administration was characterised by the modernist tendencies of the state and the dominance of the top-bottom patronage.

The management of water was firstly vested in the Hydraulic Branch (HB) of the Department of Public Works (DPW), which developed water supplies for mainly urban centres. As the years progressed, the water needs especially for agricultural use increased but could not be provided by the HB. This led to the formation of the African

Land Development Unit in the Ministry of Agriculture and Animal Husbandry, with the mandate of providing water for Agricultural use.

Most of the earlier water supplies started in townships where the local administrative units were located. It was near these administrative centres that the hospitals, police lines, schools, shopping centres, trading centres, dispensaries and houses for government officers were put up. The administrative units could not operate well without sufficient water supplies to the amalgamated population therein. The administrative units quickly attracted populations who settled close to them and opened up trading ventures. It was the traders who would later run into the problem of the lack of water. The rural areas, on the contrary, continued to use natural water sources like springs and rivers for their water needs. However, the health problems of using these natural sources led to the coordinated intervention by the international community through the Ministry of Health. This intervention was very successful and led to the proliferation of rural water supplies.

Religious bodies like the Catholic, Church Mission Society, and Anglican etc. settled in Kenya quite early and were engaged in missionary activity and later in education. In order for them to undertake their work, they required water. At times they put in their own water supplies or petitioned the government for water supplies. In this way they encouraged water development.

Just like in the case of water supply, sanitation problems were first experienced in the townships where the local administrative units were located. The impetus to better sanitation was prompted mostly by disease outbreaks due to poor sanitary dispensation. The advent of estates, whether industrial or agricultural, as was the case with sisal estates, brought people together. This was also the case during World War II and the emergency period when a large number of people were either forced to live in camps or were brought together in lieu of the war. This concentration of people warranted a better sanitary disposition.

The administration of sanitation has largely been in the hands of the Ministry of Health, albeit other actors like the District/Provincial Administration and the Public Works Department (PWD) were time and again called upon to help in sanitation. The provincial and district administration, for example, were mainly involved in the enforcement of the Public Health Act while the PWD were more concerned with the provision of sewerage and drainage systems. The Ministry of Health had the overall mandate on sanitary and public health matters.

Traditionally, sanitation was 'natural'; the inhabitants cared little about sanitation since the population was scanty and there was extensive unoccupied land where waste was disposed of. The inhabitants used bushes as a public convenience and there was no refuse collection. This could be construed from the fact that inhabitants found it quite inconvenient to adopt the settlers-imposed mode of sanitation and conservancy, which involved the use of pit latrines, bucket latrines and refuse collection. The growth of the townships experienced a corresponding advancement in sanitation works and as the most convenient sanitary technology was embraced, the conversion from pit latrines to single bucket latrines and vice versa dominated the 1920s and 1930s. The double bucket system was an advancement from the single bucket system. Waterborne sanitation was hampered by a lack of town planning, inadequate water supply systems, and a lack of finance, habits/culture and expertise.

New enhanced sanitation systems were resisted in the native reserves; however, coordinated intervention through propaganda (campaigns), coercion, colonial laws and informal education was used to break down the resistance. Sanitation of townships, labour camps, and commercial centres advanced towards waterborne sanitation over the years. This progress, however, discriminated according to race and economic interests, such that the Europeans, affluent Arabs, and Indians enjoyed the most convenient facilities while Africans who were the workers had the poorest sanitation services. There has been no specific legislation on sanition and the administration of sanitation has been fragmented. Although the main player in the provision of sanitation has been and still remains the Ministry of Health, other actors like the local government have also been useful.

## ACKNOWLEDGEMENTS

The authors wish to acknowledge the financial support from the Academy of Finland (210816) for the GOWLOP project in connection with which this chapter was mainly produced. The tireless efforts by Prof Tapio K. Katko of IEEB and Prof Juuti Petri of UTA in the development of this manuscript are highly appreciated.

## 21.8. REFERENCES

Acting Governor. (1917) The East Africa Township Ordinance, 1903, Rules. Kenya National Archives, PC/Coast/1/10/154. 1917.
A Draft of the Factories Ordinance. (1947) Rules. Kenya National Archives, BY/62/7, 1947.
Africa District Council Ordinance. (1950) Kenya National Archives, UY/5/1.1950.
Aide M'emoire (2000) Review of the Water Supply and Sanitation Sector. Republic of Kenya. Joint World Bank, KFW, GTZ and AFD Mission. December 2000.
Anderson T.F. (1950) Hygiene and Sanitation. Kenya National Archives, BY/29/45, Ref: San. 983/1. 22nd June 1950.
Aseto O. and Okello J. A. (1997) Privatization in Kenya. Studies in Modern Economic Policies. Basic Books (Kenya) Limited. Nairobi, Kenya.
A Supplement of the East Africa Township Ordinance. (1903) Rules. File No: Kenya National Archives, BY/11/01[ not dated].
BG Associates. (2001a) Learning from the past and present for the future: 'sustaining sanitation hygiene in Kenya. World Bank WSP. Volume I, Final report.

BG Associates. (2001b) Post Project Evaluation of Kwale Water and Sanitation Project (KWASP), Final Report. Kenya-Sweden Rural Water and Sanitation Program. March 2001.

BG Associates. (2002) Global Small Towns Water and Sanitation Initiative. Kenya Country Case Study. Final Draft Report. Volume I: Main Report – Comparative Analysis. Water and Sanitation Program, WSP – Africa. January 2002.

Bloomfield D.M. (1947) Sanitary Services-Kakamega Township. Kenya National Archives, BY/29/24, Ref: S/.3/73. 12th November 1947.

Bolony and Protectorate of Kenya. (1954) General notice No. 1681. Government printers.

British East African Protectorate. (1898) Sub-Commissioner Office. Kenya National Archives, Ref: PC/CP/1/1/124, Folio No.224, Nairobi, Kenya.

British East Africa Protectorate. (1916) Draft Water legislation. Kenya National Archives, Ref: AG/43/76, Nairobi, Kenya.

British East Africa Protectorate. (1923) The water ordinance, Crown Advocates Department. Kenya National Archives, Ref: AG/43/87. Nairobi, Kenya.

Bruce T.D.H. (1929) Report of Select Committee on Water Bill, 1929. Kenya National Archives, AG/43/75, Ref: TDHB/APS. 15th July 1929.

By Command of His Excellency the Governor. (1948a) Kisii Township [Refuse Receptacles] Rules 1948. Kenya National Archives, PC/NZA/2/18/30. 1948.

By Command of His Excellency the Governor. (1948b) Trading Centres ordinance, 1932. Kenya National Archives, DC/LDW/2/16/19. Ref: Government Notice. 1948.

Chairman's Report. (1957) Nairobi Peri- Urban water supplies. Kenya National Archives, Ref: WAT/36/24/212, Nairobi, Kenya.

Colchester T.C. (1947) Sanitary Services in Township. Kenya National Archives, PC/NZA/2/14/70, Ref: S.F.PH.76/5/2/19. 18th March 1947.

Colony and Protectorate of Kenya. (1913-1923) Muthaiga Water Supply. Kenya National Archives, Ref: AG/43/103, Nairobi, Kenya.

Colony and Protectorate of Kenya. (1929-1938) Colonial Reports-Annual. Kenya National Archives, AOO-1497, Ref: 1510-1858. 1929-1938.

Colony and Protectorate of Kenya. (1929-1949) Water supplies in Kilifi district including Malindi. Kenya National Archives, Ref. CA/17/89, Nairobi, Kenya.

Colony and Protectorate of Kenya. (1957a) Sessional Paper No. 98. The Manzoni Report on the Public Works Department. Government printers, Nairobi, Kenya.

Colony and Protectorate of Kenya.(1957b) Council of Ministers (CM (57) 92). Public works department, Manzoni Report. Kenya National Archives, Ref: ACW/32/30. Nairobi, Kenya.

Community Water Supply Management Project. (2003) Draft Completion Report (January 1997 – June 2003). Ministry of Water Resources Management and Development and Embassy of Finland, Nairobi, Kenya. June 17th 2003.

Crichton P. (1956a) To the Provincial Commissioners, Coat, Central, Southern, Nyanza, Rift Valley, Northern Frontier, and Nairobi Extra Provincial District. Kenya National Archives, CA/13/41, Ref: Town. 10, Vol. II/45. 22nd September 1956.

Crichton P. (1956b). Township Sewerage. Kenya National Archives, CA/13/41, Ref: TOWN 10.VOL.II. 15th June 1956.

Davies E.R. (1934) Nyeri Urban District Council Housing. Kenya National Archives, BY/21/73, Ref: H.4/20/3, 23rd May 1960.

Davies H.A. (1946) Conservancy –Kisii. Kenya National Archives, CA/13/39, ref: PH/10/7/130, 15th August 1946.

District Commissioner. (1929) Townships and Trading Centres. Kenya National Archives, PC/Coast/2/17/7. 4th January 1929.

District Public Health Officer. (1980) King'ong'o Prison Sewage Disposal Inspection. Kenya National Archives, BY/29/58, Ref: 11/A/67. 8th December 1980.

Dixey F. (1950) Hydrological Report on the Northern Frontier District, Samburu and Turkana and Report on the hydrology of the Uaso Nyiro. Kenya National Archives.

Donovan C.E. (1935) Extract from Report on Church Missionary Society Schools, Kerugoya Area. Kenya National Archives, BY/29/37. 20th May 1935.

Doria B. (undated) "64" Eldoret. Yesterday and today. A personal compilation of recorded events. Kenya National Archives, Ref: MSS/125/1. Nairobi, Kenya.

Dunn W. (1928) Water Law in Kenya; a Letter Written by William Dunn to the Editor East African Standard Dated July 31st 1928. Kenya National Archives, AG/34/75. 1928.

East African Standard. (1959) More water for Colonies Farms. Kenya National Archives, Ref: BV/14/420. Nairobi, Kenya.

Edward R. (1985) The Old Town of Mombasa, a Historical Guide. The friend of Fort Jesus, Mombasa. 1985.

Gichaaga Z. (1967) Environmental Sanitation Programmes. Kenya National Archives, BY/29/8. 27th September 1967.

Hall C.L. (1962) Nanyuki Sewerage Scheme. Kenya National Archives, BY/29/40, Ref: SAN 913 Vol. II/118. 26th May 1962.

Health Inspector, Fort Hall. (1953) Health Work in the Fort Hall Reserve during the Emergency. Kenya National Archives, BY/29/26. 3rd March 1953.

Hifab International AS. (1988) Report of Appraisal Mission. Minor Urban Water Supply Programme, 1988-92. Ministry of Water Development. April 1988.

Huxley E. (1935) White Man's Country; Lord Delamere and the making of Kenya, Chatto and Windus, Vols. I and II, 2nd edition, London.

Japan International Cooperation Agency. (1992) The study on the National Water Master Plan. Ministry of Water Development. January 1992. Nairobi, Kenya.

Jones E.R. (1954) Nanyuki Township. Kenya National Archives, BY/29/40, Ref: 14/B/46. 11th November 1954.

Jones C.R.O. (1956) Township Sewerage Kenya National Archives, CA/13/41, Ref: Town.4/16, 4/33. 3rd April 1956.

Karanja A.W. (1949) Market Days in Machakos Township. Kenya National Archives DC/MKS/10/1, Ref: 37/172/49 2nd February 1949.

Kenya National Archives. (1995) A guide to the contents of the Kenya National Archives and Documentation Service, Kenya National Archives, Nairobi, Kenya.

Kenya Railways head office records. The Kenya railways water supplies register.

Lewis A. D. (1926) Report on irrigation, water supplies for stock and water law. Kenya National Archives, Ref: AG/43/86. Nairobi, Kenya.

Ling A.P. (1947) Steel Imported Buckets. Kenya National Archives, BY/29/40, Ref: San.915/56/Vol.II. 17th March 1947.

Lloyd J.R.M. (1968) Housing in Karatina. Kenya National Archives, BY/21/135, Ref: HOU.534.4. 5th January 1968.

Louis K.M. (1950) Conservancy Bylaws. Kenya National Archives, DC/NKU/2/18/24.

Marchant W.S. (1932) Township Ordinance 1930. Kenya National Archives, DC/Taita/Taveta/3/7/1. Ref: 341/LG.5/1. 3rd March 1932.

Marsh Z.A. and Kingsnorth G.W. (1965) An Introduction to the History of East Africa. Cambridge University Press, 3rd Edition. British Institutes of History and Archaeology in East Africa.

Mats W. (2004) Towards a historical geography of intensive farming in Eastern Africa. Islands of intensive agriculture in East Africa. The British Institute in East Africa, 10 Carlton House Terrace, London, SWIY 5AH.

Medical Officer, Fort Hall. (1927) Report from Mr. Gaffney, Sanitary Inspector Fort Hall. Kenya National Archives, BY/2/73. 18th December 1927.

Medical Officer, Digo District. (1928) Sanitation In Digo District. Kenya National Archives, BY/29/31, Ref: 29/32/28. 4th June 1928.

Medical Officer of Health. (1926) Ref: Your Letter No.14/235/Vol.III of 8/7/26. Kenya National Archives BY/29/1 Ref:24/2/87. 12th July 1926.

Medical Officer of Health. (1934) Choo Campaign in North Kavirondo. Kenya National Archives, BY/29/24, Ref: S.5/18. 3rd May 1934.

Medical officer of Health. (1940) Annual Medical Report of Medical Officer of health for Nakuru, Naivasha, and Laikipia Districts and contained Townships for the Year 1939. Kenya National Archives, PC/RVP6A/24/1, Ref: R/40. 23rd January 1940.

Member for Health, Lands and Local Government. (1954) Sanitary Services in Townships. Kenya National Archives,-Ref: TOWN.13/11/115, 2nd March 1954.

Ministry of Agriculture Animal Husbandry and Natural Resources. (1953) Soil Conservation Section and its Dam Construction Units and Contracts, Annual Report. Kenya National Archives, Ref: RP/14/9, Nairobi Kenya.

Ministry of the Environment and Natural Resources. (2001) Country Strategy on Water and Sanitation Services. Final Draft Paper. Department of Water Development. September 2001.

Muchena F. and Jan F.D. (2003) Evaluation of Water Users Association Support Programme (Kajiado, Machakos and Makueni Districts). Ministry of Water resources Management and Development / Belgian Technical Cooperation. Mission October - November 2003.

Nganga. (1980) The establishment of National Sanitation Council. Kenya National Archives, BY/29/58, 19th December 1980.

Nippon Koei. (1998). The Aftercare Study on the National Water Masterplan in the Republic of Kenya. Final Report, Executive Summary. Japan International Cooperation Agency, Ministry of Water Resources, Republic Of Kenya. November 1998.

Nyanchaga E.N. and Owano A. (2001) Evaluation of Sector Performance. Kenya Country Report of Evaluation Study. Finland's Support to Water Supply and Sanitation 1968-2000. Unit for Evaluation and Internal Auditing. Ministry of Foreign Affairs, Department of International Development Cooperation. Helsinki, Finland. February 2001.

Pedraza R. (1936) Conversion of Bucket Latrines. Kenya National Archives, BY/29/1, Ref: PH.7/7/1. 27th July 1936.

Plat R.P. (1936) Kakamega Township Rules. Kenya National Archives, BY/21/98, Ref: P.H.2/9/5/12, 21st November 1935.

President, African District Council. (1953) African District Councils Ordinance, 1950, Nandi African District Council [Cleanliness of Villages], Bylaws of 1952. Kenya National Archives, BY/11/4.

Provincial Administration Water Supply Schemes. (1966-68) A Guide in Brief and Simple Terms of the Broad Meaning of the Water Act Cap 372, Laws of Kenya. Kenya National Archives, CF/13/3. 1966-1968.

Provincial Commissioner. (1941) Sanitation-Nyeri Township. Kenya National Archives, BY/29/40, Ref: PH.13/7/78. 13th June 1941.

Provincial Commissioner, Central Province. (1947) Sanitation –Machakos Boma. Kenya National Archives, BY/MKS/10/1, Ref: PH.13/2/74. 27th June 1945.

Provincial Medical Officer, Nyanza. (1952) Sanitation-Kisii Township. Kenya National Archives, CA/13/39, Ref: SN/4/52/255. 10th March 1952.

Republic of Kenya. (1965) Sessional Paper No. 10 of African Socialism and its application to Planning" in Kenya. Government printers.

Republic of Kenya. (1975) Kenya Gazette supplements No 39, Legislative supplement No 28, Water Act Cap 372. The Water (Water Development department) (Tariff) regulation. Government printers.

Republic of Kenya. (1979) First National Water Master Plan. 1979.

Republic of Kenya. (1981) The Water Act Cap 372. The Water (Water Development Department) (Tariff) (Amendment No. 2) Regulations. Legal notice No. 194. 1981.

Republic of Kenya. (1983) Sanitation field manual for Kenya. Ministry of Health, Division of Environmental Health. Edited by J.G. Mathenge, J.M. Broome, P.I. Njue, S. Stoveland, J.M. waithaka, S. Murray-Bradley, H Bhaiji, O. Osoro under the general direction of N.M. Masai, the Chief Public Health Officers.

Republic of Kenya. (1986) Sessional Paper No.1. Economic Management for renewed Growth. Government Printers, Nairobi.

Republic of Kenya. (1999) Sessional Paper No. 1 of 1999 on National Policy on Water Resource Management and Development. Government Printers, Nairobi.

Republic of Kenya. (2002a) National Development plan 2002-2008. Effective Management for Sustainable Economic Growth the and Poverty Reduction. Government Printers, Nairobi, Kenya.

Republic of Kenya. (2002b) The Water Act 2002. Kenya Gazette Supplement No. 107 (Acts No. 9). Government Printers, Nairobi. 24th October, 2002.

Republic of Kenya/Japan International Cooperation Agency. (1992) The Second National Water Master Plan. 1992, Nairobi, Kenya.

Rhodes G.D. (1950) Sewerage and Drainage Works. Kenya National Archives, BY/29/44, Ref: D.2435/4/1/2. 16th June 1950.

Ross W.M. (1916) Letter written to Hobly, the Provincial Commissioner, Mombasa, Kenya National Archives, PC/Coast/1/14/89. 30th September 1916.

Russell E.M. (1957) Village Sanitation-Mwea/Tebere. Kenya National Archives, DC/EMB/2/6/1, Ref: 2/L/36/57. 2nd November 1957.

Samez Consultant. (2004) The study on the establishment of sewerage system database in Kenya. Ministry of water resources management and development/ Japan International Cooperation Agency (JICA). Volume 3, Data sheets.

Sikes. (1922). Modern Water Legislation. Public Works Department, Nairobi. 1922.

Sikes. (1923) Water Ordinance, Kenya National Archives, AG/43/87, 26th February 1923.

Sikes H.L. (1926a) Comments on the Lewis report on irrigation, water supplies for stock and water law. Kenya National Archives, Ref: AG/43/87.

Sikes H.L. (1926b) Report by Mr. A.D. Lewis on Irrigation etc., in Kenya Colony, December 1925, Kenya National Archives, AG/43/87, 18th June 1926.

Sikes H.L. (1928) Sewerage Scheme-Kisumu Township. Kenya National Archives, BY/29/29, Ref: San.686/17. 10th March 1928.

Simpson W.J. (1914) Report on Sanitary Matters in the East Africa Protectorate, Uganda and Zanzibar. Kenya National Archives, GP.363.7SIM. 2nd July 1914.

Skytta T. K., Ojanpera S. & Mutero J. (2001) Evaluation of Sector Performance. Report of Evaluation Study. Finland's Support to Water Supply and Sanitation 1968-2000. Unit for Evaluation and Internal Auditing. Ministry of Foreign Affairs, Department of International Development Cooperation. Helsinki, Finland. April 2001.

Soil conservation Engineer. (1953) Annual Report 1953: Soil Conservation Engineer's Section and its Dam Construction Units and Contracts. Kenya National Archives, RP/14/9. 1953.

Squires H.J. (1950) Inspection Report. Kenya National Archives, BY/29/40, 25th October 1950.

Taylor A.H. and Silveira J. (1962) Report on Elgeyo Marakwet District Environmental Programme. Kenya National Archives, BY/29/7. 1962.

The Government of Kenya, World Health organization & United nations Children's Fund. (1964) Second Appendum to the Plan of Operations for an Environmental Sanitation Project in Kenya. Kenya National Archives, BY/29/6. July 23, August 7, &11, 1964.

Todd W.G. (1949) Estimates for 1950. Kenya National Archives, DC/MKS/10/1. Ref: 39/17/49. 20th May 1949.

Tope E.T. (1960) Model Bylaws [Refuse]. Kenya National Archives, BY/11/4/, Ref: 2403/1. 19th July 1960.

Tord W.G. (1954) Medical Report-November, 1954. Kenya National Archives, BY/29/58. 14th December 1954.

Town Clerk. (1923) By-Laws: The Municipal Corporation Amendment Ordinance, 1922. Kenya National Archives, AG/23/150. 3rd January 1923.

United Nations Economic and Social Council. (1959) Recommendation of the Executive Director for an Allocation Kenya, Basic Maternal and Child Welfare Services: Environmental Sanitation. Kenya National Archives, BY/29/2, Ref: E/ICEF/R.790, 13th August 1959.

Urban Water and Sanitation Management (UWASAM) Project. (2001) Plan for the second half of phase V. Summary of the proceedings and outcomes of the workshop. 13th -14th November 2001. Nairobi.

Water Development Department. (1965) A Memorandum by the Director, Water Development Department, Ministry of Agriculture Animal Husbandry and Natural Resources. Kenya National Archives, Ref: BY/35/17, Nairobi, Kenya.

Whitehouse L.E. (1951) Night soil Disposal Pits. Kenya National Archives, DC/LDW//2/16/9. Ref: PH10/8/1/115. 28th February 1951.

WHO. (1971) Sectoral Study and National Programming for Community and Rural Water Supply Sewerage and Water Pollution Control. Report No 1. General Community Water Supply Problems. Brazzaville.

WHO. (1972a) Sectoral Study and National Programming for Community and Rural Water Supply Sewerage and Water Pollution Control. Report No 2. Recommendations on National Programme for Community Water Supply Development. Brazzaville.

WHO. (1972b) Sectoral Study and National Programming for Community and Rural Water Supply Sewerage and Water Pollution Control. Report No 6. Water Legislation. Brazzaville.

WHO. (1973) Sectoral Study and National Programming for Community and Rural Water Supply Sewerage and Water Pollution Control. Report No 10. Recommendations on Administration and Organization structure for water supply development. Brazzaville.

WHO. (1975) Report of the Review Mission for the Water Supply, Sewerage and Pollution Control "Sectorial Study" for Kenya. Kenya National Archives, BY/33/72.

Wignot R.E. (1974) A Report on the Condition of UNICEF Assisted Demonstration Rural Water Supplies in Kenya. UNICEF Regional Office, Nairobi.

Williams G.B. (1907) Report on the Sanitation of Nairobi and Report on the Townships of Naivasha, Nakuru, and Kisumu. Kenya National Archives, GP, 363.7.BRI.

Wright F.G. (1938) Construction of Pit Latrines. Kenya National Archives, BY/29/31, Ref: 5.1. 23rd June 1938.

Wundanyi D.C. (1958) Sanitary Services in Townships-Voi. Kenya National Archives, CA/13/41, Ref: PH.5/1/Vol.III/265. 18th October 1958.

# 22
# THE HISTORY OF WATER CONSERVATION AND DEVELOPMENT IN THE MWAMASHIMBA AREA IN THE BUHUNGUKIRA CHIEFDOM AND IN RUNERE VILLAGE, TANZANIA

*Jan-Olof Drangert*

Figure 22.1 Location of Buhungukira area, Tanzania.

## 22.1. GENERAL BACKGROUND

Man and his livestock need water and in water-scarce areas people are preoccupied with securing the supply. In this chapter we follow the history of the water supply during the last hundred years in a rural area some sixty kilometres from Lake Victoria in Tanzania. We provide some glimpses into the pre-colonial period, then focus on major reclamation programmes during the colonial era, and end with an account of present day water-related activities. A major finding is that every village has been engaged in recurring large-scale interventions to try to solve the water problem. The development can be divided into phases. During the "German period" (1888-1916) all water development in the Buhungukira area was organised by the chief and headmen and focussed on man-made dams, while the colonial administration dreamed about pumping water from Lake Victoria to the dry areas in the Shinyanga region. Through the British indirect rule, the administration became more advanced and the chiefs were supported in organising huge labour-intensive interventions, for instance tsetse clearings in Buhungukira to provide land for the expanding human and cattle populations.

In the early 1930s, the drilling of boreholes was expected to provide water, but a series of failures put an end to this costly option, and history saw a return to man-made small dams. After World War II chiefs faced difficulties in turning out voluntary labour, and the administration introduced machinery and payment for the labour. Towards the end of the colonial period (1888-1961) large-scale interventions faded away.

Following the villagization programme (ujamaa from 1967) donors supported large piped schemes to provide the new villages with water. Again, planners anticipated a grand solution with a grid of boreholes from which the new villages were to receive water. The failure rates were high and the cost of diesel for pumping the water went up drastically due to the world oil crises. In the late 1970s, the strategy was changed into shallow wells dug by villagers and donor support for cement rings and hand pumps. Maintaining the pumps turned out to be a major problem, and soon the pumps were taken off and villagers drew by buckets from the wells. The century was concluded by villagers themselves being responsible for developing water sources and improved access.

## 22.2. THE EARLY HISTORY: PLACE AND POPULATION

Buhungukira chiefdom is situated in the Kwimba district in Tanzania, and forms part of Sukumaland. The area is east of the Moame river about 80 km SSE of Mwanza town on Lake Victoria (Figure 22.2). It became an independent chiefdom in the 17th century when it was separated from the old Nera chiefdom (Mange 1931, 4). The trade route from Lake Victoria to Tabora and the coast passed through Buhungukira and Nera and we therefore have some early written impressions of the area. The explorer Henry Stanley wrote in 1889:

> Before us, in the centre of a plain which three or four centuries ago, perhaps, was covered with the waters of Lake Victoria, there rose what must have been once a hilly island, but now the soil had been thoroughly scoured away. [...] (Here) were grouped a population of about 5,000 people; and within sound of a musket-shot, or blare of horn, or ringing cries, were congeries of hamlets out on the plain round about this natural fortress, and each hamlet surrounded by its milk-weed hedge. In the plain west of the isolated rock-heaps, I counted twenty-three separate herds of cattle, besides flocks of sheep and goats, and we concluded that Ikoma was prosperous, and secure in its vast population and its impregnable rock-piles. (Stanley 1890, 435)

The Rinderpest epidemic took a heavy toll on cattle in 1890-91 and the area returned to bush before the cattle numbers were restored (Kjekshus 1977). Brandström (1990, 3, 9) found that "By the 1920s the whole area (between Tabora and Msalala) had become tsetse-infested miombo, except for the immediate vicinities of denser settlements." Marius Fortie (1938, 31) supports this view when writing that he travelled along the old Kahama trail from Tabora to Lake Victoria in 1901, since abandoned to the tsetse flies. Fortie described the plains of Sukumaland as follows.

> This was the peaceful and fertile Usukuma (Sukumaland) so praised by Omar Sayid, a level country of sandy loam, well settled and so intensively cultivated that firewood must be fetched long distances. We marched for miles along millet fields dotted with square and round huts. (Fortie 1938, 35)

Early explorers made sparse comments on water accessibility, like Kollmann (1899, 138) who wrote that "the country /Sukumaland/ abounds in pools and ponds ..." and Stanley who wrote "it must be prosperous." The water consultant Clement Gillman compiled information about the population distribution and accessibility of water in mainland Tanganyika (mainland Tanzania). He found that two-thirds of the population lived in well-watered areas covering a tenth of the country, a sixth lived in fairly watered areas making up a twelfth of the total area; the remaining sixth lived in poorly watered areas making up a fifth of the country. Two-thirds of the country was poorly watered and uninhabited (Gillman 1936, 16).

Gillman compiled comprehensive maps of the population distribution (Figure 22.3) and types of land occupation in Tanganyika in 1934 and concluded:

> It seems impossible to hold tsetse responsible for the distribution of population, a very marked dependency on the availability of a domestic water source can be readily established. In fact, if one arranges the types of land occupation in their order of relative density, one can immediately parallel that order by one of decreasing reliability of water sources; when it will be seen that the 2/3 of the population concentrated to a mean density of 35 per km$^2$ on one tenth of the land which enjoy the benefit of permanent streams and springs or of easily accessible shallow ground water; that those living more scattered (mean density less than 10) depend on more sporadic and usually less voluminous supplies; and that in the uninhabited regions domestic water, through the accidents of geology, soil or topography, is not available throughout the year. (Gillman 1936, 16)

## 22.3. WATER AVAILABILITY IN BUHUNGUKIRA

The typical flatland areas with wide mostly cultivated mbuga plains are surrounded by scattered hills or ridges of granite bedrock forming inselberg (Figure 22.4). The plains are intersected by shallow river valleys with usually meandering seasonal river courses. The main river is the Moame river, which rises in the hills at Malampaka. It passes north of Ilula [...] and debouches in Stuhlmann Bay in Lake Victoria.[1]

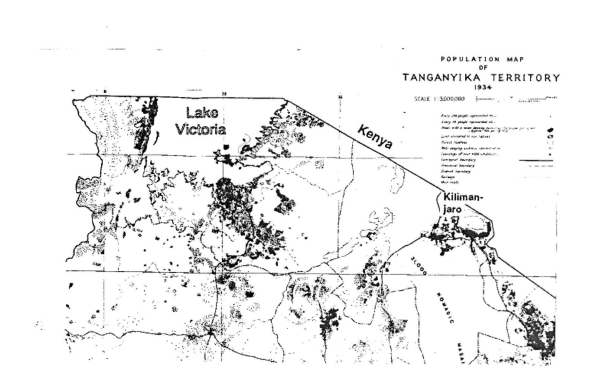

Figure 22.2 Population distribution in Sukumaland in 1936. Dots in the map idicate settlement. (Gillman 1936)

---

[1] Within the Runere area Moame catches two important tributaries, one of which is the Ndagaswa river, which passes south of Runere coming from the hilly area west of Malampaka. The other tributary, the Nyangalata river, comes from the southwest of Nyahonge, the oldest settlement established by Arabs and Indians. The confluence of all three rivers is situated east of the village Kijima. The drainage area of this river system is about 1,600 km² (WMP 1978, 41).

The climate in the area is semi-arid and the precipitation of some 800 mm per year falls during October–May. The distribution of the annual rainfall is irregular, with the extreme values ranging between 400 and 1,450 mm per year. More than 90% of the precipitation is estimated to be lost by evapotranspiration and 5% is discharged as runoff in the rivers. The remaining 5% recharges the groundwater in two ways: by infiltration in areas with favourable conditions, i.e. the sandy areas (luseni) around the granite outcrops, and by leakage from the rivers which, during the rainy season, act as influent streams (Husberg & Nilsson 1978, 1–2).

The route of rainwater suggests that water can be found around the outcrops, and in man-made dams in the mbuga soil. However, evaporation from open water bodies is high, about 1,500 mm per year. Thus, more than half a meter of the water in a dam evaporates during the dry season, and more may leak to the groundwater.

There are no lakes or wetlands in Buhungukira, and in the dry season people have to draw household water from dry river beds and a few natural springs, and cattle are driven to open man-made dams and scattered natural depressions.

In 1932 the District Officer P. M. Huggins portrayed the availability of water and the prospects of improvement as follows.

> a) Except in few sandy river beds, the supply of water is meagre and of disgusting and unhygienic composition, especially at the height of the dry season.
> b) Supplies in sufficient quantity to water cattle are few and far between in the dry season. Tramping out and erosion are therefore to be noted around available supplies.

The problems under a) can be remedied by sinking wells and boreholes whereas that under b) can be mitigated by enlarging native-made earth dams, digging new ones and sinking boreholes. All three methods have been tried experimentally and so far the following points have been noted:
(1) Wells can only be sunk round the base and on the slopes of the granite hills where there is

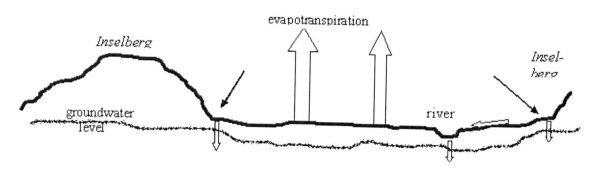

Figure 22.3 Landscape profile with infiltration areas around outcrops, and runoff and evapotranspiration from mbuga plains.

> lateritic supply. It is unlikely that many such wells would yield enough water for cattle, (2) native made dams, to be of any use in the dry season, should as far as possible be constructed so as to benefit by an overflow from a wet season spring. The majority of these continue to exude water until the end of June thereby keeping the dam full to its capacity to that date. I do not believe that a hole dug in dry earth, such as in the middle of an *mbuga*, is of any use at all as the water therein is finished before the real dry season begins. (3) All native dams should be cleaned out at the end of the dry season. (4) It is found that it is imperative to protect new dams by surrounding them except at the point of approach with *minyara* or sisal hedges. Trees are also planted in the vicinity with a view of providing shade for the stock. (Huggins 1932)

Gillman reported that a change in land use may change groundwater levels whereby water may become easy to extract:

> Not only are the residents of the peneplain miombo convinced that "water follows man" but they have for time immemorial acted on this conviction not only when choosing sites for small miombo settlements but also when pushing cultivation steppes further and further into the woodlands. ... As a typical example I can quote my own investigations at the concentration of Nyonga in south-western Tabora District. ... From the medical records it appears that when the site was chosen in 1924 there was only one poor waterhole serving ten people from which it took an hour to fill a four-gallon tin. The growth of the concentration in the first few years is shown by the following figures:
>
> | Year | Area | No. of people |
> |---|---|---|
> | 1924 | ... a few acres ... | 10 |
> | 1925 | ... 3.5 square miles ... | 1,400 |
> | 1926 | ... 9 square miles ... | 2,300 |
>
> Already one year after clearing had started shallow groundwater appeared and by November 1926 "large quantities of water were found quite near the surface" and "there has been no shortage after 1925". At the time of my inspection (August 1938) the groundwater table had risen in places so high that several huts had to be removed because the ground under them had become too wet! (Gillman 1943, 75)

Whether clearing of vegetation results in a higher or lower groundwater level has been discussed and the present standpoint of research indicates that under favourable landscape and soil conditions there may be a rise (Sandström 1995).

The history of water development in the Mwamashimba area during the 20th century is contained in contemporary documents some of which are extensively referred to below. The purpose is to give an idea of the kind and number of water interventions that the people in the area have experienced over the years.

## 22.3. TSETSE RECLAMATION IN BUHUNGUKIRA 1930-1933

In the late 1920s the colonial authority embarked upon a scheme to reclaim the forested area in order to be able to open up 3,000 square miles to settlement which would absorb perhaps 450,000 stock units and 300,000 humans from the densely crowded parts of Sukumaland (Huggins 1934). Swynnerton inspired this reclamation programme aimed at the systematic isolation of existing fly belts. The programme centred on two previously deserted areas, Buhungukira in the Kwimba federation (70 sq miles indicated in Figure 22.2) and Huru-Huru in the Shinyanga federation. District Officer Huggins wrote in a report of 1934.

> As far back as 1930, it was patent that the inhabitable parts of Kwimba District were being strained to their outmost to support an ever increasing amount of human and animal life. Most of the available land which was sufficiently well watered to support life had already been reclaimed. Notably some 50 square miles along the banks of the Simiyu River and a further fourteen square miles at the North end of Buhungukira. The Binza Tribals of the Maswa District even went so far as to cede land to the Bukwimba (people) in the neighbourhood of the Ididi River.
>
> All eyes naturally turned to that enormous waterless, fly infested, half mbuga half forest country which comprises the greater part of Buhungukira and which is nothing but the Northern extension of the Nindo mbuga system lying to the South of Shinyanga. From time immemorial cattle had skirted this area trekking down to the mbugas on the Shinyanga boundary in search for pasturage at that time of the year when all pasturage in the inhabited part of Kwimba District was eaten out.
>
> It was obvious that if this enormous waste area was to be made use of the fly and water problem would have to be tackled on a large scale, in fact that a new type of reclamation would have to be put in practice.
>
> The economic crises and the consequent removal of the District Reclamation Office precluded any immediate steps being taken whereas the visitation of locust and the consequent wasting of the pasturage taken in conjunction with the drought of 1931 so decimated the stock of the district that the finding of new pastures ceased temporarily to be such a pressing question. (Huggins 1934)

Margery Perham reported from the reclamation camp in 1930:

> We have come with 1,500 natives who have been called out by their chiefs to make a massed attack upon the sleeping sickness belt where the fly breeds in thick bush. [...] The bush around us, being thick and virgin, is expected to be teeming with wild life. [...] The attack was held up at the streams whose channels were a dense mat of undergrowth, knit closely together by the creepers known as monkey-ropes. ... The bush was soaking, the grass often higher than one's head and I was drenched up to the waist in a few minutes. You have to imagine this sort of thing on a front of eight miles. (1976, 81-82)

The bush clearing was successful[2] but difficulties were encountered in developing new water sources.

---

[2] Kapalaga (1946, 3) described the remaining bush area as follows. "... the land cover consists of open *acacia* bush with an undergrowth of grasses in the *mbuga* areas, while the *albizzias, grewias combretums* and various other trees and shrubs grow on the high ground above the *mbuga* and on the hills. Some of the hills are partially covered with *lodotia* grass which is very much used by the people as a thatch grass. *Panicum* is another type of grass found within that region. The predominant grasses in the *mbuga* areas are the *hyparrhenias, setarias*, and *cenchrus. Cynodon* has colonized a certain amount of land in the occupied areas."

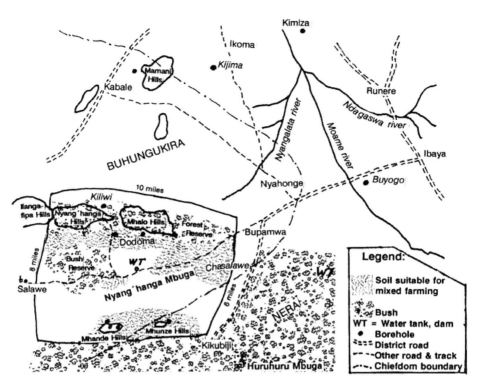

Figure 22.4 Buhungukira clearing and tributaries to the Moame river. (Based on Location map of Kabale, Nyahonge and Mwamashimba, 1948 and a sketch map of the clearing by D.O. Mr. Thornton, 1936.)

## 22.4. DEVELOPMENT OF WATER SOURCES FOR MAN AND BEAST

Huggins described the development of water sources in the Buhungukira area in 1932-33 (here *in extenso*):

> Towards the end of 1931 the Boring Machine of the Geological Department was at work in the Shinyanga District and upon completion of its labours there the opportunity was taken by the Kwimba District Native Treasuries of utilizing it in the Kwimba District. The first hole sunk, some twelve miles from Ngudu in March 1932, was successful[3] and on the strength of this it was decided to seek Geological advice and assistance in the matter of boring for water in Buhungukira. This led to the seeking of advice of the Tsetse Research Department as to the best methods of ridding the country of fly. Sites were selected for tribal boreholes and the Director of Tsetse Research at the request of the Provincial Commissioner toured the area in company with the District Officer and submitted a report. At about the same time the Director of Veterinary Services and the Pasturage Research Officer were also invited to do likewise. The reports of these officers were all satisfactory with the result that it was determined to undertake a large reclamation in 1933 under the directions of Tsetse Research Department. The Chiefs of Bukwimba Federation willingly voted the money for both development of water and the cutting of the fly infested bush. By the time plans had reached this stage two unsuccessful boreholes had been drilled at Chasalawe and

Dodoma villages[4] and realising that the whole success of the scheme depended upon the successful provision of water it was decided to dig large earth tanks in the *mbugas* before commencement of the rains. This was the state of affairs when (Huggins) proceeded on leave in October, 1932.

As for execution of the work in 1933, Huggins wrote (Figure 22.4):

The Director of Tsetse Research had drawn up a plan of campaign for the reclamation of Buhungukira covering a period of years. The first year's work covered that area bounded on the North by the Mhalo-Nyang'hanga-Ilangafipa range of hills and extending South to Mhande Hill and the Nindo *mbugas*. On the East the area is bounded by cultivation at (B)upamwa and the West is defined by the arm of *mbuga* extending from Ilangafipa to Mhande. The whole area comprises some seventy square miles but the area affected by the elimination of Tsetse bush is considerably greater.

In January an earth tank some 90'*40'*12' was dug at the Dodoma village on a Geological plan.[5] This rapidly filled and was practically the only source of water during 1933 reclamation operation. Two more earth tanks of half size were dug at Mhalo and Maboko in February but unfortunately the cessation of rain precluded them filling up.[6] In February after one more unsuccessful borehole the Geological Department struck a stream giving 1,200 gallons per hour on the high ground under Nyang'hanga Hill.[7] An order was immediately placed in England for a Lister Diesel engine and pumping plant in the hope that it would arrive in time to be erected and provide water for the working parties.

In April preliminary work was carried out by way of constructing a road from the main Nyahonge-Upamwa road to Nyang'hanga and the erection of labour camps. Bush cutting commenced in earnest on May 11th when 3,600 men were drafted into camps at Upamwa and Mhalo. Work progressed under the direction of an officer of the Tsetse Research Department aided by a casually employed European working in conjunction with the Provincial and Native Administrations who provided two more Europeans. Both the Veterinary and Forestry Departments provided additional European supervision for varying periods. The labourers, who were provided with meat and salt rations were entirely voluntary and worked for ten days under their own Chief and headmen.

As time passed, Upamwa camp was left and 4,000 men were drafted into Dodoma village camp whilst a further 2,000 continued to operate from Mhalo. By this time the local supply of water at Upamwa was exhausted and the sole source of supply was the Dodoma earth tank whence water had to be transported by porters to the labourers at Mhalo over three miles of *mbuga*. All the

---

[3] Kimiza borehole (Figure 22.4): The site chosen was in the depression, on the edge of an *mbuga* where River Magogo commences its wet season course. 60 gallons per hour were struck at 48 feet. It increased to 150 gallons at 123 feet and 465 gallons at 196 feet. Drilling ceased at 212 feet. The Geological Department boring plant was hired on a three-year payment agreement to drill for water for domestic and stock purposes.

[4] The Geological Survey Department went on boring on the East side of the Nyang'hanga Mbuga. The borehole reached a depth of 333 feet, after which it was abandoned. The formation consisted of broken schists and clay. The cost was Shs. 3,771/19, paid by the Native Treasury. After this failure, the Department started drilling in Dodoma on the West side of the Nyang'hanga Mbuga. After a depth of 506 feet the hole was abandoned. The formation was broken schist and clay. The cost was Shs. 3,373, again paid in full by the Native Treasury. After two dry holes, their drilling rig was moved to the North of the Nyang'hanga Mbuga and close to Mhalo Hill. At a depth of 337 feet, the hole was abandoned. The formation was broken schists with clay. The cost was paid by the Native Treasury (Shs. 1,000).

Figure 22.5 Chief Makwaia at one of his surface water catchment tanks.

bush of the East of a lune from Nyang'hanga to Mhande was now felled and there but remained the dense bush at the base of Nyang'hanga Hill and the light bush extending some six miles therefrom via Dodoma to Mhande Hills. It was however impossible to draft any large gang of men to Nyang'hanga camp before the installation of the pumping plant, the non-arrival of which was by now causing considerable anxiety. The Dodoma earth tank had exceeded all expectations having supplied most of the daily wants of an average of 3,500 natives and five to six Europeans over a period of 48 days. By the end of June the supply was so low that it was decided to employ only 2,000 men. In the meanwhile information was received that the pumping plant had arrived in Dar es Salaam. It arrived at Bukwimba on 9th July and hoping to have it erected by July 15th the last batch of over 6,000 men from Usmao were drafted in to the outskirts of the reclaimed area on 14th. But, *alas*, as the plant was erected, mechanical faults caused a breakdown and the long looked for water supply was not forthcoming with the result that on the morning of the 17th this last gang had to be returned to their homes.

---

[5] The Nyamiselya earth tank was begun in February 1933 in the mbuga close to the unsuccessful Dodoma borehole. When it was finished (before the end of the year) it was found to have a capacity of 500,000 gallons. Two shallow ditches led to it across the catchment area, and it was filled rapidly.

[6] Two more such tanks were dug in Mhalo and Maboko in March and finished before the end of the year 1933. These were dug out further to a depth of 15 feet in November and December the same year. The cost of all these tanks was 54 head of cattle supplied to labourers and 120 shs as wages to the overseer. Two more tanks were built in Sanjo and Chasalawe in the same way 1933. Also a tank was dug close to the Nyang'hanga borehole no 2 which can be filled, if necessary from the borehole via an earth furrow running from some 300 yards from the borehole to tank. This tank was divided into portions by a wooden barricade, termed a stop connexion by its builders through which the water passes easily from the top or drinking section into the bottom or cattle-watering section. The wooden barrier effectively prevents the cattle from invading the water set aside for human consumption.

[7] It was decided to make a final attempt to find water near the hill known as Ilangabafipa and about 200 yards from the previous hole. At three feet laterite formation was struck; at 38 feet a broken ironstone formation was struck and this continued up to 324 feet where water was found. This supply at Kiliwi yields 1,200 gallons per hour.

The total labour turn-out had been over 23,000. The next day the machinery was repaired and an excellent supply of water has been forthcoming since.[8]

Huggins wrote in conclusion:

It was unfortunate that the bush clearing could not be completed but over 3/4th of the original programme was carried out with the result that a large area is now thrown open for settlement. Settlers in small numbers have already arrived and cattle are now to be seen watering in hundreds where they have never dared to tread before.

There is however much to be done both practically and experimentally before the area can be finally settled. The Dodoma tank having been such a success many more such must be dug especially in view of the fact that three more boreholes in the neighbourhood of Mhande were sunk without success[9]. The rival merits of boreholes versus tanks have caused much controversy but both have claims to superiority. A borehole with machinery and running costs is expensive but a good one such as Nyang'hanga, properly maintained, has limitless possibilities no matter the original cost. We have to pay for our experience and we can but be guided by the excellent results obtained in other countries. The great advantages of earth tanks are the cheapness of their construction and the fact that they can be dug without limit to their numbers wherever and whenever they are required and there can be no denying that the greater the number of sources of water supply the greater is the boon to both man and beast. With regard to Dodoma it must be recollected that most of the water was used up ever before it was affected by the period of intensest evaporation commencing in July and ending with the break of the rains. It would be unwise to endeavour to rush settlement. To the native eye the area is still a wilderness with one borehole not yet properly equipped and an inadequate series of empty earth tanks. By this time next year the countryside should present an entirely different picture. (Huggins 1934, filmtape 24:232; information in the footnotes is from an appendix named Table of Water Development Schemes 1929-1948)

Despite Huggins' conclusion that boreholes and tanks "both have their own individual claims to superiority", no drilling whatsoever was undertaken between 1934 and 1942 (McLoughlin 1971, 25). The emphasis on tanks and dams was maintained and during the subsequent years a number of them were dug by the residents. In 1937 the Tsetse Department embarked on a scheme which made annual cleaning of the tanks possible. The four main tanks in Buhungukira (Chasalawe, Mhalo, Sanjo and Kahuga alias Nyang'hanga no 2) were duplicated, i.e., four new tanks were dug out in close proximity to the old.

---

[8] "The cost of the plant bought through the Crown Agents was Shs. 4,448. Railway freight amounted to Shs. 499/20. The erection costs paid to an Engineer were Shs. 700 but to them must be added Shs. 805 being the cost of materials and labour for the erection of engine shed and tank tower. A claim for Shs. 482 has also been submitted by the Engineer for further work on account of repairs to alleged defects in the machinery supplied. The total cost therefore of this borehole and its machinery and housing has been Shs. 12,057."

[9] In 1933 the drilling plant was moved due South some six miles and set up at Mhunze Hill. Boring was stopped at 408 feet although a supply of 40 gallons per hour was struck at 340 feet. This was a free borehole. After this attempt the machine was moved along the ridge to the west about 2 miles and set up at Mhande. At 172 feet schists were struck and the machine moved about 50 yards South where decayed schist was again struck and the machine lifted at 42 feet. Boring again commenced to the North but was stopped at 36 feet by order of the District Officer and the boring plant was returned to Dodoma in June. The cost was Shs. 665, which was paid by the Native Treasury.

> Of these pairs one was set aside for human use and one for cattle; suitable signboards were erected, one bearing the figure of a cow, the other a figure of a woman drawing water. The tanks used by cattle will be dry before the year is completed and can be cleaned out. Next year the clean tank will be used by humans so that the other may be drained by cattle. If the cattle drain their tank dry too early, they can either be accommodated at the borehole or by a water furrow led from the 'human' tank. (Huggins 1934)

A review of the condition of the dams in Buhungukira conducted in 1947 showed that most of them were in good order but in need of cleaning. The Dodoma tank was believed to be empty while the Nyang'hanga borehole and pump needed maintenance (there was no one in the Province who could provide the necessary services).

## 22.5. THE ORGANISATION OF WORK

The organisation of clearing work was briefly described by Margery Perham on a visit to the site in 1930.

> The natives all rolled up yesterday and they are busy building themselves grass shelters. ... We went along to Camp No. 11. ... Men were flowing in from all directions like a stream of ants-threading the bush in a single file. Each man brought rations for ten days, his big knife and his little axe. His goods were strung on a pole carried on his shoulder, his cooking pot, a basket of meal, a bundle of sweet potatoes, and a calabash of water or beer. Clothes were the fewest and the oldest. To the belt is tied a little calabash of snuff; round the neck a few shells, and perhaps pincers for picking out thorns: a couple of goat-skins as a cape or a ground-sheet.
> 
> As soon as they get in, having walked anything from ten to forty miles, they must set to work to cut down branches and grass and build themselves huts. Fires were soon going and pots of mixed meal, millet, maize and beans, were being stirred with sticks. ... Thousands of men were attacking the forest with axes, great knives and bill-hooks, shouting and singing as they struck. Five or six men would fall upon one tree and hack at it, until it came down with a long rending crash in a ruin of branches and amid screams of triumph. Then the whole group would rush at the next one. ... Last year they cleared twenty-two square miles and hard on the track of the axes the natives come in out of the crowded lands of the Wasukuma (people) to settle and, as a reward of hard work on virgin soil, to get bumper crops.
> 
> I am more than impressed by the qualities of the African and by Tanganyika's administration. Here we have 15,000 men who at a word from their chiefs have left their homes to come to work for the good of others, for very few of them can directly benefit. This is work for which they get no pay and have to provide their own food, or most of it, for they get a feast of meat three times a week. They work from 6 to 12 and all the afternoon they work away making things of wood, pots, hoe-handles, milk jars, etc. You see them working, perhaps under the tuition of some special expert. They will go back with a year's store of utensils, for wood is scarce round their own homes. At dusk they stop work, light fires and sit round in groups of five or six cooking their mealies. Then comes the dance. And they say the African is lazy! (Perham 1976, 81-83)

Here Perham gives important detail of how the men worked and lived. It is evident that the workers catered for themselves and there was ample time for private activities in the afternoons and evenings.

Figure 22.6 Excavation of dam by an ox-drawn scoop.

Figure 22.7 Construction of earth wall of a dam in the 1940's. Oxen trample to make it impervious.

The work of constructing *lambos* and tanks was organised in the same way as the bush clearing. Mzee Sendo of Bupamwa, who was (1990) the chairman of this village, gave the following account of the work in those days.

> The Bupamwa dam was built in some three months. A European did the siting and the assistant to the chief organised the construction work. Initially the recruits were not given any individual or group assignment, but after facing difficulties each one was assigned his own task that was estimated to last some ten days. If it was not completed by then, the men had to continue till it was finished. The first step in the construction work was to clear the bush in order to make the necessary levelling. Then some three pits were dug to check the soils underground. After it was found satisfactory, the topsoil was removed. The excavation of the sump or core-trench started and this soil was put to use in the building of dam walls. It was important to soak this soil - if enough water was available - and to pond the wall thoroughly in order to make it impervious. (Mzee Sendo 1990 in Drangert 1993)

Figure 22.8 A gang of villagers digging a spill-way.

A few years later "the beef and beer supplied on these occasions, as well as part of the work force, was now paid for out of departmental or Native Treasury funds." (Austen 1968, 245). The growing problem of recruiting people is mentioned in several reports. An example was given by the District Officer H. Harrison in his explanation of the managerial point of view on the clearing work at the neighbouring Madusa *Mbuga* in Shinyanga 1943.

> Operations were started with a gang of 330 only. The labour were very slow in turning out. However, on Saturday the 14th we managed to get nearly a gang - 1,115 men instead of the correct 1,250. This gang was from six different chiefdoms, which, had I known before leaving Shinyanga, I should certainly have asked for four extra African Assistants to deal with them. The local chief Ndalawa, whom I was relying on to handle any trouble with the gang, could do nothing with these people; they were not his men. Finally we had to work them separately, viz. each chiefdom on its own.
>
> The second gang of 1,065 out of 1,250 men came in very slowly, in fact, the Usmao natives of about 360 men arrived when the others had finished. This gang was the worst on record, they either never arrived at the clearing or sneaked away to cut jembe-handles, etc. After a few days of this, I brought in the clearing gang and stopped them outside the camp while it was cleared out, well over 100 men were caught who had never been near the clearing and a huge pile of jembe-handles, etc. were collected and burnt. This did not occur again. The gang was counted before it left camp and in the field when work was finished, each *manangwa* (headman) being responsible for his men. The last week was not too good as eight cases of C.S.M. were sent away. The whole camp was burnt down on September 18th when someone lighted up the *mbuga* north of the camp. (Harrison 1943)

Many water supplies were constructed under technical advice from district officers, some of which faced such management problems. But a number of village and chiefdom councils took upon themselves the task of constructing schools and dams through communal efforts without waiting until Native treasury funds were available. The following excerpt is from the Annual Report (Kapalanga 1946, 8) from the Agricultural Officer in Mwanza District: "There is a growing demand in many areas for improved water supplies and the Chief of Massanza I in particular continues to lay out and dig new dams entirely on his own."

The planning of settlements started out from human, agricultural and livestock water requirements. The carrying capacity of the land and the water requirements were calculated by using the famous "Sukumaland equation" which was developed by the well-known agricultural officer R. V. Rounce. The optimum density of 40 people per km$^2$ (100 people per square mile) dependent on one water supply in the centre of a 30 square mile area is arrived at as follows:

> At one homestead there are on average two taxpayers or a total of 17 people with an average of 14 cattle and ten small stock units (at 5 small stock equalling one stock unit) equals 16 stock units which produce altogether 16 tons of manure per annum. This manure is enough to manure eight acres every other year which is one acre more than the average acreage of arable for Sukumaland but the stock require two acres each of pasture (the average for Sukumaland is 2½ acres) equals 32 acres plus eight of arable equals 40 acres equals 16 homesteads per square mile equals 112 people per square mile, say one hundred. Three miles to walk to water is about 30 square miles equals 500 homesteads equals 8,000 stock * five gallons of water * 120 days (August to November) equals five million gallons = 10,000,000 gallons (to allow for evaporation) per 500 homes and 30 square miles. (Rounce 1949, 105)

## 22.6. THE SUKUMALAND DEVELOPMENT SCHEME 1946-1956

After the Second World War a new large integrated scheme, the Sukumaland Development Scheme, was launched.[10]

> Under this scheme over 300 large and small dams, hafirs and catchments were built, the majority by tribal turn-out labour, but the largest three dozen were built by mechanical equipment. ... In Kwimba, which is more suitable for agriculture in its north, and for grazing in the south, over 45 dams were built by hand during this period (tribal turnout, paid for at the rate of one head of livestock per every 500 man-days), and 6 more by machine. Many of the larger dams were used as reservoirs for rice irrigation. ... From 1943 to 1964, some 60 boreholes were drilled (at ginneries etc.), about half being successful. (McLoughlin 1971, 24)

During the first two years of the plan, some fifty small catchments were built or improved by using manual labour only. In 1949, for example, a second earth tank, Mwabayanda (Buyogo), was built with a standard tank 50*15*5 yards. This tank was completed in 8,729 man-days. Another earth tank, Soli in Buyogo, with a size of 35*10*6 yards, was completed in 7,845 man-days.

The first use of mechanical units in association with communal labour was introduced in 1948. Machinery, in the shape of tractors, carryall scrapers, rippers and rollers, was provided

---

[10] The estimated cost of about £230,000 grew, with the inclusion of territorial housing and water development votes, until the approved expenditure finally amounted to £472,000 to be spent over a period of ten years.

by the Water Development Department for the building of surface rain-catchment dams in sites found, surveyed and approved by the engineers. The idea was to build large dams with a capacity of some twenty million gallons, every ten miles to avoid over-concentration of stock at any point. Smaller mechanical units in conjunction with manual labour were used to improve water supplies in the occupied areas.

In the annual report from the department of agriculture in the Lake Province (1952, 30) we find the following mid-term description of the development plan. "A flight by air over Sukumaland quickly gives one a picture of the close network of artificial water supplies, which must in aggregate have helped considerably towards reducing soil erosion and improving the pastures, if only in terms of the reduction of tramping by stock."

The large Mwamashimba Dam was constructed in Bupamwa in 1956 after being surveyed the year before. The dam wall is 50 ft wide and stretches 1,148 ft to catch the runoff water from an area of one square mile. Its capacity is 750,000 gallons and the depth at the centre is 10 ft. There are two spillways (31 inches of annual rainfall) and below the dam wall are some troughs for watering cattle.

The dam contains water all year round (1990), but it is badly silted up. The villagers have tried to hire a caterpillar to excavate the silt, but this has proved impossible since such a heavy machine is expected to sink in the silted dam. The use of shovels and sledges pulled by oxen has not so far been pursued.

The demand placed on chiefs and headmen to provide labour-gangs was reduced compared to the tsetse campaigns, but it seems as if the task was not substantially easier despite payments to the workers, unless the following account from the construction of the Mkula Dam was uncommon. The Divisional Engineer of the Lake Province wrote to the Tributary Officer on 1/10/59 under the heading of Shortage of Labour.

> Work at Sapiwi (Mkula) dam is being seriously hindered by lack of labour, and I would request your assistance in obtaining 40 labourers for work here. The embankment is almost complete, and the tractors will be ready to move to the water holes, and then onto Kisesa dam within a week or so.

In a telegram 19/10/59 the Political Officer in Mwanza wrote to the Tributary Officer which indicates a weakened administration:

> I have seen the chief myself on the 9/10/59 and impressed on him the vital importance of a good turnout of labour and he has promised to bring pressure on his Parish Headmen. There should be more labour available within striking distance of Kisesa (Rutubiga) dam site, as it is nearer population centres. Naturally I cannot guarantee that full labour requirements will be met, as I have no physical control over the people concerned, but you may be assured that I will continue to bring all possible pressure to bear. I have explained to the Chiefdom Council that, failing a proper turn out, they need expect no further assistance from Government. However, it is natural for the few people already settled in such bush areas to be reluctant to help construct a dam which would attract thousands more settlers to the vicinity and restrict available land for grazing and cultivation. Would it be worth seeing the Chief about getting labour from other gungulis/parishes/? Am I right in assuming you are paying Shs. 2 a day? [11]

## 22.7. MWAMASHIMBA WATER SUPPLY IN THE 1970S

Shortage of water was experienced again in the late 1960s and the authorities in independent Tanzania embarked on a new scheme to provide humans and livestock with water. The initial idea was to pump water from Smith Sound (Lake Victoria) some 40 km away at Mbarika or from Magu Bay 70 km away. This idea was abandoned due to high costs of operation and maintenance and the problems of dealing with two regions. Handily, donor money was secured to use underground water from boreholes along the Ndagaswa river near Runere village. This Mwamashimba Water Supply was planned to supply 35,000 people in 19 villages in the Buhungukira and Nindo areas from 156 domestic points along a 120 km pipeline.

The work started in 1974/75 and was expected to be finished by 1977. In all, 16 boreholes were drilled by the MAJI [Water] department in the Runere area as preparation for the Mwamashimba water scheme.[12] The pace of the scheme was slow due to technical and managerial problems. In a letter to the District Water Engineer 18/9/1981 the Regional Development Director stated:

> The management of the work in the project is not good. Often the villagers are promised to be picked up by a truck and transported to the site where they are to dig the trench. Usually these villagers have to spend the whole day waiting for the transport without ever seeing the truck. This has happened many times and the villagers are losing their time for nothing.

More than half of the pipes and domestic points were there in 1990, but only the closest four villages had received any water. During the election campaign in October 1990 a revival of the scheme was suggested. A preliminary estimate from the Regional Water Engineer suggested that the replacement of pipes and a new booster station would cost some 45 million shs. The cost of diesel was estimated at 34 million shs per year or some 5,000 shs per household, i.e., four or five times the present taxes paid by the average household. The scheme was not revived since no one was prepared to invest the money.

---

[11] Payment had recently been introduced, and before that the Native Authority usually paid for food but not shelter. In the 1930s people brought their own food and built their own shelters when working in the tsetse eradication program.

[12] A study of the water strikes during the drilling of the wells shows that along the rivers water is stored in the whole profile from within the clay layer down to fractures in the fresh bedrock. In the remaining areas water is stored only in the fractured and, occasionally, the weathered bedrock. The river aquifer is classified as a semi-confined aquifer whereas the aquifer in the remaining areas is classified as a confined aquifer (Husberg & Nilsson 1978, 2).

## 22.8. CONCLUSIONS: ANOTHER TWO SHIFTS IN THE WATER SUPPLY STRATEGY

The failure of the sophisticated piped water supply systems led to a revision of the strategy (Therkildsen 1988). A period of shallow wells dug by villagers and equipped with hand pumps was initiated in the late 1970s. The idea of village ownership and contributions by the users had an impact on the provision of household water, but not on water for livestock (Drangert 1993).

Again, problems with maintenance became critical, and the Department of Water did not manage to set up a structure to provide outlets for spare parts for the hand pumps. Villagers were not keen to contribute to maintenance costs, and they ended up removing faulty hand pumps and started to draw water by buckets. The 1980s and 90s represent a period with collapsing government services. Villagers experienced both embezzlement of collected money and government staff sold diesel instead of using it to run the water pumps. Thus, technical problems were aggravated by administrative ones to the extent that villagers had to consider making their own water arrangements again (Drangert 1993). The following case gives a glimpse of the local possibilities for household-centred arrangements to improve access to water.

The Sukuma norm tells that each spouse does what he or she is expected to do; women fetch the water daily while men may develop a water source or improve on the technical design. Interviews with a head of household, Bwana Mfugaji, a retired agricultural officer, provide details from his household. He rated the water conditions in his homestead as follows:

> The water conditions have always been problematic in this area [Buhungukira]. The construction of shallow wells with hand-pumps a decade ago and, later, the installation of a windmill to provide the hospital with water from a drilled well improved the situation. The time prior to that was one characterized by "waiting for the water".
>
> Soon the hand-pumps broke and we did not know who was responsible. You asked whether the village council has discussed this matter. I don't know, you better ask the chairman. I suppose some people have complained. We expect the leaders to deal with the problem and they can mobilize the villagers to do what is necessary. We are not a "true" village, however, in the sense that it is inhabited by a mixture of people from all over Sukumaland. (Df1a230+370 in Drangert 1993)

Bwana Mfugaji had experimented on his own to improve access to water. He had dug a small pond for fish breeding and a shallow well for gardening. These water sources, which were far away from his homestead, became common-pool resources in line with another Sukuma norm despite the fact that he and his sons received no assistance during construction from any neighbours. He was reluctant to develop another water source due to free-riding neighbours but, at the same time, felt the need for a water source closer to the homestead. The wet season source was 400 feet away and the dry season source was more than a kilometre away. Water for his cattle was hauled with an ox-cart and at the same time household water was brought home.

> I do not face any problem and my wife does not fetch water. That is not to say that we have solved all problems, oh no! What helps me are my cattle. I have always been interested in cattle and my desire has been to stay on the farm. (f1a110 in Drangert 1993)
>
> My family is large and water consumption is high. Not all cattle are out on the grazing area. The sick ones stay here and all the calves are kept in the kraal in the yard.
>
> I need a well here at home. And there is water in the sandy slope over there. I have wished for a well for years. And it will not be long before I have one of those. ... Another idea is to build a rainwater tank. (Df1b225 in Drangert 1993)

The discussion on what was needed to implement the above ideas raised no serious obstacle. Bwana Mfugaji said he had the necessary knowledge from earlier work and it only took using the brain (*akili*). Only a few tools were required, like a *jembe*, shovel, crowbar and buckets. He also needed a cover to prevent children and cattle from falling into the well. He intended to use the ox-cart to fetch stones for lining.

> We have the ability if we get some assistance and after this inventory it is clear that we can manage. In fact all of us are concerned about water but implementation is poor. I can perhaps do something and become a good example for my neighbours. Next time you come here you will find an excellent water source! We shall expend all our efforts to produce a good example. (f1b440 in Drangert 1993)

On visiting Bwana Mfugaji a year and a half later, he had a medium-scale *lambo* [small dam] some fifty metres from the house. He told the author that he happened to know a person who was working with road maintenance. He hired him with a grader, a kind of bulldozer, over a week-end to excavate the *lambo*. The *lambo* was some twenty by fifteen metres and more than three metres at the deepest point. It was dug in an almost level section of the village and its catchment area is several thousand square metres. The *lambo* would fill to the brim early in the rainy season, and a ditch had been dug to divert excess water from entering. In a letter of March, 1993 Bwana Mfugaji wrote:

> When the rains are good, the *lambo* will be dry for only some days or a month. In 1991 there was water up to September 19, and the new rains commenced on October 13 and refilled the *lambo*. These rains ended early, April 12, 1992, and because of extensive water use the *lambo* was empty by August! The next rainy season started in November and this year we have had plenty of it. I hope that the *lambo* will not dry up at all this year.
>
> Furthermore, I hired some people to dig a second, adjacent *lambo* last year. They dug a 17 by 6 metres and 1.5 metre deep excavation by hand using *jembes*, crowbar, shovels, buckets and a wheelbarrow. So far the expenses have reached the value of one bull, and the work will continue next dry season.

The first *lambo* could provide neighbours with water during the rainy season without causing depletion. A dry season of some four months is different. We may assume a loss due to evaporation and seepage of about one metre during these four months. The total volume available is 20 by 15 by 2 metres, that is, 600 $m^3$ or 5 $m^3$ per day in the dry season. Bwana Mfugaji's big household may use 200 litres daily and 10-15 calves consume about the same amount. Some 4 $m^3$ would be then available to the neighbours. This theoretical estimate proved to be too optimistic for the first two seasons, since the water was depleted. Possibly the seepage into the ground is greater than anticipated.

# THE HISTORY OF WATER CONSERVATION AND DEVELOPMENT ...

Figure 22.8 Cooperative effort to clean overgrown *lambo*.

This shortage of water led Bwana Mfugaji to allow only a few neighbours to draw water. This is in line with the Sukuma norms, since the *lambo* is used as a water source for cattle. According to him

> [...] there are many who steal water in the afternoon and in the dark hours. Even some livestock are watered there illegally. I would like to put up a fence of barbed wire to fence off thieves.

The water quality is not good enough for drinking since overland runoff washes down pollution from the catchment area where animals graze. Therefore the drinking water is still fetched from the distant source with safe water. One obvious cost of having a *lambo* nearby is the increase in malaria mosquito breeding in the open water. Another cost is the possibility of *schistosomiasis* if the surrounding area is not kept clear of bushes. However, Bwana Mfugaji has observed no increase in the incidence of malaria and the mosquitoes do not seem to breed in the *lambo*, only in small puddles. Nor were there any signs of snails housing schistosomiasis, possibly thanks to regular removal of the grass along the *lambo*. He had also constructed a bund to prevent overland flow from the cattle kraal from entering the *lambo*.

The benefit of having water close by for household purposes is stressed by Bwana Mfugaji. In the letter he also mentioned that the survival rate of his calves has increased markedly. The benefits of more calves soon repaid the outlay for the first *lambo* of three bulls for the grader.

This supports Donald Malcolm's statement that cattle-keeping is very profitable (Malcolm 1953). Bwana Mfugaji identified some reasons for his success.

> It is not luck. Since childhood I have been interested in cattle and farming. First and foremost it takes a keen interest and enthusiasm and, secondly, patience. However, a friend of mine hired the grader to excavate an impoundment in the seasonal stream down in the valley. Unfortunately the soil was not stable enough and it collapsed when the rains began. (f2b310 in Drangert 1993)

Only smaller changes are anticipated to arise out of need or desire. Major innovations would be carried out if a favourable opportunity appears. The event of a grader in the vicinity made it possible for Mr. Mfugaji to excavate a *lambo* instead of only digging an ordinary well. The second *lambo* was built with manual labour after having been assured that it would work. The case also shows how important it is to take into account the need for water for cattle.

## 22.10. REFERENCES

Annual report from the department of agriculture in the Lake Province (1952)
The Divisional Engineer of the Lake Province wrote to the Tributary Officer on 1/10/59 under the heading of Shortage of Labour.
Telegram 19/10/59 the Political Officer in Mwanza wrote to the Tributary Officer.
In a letter of March, 1993 Bwana Mfugaji.
Based on Location map of Kabale, Nyahonge and Mwamashimba, 1948 and a sketch map of the clearing by D.O. Mr. Thornton, 1936.)
Information in the footnotes is from an appendix named Table of Water Development Schemes 1929-1948.

PRIMARY SOURCES
Harrison H. (1943) The Madusa Mbuga Clearing 1943. Report to the Director of Tsetse Research Department in Shinyanga. Letter Ref. No. 1155/217.0.
Huggins P. M. (1932) Water Conservation and Development. Report dated 23 June 1932 in Musoma District Book. Film Tape 24. Dar es Salaam National Archives.
Huggins P. M. (1934) Reclamation South Buhungukira and Nera. Filmtape 24, Kwimba District Book. Dar es Salaam. National Archives.
Kapalaga C. M. (1946) Settlement of the Remaining Bush area in Buhungukira. Report to the Agricultural Officer in Mwanza 1/10/1946.
Mange K. (1931) The Origin of the Bukwimba Country. Edited and cyclostyled at the Nyegezi Social Research Institute, Mwanza, by the Mwanza Research Team in January 1967. Dar es Salaam. Africana Section, University Library.

LITERATURE
Austen R. (1968) Northwest Tanzania under German and British rule, colonial policy and tribal politics 1889-1939. Yale University Press, New Haven.
Brandström P. (1990) Boundless Universe – the Culture of Expansion among the Sukuma-Nyamwezi of Tanzania. (Diss) Dept. of Cultural Anthropology, Uppsala University, Sweden.

Drangert J-O. (1993) Who cares about water? Household water development in Sukumaland, Tanzania, PhD Thesis, Linköping Studies in Arts and Science, No. 85, Linköping University, Sweden.

Fortie M. (1938) Black and Beautiful – a Life in Safari Land. New York. Rural Development Committee. Cornell University.

Gillman C. (1936) A Population Map of Tanganyika Territory. Geographical Review 26, 353-375.

Gillman C. (1943) A Reconnaissance Survey of the Hydrology of Tanganyika Territory in its Geographical Settings. Water Consultant´s Report, No. 6. 1940. Dar es Salaam. Government Printers.

Husberg & Nilsson Å. (1976) A Groundwater Assessment Study in the Runere Area in Tanzania. Stockholm Royal Institute of Technology.

Kjekshus H. (1977) Ecology Control and Economic Devekiopment in East African History. The Case of Tanganyika 1850-1950. Heinemann, London.

Kollmann P. (1899) Victoria Nyanza. Swan Sonnenschein & Co. Ltd, London.

Malcolm D.V. (1953) Sukumland. An African People and Their Country. London. (International African Institute). Oxford University Press.

McLoughlin P. (1971) An Economic History of Sukumaland, Tanzania, to 1964. Field Notes and Analysis. New Brunswick. PMA, 305 University Ave, Fredricton.

Perham M. (1976) East African Journey – Tanganyika and Kenya 1929-30. Faber and Faber, London.

Rounce N.V. (1949) Agriculture of the Cultivation Steppe. Cape Town.

Sandström K. (1995) Forests and Water – Friends or Foes? Hydrological Implications of deforestation and land degradation in Semiarid Tanzania. PhD Thesis, Linköping Studies in Arts and Science, Linköping University, Sweden.

Stanley H. (1890) In Darkest Africa. Vol. II. New York. Charles Scribner´s Sons.

Therkildsen O. (1988) Watering white elephants? Lessons from donor funded planning and implementation of rural water supplies in Tanzania. Scandinavian Institute of African Studies, Uppsala.

WMP. (1978) Mwanza Water Master Plan. Vol. 6. Hydrogeology Studies. Brokonsult, Stockholm.

# 23
# PROVISION AND MANAGEMENT OF WATER SERVICES IN LAGOS, NIGERIA, 1915-2000

*Ayodeji Olukoju*

Figure 23.1 Location of Lagos, Nigeria.

## 23.1. GENERAL BACKGROUND

This chapter provides a history of water supply since 1915, with an emphasis on current problems of water supply in the city of Lagos. It presents results of recent research into efforts by state and private actors to deal with inadequate water supplies in contemporary Lagos. The chapter highlights the activities of water tanker operators and the producers and retailers of sachet water ("pure water"); citizens' self-help efforts, such as the sinking of boreholes and wells; and the role of government in water supply and regulation of the quality of non-state supplies, especially "pure water." In spite of these efforts, this city of over twelve million persons continues to experience a general scarcity of water, especially during the dry season (November to March).

The provision of urban infrastructure facilities and social amenities has been a major task of local, state and the national governments in Nigeria since the era of British colonial rule (c.1900-60). Since independence in 1960, all tiers of government have grappled with this problem with largely unsatisfactory results owing to a combination of poor funding, lack of maintenance of public facilities and an unabating rural-urban drift, which has stretched urban facilities to the breaking point. Nowhere is this crisis of population pressure and inadequate urban facilities more glaring than Lagos, Nigeria's capital from 1914 to 1991, and West Africa's leading industrial, commercial and maritime centre. This chapter focuses on water supply in Lagos since the colonial era with an emphasis on current developments in this critical sector of the urban society and economy. It dwells on a period and on issues that have either been neglected or had not been adequately covered in the literature (Onakerhoraye 1984; Olukoju 2003). The discussion in this chapter is based upon recent research in selected parts of the city.

## 23.2. SCOPE AND METHODOLOGY OF THE RESEARCH

Lagos, the subject of this chapter, is an urban sprawl that is home to over twelve million persons. Consequently, the research was carried out in selected wards of the city: Ikoyi, Lagos Island, Mushin, Oshodi, Isolo, Agege, Ogba, Somolu, Ketu, Kosofe, Ajegunle, Orile, Festac, Ikeja, Surulere and Yaba. Even so, the fieldwork was focused on randomly selected streets in each of these areas, the depots and production sites of water vendors and the offices of the Lagos State Water Corporation and the National Agency for Food and Drugs Administration and Control (NAFDAC).

Four different but complementary questionnaires were administered respectively to selected households, the Lagos State Water Corporation (LSWC), NAFDAC and water vendors. A very important strategy in the peculiar circumstances of Lagos was the use of field assistants who reside in and have first-hand information on the areas of study. Their personal observation and the structured interview approach proved rewarding. Another method that enriched data collection was the decision to appoint family contacts in the two government establishments as field assistants. Yet, the returned questionnaires from the two public institutions were below average. In all, out of the 650 questionnaires that were sent out in the household category, over 500 were completed and returned. This was possible because the field assistants had to assist many of the respondents—who were themselves illiterate—in completing the questionnaires. Such respondents found it convenient to provide the answers as the questions were read, which the research assistants duly registered on the questionnaires. This method also expedited the administration of 60 questionnaires to water vendors at three of their bases in Oworonsoki, Gbagada and Ikorodu (a major settlement on the outskirts of the city).

## 23.3. HISTORICAL BACKGROUND: POTABLE WATER SUPPLIES IN LAGOS SINCE 1915

The supply of potable water to the city of Lagos was officially commissioned in 1915 after the first waterworks in Nigeria were completed at Iju on the outskirts of the city. For several decades after this event, the supply was limited to the Island of Lagos, including the European settlement at Ikoyi. It was only from the 1950s onwards that the supply was extended to the mainland, specifically communities like Agege, Oshodi and Mushin, which were then outside the municipal boundary. The spatial expansion of this facility was rather slow partly because of the cost of the project and partly because of the political complications of inter-governmental relations between the Colony (as Lagos was until independence in 1960) and the adjoining Western region.

From independence till 1967, Lagos was a Federal Territory administered by the Federal Ministry of Lagos Affairs. Thus, statutorily, the mainland settlements beyond the municipal boundary at Igbobi were outside the jurisdiction of the Federal Territory and were thus not entitled to amenities provided by it. Nevertheless, water supplies were extended to Agege and Oshodi in the late colonial period though this was initially by means of water-retailing stations. The creation of Lagos State in 1967 solved this problem by bringing the mainland settlements of Ajegunle (Ajeromi), Mushin, Oshodi, Ikeja, Shomolu and Agege into the same political entity and, more specifically, into the municipality. By the 1980s, Lagos city had expanded to incorporate all these outlying settlements, with serious implications for physical planning and the provision of public infrastructure and urban facilities.

In the meantime, the rapid increase in the population of Lagos, which had been the commercial, political, industrial and maritime nerve centre of Nigeria, had meant that the expansion of the pipeline network could not keep pace with the physical expansion of the city. First, not all parts of the expanding city were covered by public supplies. Second, even those areas within the network did not receive steady water supplies. Consequently, Lagosians had had to devise alternatives to public water supplies. This meant in essence reliance on supplies from wells, many of which were relatively shallow, and which, therefore, yielded water of dubious quality (Duncan & Olawole 1969). With the development of technology for constructing boreholes, deeper and more sophisticated "wells" were constructed by the more affluent members of the society. Still, the gap between supply and demand continued to increase, as is indicated in the following table on the projected population and water demand in Lagos between 1985 and 2005.

Table 23.1 Projected population and demand for water in Lagos, 1985-2005 (Source: Olaosebikan, 1999, 14.)

| YEAR | PROJECTED POPULATION | WATER DEMAND |
|---|---|---|
| 1985 | 6 million | 160 million gallons per day |
| 1990 | 8 million | 199 million gallons per day |
| 1995 | 10 million | 256 million gallons per day |
| 2000 | 12 million | 295 million gallons per day |
| 2005 | 15 million | 332 million gallons per day |

Meanwhile, the State government had been making efforts to bridge the widening gap. In the 1970s and 1980s, facilities of various capacities were built or expanded. The civilian administration of Governor Lateef Jakande (1979-83) constructed several micro- and mini-waterworks to supply the needs of communities across the city and state. Second, the Iju waterworks were expanded while major dams were constructed at Ishasi and Adiyan. This increased the volume of potable water available to Lagosians. Still, at the best of times, only 47 per cent of the inhabitants of the city had access to potable water "at reduced level of service" (Lagos State Water Corporation 2000). Apart from the acknowledged shortfall in supplies, the prevalence of broken pipes further reduced supplies to consumers. In any case, many of those who received supplies did not pay their bills and the Water Corporation has had to bear the burden of crippling debts. Being handicapped by a combination of adverse circumstances, it failed to meet the yearnings of a growing number of consumers. The present Governor, Bola Tinubu, has tried to solve the problem by engaging private sector technocrats to manage the corporation, with the ultimate aim of privatising it. Attempts have also been made in the meantime to secure external funding, including foreign loans, to finance the operations of LSWC.

The increasing incapacity of the state has thus created a yawning gap in the water supply sector that is being filled to some extent by private operators of various descriptions, mainly, the retailers of water: the "Hausa" porter, the water tanker driver and the sachet water ("pure water") producer and hawker. (Figure 23.2) The emergence of these retailers and the concomitant need for regulation of quality led to the establishment of NAFDAC, the government agency responsible for quality control in the food and drugs sector. In effect, the management of this important public facility has involved state and non-state actors with ramifications that are highlighted in this chapter.

A key issue in the management of water supplies in Lagos is the maintenance of quality. Allusion has been made to the low quality of well water because the wells are not deep enough

Figure 23.2 Hawkers of "pure water" and water vendor using wheelbarrow in Lagos. (Photo: Ayodeji Olukoju)

to go beyond the contaminated sources of water which are close to the surface. Moreover, the sanitary conditions of the city, especially blocked drains, as well as the dumping of refuse in drains and other unauthorised places, pose a great hazard to water supplies given the prevalence of burst pipes. Not only do these broken pipes reduce pressure and supplies, they also make possible the contamination of water supplied to surrounding areas. Consequently, there were cases of epidemics which resulted in fatalities in the 1970s and 1980s. Even in the 1990s, Lagos witnessed a wave of typhoid fever, which was commonly attributed to contaminated water.

Worse still, the public water supply fails too often in Lagos for a variety of reasons. First, the Lagos State Water Corporation does not have as much money as it needs to procure equipment and material required to adequately supply water to the city and Lagos State. The government has invested as much as N2.2 billion in the water supply (Ugwuanyi 2003) but this appears to be far less than required. It is said to need as much as N320 billion ($1.8 billion) to meet the water needs of an estimated 12 million consumers in the state (Imonikhe 2001). In any case, revenue accruing from the operations of the Corporation has consistently fallen behind outlays (Ekanem 2003). Fieldwork and responses to questionnaires reveal that not only do many consumers not know the rates charged by the Water Corporation, many of them do not pay their bills. Non-payment of water bills or poor record of revenue collection is a major problem affecting LSWC. However, since 2001, the management of the Corporation has been privatised and revenue collection has improved with the use of incentives. Still, attempts have been made to raise the huge capital required for the optimal performance of the LSWC through internal and external loans. The option of taking out external loans has, however, been criticised in some quarters

(Olaopa 2002). A complementary strategy has been the privatisation of LSWC, which has been declared to be the aim of the administration of Governor Tinubu since 1999 though this has not been achieved. To be fair, the new management of the LSWC, comprising technocrats from the private sector, appears to be achieving some results but the issue of the financial weakness of the Corporation remains unresolved. Though foreign companies have expressed interest in investing in the sector, no concrete results have been achieved (Odita-Fortun, 2002; Uwaegbulam 2002, 2003)

Second, erratic electric power supply has compounded the water supply situation in Lagos (Ajibade 2000; Ugwuanyi 2003; Olukoju 2004). Any major outage by NEPA affecting the Water Corporation's pumping stations at Iju and Adiyan automatically affects the supply of water to the city. Though such incidents might be accidental, they could also be deliberate since the Corporation is indebted to NEPA or even because of disagreements between NEPA and the state government! (Azuora, 1999) A newspaper report alluded to the rift between the two as a cause of the irregular power supplies (*National Concord* 2000, 3).

However, whatever may be said for or against private sector intervention in water supply in Lagos, it must be admitted that a considerable proportion of the water supply is at least maintained through various categories of water vendors: the "Hausa" retailer, the tanker driver and the "pure water" hawkers. Apart from the issue of adequacy of supplies, the quality of water from these sources has also been questioned and the scepticism has been justified by scientific analysis, with dire implications for public health (see, for example, Nwokocha & Nwankwo 2000; Ekpunobi 2001; Abdullahi 2002; Okeugo 2003; Olukoju 2003; Shofuyi 2003). However, whether the tacit encouragement of such private sector actors is the best option remains to be seen. To be fair, there is a growing clamour for greater private initiative in almost every sector of the economy and the water sector is definitely not an exception. Even without the call, it is clear that an alternative arrangement is required in a situation where the public water supply is seen to be both inefficient and inadequate. However, the consequence of unregulated exploitation of the city's underground water resource has not been fully grasped. Undoubtedly, a more efficient public water supply is critical to the sustainable management and conservation of the available water resources for present and future generations.

The spatial dimension and consequences of the water supply crisis in Lagos deserve some comment. Contrary to what might have been expected, the geographical distinction among the various areas of Lagos in terms of the water supply is more apparent than real. Lagos Island, where potable water made its debut in Lagos in the mid-1910s, is as disadvantaged as many areas on the mainland. Most inhabitants of the area, it was reported in 1999, "now patronise water contractors called MAI-RUWA or use well-water where available. Otherwise, it is a long queue lasting up to two hours to collect a bucket full of water barely sufficient for one individual's morning bath. Today, March 21, 1999, the going price for [a] 25 litre keg of water is N50" (Sobowale 1999). The cost of getting sufficient water for family use by this means for bathing, washing, flushing toilets and other domestic activities is thus bound to be considerable.

As for drinking water, even "pure water" (sachet water) could not be trusted in the face of the recurring outbreaks of typhoid fever in the city in the late 1990s.

Having stated that water scarcity is common to all the study locations, it must be admitted that the coping strategies necessarily differ from place to place. A critical factor in determining the dominant response to the water supply crisis is the income level of the consumers. Those who are rich enough to sink boreholes or wells, or to patronise water tankers, are not as vulnerable as the poorer people who cannot afford such interventions. To that extent, the inhabitants of high-income, low-density, planned neighbourhoods have the means or resources to cope comparatively better with water shortages and the accompanying threats of epidemics than those elsewhere. At the other end of the scale, inhabitants of the high-density, low-income parts of the city have come to terms with their predicament. They either resign themselves to their fate and patronise local retailers, or seek to tap supplies from better-served areas. In Ajegunle, the response to water scarcity consists of reliance on well water, scavenging from pipes in gutters or tapping from the nearby Apapa Government Residential Area (Nwokocha & Nwankwo 2000).

It is clear from the foregoing that official sources of water supply have always fallen short of demand, though the situation became desperate only since the 1960s, especially with the massive increase in population. The following section examines the contributions of water tanker operators to remedying the shortfall in the supply to the city.

## 23.4. WATER TANKER SUPPLY BUSINESS

Although a spatial disaggregation of responses is difficult to achieve, given the broad similarity across the city, evidence from fieldwork, especially analysis of the questionnaires, revealed the following. First, the supply of water through tankers (Figure 23.3) is necessitated by the inefficient water supply in Lagos. This has led to an informal private sector initiative of a significant magnitude. The capacity of the tankers used in the supply ranges from 250 to 1000 litres, and the prices for the full tank supply range from N500 to N1800, depending on the particular size of the tanker, the distance to be covered by the supplier and seasonal variation. Though precise figures are not available, the financial worth of the private water supply business must run into tens of millions of Naira per annum, with considerable demand being fuelled during the dry season when the scarcity of water is often acute, or by damage to the mains supplying water to an area during road construction.

Second, the business of water vending through tankers is subject to seasonal variations. During the dry season (November to March), the demand for water tends to be high and the price slightly higher than during the wet season. This is understandable as the ground water tends to be at a low ebb during the dry season. Third, the tanker operators source their water mainly from boreholes, water corporation reservoirs, and streams. To ensure that the water supplied is clean and safe, the tanker operators clean their tanks regularly. Still, the quality of their supplies cannot always be guaranteed, given the fact that tanker operators cannot always secure treated water from the government-owned waterworks while, as has been indicated above, many of the boreholes are not deep enough to reach the aquifers that yield clean water. Fourth, the clientele is varied—individuals, households, banks, companies, building sites and cement block-making sites.

Fifth, the operators face various problems, including bad roads in the city and irregular electricity supply. The latter hampers sourcing water from boreholes, which are dependent on electrical energy. Sixth, the operators have a minimal level of organization in the form of associations. However, not all operators belong to these associations. Also, these associations do not seem to wield any influence in government circles. Still, they collect dues and fees from their members, and also offer them financial aid through loans/advances. The associations play a role in the renting of tankers from their owners by the tanker drivers. The amounts payable by the tanker drivers to the tanker owners also vary with the size of the tankers and the season. For a 1000-gallon tanker, the rate is some N3000 while up to N4000 is paid for a tanker with a capacity of 2000 gallons. For tankers with a capacity of 3000 gallons, the rent delivered at the end of each day is stated to be N8000 but it is clear that the amounts vary. These figures are subject to adjustments because of the rising rates of inflation.

While the water tanker operators supply treated water to consumers in sizeable quantities, sachet water producers and hawkers do so in smaller consignments. The latter are, however, subject to quality control by NAFDAC, an agency of the federal government. The next section examines the activities of that organisation.

Figure 23.3 Water tanker owned by University of Lagos, and borehole owned by a local government in Lagos. (Photo: Ayodeji Olukoju)

## 23.5. NAFDAC AND WATER MANAGEMENT IN LAGOS

The National Agency for Food and Drugs Administration and Control (NAFDAC) was established in 1993 through an enabling Law – Decree 15, of 1993. It had been mandated to regulate and control the import, export, manufacture, distribution, sale, and use of packaged food, drugs, cosmetics, medical devices, packaged water, detergents and chemicals. For our purpose, NAFDAC plays a regulatory role in the manufacture, distribution, sale and use of packaged water across the country. This entails enforcement of compliance with set standards that ensure the provision and distribution of sanitary packaged water, an increasingly important source of supply to Lagos city.

There are two major forms of packaged water in Lagos as in other parts of Nigeria: sachet water, popularly known as "pure water", and bottled water, the latter sold in plastic bottles or larger plastic containers supplied mainly to companies and establishments. To date, NAFDAC has licensed thousands of pure water manufacturers to operate. In order to be licensed, every manufacturer of sachet water is required to meet certain requirements. First, the firm must be formally incorporated, under the appropriate Act, to produce sachet water. Second, the sachet water must have a trademark registration, which identifies its products in the market, especially in the event of any breach of the regulations. Third, production must not be done in a residential environment, and must attain the general manufacturing practice standards. Fourth, the sachet water upon laboratory analysis must satisfy the general manufacturing standards and must be safe for consumption. Fifth, the manufacturer must institute quality control measures and retain a water analyst to ensure continued satisfactory production of the sachet water.

In addition to these requirements, the manufacturer has to complete the necessary documentation and pay a registration fee ranging between N30,000 and N35,000. The licence is renewable every two or five years depending on the type of registration. Ordinary listing is for two years while full listing is for five years. On renewal, the fee paid is usually not up to that paid for registration and initial licensing.

NAFDAC usually monitors the operation of these manufacturers, to ensure compliance with manufacturing and distribution standards. This is done through spot or routine compliance visits to these factories. However, in spite of the best efforts of NAFDAC, laboratory analyses have shown that the quality of "pure water" sold in Lagos is dubious though no epidemic has been reported (Olukoju 2003).

## 23.6. SACHET WATER AND ENVIRONMENTAL SANITATION IN LAGOS

Although Lagos has a long history of poor environmental sanitation (Olukoju 1993; 2001), the city has faced a fresh menace in the poor disposal of used sachets of 'pure water' (Ekpunobi 2001; Akpabio, 2003). While this has attracted the attention of NAFDAC and other government agencies, nothing concrete or effective has been done to check this. It is acknowledged that these sachets end up blocking drainage systems and engender erosion/flooding in many areas of the metropolis.

To check the littering of the streets by the indiscriminate disposal of sachets, NAFDAC is considering a systematic replacement of the nylon sachets with plastic packs. However, this can only be done through appropriate consultation with the manufacturers. The option of penalizing the manufacturing companies for the indiscriminate littering of nylon sachets is problematic, as detecting the erring manufacturer may be difficult. Also, the responsibility for arranging for the safe and sanitary disposal of nylon sachets of 'pure water' was not stated in the licensing guidelines. Thus, NAFDAC is not yet justified in penalizing these manufacturers. One other option, which is feasible, is that of educating the public on the proper disposal of their nylon/polythene sachets. NAFDAC is also exploring this option. Still, it needs the cooperation of the users themselves, the local government authorities, religious and community leaders, and schools. Each of these actors has a role to play in educating, correcting and sanctioning errant consumers of packaged water.

On the issue of the quality of water provided by whatever means for consumption by Lagosians, NAFDAC is aware of the inadequate and relatively unsafe water supply in Lagos. However, the Agency believes that the appropriate and regulated involvement of private operators in the business of water provision may offer a sustainable solution. NAFDAC is nonetheless committed to regulating and controlling the activities of these companies to ensure that the water supplied is safe for human consumption. The Agency, in the bid to achieve its set objectives, relates effectively to all stakeholders in the business of water supply and the relevant supportive government agencies like Local Governments.

## 23.7. CONCLUSIONS

This chapter has highlighted major developments in the supply of potable water to the inhabitants of Lagos, Nigeria since the colonial period, with a focus on recent developments. The following are the notable features of water supply in the city during this period.

Older neighbourhoods in the original municipality of Lagos are generally covered by a network of pipelines though most are also aged, leaky or damaged. Places like Ikoyi and Lagos Island, Mushin and most parts of Shomolu have public supplies but Ajegunle, a densely populated section of the city, is largely outside the network of public mains! In general, Lagosians depend more on boreholes and wells as sources of water than on the public water supply. The usage of water is not limited to the domestic domain as demand also comes from commercial consumers, especially for construction purposes.

The public water supply even in highbrow areas such as Ikoyi and Victoria Island is generally erratic, compelling residents to depend on water vendors for their supplies. Shortages in these areas are said to last for several weeks. But FESTAC Town (a large estate) provides a striking exception, as it seems favoured with functional public water taps. This, however, did not stop the inhabitants of the area from digging boreholes and wells as reliable alternatives to the public water supply. Likewise in Surulere, the water supply does not pose as serious a problem as is the case in Orile and Ajegunle, where people depend mostly on water vendors. According to respondents, for most residents of tenement buildings, water pipes are merely for decoration.

People in this part of Lagos wait eagerly for the rainy season—when they can collect rainwater—to ease their dependence on the purchase of water. In comparative terms, inhabitants of high-density areas like Ajegunle, Mushin and Shomolu suffer more from irregular supplies than those in low-density, high-income areas like Victoria Island. Again, the densely populated areas on Lagos Island are as disadvantaged as similar areas on the mainland.

The pattern of water consumption in most Lagos households is generally similar though the size of households and the type of building determine the quantity of water consumed per day. The average daily consumption per household is accounted for by cooking, bathing, the flushing of toilets and washing. Most people boil water and this is the general treatment for bad water; during the dry season, most rely on the purchase of "pure water" (Ekpunobi 2001). Indeed, the survey reveals that Lagosians have become more dependent on packaged water, especially "pure water," in their daily activities as it becomes the alternative source of water each time (as is often the case) public supplies fail. Still, the dependence on "pure water" is more acute on the mainland than in the low-density areas of Ikoyi and Victoria Island, and other high-income residential estates in other parts of the city, where affluence and status permit access to bottled water, potable water, borehole water and supplies by tankers. In effect, "pure water" is the preserve of the lower classes.

On the whole, the water business is big business in Lagos (Magbor & Mustapha, 2001). The big-time retail business is dominated by retired or retrenched workers and other persons who eke out a living, as water tanker drivers, selling water to individual homes and for construction purposes. However, a newspaper report claimed that some unscrupulous water tanker operators deliberately sabotage public supplies to boost their own business (*The Punch* 2001)! But, it is difficult to verify the veracity or extent of this practice.

## ACKNOWLEDGEMENTS

I acknowledge the financial support of the French government-sponsored programme on urban development (PRUD), managed by IMSTED-GEMDEV, France, towards the research leading to this publication and for my participation in conferences in Dakar and Paris in 2003 and 2004 in the context of the project. I am grateful to Professor Alain Dubresson and Dr. Laurent Fourchard, who led the Global South and Nigerian teams, respectively, to which I belonged. The assistance of my research assistants, especially Paul Osifodunrin and Lanre Davies, is hereby gratefully acknowledged.

## 23.8. REFERENCES

Abdullahi F. (2002) How pure is our water? ThisDay (Lagos), 28 April 2002.
Adebusuyi D. (1999) Persistent outages causes water scarcity in Lagos. The Punch (Lagos), 16 August 1999.
Ajibade A. (2000) Obstacles against provision of potable water in Lagos. Daily Times, 19 June 2000.
Ajunwa C., Ebhodaghe S. & Olaiya A. (1999) Lagos taps dry up, residents thirsty. The Guardian, 4 October 1999.
Akaraogun O. (1999) Water and public health neglect. Daily Champion (Lagos), 23 May 1999.
Akitoye H. (1999) Water shortage worries Lagosians. National Concord (Lagos), 4 July 1999.
Akpabio R. (2003) The menace of 'pure water' sachets. The Guardian, 24 March 2003.

Azuora C. (1999) Lagos water shortage blamed on black-out. New Nigerian (Kaduna), 1 October 1999.
Bamidele R. (1999) Water shortage claims 17 lives in Lagos. National Concord, 15 December 1999.
Banjoko D. (1999) When the trickle dries up. The Punch, 21 May 1999.
Duncan J.W.K. & Olawale A.O. (1969) Properties of Water From 9 Existing Wells in Shomolu, A Suburb of Lagos. The Nigerian Engineer 6(2) 17-19.
Ekanem W. (2003) LSWC needs N2b to supply water annually … earns N496m. The Punch, 18 January 2003.
Ekpunobi B. (2001) Poor or pure water, the beat goes on. Daily Champion, 22 March 2001.
Ijediogor G., Ekunola T., Olaiya A. & Adebanwo K. (1999) In search of the fountain of life. The Guardian, 8 May 1999.
Imonikke T. (2001) Corporation needs N320bn to supply Lagos water. Daily Champion, 20 April 2001.
Inyama N., Ogah D., Nwannekanma B., Ebosele M. & Nwadike C. (1999) Again, Lagos taps run dry. The Guardian, 14 December 1999.
Isiguzo I. (1999) Water sir, not words. Vanguard (Lagos), 21 May 1999.
Lagos State Government (1987) Lagos State Handbook 1987, Ikeja: Ministry of Information.
Lagos State Water Corporation, 2000:28.
Magbor D. & Mustapha Y. (2001) Water supply is big business. Daily Times, 12 January 2001.
National Concord (2000) Water corporation, NEPA settle rift. 3 January 2000.
Nwokocha J. & Nwankwo T. (2000) In Lagos, safe water is a swan song. The Vanguard, 19 March 2000.
Obike U. (2001) Lagos repositions corporation for efficient water supply. This Day (Lagos), 19 December 2001.
Odita-Fortune S. (2002) Lagos, British firm team up to improve water supply. The Guardian, 13 February 2002.
Ogah D. (1999) Lagos shops for N142 million to tackle water scarcity. The Guardian (Lagos), 14 September 1999.
Ogbu C. (2001) LSWC reshapes for privatisation. The Punch, 23 December 2001.
Okanlawon S., Ijediogor G., Ajunwa C., Olaiya A. & Nduka H. (1999) City of water is thirsty. The Guardian on Saturday (Lagos), 24 April 1999.
Okeugo C. (2003) Epidemic imminent as water scarcity hits Lagos. ThisDay, 8 March 2003.
Olaopa T. (2002) Water supply without external loans. The Guardian, 8 March 2002.
Olaosebikan Bola (1999a) Lagos State Water Corporation: Dawn of a New Era, Lagos: Lagos State Water Corporation, 1999.
Olaosebikan Bola (1999b) Lagos water emergency. National Concord, 19 December 1999.
Olukole T. (2000) Water epidemics loom in Lagos. Nigerian Tribune (Ibadan), 8 January 2000.
Olukoju A. (1993) Population pressure, housing and sanitation in West Africa's Premier Port-City: Lagos, 1900-1939. The Great Circle: Journal of the Australian Association for Maritime History 15(2), 91-106.
Olukoju A (2001) The Pluralisms of Urban Waste Management: A Comparative Study of Lagos (Nigeria) and Tokyo (Japan) in Agwonorobo Eruvbetine (ed.), The Humanistic Management of Pluralism: A Formula For Development in Nigeria, Lagos: Faculty of Arts, University of Lagos, 2001, pp.508-525.
Olukoju A. (2003) Infrastructure Development and Urban Facilities in Lagos, 1861-2000, Ibadan: IFRA, 2003.
Olukoju A. (2004) 'Never Expect Power Always:' Electricity Consumers' Response to Monopoly, Corruption and Inefficient Service in Nigeria. African Affairs 103(410), 51-71.
Oluwafemi S. (1999) For Lagosians, no water to drink. National Concord (Lagos), 9 July 1999.
Sanni L. & Ajulo S. (1999) Water rates to go up as LSWC seeks N2b. The Guardian, 20 December 1999.
Shofuyi S. (2003) Study x-rays poor quality of 'pure water'. The Punch, 4 February 2003.
Sobowale D. (1999) Potable water: Marwa's waterloo. Vanguard, 24 March 1999.
The Punch (2000) Water supply in Lagos. (Editorial), 19 June 2000.
The Punch (2001) Lagos State Water Corporation decries destruction of equipment. 25 December 2001, 6.
Ugbodaga K. (2001) Why private sector participation in water supply. Daily Times, 20 April 2001.
Ugwuanyi E. (2003) Lagos water corp. invests N2.2b to boost water supply. The Vanguard, 18 February 2003.
Uwaegbulam C. (2001) Lagos kicks off fresh bid for plum water supply scheme. The Guardian, 25 March 2001.
Uwaegbulam C. (2002) SA firm joins bid for Lagos private water scheme. The Guardian, 28 January 2002.
Uwaegbulam C. (2003) Lagos drafts law to privatise water supply. The Guardian, 13 January 2003.

# 24

## EXPANDING RURAL WATER SUPPLIES IN HISTORICAL PERSPECTIVE: SIX CASES FROM FINLAND AND SOUTH AFRICA

*Petri S. Juuti, Tapio S. Katko, Harri R. Mäki & Hilja K. Toivio*

Figure 24.1 Key places mentioned in the chapter.

## 24.1. GENERAL BACKGROUND

Although the greatest problems in water supply and sanitation are linked to urban population growth, the majority of the world's population still in the first decade of the new millennium live in the countryside. Thus, billions of people, particularly in developing countries, live largely in agrarian or other rural-like conditions.

This chapter describes altogether six cases from the Finnish and South African countryside — with various types of small-scale and expanding systems in the rural environment at various times. Even long after WWII, Finland is an agrarian country, where traditional bucket and well systems are still in use outside the networked systems in dispersed rural areas. In the case of South Africa, there is not much long-term research in this field because, for obvious reasons, other aspects of its political and economic history have been in the forefront. While Finland has been placed among the top positions in several water and environment-related international comparisons, South Africa has also provided some technological breakthroughs in water management. In any case, examples from history can on their part illuminate current local water crises.

## 24.2. OBJECTIVES AND STRUCTURE

This chapter discusses the evolution of rural water supply in Finland and South Africa in historical context. The distance between the case countries is very long, and there are no colonial ties, either. Still, some similarities or analogies can be found. Surprisingly, the annual rainfall average is almost the same. In Finland it is between 600 and 700 mm. In northern Finland the annual rainfall is about 600 mm, half of which falls as snow. In South Africa the average annual rainfall is 464 mm, ranging from less than 10 mm in the western deserts to 1,200 mm in the eastern part of the country. However, some 65 percent of the country does not receive enough rainfall for successful dryland farming.

Since 1960, rapid growth in the human population of Africa has created a great demand for water and food. This population expansion has caused many problems in land use and erosion - particularly within catchment areas - that is the areas drained by a river or body of water. The overall expansion of human settlements has increased organic loadings to water bodies and groundwater. Especially small lakes could not take the effluent load without damaging the ecosystem.

Based on six case studies — three from Finland and three from South Africa — the chapter provides an overview and analyses the key principles that drive the evolution of rural water services. Can we find certain patterns, principles and practices that have possibly some analogies and similarities? Finally, we will discuss what kind of possible principles of sustainable and viable water services and their management in rural areas can be used successfully. And what have been and what are possible present linkages between rural and urban systems and development.

Petri S. Juuti, Tapio S. Katko, Harri R. Mäki & Hilja K. Toivio

The chapter concentrates on water supply but also deals with sewerage services to some extent. The first case concentrates on the overall rural water and sanitation development in Finland, the second case on the small municipality of Kangasala and further its Sahalahti region (3rd case), the fourth case on South Africa and a historical water conflict in Southern Bechuanaland, the fifth case on the Mphahlele Rural Area, and the sixth case on wind power as a new innovation for watering Karoo, a dry plateau area in South Africa.

## 24.3. CASE 1: DEVELOPMENT OF FINNISH RURAL WATER SUPPLY

The earliest sources of water supply in Finland were springs and wells, but surface waters also had to be used in times of drought. Then, various types of simple piped systems started appearing on manors and larger farms in the end of 1800s. House owners also started to establish small associations in the early 20th century for the purpose of building wells. (Juvonen 1998, 29; Juuti et al. 2003, 22–25)

The oldest examples of water services in Finland are from the countryside, whereas similar rural technologies were also used in earlier urban centres. Traditionally so-called witchers sited wells. It was believed — and some people still believe — that these local experts can locate water veins with the help of a forked willow branch.

In 1949 to 1950, some 40 of the most famous well witchers were invited to locate the "water veins" in the Botanical garden of Helsinki University. The eyes of the witchers were covered, and the result was as many different maps on water veins as there were witchers. Yet, a considerable number of the citizens still seem to rely on these witching methods. In reality, well witchers are practical geologists who are able to locate the most potential sites simply through vegetation, other signs in nature, and their earlier experience. Besides, small amounts of groundwater can be found in Finnish conditions almost everywhere. (Katko 1997a)

The earliest written proposal for constructing a common piped water supply appeared in the Wasa Newspaper in 1863. (Hn. 1863) It was suggested that a drilled wooden pipeline be built to carry water by gravity from a natural spring. The first documented case of such a common piped water supply with several users was constructed in Ilmajoki in Ostrobothnia in 1872. This may be compared with the first urban water supply and sewerage system that started operation in 1876 in Helsinki, the Finnish capital.

The rural system in Ilmajoki was soon followed by other small systems in the same area. These water pipes were constructed out of wooden pipes that were drilled manually from pine trees. A skilled driller could produce some 40 to 50 meters (130 to 160 feet) of pipe in a day. (Figure 24.2)

Figure 24.2 A manual driller of water pipes made of pine wood in 1913 in Sumiainen, Central Finland. (Photo: Vanhan Äänekosken kotiseutuyhdistys.)

In the early 1930s, a special machine was developed for drilling wooden pipes. It is difficult to know the origin of the use of the wooden pipes, but according to one professional article (Mäkelä 1945), the idea might have come to Ostrobothnia from the United States along with returning emigrants. However, the earliest single wooden water pipe in Finland has been found in Turku, dating back to the mid-1600s. Besides, wooden pipes were also largely used in the European continent like Germany (Bärthel 1997), and many of the structural and civil engineering codes and practices in Finland were borrowed from Germany (Katko 1997b, 53; Katko 2000).

Traditional wooden well structures were considerably improved once concrete rings came into the market in the 1930s. The traditional counterpoise lift and winches, and later also hand pumps, were introduced. Local experts manufactured the latter from wood before the factory-made models out of cast iron came to the market. After World War II, the Work Efficiency Institute of Finland introduced a developed model of the women's double yoke with padding for the shoulders. The need for this model, at that time a new innovation, was quite real. In 1951 it was estimated that the Finnish women walked daily the distance from the earth to the moon and back while carrying water from wells to the cow shed and house (Wäre 1952).

The evolution of rural water supply is largely based on cattle farming. At first, electricity was introduced to farmhouses in Finland. Second, a water pipe was constructed to the cow shed, followed by a sewer constructed from the house. Finally, a water pipe was drawn to the house (Figure 24.3). (Myllyntaus 1991, 248-253; Juuti & Wallenius 2005, 46-47)

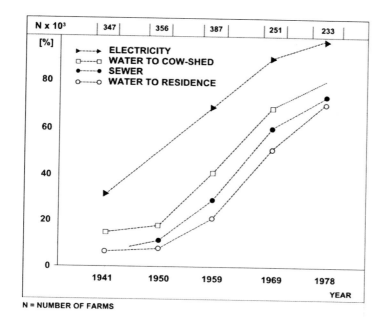

Figure 24.3 Diffusion of electricity, water and sewers to Finnish farmhouses. (Source: Katko 1997a)

The Finnish government started to support the rural water supply through the first financial support act in 1951. This was preceded by the work of the Parliamentary Committee for the Rationalizing of Households, having also three female parliament members — thus highlighting the role of women in daily household matters like the supply of water (Kotitalouden rationalisoimiskomitea 1950).

## 24.4. CASE 2: EXPANSION OF SYSTEMS IN AND FOR KANGASALA MUNICIPALITY

The next case presented is from the municipality of Kangasala, Finland – some 150 kilometres north of Helsinki, the capital, and just east of Tampere, the third biggest city of Finland. It shows how the first individual water supply systems for single houses or a few houses were used. This was followed by several consumer-managed cooperatives that were gradually also established for the central village and later taken over by the municipality. The systems expanded further and inter-municipal cooperation was started – first in sewerage services and later in water supply.

The conditions in some parts of Kangasala did not allow building wells. One such area was the central part of the village around the church, where different types of well associations sprung up; another was the Ilkko area further away from the municipal centre where groundwater lay quite deep. The well-based system remained the primary form of water supply for the first few decades of the 1900s, although primitive piped systems also existed. Manors, hospitals, schools and other large consumers of water had their own small works or highly productive wells. (Ilkko Water Cooperative; Juuti et al. 2003, 24)

The Tulenheimo manor in the central village had its own small-scale piped water system already at the turn of the 20th century. It drew water from a driven well on its own land. The large Lihasula farm in the northern part of the municipality led water by gravity via wooden pipes to the cowshed, stable and the main building of the farm already in the mid-1890s. As consumption increased on Lihasula farm, water was drawn from Lake Vesijärvi starting in the 1930s. A wind engine pumped the water into a tank on top of a hill. Both Lihasula and Tulenheimo used windmill pumps: in the former a natural well was the water source and in the latter bore holes were installed and the best available technology and specialist were used. (Vesiyhtiöpapereita 1927–1928, 21.6.1927, 28.9.1927, 8.12.1927, 26.10.1928; Raitio 2000, 74)

Kangasala municipality started first to supply water to its own units and facilities. Various regulations had already earlier been enacted to promote fire safety, which was the first concern immediately followed by the water supply, sewerage and public hygiene. Wells played a central role in school development projects. One task of the Municipal Building Board was to oversee that public building projects, including their wells, were properly implemented in the early decades of the 20th century. The work was contracted out and its quality was monitored closely. The quality of the well water was also tested regularly. (For example, Municipal Building Board, 11.11.1929, 4.12.1929, 17.12.1934; Virkkala et al. 1949, 495; Anttila, 26, 208–210)

In 1927 the Kirkonkylän Vesijohto Osakeyhtiö (Church Village Water Pipe Co.) was established. Later it became known as the Tulenheimo Water Co., in which the municipality was also involved. The waterworks comprised a small pumping station on Lake Ukkijärvi, a 2-inch galvanized water pipe, a small water tank and a distribution network. Initially water was drawn directly from the lake, but later a well was built on the bank. (Figure 24.4) Part of the water issued from an esker while the rest was absorbed from the lake through bank side gravel strata. The Tulenheimo Works served large water consumers and some houses. It played a major role in the development of the municipal water supply and sanitation. The time of the modern conveniences of central heating, piped water and inside toilets was still far away. Only in 1936 did the first house receive such conveniences. (Minutes of counselors, 19.9.1927 § 18; Vesiyhtiöpapereita 1927-1928, 21.6.1927, 28.9.1927, 8.12.1927, 26.10.1928; Raitio 1992, 55)

The population of Kangasala increased slowly until the 1940s, but in the aftermath of WWII over a thousand Karelian immigrants moved there. In 1949 the population exceeded 10,000 – it more than doubled in 50 years. Between 1940 and 1975 the population increased 2.5-fold. The accelerating population growth of the 1950s increased pressure to provide water supply and sanitation services. (Anttila 1987, 26, 44–45, 484–485; Äikäs 1999, 8) In densely populated communities, the lack of sewerage contaminated wells, which also ran low. The isolated problems that occurred in the early decades of the 20th century, which the Board of Health attempted to control and eliminate, had grown into a serious problem. As the population continued to grow, it became a major issue of municipal administration. The requirements and needs of fire safety also become clear from the articles written for Kangasalan Sanomat during that period. (For example, Board of Health, minutes 21.1.1928 §1, 17.3.1928 §1, 19.1.1929 § 6; KS 1921, 1953, 1954, 1955a, 1955b)

In 1952, the municipal council decided that something had to be done about water supply and sanitation. The Board of Health also started paying attention to the contamination of groundwater. The need for sewerage services was apparent since Tulenheimo Water Co. did not provide them and it was feared that the source of water supply, Lake Ukkijärvi, would be contaminated by the waste carried by surface waters and discharged by the sewers of buildings on the shore. (Minutes of counsellors, 31.5.1952 § 20; Juuti et al. 2003, 66–67) In 1953 a decision was made to establish the Kangasala Water and Sewerage Cooperative, or the so-called "smelly cooperative", the municipality being the majority shareholder. The water supply and sanitation plan of the municipality was completed the next year. At that time the population stood at 13,000. The first stage of the groundwater intake plant using bank filtration from Lake Ukkijärvi was also completed in 1954, and work on the Kirkkoharju elevated tank was launched. It was decided to sell the Tulenheimo Water Co. to the Kangasala Water and Sewerage Cooperative. The distribution network of the cooperative was preserved and it started drawing water from the well of the new cooperative. (Water Cooperative, annual report 3.9.1953–31.12.1954; KS 1955a)

In addition to a water pipe, the cooperative started building a primitive mechanical wastewater purification plant, a so-called Emscher tank. The construction of a main sewer line was hastened on by the completion of a secondary modern school building and the need for sewer connections for other new buildings. *Kangasalan Sanomat* (local newspaper) considered a sewer system indispensable. Financing for water supply and sanitation projects was drawn from different sources: own funds were complemented by state subsidies and the municipality's unemployment reserves. Loans were also taken from local banks and the Social Insurance Institution. (Water Cooperative, annual report 3.9.1953-31.12.1954; KS 1953, 1954, 1955a; Räsänen 2002)

The first stage of the water system was inaugurated in January 1955. Water was drawn from two wells on the shore of Lake Ukkijärvi. The new water pipes were dimensioned according to the fire hydrants. The feeder pipes of the former Tulenheimo Water Co. to the Kirkkoharju elevated tank had to be replaced with larger ones. In order to meet fire safety requirements, a new water reservoir was built atop the esker. It was later expanded to twice its original size. (Räsänen 2002, Water Cooperative, annual report 3.9.1953-31.12.1954, annual report 1955; KS 1955a)

Figure 24.4 Lake Ukkijärvi, first water intake place in Kangasala. (Photo: Petri Juuti)

The main construction phase of the actual municipal water supply and sewerage works coincided with the major relief work periods in the 1950s and 1960s. Public works eased the severe unemployment in the municipality and provided infrastructure. (KS 1957; Anttila 1987, 511; Lumme 2002; Kouhia 2002) By a decision of the municipal council on 7.3.1959, the main responsibility for water supply and sanitation was transferred to the municipality, and a municipal water supply and sewerage works was established. The works implemented major water supply and sewerage projects in the western part of the municipality, although the Ilkko Water Cooperative remained active alongside the municipal works until 1966. Several people worked for both the municipal and the cooperative organisation simultaneously. However, the cooperative society was financially unable to deal with the tighter wastewater purification requirements or build long transfer lines. (Minutes of counsellors, 7.3.1959 § 73; Almonkari 2002; Juuti et al. 2003, 78–79)

In 1959 the municipal council decided that a separate biological wastewater treatment plant would not be built at the eastern end of Lake Pitkäjärvi, but that instead the sewage of several population centres would be led to a trunk sewer leading to Lake Kirkkojärvi. The aim was to keep other lakes clean of sewage, as they were the only lakes suitable for recreational use in areas of population concentrations. The City of Tampere also adamantly opposed leading sewage into Lake Pitkäjärvi or Lake Kaukajärvi. It was also deemed that Lake Kirkkojärvi was already so contaminated that it did not have the same recreational value as the other lakes. Consequently, it could receive even more sewage. (Minutes of counsellors, 7.3.1959 § 72; KS 1959)

The municipal council approved the leading of sewage to Lake Kirkkojärvi on the condition that a mechanical treatment plant was built on the shore immediately. It also insisted that the plant build a pumping station to carry the treated water into Lake Roine. This idea was, however, never realised since the City of Tampere was planning to take its raw water from the same lake. (Minutes of counsellors, 7.3.1959 § 72; KS 1959) A plan for conducting sewage was also on the agenda of the council meeting on 7.3.1959. Mr Väinö Pälli, a bank manager, predicted that Lake Kirkkojärvi would become polluted and was right on the mark, unfortunately. Studies supported his views: the first signs of water pollution were apparent already at the end of the 1950s. In the early spring of 1961 oxygen depletion was detected in almost every part of the lake. Fish-kill occurrences were also noted. (Minutes of counsellors, 7.3.1959 § 72; KS1959; Juuti et al. 2003, 80–81)

The final decision to build a wastewater treatment plant on Kirkkojärvi was made in June 1959. The plant was dimensioned for the sewage of 4,000 people. It treated sewage only mechanically; it was then led into Lake Kirkkojärvi. Septage was also transported to the plant. Yet, Lake Kirkkojärvi could not take the effluent load and was consequently damaged, partially irreversibly. However, the municipality acted lawfully and the practice was not exceptional in Finland at the time. (Anttila 1987, 511–512; KS 1960; Juuti et al. 2003, 81–82) From 1980 onwards all Kangasala municipal wastewaters have been pumped to the main wastewater treatment plant of the City of Tampere completed in 1973. (Almonkari 2002)

While the decision-makers of Kangasala pondered sewerage solutions, those of the City of Tampere faced problems related to water supply: Lake Näsijärvi, the old raw water basin, had been contaminated especially by forest industry wastewaters which necessitated drawing water from some other source. At the end of the 1950s, all eyes turned to Lake Roine in Kangasala. Its use was first suggested in December 1959. The project was given impetus by the water quality survey of Lake Näsijärvi released in 1962, which showed that the southern part of the lake was contaminated by sulphite lye. (Almonkari 2002; about Tampere, see Juuti 2001 and chapter by Juuti & Pál)

The construction of Kangasala's water supply and sanitation culminated in the early 1960s and continued until the late 1970s. (KS 1972a, 1972b) Figure 24.5 presents the overall expansion of the water supply in Kangasala, first based on individual wells. In second phase came the gradual shifting to piped rural cooperatives, and then to municipal systems. Finally the wholesale-based intermunicipal system is planned to be expanded for a wider region by 2010. Yet, we should remember that even in the first decade of $3^{rd}$ millennium we have and will have many on-site systems in dispersed rural areas where connecting to networked systems will not be economically possible or feasible. Cooperatives in this case have merged over time. Thus, various systems have developed from on-site to small systems, key planned areas taken over by municipal systems and finally inter-municipal cooperation both in water supply and sewerage have merged.

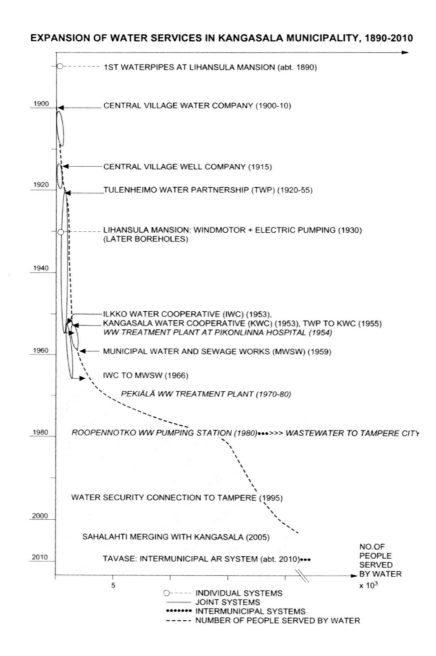

Figure 24.5 Expansion of water services in Kangasala municipality, 1890-2010.

## 24.5. CASE 3: THE DEMAND FOR WATER ON BROILER FARMS IN SAHALAHTI

The area of Sahalahti is located in Häme, circa 30 kilometres east of the city of Tampere. In 2005 Sahalahti municipality merged with the municipality of Kangasala. In spite of the consolidation of local government areas, the area of Sahalahti has some unique characteristics. For example, the prepared food industry of Saarioinen Ltd. in Sahalahti is the main source of livelihood in the area. One of the best-known products of Saarioinen is a type of broiler that has a long tradition in Saarioinen. The first chicken house on the Saarioinen farm was founded in 1945. In the 1950s, Saarioinen began to breed broilers and bring them into the market first in Finland. (Raitio 2001, 220, 208, 213) Thanks to Saarioinen, there are nowadays several farms specialized in breeding broilers in the area of Sahalahti. Because of this specialization, there has been a great demand for water services in Sahalahti, which has been important for broiler breeding itself but also for the environment.

One of several farms in the village of Pakkala in the Sahalahti area is the Kunni farm. The development of water services on Kunni farm has been mainly based on the need for water for cattle. The new methods replaced the old ones on Kunni farm before the 1950s, which means that there was no longer a need to carry water to the cowshed. Water was pumped by hand or by electric energy to a water tank in the cowshed. A water pipe was laid to the cowshed and the house in 1951, and it did not take even ten years before both hot and cold water ran to the house. Water was taken from nearby Lake Pakkalanjärvi via pipes and only drinking water was carried from the well. Water services were well organized before the municipal system of water pipes was built in the 1970s. (Mäkijärvi, H. 2003; Mäkijärvi E. 2003; Mäkijärvi E. 2004a)

By 2005 the need for water had increased remarkably on Kunni farm, caused by breeding broilers with farming. Water consumption has increased parallel to increased broiler breeding on the farm. Annual consumption is now circa 5,000 cubic metres, which is much more than a usual farm's consumption. Though water comes from the municipal system of water pipes, it isn't sufficient in case of an emergency, for example a major fire. For safety's sake, Kunni farm also has its own well. (Mäkijärvi E. 2003)

After all, water supply is one part of the water service and the other important thing is waste management. Its development has been significant in recent years. First wastewater was put in a septic tank but during the last ten years it has been put in an enclosed well. From the enclosed well, sludge has been taken to the fields during the growing season and it has been taken to the sewage treatment plant of Sahalahti at other times. Nowadays the situation is changing. The municipality has already built a drain to which Kunni farm will probably be linked during the years 2005–2006. This means that all wastewater from the farm will run via pipes to the sewage treatment plant. (Mäkijärvi E. 2003; Mäkijärvi E. 2004a; Mäkijärvi E. 2005)

Another farm in Sahalahti, owned by Juhani Mäkijärvi, is located in the village of Pakkala like those of many other broiler farms. This farm was founded in 1983. Water supply had to be organized at the same time because the municipal system of water pipes ended one kilometre from the farm. The only choice was to continue it. If the municipal water service is out of order, a spring and a well near the farm help in getting water. One precautionary measure is a five cubic metre tank through which water runs all the time. There is also a booster station on the farm because the farm is still located at the end of the networked system of water pipes. (Mäkijärvi J. 2003)

Compared with other broiler breeding farms in Finland, Juhani Mäkijärvi's farm is also a big one. Therefore its water consumption is great — 5,000 cubic metres for the broilers' drinking water and 300 cubic metres water for washing and household use per annum. Wastewater comes mainly from breeding establishments' washing water. J. Mäkijärvi says that one important function of water is to maintain the level of hygiene in breeding establishments. The big breeding establishments have to be washed many times a year. This is a big change since chickens have been bred in a breeding establishment instead of being outside. The major part of farm's wastewater, breeding establishments' washing water, has been taken to the sewage treatment plant of Sahalahti. Before the municipal sewer system only household wastewater has been put into the ground through the septic tank. J. Mäkijärvi's farm will probably be linked to the municipal sewer system during the years 2005–2006. After this, all wastewater will run via pipes to the sewage treatment plant. (Mäkijärvi J. 2003; Mäkijärvi J. 2004a; Mäkijärvi J. 2004b; Mäkijärvi J. 2005)

Water services had been well organized in Sahalahti even though it was a small municipality. Development of water supply and sewerage began already in the 1970s. A system of water pipes was constructed in the scattered settlement areas in 2003 and in 2004. It was proposed to extend the system of water pipes before Sahalahti merged with Kangasala municipality. The main reason for the rapid development of water services is the farms' specialization in broiler breeding. (Mäkijärvi E. 2003; Mäkijärvi E. 2004a) In the future, the project TAVASE (the inter-municipal project for forming artificial groundwater) will also help in getting pure water to the farms. (Mäkijärvi J. 2004a) All in all, a broiler farm consumes much more water than the usual farm without any specialization. Because of the great need for water, there is a demand for well-organized water services on the broiler farms. Pure water has an important role in broiler breeding and its sufficiency has to be guaranteed also in unexpected situations.

Sahalahti will take part in the inter-municipal project TAVASE and one of the sequences is transferring wastewater from Sahalahti, including from Kunni farm, to Tampere. Firstly after pre-processing, the wastewater will be transferred via pipes from Sahalahti to Tampere. In the sewage treatment plant of Tampere, the solid part of the waste will be separated from the wastewater. The solid part of the treated sludge of the plant will be transferred from Tampere to Sahalahti for final composting. The runoffs of the composting plant will not be put into the ground but on the contrary they will be collected and transferred after pre-processing to the sewage treatment plant of Tampere. This back and forth circulation of wastewater will be possible in the future. (Mäkijärvi E. 2004a; Mäkijärvi E. 2004b)

Farmer and broiler breeder Erkki Mäkijärvi says that nowadays when the need for water has increased at the same time as the size of cattle, the municipal system of water pipes has been very significant and necessary to farmers and the municipal sewer system to the environment. Even though the water services have caused costs like new investments and costs of linking to the municipal water services, E. Mäkijärvi says that the water services have to be well organized especially on big breeding farms. (Mäkijärvi E. 2004b) Juhani Mäkijärvi agrees with his brother Erkki Mäkijärvi on the importance of the municipal system of water pipes. He says that without the municipal water services it would be more difficult to maintain operation of the farm. (Mäkijärvi J. 2004b)

This case of broiler breeding is an example of great water consumption without a high amount of waste. The environment has not been forgotten. The fact is that the wastewater from broiler farms is only a minor part of the amount of farms' water consumption. Even though the broiler breeders' one main concern is to guarantee the sufficiency of pure water for broilers, they have also attended to waste management. All in all, the development of water services has depended on the great need for water. Thanks to the municipality of Sahalahti, the farmers have not been alone in developing water services.

Figure 24.6 Chicken house in Sahalahti, Finland - a building in which broilers are reared in confined conditions (Photo: Annele Mäkijärvi)

## 24.6. THE DEVELOPMENT OF RURAL WATER SUPPLY IN THE REPUBLIC OF SOUTH AFRICA

The Republic of South Africa (Figure 24.7) has a total area of 1.22 million km² out of which only 13 percent is considered as cultivable. The population is estimated at 44.8 million (2001), of which 49 per cent is urban. The average population density is about 37/km², ranging from more than 500/km² in some rural areas to only 22/km² in other parts.

The importance of water was recognized early in South Africa by politicians, writers, and scientists, for example in 1886 John Noble wrote: "Where water is abundant, foliage is plentiful, and the town or village has a cheerful and comfortable appearance; where it is absent, the town has dry and desolate look." (Noble 1886, 177)

This assessment is still valid today.

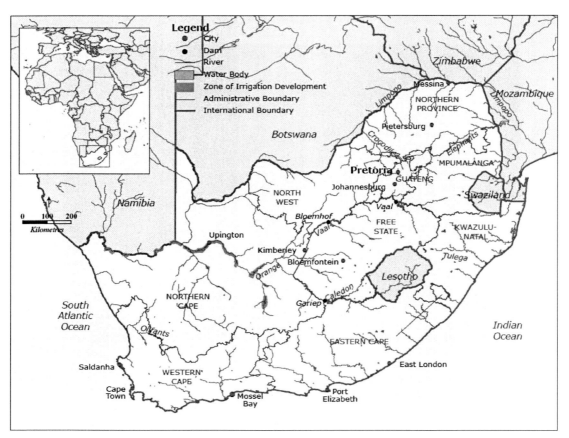

Figure 24.7 Map of South Africa. (Source: http://www.fao.org/ag/agl/aglw/aquastat/countries/south_africa/index.stm)

Figure 24.8 Harts River and the surrounding area. (Source: Mäki 1991, Appendix 2, map 2)

## 24.7. CASE 4: HISTORICAL WATER CONFLICT IN SOUTHERN BECHUANALAND, 1881-1883

This case shows how water issues can develop into the stage of armed conflict and the consequences for those on the losing end.

Southern Tswana occupied Southern Bechuanaland around Harts River in the 19th century. This area was dry: precipitation was seasonal, unreliable and sparse, and falling mostly in heavy thunderstorms during the summer. Riverbeds were dry, for the most part, with a visible flow for only a few months. The soil retained little moisture. There was, however, some geographical compensation. Firstly, water could be obtained by digging wells in the dry riverbeds. Secondly, there was a limestone formation, which gave rise to numerous springs, some of which retain their flow throughout the year. Thirdly, in the eastern parts of the Harts valley, the soil was richer. (Shillington 1985, 4-6)

The main source of livelihood for Tswanas was cattle grazing. The grasslands of the area were especially suitable for this because there were no tsetse flies. Before the Europeans came, the power of the strongest chiefs was based on their huge herds of cattle. By 1880 Tlhaping herds in the area could be counted by the tens of thousands (Agar-Hamilton 1937, 215), and taking care of these huge herds demanded the control of water sources. Some of the largest cattle herds in the world are still found in this area. (South Africa Yearbook 2003/2004, 21) The competition for the pastures, arable land and springs between native groups became acute quite easily. Conflicts occurred particularly during the dry season. Quite often a whole chiefdom had to move to get water and usually this meant going to another chiefdom's area. Battles happened mostly during the cattle raids and usually ended quickly. (Maylam 1980, 15-16; concerning Tswanas, see also Schapera 1959 and Sillery 1952)

In the spring of 1881, war broke out in the southern parts of the area between Tlhaping chief Mankurwane and Kora chief Mossweu. Some months later, fighting began between Tshidi Rolong chief Montshiwa and Rapulana Rolong chief Moswete in the north. The white "volunteers" soon began to arrive in the area to fight for the native chiefs and the situation got out of control. (Schreuder 1980, 89-90)

In the southern parts of the conflict area reasons for the conflict were cattle and timber, while in the north it was arable land and water. (Report of Commissioners 1882, 28-35) In both areas, conflicts increased the chances for cattle thievery. This became the most important economic activity in the area and was one reason for the continuation of war. White cattle thieves came mostly from Kimberley's diamond fields and took their loot there so that the natives got the blame. (Agar-Hamilton 1937, 199)

In his letter to Colonial Secretary Kimberley, Governor Sir Hercules Robinson said that the main reason for the difficulties in the area was not the quarrels between the chiefs but the action of unprincipled European adventurers who, eager for land and cattle, were attacking Mankoroane and Montsioa without a cause (Governor Robinson to Colonial Secretary Kimberley 1882). Most of the whites were only advancing their own interests. First they took cattle for compensation, but very soon they began to demand land rights. This led to gross misunderstandings because of divergent conceptions of property. Chiefs thought that they only gave away the right to use land. Whites assumed that having obtained a mark on a paper from an illiterate chief they had acquired absolute ownership. Most of these whites never intended to cultivate the land themselves or raise cattle there. Their purpose was to sell the land rights when the situation calmed down. (Report of Commission 1886, s. 76-84, Annexure F, Schedules B & C; Sillery 1952, 47-48; Mackenzie 1887, 233, 249; Shippard 1901, 58-59)

Most of the whites were cattle- or horse thieves, were supplying ammunitions to natives, or were selling rights to cut timber in the native territory. Most of the monies so collected ended up in the pockets of the whites and the Tswana chiefs got only a fraction of the profit from the woodcutting licences or the ammunitions they authorized the whites to buy. It is also very probable that the stolen horses and cattle ended up in Kimberley and were not delivered to Tswanas as was planned. (Mäki 1998, 157)

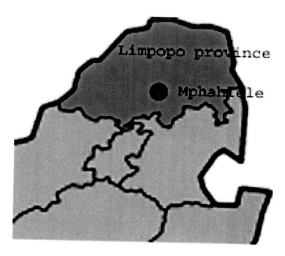

Figure 24.9 Situation of Mphahlele area.

When the area had calmed down in 1883, most of the best-watered lands were in the hands of whites and the situation of the natives was worse than it had been earlier. Conflicts started as normal native fighting but soon escalated into a real battle for the best lands after the white "volunteers" began to take part. The nature of the war changed from cattle rustling to conquering the opponent and taking over his lands. Behind the "volunteers" loomed the mining industry of Kimberley that protected its own interests by taking over the fuel and labour resources of the area.

## 24.8. CASE 5: MPHAHLELE RURAL AREA

In South Africa, the Mphahlele Rural Area (Figure 24.8) in what is nowadays Limpopo province has a long history of water shortages and in the past many attempts were made to solve the problem; however, financing had been a barrier. This includes financing the works and as well the ability of the users to pay for the development. Therefore, the state assistance for the communities had played a major role. The first attempt was made in 1937 with the digging of trenches to supply water from the river, and the second in 1939 when some boreholes with windmill pumps were installed. (http://www.delportdupreez.co.za/html/html/dpa_osab.htm#History)

The Department of Health found in 1960 that drinking water in the area was unhealthy. After that, some attempts were made to improve the situation and boreholes were drilled. From 1968 to 1972 the government started an extensive drilling programme and the situation improved.

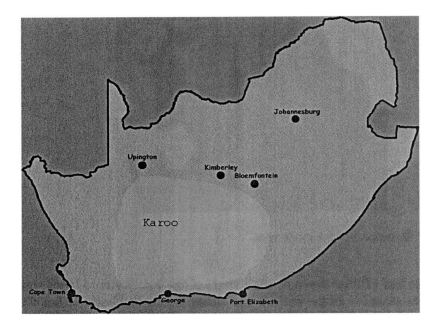

Figure 24.10 Karoo.

There are still many of these boreholes in use, nowadays equipped with diesel-driven pumps. They are the major sources of potable water to the areas.

In the 1980s the Department of Water Affairs of the then Lebowa Government Service assisted the installation of networks in a few villages. Development Forums were organised in 1994 with some areas for an improved water supply system. At the same time, the communities asked for assistance from the Board of Northern Transvaal Water (Later Lepelle Northern Water). Some 34 villages elected a Zonal Development Forum to negotiate on behalf of the communities. This Board approved giving 12 million rands for the improvement of the bulk water supplies to villages in the Mphahlele areas. In 1995 the Project Engineers received instruction to proceed with this project as part of the Olifants Sand Water Transfer Scheme. (http://www.delportdupreez.co.za/html/html/dpa_osab.htm#History)

## 24.9. CASE 6: WIND POWER AS A NEW INNOVATION FOR WATERING KAROO

It is not entirely certain when the first windmills appeared in Karoo (a dry plateau area in interior South Africa), but Civil Commissioners of the Cape government confirm their existence in the arid western Karoo in the 1870s. The earliest recorded example of the imported mill is from the year 1874 in Hopetown. (Archer 2000, 681-682)

Windmills, most commonly connected with boreholes, are mentioned for the first time in the Cape census in 1904. There were said to be 1,275 wind motors and 364 water wheels in use in the Cape Colony. The agricultural census sources show that the number of windmills on South African farms in 1926 was 44,000, in 1946 about 101,000 and in 1955 as high as 151,000. In recent times, technology has evolved further and the use of submersible pumps in boreholes and solar power spread into Karoo during the 1980s and 1990s. (Archer 2000, 682)

By 2003 there were still some 300,000 windmills with boreholes on farms across South Africa, second in number after Australia. They are primarily used for watering livestock and supplying communities with water. (http://www.africaguide.com/facts.htm; South Africa Yearbook 2003/2004, 481)

## 24.10. DISCUSSION AND IMPLICATIONS

### Current situation in South Africa and in Finland

In South Africa some 6.2 million people have access to a basic supply of water, obtaining it from sources that are further than 200 metres away. According to the 2001 Census, five million people still need access to a basic supply of water or are without the bare minimum supply. These take water directly from dams, pools, streams, rivers or springs, or purchase water from water vendors. Households with access to water increased from 80% in 1996 to 85% in 2001. In 2002/03, some 1.2 million people received water-supply infrastructure and 65,105 toilets were built. The Department of Water Affairs and Forestry had intended to eradicate the bucket system in 430,000 households by 2006.

Still there are problems in water management in South Africa. For instance, adjacent communities could pay different amounts depending on the systems installed. Rural households pay for water from standpipes, whereas households in Durban, getting water on site, have been getting the first 6 kilolitres per month for free. Such inequities have led to tensions inside and between villages. And vast areas are still not receiving water services at all. (For a very critical assessment of the current situation, see Bond 2002)

### Community and Resident Participation

Residents' associations have been active, thereby forcing water supply and sanitation onto politicians' agendas. This is an important factor in every project wherever it might be. For example, at the end of the 1960s and early 1970s in Kangasala, sewers were also demanded in addition to water pipes. People started opposing the contamination of Lake Vesijärvi since wastewaters of the Ruutana residential area about 10 kilometres from the centre of the municipality ran into it. The situation was much the same as in cities at the beginning of the period of strong growth: human habitation polluted wells since appropriate sewerage did not exist, and what little water there was threatened to dry up. Although the time and place have changed, we are talking of a similar phase in communal development: the population density and standard of living increased along with consciousness about the state of the environment. (KS 1968, 1972a, 1972b, 1973; Almonkari 2002)

In South Africa there have emerged a vast amount of governmental and non-governmental Development Forums for improved water supply systems since 1994. Consciousness of environmental problems and threats has also increased at the same time.

About four-fifths of the population of Kangasala live in built-up areas at the start of the 21st century. A ribbon-like settlement developed along the road to Tampere and thus ribbon-like water supply and sewerage trunk lines were completed towards the end of the 1970s. A major trunk line was built in 1995 in order to provide a connection to the network of the City of Tampere primarily for emergencies. The municipality promotes ecologically sustainable human habitation: for example the residents of the Riku Eco Village are responsible for their wastewater treatment on site. (See, for example, http://www.kangasala.fi/matkailu/english/index.htm; Juuti et al. 2003, 131–132)

The numerous lakes and long shoreline of the Kangasala municipality have influenced its settlement greatly. It is in the best interest of the municipality to see to it that the lakes remain in good condition. With this in mind, different solutions have been developed for water supply and sanitation in dispersed settlements including the so-called mobile sewer: an information technology-based septic tank emptying system. In an extensive and dispersedly populated area it is not possible to connect each property to a centralised water supply and sewerage system at a reasonable cost. Now Sahalahti region is a part of municipality of Kangasala, the merger happened in the beginning of the year 2005.

In South Africa, our other case country, play pumps that use the energy of children at play to pump water were installed in 2001 in 40 rural villages throughout South Africa with the funding of the Kaiser Family Foundation. An additional 60 play pumps were installed in 2002. In April 2004 a new six-year programme began for 2800 rural schools in Kwazulu-Natal. It includes the installation of 40 play pumps at these schools. A local water utility, Umgeni Water, is also involved in this project. Water tanks are installed a few metres away from the play pumps. The pumps can generate some 1,400 litres of water per hour, saving young women time and energy they would otherwise be spent walking to and from more remote water resources. Play pumps also help prevent diseases such as cholera that can stem from open-water supplies. (South Africa Yearbook 2004/2005, 601)

## Implications

Lessons learned from South Africa and Finland might be useful in developing countries. But what are these lessons?

In the first phase, the role of natural resources such as water, rich soil, wood and minerals, was central when permanent settlements were built for the first time. Cattle and farming created a great need for water, and affected where wells were dug. Cattle need a large amount of water: one cow's average need is about 50 litres in day.

In South Africa, water and other natural resources were often behind conflicts in the early years. Materials used for the pipes and pumps were the same in both countries. The same goes for the innovations; in both countries they used wind power for pumps before electricity. The drilling techniques and methods for finding water were similar.

On the whole, the common rural water pipelines in Finland developed gradually from small to larger systems. These systems evolved especially in the Ostrobothnia region based on the needs of cattle farming and rural households. Water was taken from natural springs, located often at higher elevation than the service areas and led by gravity. The settlements were concentrated along the riverbanks, and there was also a long tradition of joint efforts in the region. It is believed that this tradition of constructing joint water pipes by consumers themselves originates partly from the lake drainage associations that were very common in the 18th and 19th centuries (Anttila, 1967).

In South Africa, water pipelines in rural areas are still in the development stage. There are vast areas where water is taken directly from dams, pools, rivers or springs. There are large governmental programmes, for instance, for drilling wells to improve the water supply situation. Consumer-built water pipes are nearly unknown in South Africa. In Finland, rural areas still, in 2005, relied largely on traditional wells and on-site sanitation. About 80% of Finns are connected to sewer networks and about 90% to water supply networks. Those without connections are located mainly in the countryside. Before 2004, sanitation in the countryside was not controlled well enough. Since then, new legislation has required much more precise plans than earlier and every house owner is responsible for the construction of the on-site system. (Mattila 2005, 137-151; Juuti & Wallenius 2005, 148-149)

Local expert knowledge was used amply and the adaptation was tailored to the conditions of Kangasala. The dimensioning of the 1953 waterworks was a success; the needed capacity and extension possibilities of the waterworks proved to be correct.

The combining of water acquisition, sewerage and environmental protection started long after the crisis of Lake Kirkkojärvi in 1961. Since 1980, wastewaters have been pumped via a transfer sewer to the main wastewater treatment plant of the City of Tampere.

It was decided to finance the activity of the cooperative system of 1953 on the basis of non-metered consumption, but luckily by a decision of the municipal council in 1959, the main responsibility for water supply and sanitation transferred to the municipality and municipal waterworks were established. The decision to base the financing of the activity on metered consumption was then also made. This has made possible the sensible development of the utility, which probably would not have been possible in light of the examples with other charging principles.

Preference was then not given to better quality groundwater in spite of various warning signs, but the decision-makers stuck with untreated surface water, which contained unclean wastewater. Only decades later did groundwater become part of the water supply of Kangasala. The decision to build a wastewater treatment plant on Kirkkojärvi in 1959 was a catastrophe, which in fact destroyed rather good quality of waters of Lake Kirkkojärvi.

These principles are mainly related to the municipal water supply and sanitation. Bigger industries like pulp and paper mills in the neighbouring municipality of Tampere have traditionally had their own systems although they used to have some connections to the city water works at a certain stage.

In Kangasala, water supply and sanitation has been implemented through different types of organizations over the years. The sector has developed gradually as have the organizations responsible for it. Initially there were farm-specific systems, small companies and cooperatives in several areas. A water and sewerage cooperative came into being in the central village; it later became the municipal water and sewage works. However, cooperatives operated alongside them and merged or otherwise disappeared from the scene as municipal water and sewage works evolved. Gradually inter-municipal cooperation also emerged, at first in water pollution control, and later in water supply.

Good results have been achieved in water protection as a whole. Yet, earlier discharges of wastewater, e.g. into Lake Kirkkojärvi in Kangasala, or measures such as lowering the water level of lakes, unavoidably affect their present and future state. The incorporation of water and sewage works also introduces new challenges. Historically, inadequate purification of wastewaters and their discharge into small lakes has been opposed – even on quite good grounds.

## 24.11. CONCLUSIONS: POSSIBLE PRINCIPLES OF SUSTAINABLE WATER SERVICES

Management of water services should be demand-based, so that the solution is up to "water demand". In different areas, the difference between water and need can be great. Because of this the solutions to water demand must apply according to local variations and conditions.

In the countryside new innovations and techniques can be taken in use more easily than in cities because there is not so great technological path dependence. For example, wind power was used largely to lift water in the Finnish countryside up till the 1950s before electricity, and it is now used successfully in the South African countryside when electricity is not easily available.

Most of the solutions in both countries were originally started in individual systems that gradually have expanded. Since then, the local government has become involved gradually in bigger villages and townships.

In many areas cattle farms earlier created the demand for water while later, e.g., food-preparing industries took the lead. With the expansion of piped water, the need for proper water pollution control, including wastewater treatment, became a must.

Small systems still survive in individual households in both case countries; even water cooperatives exist and they have not vanished as predicted in Finland. Municipality-owned utilities are developed based on modern principles, while inter-municipal systems are also taken into use both in water supply and sewerage services. Thus at the same time we have the whole spectrum of diversity of systems, instead of just jumping merely from one to another and leaving others aside.

The example of using play pumps to get water in the poorer areas of South Africa also shows that with the new, innovative solutions you can easily make the water supply situation better in rural areas. The same also applies also to the old, proven methods like windmills; even with new, improved apparatus, the old methods can still be the best solution in some areas.

There are also dangers in water-scarce areas. There is always a possibility that water becomes an issue for armed conflict, as has happened earlier. The difference is that nowadays conflicts are much more destructive than they were in the late 19th century.

Solving the major issues concerning water supply and sanitation will require major investments including organising the services for sparsely populated areas. Various development projects provide a good example. Clean water and wastewater, or sanitation, are nevertheless tied together. Inadequate sewerage may also lead to the contamination of clean water.

Every case and environment is of course unique and there are no global rules or principles taken as a handbook solution in rural-like conditions. Local conditions, natural resources and traditional solutions vary a lot even in one country. Community water supply and sanitation probably touches the lives of more people than any other public service. In every case and every area, cultural traditions are different and decision-making traditions, too. One must know these local conditions and traditions and also local history when plans are prepared. Local people have this knowledge and their lives are involved.

That is why the residents must be taken very carefully into account when decisions are made. It is also good to know how the same kind of problems are solved elsewhere. What is needed is not copying the solution directly but looking for good possible strategies and ideas and maybe even learning from mistakes that others have made earlier.

Figure 24.11 Cows from Finland with "EU - earrings" and President Tarja Halonen with Professor Pentti Arajärvi. (Photo: Petri Juuti)

## 24.12. REFERENCES

PRIMARY SOURCES

Board of Health (Kangasala), minutes 21.1.1928 §1, 17.3.1928 §1, 19.1.1929 § 6

Governor Robinson to Colonial Secretary Kimberley, 9.7.1882, Colonial Office Confidential Papers 879/19/241, No. 33, microfilm in the History Department of the University of Tampere.

Ilkko Water Cooperative, Kangasala Municipal Archives.

Minutes of counselors (Kangasala Municipal Archives) 19.9.1927 § 18, 31.5.1952 § 20, 7.3.1959 § 72, 7.3.1959 § 73

Municipal Building Board (Kangasala) 11.11.1929, 4.12.1929, 17.12.1934

Report of Commission appointed to Determine Land Claims and to Effect a Land Settlement in British Bechuanaland, British Parliamentary Papers, C. 4889 (1886).

Report of Commissioners appointed to Inquire into a report upon all matters re. settlement of Transvaal Territory, Part II, British Parliamentary Papers, C. 3219 (1882).

Vesiyhtiöpapereita 1927–1928, 21.6.1927, 28.9.1927, 8.12.1927, 26.10.1928, Archieves of Minna & Anssi Apajalahti.

Water Cooperative (Kangasala), annual report 3.9.1953-31.12.1954

Water Cooperative (Kangasala) annual report 1955.

## LITERATURE

Agar-Hamilton J.A.I. (1937) The Road to the North, South Africa, 1852-1886. London.

Anttila O. (1987) Kangasalan historia III. v.1865–1975. 110 kehityksen ja kasvun vuotta. Kangasala.

Anttila V. (1967) Järvenlaskuyhtiöt Suomessa. Doctoral dissertation, University of Turku. Kansatieteellinen arkisto 19. Turku.

Archer S. (2000) Technology and ecology in the Karoo: A Century of Windmills, Wire and Changing Farming Practice. Journal of Southern African Studies 26(4), 681-682.

Bond P. (2002) Unsustainable South Africa: Environment, development, and social protest. Pietermaritzburg.

Bärthel H. (1997) Wasser für Berlin. Herausgeber Berliner Wasser Betriebe. Verlag für Bauwesen Berlin,Germany.

Hn.(Pseudonym) (1863) Wasabladet 7.3.1863.

Juvonen K. (1998) Kangasalan Joulu.

Juuti P. (2001) Kaupunki ja vesi. Doctoral dissertation. Acta Electronica Universitatis Tamperensis 141. Tampere.

Juuti P. & Wallenius K. (2005) Brief History of Wells and Toilets. Pieksämäki.

Juuti P., Äikäs K. & Katko T. (2003) Luonnollisesti vettä. (Water naturally. Kangasala water utility 1952–2002) Saarijärvi.

Katko T. (1997a) Water! – Evolution of water supply and sanitation in Finland from the mid-1800s to 2000.

Katko T. (1997b) The use of wooden pipes in rural water supply in Finland. Vatten, 53, 267–272.

Katko T. (2000), Long-Term Development of Water and Sewage Services in Finland. Public Works Management & Policy 4(4), 305–318

Kotitalouden rationalisoimiskomitea (1950) Maaseudun vedenhankinta- ja viemäriolojen parantaminen. Mietintö no. 1.

Mackenzie J. (1887) Austral Africa, Losing it or Ruling it, being incidents and experiences in Bechuanaland, Cape Colony and England. London.

Mattila H. (2005) Approriate Management of On-Site Sanitation. Tampere University of Technology.

Maylam P. (1980) Rhodes, the Tswana, and the British: Colonialism, Collaboration, and Conflict in the Bechuanaland Protectorate, 1885-1899. London.

Myllyntaus T. (1991), Electrifying Finland. The Transfer of a New Technology into a Late Industrialising Economy, London: Macmillan & ETLA 1991, xvi, 248-253

Mäkelä U.O. (1945) "Painevesijohdoista Toholammilla". In Maanviljelysinsinööriyhdistyksen vuosikirja 1944–45.

Mäki H. (1991) Gladstonen hallitus ja Betsuanamaa 1880–85. MA thesis, University of Tampere.

Mäki H. (1998) Valkoiset seikkailijat: Betsuanamaan sodat ja valkoisen vallan ekspansio Etelä-Afrikassa 1881–1883. In Kansa ja kumous: Modernin Euroopan murroksia 1880–1930. Historiallinen Arkisto 111. Helsinki.

Noble J. (ed.) (1886) History, Productions, and resources of Cape of Good Hope, Official Handbook.

Raitio R. (1992) Kangasalan kirkon seutu vuosisadan vaihteesta 1950-luvulle. 2nd ed., Kangasala.

Raitio R. (2000) Lihasula, Palko ja Lisko. Historiaa ja tarinoita Kangasalan Lihasulan ja Palon kyliltä, Jyväskylä.

Raitio R. (2001) Maatalouspitäjästä Suomen teollistuneimmaksi kunnaksi. Sahalahden historia III 1869–1999. Jyväskylä.

Schapera I. (1959) A Handbook of Tswana Law and Custom. London.

Schreuder D.M. (1980) The Scramble for Southern Africa, 1877-1895. The Politics of Partition Reappraised. Cambridge.

Shillington K. (1985) The Colonisation of the Southern Tswana 1870-1900. Pretoria.

Shippard S.G.A. (1901) Bechuanaland, in British Africa, The British Empire Series II. London.

Sillery A. (1952) The Bechuanaland Protectorate. Cape Town.

South Africa Yearbook (2003/2004).

South Africa Yearbook (2004/2005).

Virkkala K., Luho V. & Suvanto S. (1949) Längelmäveden seudun historia I. Kangasalan historia I. Forssa.

Wäre M. (1952) Maaseudun vesihuolto. Mitä-missä-milloin, kansalaisen vuosikirja 1953, 216–219.

Äikäs K. (1999) Kangasalan vesihuollon strategia vuoteen 2020. Diplomityö, TTKK.

INTERNET

http://www.africaguide.com/facts.htm, accessed 1.2.2004

http://www.delportdupreez.co.za/html/html/dpa_osab.htm#History, accessed 19.3.2004.

http://www.iied.org/sarl/dow/tanzania/chapter1.html, accessed 19.3.2004.

http://www.kangasala.fi/matkailu/english/index.htm;

http://www.realgoods.com/calsolar/downloads/casestudy_masai.pdf, accessed 28.4.2004

INTERVIEWS

Almonkari J. (2002), 15.5.2002
Kouhia E. (2002), 26.4.2002.
Lumme H. (2002), 24.4.2002.
Mäkijärvi E. (2003), 28.3.2003
Mäkijärvi E. (2004a), 24.5.2004
Mäkijärvi E. (2004b), 10.9.2004
Mäkijärvi E. (2005), 3.5.2005.
Mäkijärvi H. (2003), 28.3.2003
Mäkijärvi J. (2003), 27.3.2003
Mäkijärvi J. (2004a), 24.5.2004.
Mäkijärvi J. (2004b), 10.9.2004
Mäkijärvi J. (2005), 3.5.2005
Räsänen A. (2002), 7.2.2002.

NEWSPAPERS

KS, Kangasalan Sanomat

| | | |
|---|---|---|
| 22.1.1921. | 10.1.1959. | 5.9.1953. |
| 22.10.1960. | 18.12.1954. | 6.4.1968. |
| 5.3.1955a. | 10.3.1972a. | 9.5.1955b. |
| 10.11.1972b. | 13.7.1957. | 26.10.1973. |

# 25

## SISTER TOWNS OF INDUSTRY: WATER SUPPLY AND SANITATION IN MISKOLC AND TAMPERE FROM THE LATE 1800S TO THE 2000S

*Petri S. Juuti & Viktor Pál*

Figure 25.1 Location of the two case cities: Miskolc in Hungary and Tampere in Finland.

## 25.1. GENERAL BACKGROUND

This chapter discusses the development of water supply in Hungary and Finland in two case cities, Miskolc and Tampere, from the late 1800s to the 2000s. Industrialization was one of the main joint drivers towards establishing water and sewerage services in the two cases. Demands grew and several water crises emerged in both cases throughout the period. After long struggles, different solutions were found and problems in water quantity and quality are now mainly over.

Rapid industrialization and population growth were remarkable in the period and new constructional solutions were demanded. Densely built apartment houses brought new challenges both for water supply and sewerage. Today, both countries are members of the EU and follow EU regulations on water-related matters, however, the evolution of water-related services has been rich in problems and controversies in both cases.

The study is based on the different archive materials, newspapers, technical journals, conference reports and literature. A specific research question was how local water services developed in these cities, under the influence of rapid urbanization and industrialization. Methods used combined statistical and archive data of the case studies.

Hungary is situated in the Carpathian Basin of Central Europe, along the rivers Danube and Tisza (Figure 25.1). The Hungarian Kingdom was established in 1000 and until the partial Turkish occupation in the 16$^{th}$ century, it played an important role in Central Europe. Later on the Habsburgs ruled the country and Hungary became an equal partner of Austria, in the new dualist empire, by the year 1867. In 1920, the present borders of the country were drawn; therefore most river flows entering the country from neighboring states have been causing several problems of floods and trans-boundary water pollution in recent decades.

In Hungary, Miskolc, 160 km northeast of the capital, Budapest, and its industrial area, Diósgyőr, have been industrial centers from the 1860s. The surroundings of Miskolc and Diósgyőr, once independent, but nowadays functioning as one united town, have been settled since Paleolithic times. During the Middle Ages, Miskolc was a small agricultural and commercial centre. The first iron furnace of the region was established in the 1770s. The rise of industry and the significant growth of the town started during the second half of the 19$^{th}$ century. Soon Miskolc became an important industrial and commercial centre, based on her iron- and engineering works.

By the end of the 19$^{th}$ century, the town had approximately 40,000 inhabitants, but did not have a water supply system. Wells were inadequate for citizens and the only pipeline was built for the needs of the Diósgyőr Ironworks. Therefore cholera and other epidemics appeared frequently. One of the biggest cholera epidemics swept through the town in 1873. At the same time, the town was endangered regularly by floods because of its geographical situation. By the end of the 1880s, a decision was made to improve water services and canalisation by the town council. In 1913 a water intake plant at the Tapolca charts springs, an almost 40 km long water supply and wastewater system and a water tank on the top of Avas hill, were constructed.

Water services concentrated only on the downtown area and 18 per cent of total dwellings were connected. At the same time, a wastewater plant was constructed on the banks of the river Sajó. (Teslér 1963, 13.)

Between the two World Wars, a modest development took place in the water supply system. Modernisation was carried out with the help of foreign bank loans. The unfavourable economic situation was caused by the loss of important territories in 1920 and later on the Great Depression. During World War II, the town was bombarded several times because of its economic importance. Therefore, a significant length of pipelines was damaged and the waste water treatment plant was not reconstructed until 1972.

Tampere is an industrial inland city situated approximately 140 kilometres northwest of Helsinki, the capital of Finland (Figure 25.1). The city was established in 1779 along the rapids between two lakes, Näsijärvi and Pyhäjärvi, with an elevation difference of 18 meters. The centre of the city is traversed by the nearly one kilometre long Tammerkoski Rapids around which the city and its industry originally grew up. The rapids, running from north to south, break an esker formation separating the lakes. The esker is one of the highest traverse ridges in the world formed during the Ice Age. Nowadays, Tampere is the largest inland city in the Nordic countries and the third largest city in Finland.

Except for some cases in Hungary and Finland, water supply was a major problem for these settlements but the above-mentioned rapid economic development and urbanization allowed for and made a necessity of the introduction of modern water supply and sanitation systems.

City fires were great problems in Finland during the 1800s; even the capital city of that time, Turku, burned down in 1827. This was due to wooden houses and the lack of pressurized water for putting out fires. Yet, fires were not major threats to urban dwellers in 19th century Hungary, since most buildings were built out of bricks or stones in rocky areas. Wood was an expensive and scarce raw material, and was therefore used only for certain construction purposes.

In **Tampere** industrialization was one of the key drivers for developing and establishing water and sewerage systems. This happened at quite an early stage in Tampere - compared to the development in the whole country. The history of the community water supply and sewerage services goes back to 1835, when the first water and sewer pipes were constructed. The system was quite simple and constituted a so-called bucket system. The first water-protection regulation in Tampere concerned this system. (Tampere city record office, 1838) The rapid growth period in Tampere started a few years later. (Rasila 1984, 131) Most of the industries were established along the rapids utilizing the river water, e.g. by using water wheels in this phase.

Along with fast industrialization, the city grew rapidly; during the period of 1835-1921 the population rose from about 1,600 to over 40,000. The first common water pipes serving the City of Tampere were taken into use in 1882, taking water from Lake Näsijärvi until it became polluted due to industrial loadings from the surrounding areas. Due to this decreasing water quality, in the 1960s a new source of drinking water needed to be found. It was then that Lake Roine, situated 20 km to the east of Tampere, was discovered and taken as the major raw water source. (Juuti & Katko 1998)

In the beginning of the 20th century, the raw water basin of Tampere, Lake Näsijärvi, was already partly polluted and typhoid epidemics were common until 1917. For a discussion of these problems, see Chapter 10 by Juuti & Katko. The pollution of the lake became more and more serious during the following years.

The wastewater matter was taken up again only in the 1950s, and in 1962 the first wastewater treatment plant was completed in Rahola, for the western suburbs of the city. Then, finally, Tampere had a modern water and sewerage system in every respect, although it did not cover the entire city. During severe typhoid epidemics in 1915-16 local newspaper Aamulehti was very interested in the wastewater question. Later on it mostly reported the decisions of various administrative organs. Interest was aroused again after the lakes became indisputably polluted by the turn of the 1950s and 1960s. (Juuti 2001; Katko, Luonsi & Juuti 2005) Milestones of both case cities' water and sewage works are shown in Table 25.1.

## 25.2. FROM WORLD WAR II TO THE 1960S

Miskolc was faced with accelerated urbanization and industrialization after World War II. Water-related investments could not keep pace with the increasing demands of citizens and local industries. Therefore, a significant need for housing had to be satisfied. New blocks of flats were built with bathrooms and WCs in different areas of the town starting in the early 1950s. In these new buildings dwellers wasted tap water because it was to be found in abundance, often without water meters and for a very low, state-subsidized price. Creating a water shortage by wasting water was a significant danger already in the late 1950s and experts kept on warning the public of the future consequences. (Észak-Magyarország 1959, 4) In fact, such wastage of water has been a global problem until the present.

Before World War II most consumers had water meters in Miskolc, but after the late 1940s the installation of meters was neglected. Responsible experts at the water company complained about the shortage of spare parts for old meters. Neither the manufacturers nor the suppliers were ready for the accelerated need for meters. In addition, the state policy of social care included cheap and abundant running water. Besides free education and free health care, cheap housing was also guaranteed by the government. Furthermore, most new residents of Miskolc had been born and brought up in the countryside, lacking adequate water supply, drainage, electricity and heating. For new residents, moving into new housing blocks often meant the first time experiencing a WC, bathrooms and central heating. Therefore, their personal connection towards water was an exploiting and materialistic attitude. (Észak-Magyarország 1962, 2)

At the same time, water pipes were constructed and community wells were set up in the suburban zones. The areas of Hejőcsaba, Martintelep and the Western outskirts were supplied by water, but no efforts were taken to build drains. (Észak-Magyarország 1949a, 9) In addition, suburban sprawl and territorial expansion took place, which caused further problems.[1] Citizens

---
[1] In 1945 the community of Greater Miskolc was established.

Table 25.1 Milestones of the waterworks in Miskolc (Pál) and Tampere (Juuti & Katko).

| MISKOLC | YEARS | TAMPERE | YEARS |
|---|---|---|---|
| 1. Hámor artificial dam (for industrial needs) | 1812-present | | |
| | | 1. Head of rapids, iron pump | 1835 to 1880s |
| | | 2. Head of rapids, gravity system | 1882-1906 |
| | | 3. Sewerage system, 1st phase | 1876-1894 |
| | | 4. High-pressure water supply system | 1898-1931 |
| | | - Myllysaari high-pressure pumping station | 1898-1931 |
| | | - Pyynikki water reservoir | 1889-present |
| | | 5. Pispala water cooperative | 1907-62 |
| 3. Felső- spring intake; gravity system (industrial) | 1908-present | | |
| Gallya-spring intake; gravity system (industrial/citizen) | | | |
| | | 6. Groundwater inventories, western side | 1911-20 |
| 4. High-pressure water supply system | 1913- | | |
| - Tapolca high-pressure karsts water intake station | 1913-present | | |
| - Avas water tank | 1913-present | | |
| 5. Sajó wastewater treatment plant | 1913-44 | | |
| 6. First water supply main | 1913-present | | |
| | | 7. Groundwater inventories, eastern side | 1916-20 |
| | | 8. Kaupinoja surface water treatment plant | 1928-90 |
| | | 9. Old water tank of Kaupinkallio | 1928-present |
| | | 10. Mältinranta surface water treatment plant | 1931-72 |
| 7. Second water supply main | 1939-present | | |
| 8. Avas tunnel | 1939-present | | |
| 9. Tavi-spring intake (industrial) | 1940-present | | |
| | | 11. Mustalampi groundwater intake | 1950-present |
| 10. Avas dual water tanks | 1954-present | | |
| 11. Anna spring intake (industrial/citizen) | 1955-present | | |
| | | 12. New water tank of Kaupinkallio | 1958-present |
| | | 13. Rahola wastewater treatment plant | 1962-present |

| MISKOLC | YEARS | TAMPERE | YEARS |
|---|---|---|---|
| 12. Szinva spring intake (city) | 1963-present | | |
| | | 14. Hyhky groundwater intake | 1966-80, 97-present |
| | | 15. Messukylä groundwater intake | 1967-present |
| | | 16. Tesoma water tower | 1969-present |
| 13. Intermunicipal network (boldva+sajóecseg intake) | 1970-present | | |
| 14. Sajó wastewater treatment plant | 1972-present | 17. Viinikanlahti wastewater treatment plant | 1972-present |
| | | 18. Peltolammi water tower | 1972-present |
| | | 19. Rusko surface water treatment plant | 1972-present |
| 15. Alsózsolca groundwater intake | 1974-present | | |
| 16. Tetemvár water tank | 1974-present | | |
| 17. East main | 1974-present | | |
| | | 20. Works in Polso and Kämmenniemi | 1976-present |
| | | 21. Pinsiö groundwater intakes | 1976-present |
| | | 22 Julkujärvi groundwater intake | 1979-present |
| 18. Hernád groundwater intake | 1977-present | | |
| | | 23. Hervanta water tower | 1982-present |
| | | - improved and expanded Rusko treatment plant | 1989-present |
| | | 24. Oxidization of Pyynikki depression | 1983- |
| 19. Reintroduction of water meters | 1990-present | | |
| 20. Improved Sajó wastewater plant (biological treatment) | 1994- present | | |
| 21. Improved Sajó wastewater plant (sludge dry plant) | 1998- present | | |
| 22. Improved Sajó wastewater plant (recycling sludge) | 2007?- future | | |
| | | 25. Possible artificial recharge | 2012? |
| | | 26. Relocation of Viinikanlahti wastewater treatment plant | 2020? |

in suburbs used brick waste water tanks, which polluted ground water on a widespread basis. Although the importance of waste water and rain water pipes was emphasized in theory, the priority of water-related investments was building pipelines until the late 1980s.[2] In addition to community water needs, industrial water demand accelerated. During the 1950s and 60s, Miskolc became the second largest industrial centre of Hungary after the capital Budapest. Beside traditional ironworks, machine works and paper manufacturing, as well as new branches of the processing industry such as glass, medicine and alcohol manufacturing, were set up.

A water shortage emerged first by the early 1950s. At that time, the communist town council was debating the priority of industrial and community water supply. Local industrial plants supplemented their own water pipes from community networks even during the time of water shortage. (Borsodi Műszaki Élet 1961, 23- 24)

Later on, the water shortage worsened and different projects and seminars were set up in order to solve the problem. In summer 1964, around 30,000 inhabitants did not have access to tap water for months and even the county hospital had to delay surgeries because of the lack of water. Therefore, some of the people started to use old, polluted wells. For example, top floor citizens had to use their neighbours' taps downstairs in order to get water. (Észak-Magyarország 1966, 5) Therefore, water problems encouraged good social relations among citizens, but the negative effects were undoubtedly more harmful. Water intake plants ran on maximum capacity but after dry winters, springs dried up and provided 2-3 times less water than normally. At the same time, industrial plants were running on full capacity, because an adequate amount of water was provided for them under the decision of the town council.

After the terrible draught of 1964, a new level of politicisation of the water issue was about to be reached.[3] The reason for the seasonal water shortage was the karsts springs of the nearby Bükk Mountains. At that time, the city water network was based on karst sources. Karst water is created in extended limestone cave systems by water inflow from rain. While water is infiltrated through the rocks, it is naturally purified. The only problem was the low quantity of this high quality drinking water.

After the mid-1950s Polluted local rivers could not be used for water intake. In addition, most rivers enter the region from upstream countries (abroad). The possible intake sources of Miskolc, the rivers Sajó and Hernád, already entered the country carrying the pollution of Czechoslovakian industrial plants. What is more, the river Sajó runs through the heart of the Borsod industrial area, which resulted in additional pollution from steel, engineering, chemical and glass industrial plants. Finally, attention was not paid to the proper cleaning of sewage in any of the socialist countries during the 1950s and early 60s. For example, the waste water plant of Miskolc was not rebuilt until 1972.

---

[2] Almost all annual reports of the City waterworks in the 1960s emphasize the urgent need for wastewater and rainwater pipe construction and a modern waste water plant. (Miskolc Waterworks papers ,No. 29-35)

[3] Numerous seminars and scientific meetings were held in order to solve the water problems of Miskolc in 1958, 1964, 1967 and 1973.

Figure 25.2 The Szinva spring situated in Lillafüred, in the Bükk Mountains, was has been connected to the water supply system of Miskolc since 1962. (Photo: Viktor Pál)

All in all, the excreta of about 500,000 people and the waste water of dozens of highly polluting, mammoth-sized industrial plants mixed into a deadly cocktail for fish, which died out in the Sajó river by the early 1950s. Another local water source, the river Hernád, was regularly endangered by the waste water of the Kosice ironworks, in Slovakia. The above-mentioned facts made the idea of a river intake plant on the Hernád very risky and expensive. The 1964/IV Water Act stated:

> "Those industrial plants which pollute water bodies have to be equipped with waste water treatment machinery." (1964/IV Water Act, 14 §)

Despite laws, industrial plants kept on polluting, because water filters were more expensive than annual bans, even in the long run. All in all, the main rivers of the region died out constantly or occasionally for many decades, as a result of industrial activity starting from the late 1950s. Local authorities and the government did not make much effort to solve or improve the situation. River pollution was rather considered as a threat to the national economy, than an environmental hazard. Officials and experts were worried about the economic consequences of pollution. Fish kills and the reduce of habitat caused significant losses for state-owned companies. Probably the economic loss caused by pollution generated more interest in a different, more environmentally conscious way of thinking in the early 1970s. (Miskolc Waterworks papers, No. 40-42-a.)

The water-related problems show many advantages and disadvantages of the socialist model. In Miskolc, centrally planned blocks abolished the slums of the town and created a modern, although unsustainable way of living. Workforce was needed in the many factories of the region and the local authorities, under governmental pressure, granted permission for the new housing estates of the town, although it was known that the water situation was worsening significantly and there would not be enough water for everyone. Therefore, economic development was a priority, just like in other industrial strongholds of Hungary. While a significant number of new workers came from the satellite municipalities of Miskolc, the majority of them moved from other parts of the country. New residents needed comfortable housing, within a short time and with a limited budget. Therefore, environment- and sustainability-conscious opinions were neglected not only by the government but also by the residents trying to reach a better standard of living. (B.A.Z, No. 40)

**In the case of Tampere**, public concern about water arose in the 1950s. The local newspaper *Aamulehti* reported on planned water quality surveys on Lake Vanajavesi near the forest industry centre of Valkeakoski in March 1953. The concern of professionals could be detected clearly: The article was titled: *"Bad water management is worse than losing a war".* The anonymous writer, clearly a water expert, pointed out how water bodies next to the biggest communities had become alarmingly dirty. The article also referred to toxic compounds that make water totally unusable. (Anon. 1953)

Lakes Näsijärvi and Pyhäjärvi had become heavily polluted by the pulp mills in spite of restrictions and industries´ own but insufficient actions. The traditional intake water body of the Tampere City Water Works, Lake Näsijärvi, had badly deteriorated. Its water quality was so low that it was not possible to treat it with conventional chemical methods. (Murto, 1964a, 11–12; See: Lehtonen, 1994, 83) A survey carried out in 1960-61 showed that the major polluters were sulphite pulp mills, like the one in nearby Lielahti, which had operated since 1914. *Aamulehti*, the major newspaper, did not criticize the mill when it was constructed, rather writing about its construction in a neutral way. *Kansan Lehti*, the working class newspaper, paid at the same time attention to air pollution and low housing standards. Aamulehti argued that it was beneficial for the city to reserve industrial sites along the watercourses. (Murto 1964b, 1–11; Juuti 2001, 216; Aamulehti 1961b) However, in the 1960s the tone was totally different.

At the turn of the 1960s, Aamulehti started to write about water pollution. Mr. Matti Murto, the managing director of Tampere City Water and Sewage Works, had several articles published on the threat of industrial water pollution. (Aamulehti 1959; Aamulehti 1961a) In 1962 a Water Act came into force enabling the authorities to set legal requirements and time schedules for water polluters. The quality surveys in the 1960s showed the worsening condition of water bodies. Therefore, the relocation of raw water intake to lakes in Kangasala, east of Tampere city, was discussed.

## 25.3. EYES WERE OPENED, FROM THE 1960S TO THE LATE 1980S

In Miskolc, after the water supply crisis of the mid-1960s, conferences and project meetings were held more often and the public debate became more intense. All available possibilities were examined as to how new sources of water could be exploited. Artesian wells, a gigantic 100 km long water pipe from the river Tisza and other plans were discussed. The main aims of all these studies were how to find new water resources and to encourage city dwellers to consume less tap water. But in reality, the water needs of industry still accounted for about 65 per cent of water needs in the region. (Miskolc Waterworks papers, No. 40-42-a) In spite of changes, water was handled as a raw material, a necessary good for industrial needs and a conductional good for the community in contemporary rhetoric until the mid-1960s.[4] The worsening condition of local waterways and water resources was not displayed as an environmental problem primarily in the media until the end of the 1960s.

By the early 1960s, different ways of wasting water were criticized. Journalists pointed out that the local community should be more conscious about the use of piped water, so that less water would have to be pumped into the system. By the late 1960s, a new tone of environmentally conscious journalism emerged, based on foreign examples. One of these journalists was Tibor Priska, who published many articles on environmental issues in local papers. In the article "Black Rivers he writes[5]: What is a lot? And what is a little? The environmental fine, or building a waste water treatment plant? To build a waste water plant may cost millions [of Hungarian Forints], to pay the fine may cost not more than a hundred thousand. But the final damage will arise later, when water is used to irrigate lands, fish die out etc. What is more, this is happening each year. (Priska 1968)

By the early 1970s, the new, more environmentally conscious way of thinking emerged among political decision-makers and experts as well. Socialist Hungary was not completely isolated from foreign innovations and had limited relationships with countries in Western Europe and overseas. Despite the country's relative isolation, foreign experiences helped to address water problems. By the early 1970s, German and English experts examined the river Sajó and Hungarian engineers took several study trips to West Germany and the UK. A fund of one million US dollars was received from the Development Programme of the United Nations for the Sajó project by 1972. The first phase of the project was to learn about foreign experiences and adopt monitoring and laboratory techniques. The models of the Thames and Lee rivers of England, the Treen in Scotland and the and Ruhr in Germany were studied. The pollution of the Sajó became an internationally recognized environmental problem. (Észak-Magyarország 1972, 3) There were also expert exchange programmes such as the one involving the Hungarian and Finnish Governments.

---

[4] The best examples of environment-related rhetoric are newspaper articles. There one can find many features of social attitudes towards water, e.g. Észak-Magyarország1949b, 4; Észak-Magyarország 1958a, 2; Észak-Magyarország 1958b, 3.
[5] Author's Translation.

To solve water supply problems, a decision was made to construct a water intake plant on the river Hernád by 1969. Further investments took place in the karstic water intake areas. In addition, a pipeline was built between Miskolc and the Northern Hungarian regional waterworks system, in order to guarantee adequate water supply.[6] In the early 1970s, a mere 10 per cent of the town's annual water consumption was supplied from the Kazincbarcika region by an inter-city network. A small scale intake plant was built based on the pebble basin of the [river] Sajó in Alsózsolca, by 1974. In addition, a water intake plant was constructed close to the river Hernád, in 1977, after many years of delay and fierce debate. Later on waste water loadings on the Hernád emerged again from the Kosice Ironworks, Czechoslovakia and endangered the intakes of Miskolc downstream. (Miskolc Waterworks papers, No. 39)

In 1978, new intake plants were taken over by the Northern Hungarian Regional Waterworks following a political decision. Therefore, the majority of the supplied water had to be bought by Miskolc City Waterworks. Water costs were rising, but they were refunded by the socialist government until the end of the 1980s. The state refund kept water prices modest and thus consumption habits did not change. Water pipe construction remained the priority and the sewage network grew on a smaller scale. For example, in 1973, out of the total 19 million forint investment of the city waterworks, 9 million was concentrated on new water pipes, 6.5 million on construction and machinery, 2 million on drains, 1.5 million on pools and thermal baths and the rest half million went to other expenses. (Miskolc Waterworks papers, No. 37)

Figure 25.3 The wastewater plant of Miskolc with mechanical treatment was constructed in 1972. (Photo: Viktor Pál)

---

[6] The Miskolc City Waterworks and the Borsod-Abaúj-Zemplén County Waterworks were created by 1955 from the same regional water company.

In the 1970s-80s, about 65-70 per cent of the supplied water in Hungary was still consumed by industrial plants. Until 1972, industrial and public wastewaters were not treated by any means. A wastewater treatment plant was designed in 1955 and the preparation works [had] started in 1963. (Miskolc Waterworks papers, No. 42-a) Despite the need for a waste water plant, the government cancelled the construction in 1965. At the same time, a significant share of the Miskolc City Waterworks' profit went to pay water pollution fines every year. By the end of 1972, the city waste water treatment plant was constructed on the bank of the Sajó river with a mechanical phase (Figure 25.3), providing only 30 per cent efficiency. (Miskolc Waterworks papers, No. 34-36) The plant was completed by 1986. That was the time when the annual water consumption reached its peak at around 40,000 m3 in Miskolc between 1985-1988. During the late 1980s, the growth of the region slowed down and city plans concerning future investments had to be reconsidered. Migration and industrial output decreased significantly. Therefore, water supply conditions improved and summer water crises became less frequent and widespread, but wasting water and network loss have still remained significant problems up until the present.

**In the same period in Tampere,** in the mid-1960s it was estimated that water consumption would increase from about 200 litres per person/day to 250 litres in 1970 and even to 450 litres, along with a population of 300,000, by the 1990s. Yet, after the oil crisis in 1973, consumption and the introduction of a sewerage charge raising water prices to more than double, the consumption levelled off. (Figure 25.4) (Murto 1964; Juuti & Katko 1998)

Big changes in pulp mill production processes, like alcohol production as a by-product in sulphite mills already before WWII, the later transfer to sulphate pulp production with the evaporation and incineration of black liquor as well as the effective treatment of municipal wastewaters since the late 1960s, improved the condition of water bodies. In 1971, the sulphite pulp mill at Lielahti started to treat its wastewaters mechanically while in 1972 an evaporator and a recovery furnace were completed and about 85 percent of the black liquor was recovered. Still, pulping and bleaching continued to cause environmental problems and the mill was about to be closed. (Katko et al. 2005)

At this time, the left-wing working class newspaper, *Hämeen yhteistyö*, wrote on 23 December 1971 in a long article titled *Pollution Emperors Protected by City* that the city of Tampere had not taken serious steps against industry waste waters. (*Hämeen yhteistyö* 1971)

In 1985, the old pulp mill was replaced by the first so-called chemi-thermo-mechanical pulp mill in Finland. Biologically activated sludge treatment for wastewaters was also introduced. This decreased the wastewater load by 90%. However, at the same time external energy consumption increased dramatically. This was due to the high yield for pulp, of the order of 90 percent, compared to chemical pulping where the pulp yield is less than 50 percent and most of the mill's energy, even surplus, can be generated from the wood itself. The reduced wastewater volumes were now treated in an activated sludge process that was constructed by utilizing the large old clarifiers effectively for primary sedimentation and aeration. (Tampere Water 1962-

Figure 25.4 Managing director M. Murto was forecasting the future developments of water consumption in Tampere in the 1960s.

1967; Könönen 1994; Juuti 2001, 217; Juuti et al. 2000, 143–144; concerning the conditions of wastewater treatment in the whole country, see Lehtonen 1994, 71–85; Luonsi 1987, 83–91)

Industrial changes and new community wastewater treatment plants together soon brought about visible effects in the quality of water bodies. Within the Tampere water and environmental district of that time, the industrial biological oxygen demand (BOD) loading was reduced by over 50 per cent from the early 1970s to the early 1980s. Except for the Lielahti pulp mill, industries in the centre of Tampere discharged part of their wastewaters to the treatment plant of the city. (Juuti 2001; Katko et al. 2005; Tampereen vesi- ja ympäristöpiiri 1986, 27)

## 25.4. STREAMS OF OLD AND NEW CHALLENGES

The political and economic changes of the late 1980s affected **Miskolc** significantly. Ownership, profile and production scales were changed in the case of many industrial plants along the Sajó river and the Szinva creek. In Miskolc, heavy industrial production slowed down and stopped later on. At the same time, the nearby chemical works of Kazincbarcika, Sajóbábony and Tiszaújváros were transformed and flourished with the help of foreign investments. The environmental situation of Miskolc improved significantly, because of closing down major industrial plants. After having medium quality sewer water, the Sajó river and the Szinva creek became good quality water flows with rich fish stock. (Vári & Kisgyörgy 1998, 223-225) After more than 40 years, fish came back to the river Sajó in the 1990s.

At the same time, local karst waters in the nearby Bükk Mountains were protected by the almost full sewerage of mountain villages. Thanks to these developments, the quality of drinking water has been improved. From the beginning of the 1990s, rapid sewerage construction took place in the Miskolc area to provide wastewater services to all residents. In 1994, the biological phase of the waste water plant was put into order. (Vizeink, 1995, 2) The following investment was a sludge dry plant in 1998. Nowadays 99 per cent of all homes are connected to the water system, and more than 85 per cent of the city's territory is connected to the sewer system. That means 62,000 homes have joined the sewage system. Lately peripheral suburban areas have also been attached to the networks. Due to the decline in per capita consumption, the partly city-owned company pumps 50 per cent less water to the network in 2005 than at the end of the 1980s. (Environmental and Conservational Programme for the City of Miskolc 2004)

In 2004, the length of the water supply system was 650km, and the sewerage system was about 435km, the latter discharging some 18,950,000 m3 of waste waters annually. (Environmental and Conservational Programme for the City of Miskolc 2004) During the last 15 years, more than 25 per cent of the local population has emigrated because of unemployment. State subsidies decreased and water meters were introduced by the early 1990s. Therefore, municipal water consumption decreased even more. The new units of sludge drying-out and sludge incinerating are planned to be introduced in the near future. (Figure 25.6)

Water pipe construction hardly could keep pace with the accelerated development of the town and sewage lines could not be constructed due to limited construction capacity. By the 1990s, the sewage systems were upgraded and broadened. At the same time, the population of the town started to stagnate and decline in the mid-1980s. The collapse of the socialist economy, rising unemployment and social problems resulted in emigration to other regions of Hungary. Therefore, a relatively extended system, designed for approximately 300,000 citizens and several industrial plants, is currently run at half capacity for the present population of Miskolc and satellite communities. Decreasing consumption naturally causes financial problems for the local waterworks and in 2005 a partial privatisation took place. Experts hope that private capital would solve the problems of the ageing network, significant leakage and other questions. The success of this approach remains to be seen. (Èszak-Magyarország 2004, 7)

The present priorities of water legislation are to encourage the conservation of water resources. These tasks have been emphasized globally but the installation of meters, high water and sewage prices and the collapse of industry have solved environmental problems rapidly in Miskolc.

In **Tampere**, the question whether to use groundwater or surface water has been debated for a long time. Even in 2005, there is intense debate over the planned artificial ground water system in the region. Already in 1915-16, there were discussions about whether Tampere should begin to use groundwater, which in terms of healthfulness and taste was better than the water of Lake Näsijärvi. After groundwater inventories, the result was that in 1920 the city abandoned the groundwater plans. After the decision, a solution other than groundwater had to be found, and in 1921 the city council approved the building of a plant using surface water with rapid sand filtration. The idea of using groundwater in Tampere was not reintroduced until the 1950s. (Juuti & Katko 1998, 101-107; Juuti 2001, 190-194) The long-term development of water consumption and population growth in Tampere is presented in Figure 25.5.

The new Kaupinoja surface water intake was in a safe place, and opened in 1928. The water from Lake Näsijärvi was chlorinated and filtered through sand filters. Another new plant for the western parts of the city was built in Mältinranta, upstream from the rapids in 1931. After

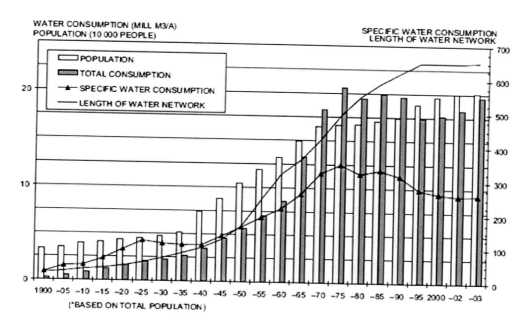

Figure 25.5 Water supply in Tampere, Finland 1900-2003. (Source: Tampere Water, annual reports 1898-2003)

World War II the water works grew rapidly. At the turn of the 1960s, the quality of raw water became a problem. The pulp industry had polluted Lake Näsijärvi, the age-old upstream raw water basin. In 1972, Lake Roine, which lies in a neighbouring municipality, became the new source of raw water. It constituted one of the best raw water basins in the country, but later some seasonal problems were faced in relation to odour and taste due to algae. Water treatment was enhanced by introducing activated-carbon and flotation treatments. Also, the alkalinity of treated water was increased to prevent the corrosion of pipes. Water treatment and the quality assurance of piped water are continuously improved. (Juuti & Katko 1998) It is planned that by approximately 2010 a regional water supply system based on ground water through artificial recharge will serve the city and the neighbouring municipalities. Thus, the question of whether to use ground or surface water has been there for a century. (Juuti 2001)

In Tampere cooperation over the city borders has been carried out for quite some time. Sections of surrounding municipalities have been incorporated into the city over the years. The city water works assumed control of that particular system in the 1960s. Two built-up areas of the Teisko-Aitolahti rural district, which nowadays is part of Tampere, have water supply and sewerage systems run by the city. In other sparsely populated areas, water supply and sewerage will continue to be provided through cooperatives or by individual property owners themselves. (Juuti & Katko 1998)

Since 1972 the wastewaters of the central and eastern parts of Tampere have been led to the Viinikanlahti treatment plant on Lake Pyhäjärvi downstream from the city. The plant is the largest environmental investment the city has ever made. It lies on top of what used to be waste masses and, due to changes in land use, is being encroached upon by the expanding core of the city. The plant was expanded into a chemical treatment plant in 1976 and into a biological-chemical plant in 1982. Now there are several plans to move it away from Tampere, for example to the neighbouring municipality, Nokia. The Pyynikki traverse ridge divides the city into two natural sewerage systems that originally depended merely on gravity, but later introduced also pumping. (Juuti & Katko 1998)

The first plan for broader intermunicipal cooperation in the Tampere region was devised in 1972. The first general plan for water supply to Tampere and Valkeakoski, TAVASE, was drafted in 1993. It covered some ten municipalities. The idea of creating artificial recharge for the needs of the entire area also emerged. (Juuti et al. 2003, 121–127, 167–173; Kangasalan Sanomat 1972)

The factories on the banks of Tammerkoski Rapids, representing the textile, leather and wood processing industries, had their own water intake plants and sewers by the Rapids from early on. Later on, starting in the 1960s, they also built water treatment facilities. The Finlayson factories in Tampere were the first in Scandinavia to introduce sprinklers in 1837, only six years

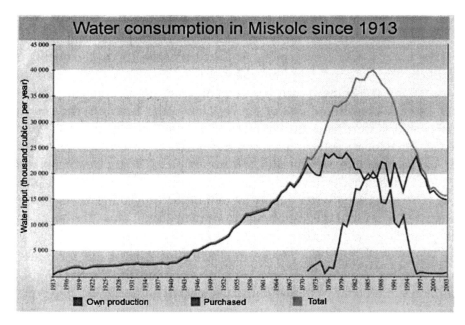

Figure 25.6 Water consumption in Miskolc 1913-2000.

after their invention. The factories were also big customers of the city water works. For example, the industry per cent share of the total water consumption was around 20-30% until the 1990s and after that around 10%. When water pollution control became topical, enterprises started to clean their effluents and gradually also to lead them to the city's wastewater treatment plant. Industry also underwent a major structural change in the 1980s, which further reduced the wastewater load. (Tampere Water, annual reports 1898-2003; Juuti & Katko 1998, 238-251)

## 25.5. CONCLUSIONS

Although the political and economic development of Finland and Hungary are significantly different, the development of water services and related environmental questions had surprisingly similar elements during the research period.

Early efforts to provide water were made during the first decades of the 1800s. Simple gravitation systems were soon replaced by high-pressure schemes. The main indicator behind the shift from wells to water pipes and from ditches to sewerage was the threat of various diseases. In addition, fire was a major danger, especially in Tampere. Industrialization caused rapid urbanization in Tampere during the 19th century and similar development happened in Miskolc after World War II. The early development of the textile industry had a leading role in Tampere, while steel and engineering factories were enlarged by the 20th century in Miskolc.

Wastewaters of the two similar-sized cities polluted water bodies until efficient waste water treatment was introduced in recent decades. The share of industrial use of the community water supply system peaked around 50 per cent in Tampere and 65 in Miskolc, after rapid industrialization of the latter. At the same time, the total community water use peaked in Finnish towns in the mid-1970s and in several Hungarian cities between 1985 and 1988. The main reasons for the decline in water use in Miskolc were deteriorating industry and newly installed water meters. The significant change of the industrial profile of Tampere has decreased local water consumption.

The rapidly accelerating need for water in Tampere was met by surface water from Näsijärvi Lake, which became contaminated fast. Later on groundwater was utilized, and artificial groundwater will likely be produced in the future. Miskolc has been served by the karst springs of the Bükk Mountains, instead of the region's heavily polluted surface and ground water resources. By the 1970s, filtered ground water had to be taken from neighbouring municipalities to meet the city demand.

In Miskolc urban water supply was often ranked below the needs of industry. This was also true to some extent in Tampere. Later, the priority of water supply was to satisfy consumer demands and to increase the pipe network between 1940s-1980s. Minor water quality issues were handled by the stagnation and decline of consumption. The paradigm shift from quantitative to qualitative and from the construction of water pipes to the construction of sewage and waste water treatment was closely connected to the collapse of communist industry in Miskolc. Changed ownership and management along with new laws and institutions have led to the successful adoption of EU regulations and norms during the past 15 years in Hungary. On the one hand, the economic depression led to financial problems; on the other, it solved major pollution-related problems. Therefore industries began to protect waters and later on, co-operation increased with the city wastewater works.

In Finland the environmental consciousness rose earlier. Industrial production systems and processes were modernized and water laws and public awareness increased public pressure. During the past 15 years, industrial production shifted towards high-tech products and services in Tampere and mostly services in Miskolc, which largely decreased water consumption and pollution. In addition, this has been a global trend in industrialized countries. A key principle in both case countries has been that municipality-owned water and sewage works have bought – and continue to buy – contracting, design and expert services from the private sector based on competitive bidding. In past years, the same development has happened also in Miskolc.

The number and significance of environmental issues have decreased in Finland and Hungary in recent years. Industrial waste water, agricultural chemicals and air pollution from manufacturing and power plants still pose a problem in Finland. Unsolved environmental problems in Hungary are frequently caused by the international pollution of rivers. In this sense, water does not know limits neither in place nor time. (Figure 25.7)

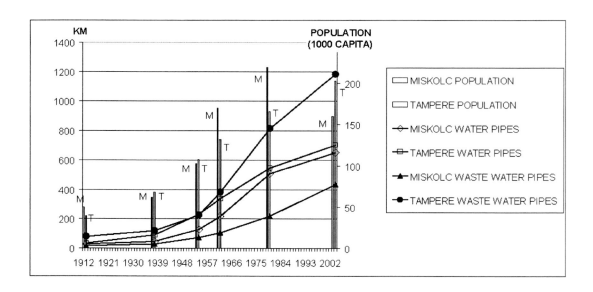

Figure 25.7 Growth of water and wastewater pipes and population in Miskolc and in Tampere 1900-2004.

The history of Tampere and Miskolc shows how water-related problems emerged slowly and remained unrecognized for long by the public. When water problems started to affect everyday life, decision-makers adopted legislation in order to solve the emerging problems. These comprehensive solutions generally require large investments.

Despite the different political, economic and social development of Finland and Hungary, roughly similar environmental problems appeared at the same time. Adequate solutions to these problems, however, were found in Hungary only after the political changes of 1989.

Nowadays the waterworks of Tampere and Miskolc operate under ISO 14001 environmental management standards and several future development plans have been announced.

## 25.6. REFERENCES

PRIMARY SOURCES

Act 1964/IV on water pollution and the related 32/1964 (XII.13) governmental order.
Miskolc Waterworks papers B.A.Z County Archives, Mezőcsát Archives, XXIX-113. No. 34-36.
Miskolc Waterworks papers B.A.Z County Archives, Mezőcsát Archives, XXIX-113. No. 42-a.
Miskolc Waterworks papers B.A.Z County Archives, Mezőcsát Archives, XXIX-113. No. 37.
Miskolc Waterworks papers B.A.Z County Archives, Mezőcsát Archives, XXIX-113. No. 39.
Miskolc Waterworks papers B.A.Z County Archives, Mezőcsát Archives, XXIX-113. No. 40-42-a.
Miskolc Waterworks papers B.A.Z County Archives, Mezőcsát Archives, XXIX-113. No. 40.
Miskolc Waterworks papers B.A.Z County Archives, Mezőcsát Archives, XXIX-113. No. 29-35.
Tampere city record office, minutes of city administrative court 3.4.1838.
Tampere Water, annual reports 1898-2003

LITERATURE

Aamulehti (1959) 18.12.1959

Aamulehti (1961) 12.1.1961.

Aamulehti (1961) 2.4.1961.

Anon. (1953) Aamulehti, 5.3.1953.

Borsodi Műszaki Élet (1961) No.2. 23- 24 pp.

Environmental and Conservational Programme for the City of Miskolc. 2004. [http://www.miskolc.hu/oldal.php?menupont_id=190]

Észak-Magyarország (1949a) Jelentékenyen bvül Miskolc vízhálózata, 17.04.1949

Észak-Magyarország (1949b) 200 százalékra teljesítik a hároméves tervet a miskolci Vízgazdálkodási Hivatal dolgozói. 02.10.1949.

Észak-Magyarország (1958a) Miskolc város közmveinek és közlekedésének fejldése. 15.10.1958.

Észak-Magyarország (1958b) A vízügyi ankét második napjának vitája. 06.11.1958. 3

Észak-Magyarország (1962) Az ivóvíz is érték ? takarékoskodjunk vele! 28.04.1962

Észak-Magyarország (1966) Ismét hiánycikk lehet a leger?sebb ital. 24.04.1966

Észak-Magyarország (1972) Megmentik a Sajót, 22.07.1972.

Hämeen yhteistyö (1971) Yhtiöiden saastekeisarit kaupungin suojeluksessa. 23.12.1971.

Juuti P. (2001) Kaupunki ja Vesi. Doctoral dissertation. Acta Electronica Universitatis Tamperensis 141. Tampere.

Juuti P. & Katko, T. (1998) Ernomane vesitehras. Tampereen kaupungin vesilaitos 1835- 1998. Tampere

Juuti P., Rajala R. & Katko T. (2000) Ympäristön ja terveyden tähden, Hämeenlinna, pp.143–144.

Juuti P., Katko T. & Äikäs K. (2003) Luonnollisesti vettä. Kangasala, Vesilaitos 1952-2002. Saarijärvi.

Kangasalan Sanomat 10.11.1972

Katko T., Luonsi A. & Juuti P. (2005) Water pollution control and strategies in Finnish pulp and paper industries in the 20th century. (in IJEP, International Journal of Environment and Pollution), IJEP Vol. 23, no. 4, 368-387.

Könönen R., Jäähyväiset Lielahden liukoselluloosatehtaalle -Tervetuloa kemihierre.

Lehtonen J. (1994) Jäteveden puhdistuksen kehitys Suomessa pitkällä aikavälillä. TTKK, VYT No. B 58. Tampere.

Luonsi A. (1987) Development of wastewater treatment technology in Finnish pulp and paper industry. Kemia-Kemi, no. 2, pp. 83–91. (original in Finnish)

Murto M. (1964a) Tampereen vesilaitos kaupungin asukkaiden palvelijana. Tammerkoski, Vol. 26, pp. 11–12; See: Lehtonen, J. (1994), p. 83.

Murto M. (1964b) Presentation for Tampere-Seura (11 November 1964), pp.1–11;

Priska T. (1968) Black rivers. Észak-Magyarország, 13.03.1968, 5

Rasila V. (1984) Tampereen historia II. 1840-luvulta vuoteen 1905. Tampere

Tampereen vesi- ja ympäristöpiiri (1986) Vesiemme hyväksi. Pirkanmaan ja pohjois-Satakunnan vesienkäytön, hoidon ja suojelun kehittämisohjelma, p.27.

Tampereen kaupungin vesilaitos, Annual reports 1962-1967.

Teslér J. & Piukovics J. (1963) A miskolci vízmvek és fürdk 50 éves története. Miskolc, ,

Vári A. & Kisgyörgy S. (1998) Public participation in developing water quality legislation and regulation in Hungary. In: Water Policy, 1 (1998) 223-225 pp.

Vizeink (1995) No. 1-2.

Worster, D. (1994) Nature's economy. Cambridge: Cambridge University Press.

# 26
# WATER SUPPLY AND SANITATION IN RIGA: DEVELOPMENT, PRESENT, AND FUTURE

*Gunta Springe & Talis Juhna*

Figure 26.1 Location of Riga, Latvia.

# 26.1. GENERAL BACKGROUND

Latvia regained independence from the former Soviet Union in 1991 and joined the European Union in 2004. During the transition period from a Soviet centrally planned economy to a market economy, attitudes towards environmental issues, such as drinking water and wastewater treatment, have changed. These changes have been influenced by European policy, which puts strong emphases on environmental protection and public health. These processes are of special importance for Riga (Fig. 26.2), the capital of Latvia, with its high population density. The total population of Latvia is 2.32 million, about 735,000 of which live in Riga. Riga occupies 307 km² at the Gulf of Riga. The city was founded in 1201, and during the medieval period developed into an important trading centre within the Hanseatic League. In the second half of the 19th century Riga transformed into a modern industrial centre of great importance within the Russian Empire. Today Riga is the largest city in the Baltic States and in its vision for the future the city is seen as a centre of European importance in the Baltic Sea region.

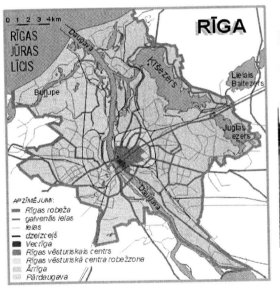

Figure 26.2 A map of Riga (www.rsp.lv/arh/teksts/ infoavoti/kartes/Riga.html) with its bodies of water (Lake Kisezers, Lake Juglas, Lake Lielais Baltezers, the River Daugava and the Gulf of Riga) and a panorama of Riga (www.eidos.lv/lv/all_other/ photo/riga/riga_1.php).

Riga is rich in water resources, including surface waters (Fig. 26.2) and groundwater. The largest river of the Baltic States—the River Daugava—crosses the city. There are many lakes located in the vicinity of Riga, and rich groundwater aquifers are available.

## 26.2. DEVELOPMENT OF WATER SUPPLY SYSTEM IN RIGA TILL THE 20ᵀᴴ CENTURY

In the Middle Ages, like most other towns, Riga obtained water for the needs of its inhabitants and crafts workshops from protection canals and shallow wells. Along with an increase in population density, these sources became scarce and polluted. Therefore the city Council had to think about a way to transport water from more distant sources. In 1582 the Polish king Stephan Batory granted the privilege to divert water from the small river located outside the city through the artificial canal to Riga. This water resource gradually dried up, but the canal system existed till 1812 when Napoleon's army destroyed it (Zeids 1978, 304, 430–431; Archives of Municipal enterprise "Riga Water").

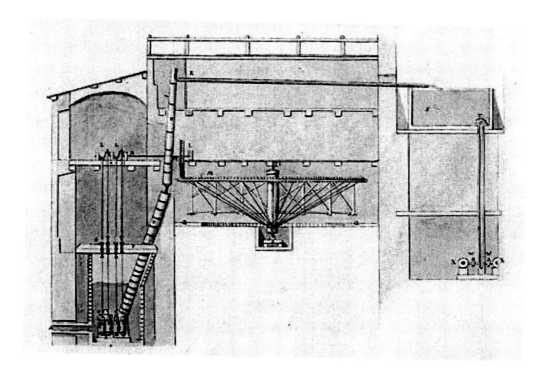

Figure 26.3 The first pumping device in Riga, built in 1663 (Source: Zeids, 1978)

# WATER SUPPLY AND SANITATION IN RIGA

The first water-pumping device (Fig. 26.3) in Riga was built in 1663 to take water from the River Daugava. Six horses were used to run eight pumps with brass pistons, which helped to pump water from the river through a special wooden pipe into a wooden storage tank located at the top of the tower. From the storage tank water was distributed though wooden pipes to houses of noblemen. This water supply system was used till 1863.

In 1863 a new pumping station with steam engine plunger pumps was put in operation (Fig. 26.3). The water was taken from the Daugava upstream from Riga sewage discharge; therefore, water was relatively safe for drinking. This water supply system satisfied the needs of the city for about 40 years. To ensure pressure in upper floors at the end of the 19th century, several water towers were built in Riga (Fig. 26.4).

Systematic sewer construction was started in the middle of the 19th century. Up to 1907, the total length of the sewer system was 19.6 km. Most often combined (foul and storm water) sewer collectors were built at the time (Pāvels 1932).

Figure 26.4 The first pump station started water intake from the River Daugava in 1863; one of the four typical water towers of Riga, built in 1897 (Source: Archive materials of Riga Water Ltd.).

## 26.3. SITUATION IN RIGA WATER SUPPLY IN THE 20ᵀᴴ CENTURY UP UNTIL TODAY

At the beginning of the 20th century, Riga developed as a large industrial and social metropolis. In 1800 there were 29,500 inhabitants in Riga, in 1863—77,467, but in 1897 there were 282,230 citizens. (Latvijas statistikas gada grāmata 1937/38)

Due to pollution from agriculture and industry, water quality in the Daugava changed for the worse, causing different diseases, for example, a typhoid epidemic (about 2000 affections cases) in 1900 (Krastiņš 1978, 215). The extension and improvement of the existing system became inexpedient. In 1883 large quantities of high quality groundwater were discovered in the surroundings of Riga (Bukulti, Rembergi and Zakumuiza). It was appreciated as the second best after Vienna, which used water from the Alps. At the end of 1904 the groundwater intake station Baltezers, located 20 km from Riga, was put into operation. The water intake and pumping station cost 3 million golden roubles of tsarist Russia (Ronis 1932). In 1913 the water intake from the Daugava was banned.

Nowadays in the engine room of this old station in Baltezers, the museum of the Riga Water Supply (opened in 1988) is located. The exhibition tells of the development of methods of water supply in Riga during four centuries. The museum displays steam boilers, produced in the early 20th century, pumps, pipes, different kinds of water meters and testing equipment (Fig. 26.5).

In 1912 installation of the sewer system within the centre of Riga - Old Town was almost completed. In 1913 a sewer installation project was started on the west bank of the Daugava (Pardaugava) totalling 6.5 km with an outlet into the river. In the last pre-war years the sewer network was expanded by 71 km and its total length reached 90.6 km (Pāvels 1932, 485–486;

Figure 26.5 Exposition Museum of Riga Water Supply in Baltezers with steam pumps and remnants of wooden pipes used in the 17ᵗʰ century (Source: Archive materials of Riga Water Ltd.)

Krastiņš 1978, 216–218). In 1914 the population of Riga had increased already to 520,000 (Latvijas statistikas gada grāmata 1938. gadā, 6).

After World War II the industrial production of Riga grew along with the size of the population. To meet the needs of increasing water consumption, artificial groundwater recharge was started in 1953. Several new groundwater abstraction plants (Baltezers I, Baltezers II, Rembergi, Gauja I, Gauja – experimental, Jugla, and Katlakalns) were developed in the 1960s and 1970s.

By the fall of 1991, most of sewerage was dumped raw or partially treated into the Daugava and further into the Gulf of Riga. It resulted in an extensive contamination of the ecosystem.

Further increase of water consumption urged the city to search for more water sources. Over-extraction of groundwater resulted in a lowering of its level in the Riga region. Thus in 1979 the drinking water treatment plant (WTP) Daugava, which was taking water from the Daugava, was built. The designed capacity of the plant at the beginning of operation was about 200,000 $m^3$ per day. However, in the middle of the 1980s, along with a change in drinking water standards, the yield of the plant was reduced.

Today providing good quality drinking water is one of the main goals of the National Environmental Policy Plans for Latvia (1995, 2003). The municipal enterprise Riga Water Ltd is in charge of drinking and wastewater handling services for Riga. The enterprise is involved in the internationally financed Riga Water and Environment Project, which is directed to the establishment of a safe and high quality water supply in Riga and treatment of all wastewaters in accordance with modern requirements. The project started in 1996, in 2001 its first stage was finished, and the second stage (2002-2006) has started. The third stage has to be completed by the year 2008.

The first activities within the project were directed to the rehabilitation of drinking water produced by WTP Daugava. The existing treatment chain (pre-chlorination - flocculation sedimentation - sand filtration - post-chlorination) of the plant was no longer sufficient for fulfilling the guidelines of European Directives. During more than two years, different process combinations were studied in the pilot plant. As a result, several improvements were considered. Primary chlorination was changed to ozonization. Sand filters were converted into with air and water backwashable multi-layer filters. The second ozonization step was introduced to guarantee complete disinfection and the removal of organic matter. Biologically active carbon filtration was introduced to increase the biological stability of drinking water in distribution networks. Although chlorine is still used before water distribution, the doses are significantly lower because the consumption of chlorine has been highly reduced.

In 1991 the first wastewater treatment plant (WWTP) Daugavgriva was finished and put into operation. This treatment plant handles wastewater from Riga and partly from neighbourhoods. It serves more than 850,000 inhabitants, and at the moment about 81% of Riga inhabitants are connected to centralised wastewater networks. The capacity of the plant is 350,000 $m^3$ of wastewater per day. The plant uses mechanical and biological treatment. Modern process of biological treatment for the removal of phosphorus was realised at the plant.

At present Riga uses a number of sources for its water supply, including surface water, natural groundwater and artificially recharged groundwater from Lake Mazais Baltezers. As a result of a decrease in industry and the installation of water meters in apartment buildings (Rubulis et al. 2001), the consumption of drinking water in Riga has dropped at an unexpected rate since the latter half of the 1990s. At the moment, daily consumption of drinking water is about 150,000–170,000 m$^3$. In 1861 the estimated daily consumption per capita was 114 litres, whereas during the last century it increased to 250 litres and more (Archive materials of Riga Water Ltd). With

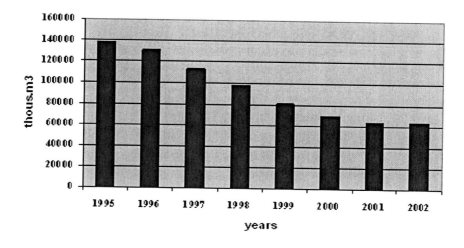

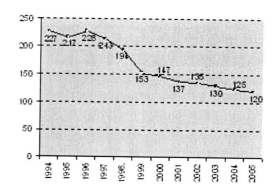

Figure 26.6 Annual drinking water consumption in Riga 1995–2002 (a) and in litres per capita per day, 1994–2005 (b). (Source: Municipal enterprise Riga Water Ltd)

the beginning of the installation of water meters in Riga, the consumption started to decrease and eventually stabilised at the level of 120 litres per capita per day. (Fig. 26.6)

At the moment, in most cases the drinking water in Riga meets the water quality standards of Latvia (State of environment in Riga 2001), and after further improvement in the water supply, it will meet the Directives of the EU. At WWTP Daugavgriva several improvements have been realized, including sludge handling and an activated sludge process (Riga Water and Environment Project).

## 26.4. PROBLEMS CONCERNING RIGA'S DRINKING WATER AND WASTEWATER AND FUTURE PROSPECTS]

During the last decade, as a result of substantial efforts in the area of environmental protection, the inland water quality of Latvia has begun to improve (Juhna & Klavins 2001). At the same time, some environmental problems have remained. One of them is an eutrophication of raw water sources used for drinking water production. It concerns Lake Mazais Baltezers, which is the main source for the artificial recharge plant supplying up to 25% of drinking water for Riga. At the beginning of the 20$^{th}$ century, Lake Mazais Baltezers was clean, and its sediments had a high degree of mineralization (Ludwig 1908). In contrast, now the sedimentary material of the Lake consists mostly of organic matter rich in nutrients. The occurrence of cyanobacterial blooms has become typical for the Lake Mazais Baltezers and also for the infiltration basins taking water from the lake, as well as in the further course of artificial recharge of groundwater (Springe et al. 2001).

The self-purification capacity of another source of drinking water—the Daugava—in the riverine part is quite high (Sprinģe, 1995). The situation was changed due to the building of three hydropower plants in the 20$^{th}$ century on the Daugava. After the dams were built, algal blooming, previously untypical for the Daugava (Kumsare 1967), with the development of potentially toxic species (Druvietis 1997, 65; Druvietis 2001), has been observed. Microcystins from *Microcystis* spp. were detected in the Daugava (Balode et al. 2001) as well as in Lake Mazais Baltezers (Eynard et al. 2000; Balode et al., 2001).

Protection of surface water bodies has to become more stringent, and implementation of Latvian laws harmonised with EU requirements as well as the basin management approach defined in the Water Structure Directive must be carried out.

In general, the quality of drinking water is good and meets EU Directives at the plants, but it deteriorates in pipelines during distribution to the consumers (Halla 1998). This is largely connected with outdated water networks. Today Riga's water supply network extends over 960 km, and at least 70 km of it are in poor condition and must be renewed. It should be noted that the oldest networks, built during the four decades from the beginning of the 20$^{th}$ century to about 1940, are in comparatively good condition with few exceptions. These systems were constructed of 300 mm to 800 mm diameter cast iron pipes with an anticipated lifespan of 50 to

90 years. In contrast, the more recent networks built between 1940 and 1990 with pipes made in Russia are in poor condition.

Another reason for the decrease in water quality is the presence of biologically degradable organic carbon allowing for bacteria to grow in water and in biofilm (Juhna, Klavins, 2001, 306–314). Therefore Riga has joined one of the European research projects that are aiming to control the microbiological stability in drinking water distribution networks (Surveillance and control of microbiological stability).

Hopefully at the second and third stages of the Riga Water and Environment Project, the problems will be solved as one of the project aims is devoted to preventing a decrease in the water quality in the water supply network.

In respect to wastewater handling, the next investments will be largely allocated to the extension and reconstruction of networks leading to environmental improvement. The former combined sewer system will be replaced with separate foul and storm networks.

## 26.5. CONCLUSIONS

The history of Riga water supply and sewerage extends over four centuries. Every period has had its problems, and they were resolved according to the knowledge and possibilities of the time for the sake of human welfare. This still will remain a priority in the $21^{st}$ century, although the technologies for reaching it will differ from those in the $17^{th}$.

## 26.6. REFERENCES

Archives of Municipal enterprise Riga Water Ltd ("Rīgas Ūdens")

Balode M., Ward C., Hummert C., Reichelt M., Purina I., Bekere S. & Pfeifere M. (2001) The first toxicological studies of harmful algal blooms in Latvian waters. In Hallegraeff, G. et al. (Eds) Harmful Algal Blooms 2000, Intergovernmental Oceanographic Commission of UNESCO, pp. 83.

Druvietis I. (1997) Observations on Cyanobacteria blooms in Latvia inland waters. VII International Conference on Harmful Algae, Vigo – Espana, pp .65.

Druvietis I. (2001) Cyanobacteria blooms in dammed reservoirs, the Daugava River, Latvia. In Hallegraeff, G. et al. (Eds) Harmful Algal Blooms 2000, Intergovernmental Oceanographic Commission of UNESCO, pp. 105-107.

Eynard F., Mez K. and Walther J.-L. (2000) Risk of Cyanobacterial toxins in Riga waters (Latvia). Wat.Res. 4 (11), 2979 - 2988.

Halla T. (1998) Water management in Riga and in Copenhagen. The Sea and the Cities. A Multidisciplinary Project on Environmental history. http://www.valt.helsinki.fi/projects/enviro/articles/articles.htm

Juhna T. & Klavins M. (2001) Water-Quality Changes in Latvia and Riga 1980-2000: Possibilities and Problems. Ambio, 30 (4-5), 306-314.

Krastiņš J. (ed.) (1978) Rīga 1860 – 1917 (Riga 1860–1917) Rīga, Zinātne. (in Latvian)

Kumsare A. (1967) Hydrobiology of the River Daugava. Zinatne, Riga, 186 p. (in Russian)

Latvijas statistikas gada grāmata (Statistics yearbook of Latvia). 1937/38, astoņpadsmitais gada gājums (18th annual set), Rīgā, 1938. gadā, pp. 6 (in Latvian)

Ludwig F. (1908) Die Küstenseen des Rigaer Meerbusens. Arb. d. Naturforscher-Ver. zu Riga N.F. 11

National environment protection policy plan for Latvia (1995) http://www.varam.gov.lv

National environment protection policy plan for Latvia 2004 – 2008 (2003) http://www.varam.gov.lv

Pāvels R. (1932) Rīgas pilsētas kanalizācija (Sewerage of Riga City). In: Rīga kā Latvijas galvas pilsēta (Riga as the capital of Latvia), Rīga, pp. 485 – 488 (in Latvian)

Riga Water and Environment Project. http://www.rw.lv

Ronis A. (1932) Rīgas pilsētas apgādāšana ar ūdeni (Riga City water supply). In: Rīga kā Latvijas galvas pilsēta (Riga as the capital of Latvia), Rīga, pp. 459 – 468 (in Latvian)

Rubulis J., Snidere L. & Briedis V. (2001) Problems with drinking water metering in apartment buildings and flats in Riga city, Latvia, Research Studies Press, in: Water software systems: Theory and application, 1, 349-356

Spriņģe G. (1995) Bacterial dynamics in the lower reaches of the Daugava River. Proc.of Latv.Acad. Sci., 1995 3/4, 108-112.

Springe G., Druvietis I. & Juhna T. (2001) Development of potentially toxic cyanobacteria and bacteria during artificial recharge of groundwater. In Hallegraeff, G. et al. (Eds) Harmful Algal Blooms 2000, Intergovernmental Oceanographic Commission of UNESCO, pp. 503-506

State of the Environment in Riga 2001 http://www.ceroi.net/report/riga

Surveillance and control of microbiological stability in drinking water distribution networks. http://www.safer-eu.com

Zeids T. (ed.) (1978) Feodālā Rīga (Feudal Riga) Rīga, Zinātne. (in Latvian)

# 27
## WATER AND ENVIRONMENT IN ONE INDIGENOUS REGION OF MEXICO
*Patricia Avila García*

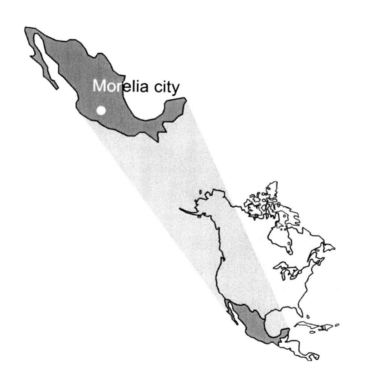

Figure 27.1 Location of Morelia city, capital of Michoacan State where study area, Meseta Purepecha indigenous region is.

## 27. 1. GENERAL BACKGROUND

The *Meseta Purepecha* is an indigenous region located in central-western Mexico (Figure 27.1), an area where certain peculiar geological and topographical conditions impede the formation of plentiful springs and waterholes. This is allocated in the borders of two important river basins: Lerma-Chapala and Rio Balsas (Figure 27.2). In stark contrast to this, the region is characterized by abundant precipitation and is covered by extensive forests, two factors that facilitate filtration and the recharging of aquifers. Thus, it is a water-producing zone that benefits the surrounding area, where springs, rivers and lakes are plentiful.

In spite of the limited natural availability of water, however, most human settlements in the *Meseta* date from pre-Hispanic times and have therefore proven capable of sustaining populations of considerable size for a long period of time. This apparent contradiction can be explained by the fact that the people have developed sociocultural strategies for water management based on the following elements:

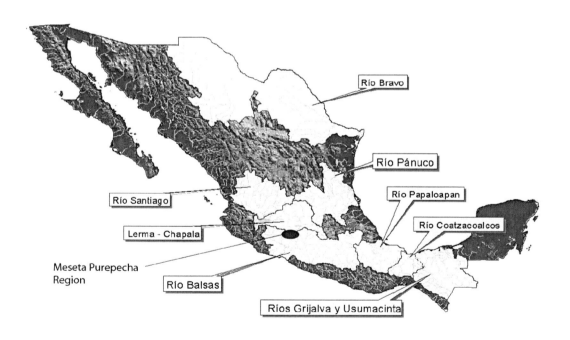

Figure 27.2 The principal river basins in Mexico (source: Water National Commission).

1. The emergence of what we may call a "culture of water scarcity", that utilizes modest volumes of water due to the lack of adequate sources of supply.

2. A form of social organization that permits "community control of water", where this resource is seen as a collective good to which the entire population must be assured access. In addition, all members of the community share the responsibility of conserving and maintaining the sources of supply and for capturing, transporting and distributing water.

3. The emergence of a "culture of ecological water use and management" associated with the *Purepecha* people's cosmovision (worldview), where water is highly-valued and must be cared for because it is a "fruit" bestowed by "mother nature" (*Cuerauahperi*). This attitude is reflected in the practices of water use and management, whose basic ecological principles are: low consumption patterns (little waste), the diversification of sources of supply (utilization of rainwater, springs, watering holes), multiple applications (productive and domestic uses) and recycling (minimal discharge).

Historically, the conditions of life that the indigenous people of the *Meseta* have experienced have been very harsh: walking long distances for water, waiting all day and night for small amounts of water and even consuming water contaminated with dirt and organic material. However, certain sociocultural strategies have decreased the inhabitants' vulnerability in the face of water scarcity. In the most critical months (March to May), for example, water is rationed by community agreement in order to guarantee each family the minimum volume necessary, while from July to September supplies are complemented by the exploitation of rainwater captured as it runs off the roofs of the peoples' houses.

Figure 27.3 Forest and land use in the Meseta Purepecha (photo: Patricia Avila).

## 27. 2. THE REGION OF STUDY: THE MESETA PUREPECHA

The geographical location of the region known as the *Meseta Purepecha* is within the transversal neo-volcanic axis of central-western Mexico. The mean elevation of the plateau areas is 2300 meters above sea level, while the mountainous parts can reach altitudes as high as 3300 meters (Avila 1996). The climate is temperate with summer rains, and the vegetation is classified as temperate forest, where pine, oak and spruce trees predominate (Figure 27.3). In hydrological terms, the region contains seven zones: Charapan, Paracho, Arantepacua, Tanaco, La Mojonera, Pichataro and Zinziro, all of which are characterized by "runoffs unrelated to any apparent surface drainage system, and that lack a permanent, organized fluvial network, flowing as underground streams." (Sanchez 1987)

Despite the region's high level of precipitation—mean annual rainfall of 1274 millimeters (50.15 in.), 95% of which falls from June to October—neither hydrological networks with superficial currents (rivers) nor permanent bodies of water (lakes) exist there. It is only during periods of intense rains that one observes water currents that run down from the higher elevations, but even these disappear after just a few hundred yards due to the high permeability of the subsoil. As a result, there are only a few springs and waterholes, characterized by modest flows and an extreme sensitivity to variations in the amount of precipitation. Flow rates are below 5 liters per second and tend to diminish even more as each rainy season recedes (November to May), with no recovery until the new rainy season and the process of recharging the aquifers begin once again.

In contrast to this reality, the *Meseta* is of tremendous hydrological importance for surrounding regions due to its biophysical and climatic conditions (abundant precipitation, high soil permeability, elevated topography, predominantly forest vegetation) that aid in recharging the aquifers (Figure 27.4). Numerous springs, rivers and bodies of water exist all around the *Meseta* and there are innumerable deep wells in the surrounding areas (Uruapan, Los Reyes, Patzcuaro and Zamora).

## 27.3. POPULATION AND WATER SCARCITY

This region is located in the central area of the state of Michoacan and includes some 43 localities, divided among eleven municipalities. Most of the communities (28) are located in the municipalities of Charapan (6), Cheran (3), Nahuatzen (10) and Paracho (9), while the other 15 pertain to the municipalities of Tangancicuaro (1), Los Reyes (3), Uruapan (4), Tingambato (1), Patzcuaro (2), Erongaricuaro (2) and Chilchota (2). The total population is perhaps 120,000, the majority of whom are indigenous. Though official statistics on the number of indigenous *Purepecha*-speakers do not confirm this, even the given figure of 35-40% certainly indicates the significance of this population in the region. Among the main productive activities undertaken we find temporal agriculture (corn), forest exploitation (timber, pine resin), the elaboration of semi-finished products (furniture, packing crates for fruits and vegetables) and a wide range of

Figure 27.4 Water and springs in the surrounding area of the Meseta Purepecha (photo: Patricia Avila).

artesanal items. This region is considered one of the poorest in the state of Michoacan, as the indicators of the population's material conditions of existence (such as income, employment, housing, health and education) are well below state and national averages (Figure 27.5).

The scarcity of water in the *Meseta* dates back many centuries, as reflected by *Purepecha* mythology, early settlement patterns and even the chronicles of the first Spanish travelers in the region. Water scarcity can be attributed to the area's peculiar biophysical conditions, which permit neither the formation of rivers and bodies of water nor of bountiful springs and watering holes. The traditional sources of supply have been rainwater and small, intermittent springs and waterholes (called '*ojos de agua*') that tend to dry up quickly as soon as the rainy season ends (the dry season runs from November to May). This has generated a situation of great hardship for the population, especially during the months from March to May (Figure 27.6).

> [in Nahuatzen]... the areas of service around sources of water are zealously guarded during the months of scarcity, when women get up at two o'clock in the morning to stand in line if they want to obtain enough water for the day. Even as late as ten at night the row of empty pails waiting to be filled reflects the town's great tragedy (Aguirre 1952, 290).

In spite of this, the region has been home to significant population nuclei, whose origins can be traced back to the pre-Hispanic era, though many of the original town sites have been abandoned and the population relocated as a function of the proximity of water, and some localities were established as a function of factors such as the availability of arable land and forests (Avila 2006).

Figure 27.5 Indigenous people and poverty (photo: Patricia Avila).

## 27.4. THE INDIGENOUS COSMOVISION OF WATER

The *Purepecha* mythology (Corona 1986) has a legend concerning the origin of the world that reveals the divine character of nature. This myth says that the god of eternal fire, *Curicaeri*, created the sun (*Tata Juriata*) and the moon (*Nana Kutsi*), and that it was from their union that nature—*Cuerauahperi*—was born, representing harmony and the mother of all that exists on earth: mountains, water, plants, animals and mankind.

Water is related specifically to the five gods of rain (called the *Tiripemencha*), who were brothers of the God of Celestial Fire, *Curicaeri*. In the *Purepecha* language, this name means "divine or precious water". It is believed that the *Tiripemencha* live in five houses in the sky and are represented by clouds located in different positions and associated with certain colors: *Ocupi-Tiripeme* or *Chupi Tiripeme*, the blue god, is in the center and has his seat on the island of *La Pacanda* in Lake Patzcuaro; to the east (in Curianguro) sits *Tiripeme-quarencha*, the red god; to the west (in Iramuco) is *Tiripeme-tutupten*, the white god; to the north (in Pichataro) we find *Tiripeme-xungapeti*, the yellow god; and, finally, to the south (in Pareo) sits *Tiripeme-caheri*, the black god.

Like *Ocupi Tiripeme*, the goddess *Cuerauahperi* also has four offspring, though hers are all daughters—clouds—who surround and accompany her. Thus, both deities appear to have similar significance or meaning which implies that the sacred character of water is comparable to that of nature itself. By the same token, the human sacrifices that were carried out in the ceremonial centers located near various springs and waterholes, as in Zinapecuaro, begin to

Figure 27.6 Water scarcity and childhood (photo: Patricia Avila).

make sense once we understand the people's belief that "death" precipitated the rebirth of water. In addition, we should note that in mythology, it was there that the goddess who created all springs, *Cuerauaperi*, dwelled.

> And they sacrificed those slaves and upon removing their hearts held their ceremonies with them, and while still hot they were taken from the town of Zinapecuaro to the hot springs of the town of Araro, [where] they were thrown into a small hot spring and were tied to planks, and they dropped blood in all the sources of water in the town, which were dedicated to other gods. And from those sources of water there arose vapors and they said that from there the clouds emerged for rain to fall, and that the goddess *Cuerauaperi* who sent them from the west was there. And it was out of their respect for her that they cast that blood into those sources of water (Alcala 1977, 50).

We can see that it was through this cosmovision that water obtained its divine origin, based on a close, ongoing relationship with the celestial god of creation, the gods of rain and the goddess of the earth. The origin of human beings, in contrast, is not portrayed as divine, but is attributed to a level quite different from that of the gods. The associations that mankind forges with the deities are based on a relationship of respect and harmony and, in the particular case of water, the population looks to "procure (look after) it" so that it will never fail them and they will be able to survive (Avila 1996).

## 27.5. MYTHS RELATED TO WATER

In the *Meseta*, myths related to the origin of water are an expression of the need to find sources of supply for a population that confronts severe conditions of scarcity. Among the many myths that mention water, we find several that associate it with orphan children who discover sources of water thanks to signs from a bird, while others relate the act of sexual intercourse with the emergence of water. One of these myths is told as follows:

> She was an orphan girl, and no one liked her. She lived in Old Paracho, but as there was no water there they had to bring it from Aranza. Maria was always very dirty and never combed her hair. People sent her to fetch water because they considered her a servant. She carried water in a clay vessel (*cantaro*) and had to make two trips every day… one in the morning and another in the afternoon. People always scolded Maria, because she arrived too soon or too late. This was because no one cared for orphans.
>
> On one occasion when Maria was going for water, a small bird appeared and sprayed her hand with water. But Maria paid no attention and continued on her way to Aranza. But then it happened again… and then again. The third time, Maria peered about to see where the little bird that sprayed her with water had come from and saw that near the place from which it emerged there was a waterhole. So Maria no longer had to walk all the way to Aranza for water, she just took it from that place. And from that time on she made three trips for water every day instead of just two.
>
> Then the townsfolk began to notice that she brought water more times every day and that it didn't take her as long as it had before, so they followed her and discovered where she was getting the water. Then they went and told the priest what was going on, and he said they should clean Maria up, comb her hair and bathe her and then take her to the waterhole, throw her in and leave her there to die. And that was what they did, and after that there was always water for them to carry to Paracho. There is the belief that if you throw a soul into a place where water emerges, it will never dry up (Acevedo 1982, 29-30).

In two variants of this myth, the principle figure is sacrificed in order to obtain more water, because while he or she is alive, it is scarce. Their death is not seen as a kind of suffering, but as a divine act that assures the renovation of the source of water. It is for this reason that the personages are associated with the god of water: the only force capable of making the vital liquid come forth. In a second version, however, it is a couple's sexual intercourse that symbolizes the act of creating and reproducing water from the recesses of the earth (mother nature).

> Many years ago, there was a young girl named Maria Lapis, who had a lover whom she always met when she went to get water. No one knew of their encounters, much less that Maria had found water in a place close to the town of Paracho. Then some people began to notice that it didn't take her very long to come back with the water and, to satisfy their curiosity, they decided to follow her. In a certain place they found the couple consummating their love and in that very place there was a waterhole. The people decided to drown her there so that water would never cease to flow. And so it was that a spring emerged that even today quenches the people's thirst (Aguirre 1952, 285).

## 27.6. FESTIVALS AND RITUALS RELATED TO WATER

In the *Meseta Purepecha* festivals and traditions related to water are many and varied (Figure 27.7). For example, on Tuesday of Carnival week in Pichataro, a festival is held in honor of water: it is celebrated at a different watering hole each year and everyone (cattle ranchers, people from certain neighborhoods or from the entire town) who uses water from that particular source is invited to participate in the *faena* (collective preparations). Though the purpose of this festival is to demonstrate just how greatly water is appreciated and valued, the collective work involved also tends to solidify a spirit of cooperation and participation among the people. Envies seem to be forgotten, and all the tasks are carried out with joy and solidarity. Women, men and children all work side-by-side to clean the spring and to hew troughs by hollowing out tree trunks with their machetes, in which they will store and transport the water.

This festival is the responsibility of a man called "the collector", the *carguero* of the god-child, though everyone cooperates with adornments, food and other details. Music is played the whole time to raise the spirits of those who work in the *faena*. Later, the women and children arrive, bringing food, soft drinks, beer, cigarettes and liquor. The men distribute the drinks, while the women make sure that everyone is fed. Men eat first, followed by the children and then the women. It is a day of partying and sharing, and everyone participates in the dancing and playing, while the sky echoes with the explosion of fireworks. Women wear their most beautiful aprons and put multi-colored ribbons and scarves in their hair.

During the festival everyone dances and throws confetti and people play jokes upon one another. Children paint their faces and run around trying to break egg-shells filled with confetti on the heads of their friends, or throw handfuls of flour at them. No one gets away, though great care is taken to assure that the water is never befouled. As evening falls, they return to the community with live music and along the way they play and dance to the rhythm of the *torito* (during *carnaval*, musicians are commonly accompanied by a dancer bearing an outfit representing a young bull, a tradition that is widely diffused in many localities of Michoacan). Upon arriving at the central square (*plaza*), a great party is held that includes a presentation in which a rancher pretends to rob a cow and is chased by everyone in town until he is finally caught. At the same time, the carnival *torito* dances while children and young people play with him, pretending to be bullfighters (Information provided by several people from Pichataro, and based on my own observations and participation in the festival).

## 27.7. SOCIOCULTURAL STRATEGIES FOR ECOLOGICAL WATER USE AND MANAGEMENT

The unusual biophysical conditions (limited availability of water) and the relatively dense indigenous population of the *Meseta Purepecha* have influenced the development of sociocultural strategies for ecological water use and management as means of confronting scarcity. These strategies are based upon an efficient, multivalent and diversified system of water exploitation, and on forms of social and community control that make it possible to regulate the population's access to water and so conserve this valuable resource (Avila 1996). Strategies vary from one community to another and, above all, are more deeply engrained in those towns where scarcity is more extreme; but it is thanks to these strategies that it has been possible to maintain a substantial human and animal population, not to mention the wide variety of flowers and plants that are found in the patios of the houses (Figure 27.8).

**Social control of water**

The predominant form of water exploitation in the *Meseta* can be termed 'collective usufruct', whose guiding principle is that water is considered a vital element of the community's patrimony. This means that it is seen as a resource that belongs to the collectivity and that the entire community is responsible for preserving it and assuring its availability today and in the future (Figure 27.9). Decisions concerning the access, use, management and distribution of water are made in communal meetings and assemblies. Access to sources of supply is free for

Figure 27.7 Water is sacred to indigenous people (photo: Patricia Avila).

all members of the community, though during the months of greatest scarcity restrictions are applied in an effort to make sure there will be enough to go around.

The volume of water destined for each family is also defined collectively and, though the amount may be quite modest, all families are guaranteed a minimum quantity to satisfy their basic needs. This internal organization prevents some people from using more water than others, as individuals are named to monitor the sources of water and to ensure an equitable distribution. In addition, communal agreements have been reached that allow certain sources of supply to be shared between neighboring towns, which come together to 'pool their resources' in order to combat the problem of scarcity (for example, Pichataro shares water with Erongaricuaro, Nahuatzen with Cheran). Such agreements also reflect the conservation of certain bonds of solidarity that are characteristic of the culture of the *Meseta*.

Another regulating mechanism that restricts the volume of water people receive is designed to assure that there is enough water available for animals (cattle) as well. People voluntarily concede part of their own water supply so that everyone can keep a few heads of cattle (normally 1 to 4), which are important for certain agricultural tasks, for transporting firewood and timber and as a kind of alimentary reserve for the family during periods of great economic hardship.

In some communities, the same sources of supply are used by both people and animals, but at different times of the day: for example, families may have access to water in the mornings and

Figure 27.8 A traditional indigenous house and the reutilization of runoff water (photo: Patricia Avila).

afternoons, while the evening hours are reserved for their animals. There are also cases in which two lines are formed, one for the population and the other for animals, while in other towns certain watering holes are reserved exclusively for the use of animals. On occasions, however, when scarcity reaches extreme proportions, the amount of water available is insufficient to satisfy even the basic necessities of the cattle and some animals "die of thirst" or are simply sold for butchering before they die.

With respect to collective water management, these communities have also developed cultural practices designed to conserve and rehabilitate the sources of supply and have implemented projects designed to capture, transport and store this vital liquid. Among these practices we find collective work projects involving labors undertaken in the collective interest of the community that do not imply any form of economic remuneration. At the root of these practices is the idea that because water is a collectively-held resource, everyone must collaborate in improving the means of exploiting, conserving and maintaining the sources of supply.

The communal authorities call people together for such tasks as dredging and cutting away the undergrowth to protect their springs and waterholes, to do maintenance work on existing infrastructure, or to undertake new constructions. Men are organized by *barrio* (neighborhood) to participate in these activities, while the women and children prepare food and bring it to the work site. Generally speaking, these projects are organized on Sundays—when people generally rest from their normal economic activities—and there is often almost a festive atmosphere as the people set to their tasks with great enthusiasm. It should be mentioned that people who do not participate in these *faenas* are looked upon very negatively in the community and on occasions may be fined or even punished by being put in jail.

Sociocultural practices of this nature are fundamentally important in assuring and maintaining adequate supplies of water, because with the passing of time, the structures designed to capture and transport water deteriorate or begin to get clogged with sediments. Thus, constant maintenance is needed to avoid reductions in the flow. Moreover, the work of cleaning the holding tanks and cisterns requires the participation of the entire town. For example, just before the onset of the rainy season in Cheranatzicurin, residents undertake the work of cleaning the cisterns in which they will store rainwater for later use.

Finally, in some localities these responsibilities are delegated to neighborhoods or to certain productive sectors that must maintain and monitor the exploitation of specific sources of supply, while others may be open to the entire populace. In both cases, the *faena* continues to be both an elemental practice and a pre-condition of conserving one's right of access to existing sources of water.

### Efficient water use and management

Patterns of water consumption in the communities in the Meseta are associated with a culture that attempts to 'optimize' use and exploit water 'up to the very last drop'. The volume of water destined for the range of domestic-productive uses is very small indeed, as it is calculated on the basis of a rationality that stresses efficient and rational use. In fact, most houses have no running water, drainage systems or sanitary installations (taps, flush toilets, showers, etc.) that would require high levels of consumption.

According to my own calculations, the average amount of water per inhabitant in the *Meseta Purepecha* was just 12 liters per person per day (l/p/d), a figure that is very close to the minimum necessary for satisfying basic necessities of 8 l/p/d. The reasons for this, of course, are scarcity and the large amount of time and energy that people must invest in obtaining and transporting water (Avila 1996). Inside the towns we also find mechanisms designed to prevent people from wasting water: when someone is seen making wasteful use of water they are likely to be reported immediately to the communal authorities who will reprimand them and, in cases of repeated offenses, fine or even jail them (Figure 27.10).

### Diversified water use and management

One way of satisfying the demand for water throughout the year and of achieving greater self-sufficiency in terms of supplies for the population is to design strategies to diversify water use and management. Basically, the idea is to combine and exploit the different sources of supply that exist during the year (rainwater, springs, waterholes, run-offs). Water exploitation is closely related to the hydrological cycle: during the months of extreme scarcity (November to May) the flow from the usual sources of supply tends to diminish and, especially from March to May, people implement a variety of mechanisms designed to combat the lack of water. During the following four months (June to October) existing supplies are complemented by the rainwater

Figure 27.9 Water is a common good for indigenous people (photo: Patricia Avila).

that falls and increased flows at the springs and waterholes, which tend to expand one or two months after the rains begin.

With a view to maximizing water supplies, measures are taken to increase diversification as far as possible by exploiting all available sources. This means that the people use water from various sources at the same time, be they springs, waterholes or rainwater. As part of this effort, some sources may be reserved for specific uses (according to the quality and quantity of water they contain), such as drinking, washing or bathing, for example; or may be assigned to a certain neighborhood or locality. In Pichataro, for instance, in some cases the maintenance and exploitation of the springs is the responsibility of individual neighborhoods, while in others it is shared with nearby communities that have access to common sources (Comachuen, Erongaricuaro, Huiramangaro). All of these means of exploiting diverse sources of water allow the people to decrease their vulnerability in the face of scarcity. Given that in the most difficult months of the dry season the flow of water from most springs and waterholes tends to diminish quite markedly, the people's ability to respond is increased if they can turn to alternative sources of supply. If they were to depend on only one source of water and had no secondary sources, in contrast, their situation would become even more difficult.

With respect to rainwater, it must be remembered that the mean annual precipitation in the region is very high, especially during the summer months. The onset of the rainy season, then, represents for the people the end of a long period of scarcity. The first rains have hardly begun to fall when people begin to appear with their clay vessels, buckets, tubs and other recipients to trap the maximum quantity of water they can as it runs off the roofs of their houses and

Figure 27.10 Indigenous women washing clothes in a communal area (photo: Patricia Avila).

is carried in wooden troughs to be stored in wooden cisterns (*canoas*). Once deposited, this rainwater is used for drinking, cooking, washing clothes and bathing (Figure 27.11).

We can see, then, that the *Meseta* is characterized by a culture of water use and administration that has allowed the population to confront the conditions of extreme scarcity that last for one third of the year. However, the problem of storing rainwater has constituted one of the main limitations that prevent people from maximizing their exploitation of this particular source of water. Most houses do not have large holding tanks or cisterns, but only a very modest storage capacity, sufficient to hold enough water for just a few days (perhaps 1 to 3). For this reason, when there is a break in precipitation during the rainy season (what the people call a *veranito*, or 'little summer'), then the towns immediately begin to suffer from scarcity once again, due to the absence of installations capable of storing large volumes of water.

Related to this matter is the fact that the *Meseta* has been the scene of interesting experiences in the development of technologies designed to increase the efficiency of rainwater exploitation. For example, at least two families in the town of Capacuaro now have large cisterns in their houses, so that the amount of water they can store during the rainy season is more than sufficient to satisfy their total annual requirement. In this way, these families achieved self-sufficiency and no longer have to purchase water or carry it from the communal holding tanks.

In summary, the strategies for diversifying water use and management have made the population less vulnerable to scarcity through the development of an efficient and rational system of exploitation.

## Multivalent water use and management

The multivalent use and management of water is a strategy that consists of re-utilizing (re-cycling) and exploiting water for productive and domestic applications. The logic of this approach is based on an ecological form of water management that looks to minimize the volume of run-off generated in the domestic sphere by optimizing water use, re-utilizing 'gray water' (dirty, used water), and attempts to maximize this resource by using the same water for different purposes. For example, the water used in such domestic activities as washing dishes, washing clothes, rinsing the cooked corn dough used to make *tortillas* and for personal care are normally carried out in the patio or yard of the houses, because—as mentioned above—most dwellings have no sanitary installations or drains.

Part of this runoff water is absorbed directly by the subsoil and reincorporated into the hydrological cycle (a fact that may contribute to the contamination of local sources of supply, such as waterholes and springs), while another part is collected in buckets, pails or tubs to be reused for other purposes, according to its quality: water from the *nixtamal* and water that contains no soap residues (detergent, chlorine) can be used to irrigate plants and gardens or consumed by animals, while soapy water is used to wet down the patio or the street before sweeping and cleaning.

Secondly, local sources of supply may have specific applications – some are used for drinking water or cooking water, while others provide water for washing clothes and bathing, and there

Figure 27.11 Rainwater is caught from roof of the house and stored in buckets (photo: Patricia Avila).

may also be tanks for cattle or for industrial uses. Other sources, however, may have multiple uses.

Cheran provides us with an interesting case, as the people there have defined a specific type of exploitation for each source and have clearly delimited each site according to type of use. Though most of the springs are exploited for domestic use, they may also have troughs that channel off water for consumption by animals. There are also places used exclusively by men or women for bathing: men use a spring near *Uekuaru*, while women bathe a few hundred meters away in a ravine where there is a modest flow of water. Further downstream, women go to wash clothes and men and boys lead their animals to drink. At another point, water is siphoned off towards a resin plant, where it is used in the processing of tar and turpentine. Finally, the remaining runoff filters into the subsoil. Standing water or the fluid that slowly runs off through this ravine tends to disappear a few meters beyond the community's borders, due to evaporation and high soil porosity.

This multivalent use and management of water has made it possible for various localities in the *Meseta* to keep fairly sizeable herds of cattle, as the people are willing to share their sources of supply or even to cede them to their animals. Thus, the development of cattle-raising has been due, in large part, to the people's strategies of managing both water and pasturelands in collective fashion. By the same token, the existence of flowers and plants inside people's homes during the entire year despite the scarcity of water (as in Cheranatzicurin, Ocumicho and Pichataro), can only be explained by their reutilization of runoff water from domestic activities for watering (Figure 27.11).

## 27.8. CONCLUSIONS

Over the centuries, the indigenous population of the *Meseta Purepecha* has had to develop a variety of strategies for water use and management that have allowed it to confront the problem of scarcity. Their strategies are based on sound ecological principles and programs of social regulation that include: a) social, community-based control of water that assures access to the entire population and the conservation of this resource; b) efficient water use and management that has given rise to rational forms of exploitation; c) diversified water use and management that has optimized the exploitation of all available sources of supply, including rainwater; and, d) multivalent water use and management that has permitted the development of a range of productive and domestic activities and maximized exploitation through recycling.

However, scarcity, in and of itself, cannot explain the high value that people attribute to water, the high esteem in which it is held, nor the logic upon which the strategies of water use and management are based. In order to understand this, it is necessary to enter the sociocultural dimension and comprehend that for the *Purepecha* people (the indigenous people of the region), water has a sacred and divine character that is revealed in their myths, rituals and cultural practices.

In the cosmovision of the *Purepechas*, human beings are seen as a 'fruit of nature' and the relationship established between humankind and nature is, therefore, based on principles of respect and harmony. This is because the goddess of creation, *Cuerauahperi*, and the god of water, *Ocupi Tiripeme*, are very similar in nature, and humans cannot act upon a level of superiority, but only through a relationship of respect, mediated by the technology and knowledge they generate.

Figure 27.12 Re-utilization of water in the houses (photo: Patricia Avila).

It is in this sense that the sociocultural practices associated with water use and management are reflections of this cosmovision: water is part of a highly-valued patrimony that must be cared for and conserved in order to assure its availability both today and for future generations. Its exploitation is based on ecological principles that include not only forms of use and management that are efficient, multivalent and diversified, but also measures designed to conserve this valuable resource. Thus, it is the binomial 'water culture/environment' that explains the continuous existence of large population nuclei in the *Meseta* since pre-Hispanic times, in spite of the severely limited availability of hydrological resources that has characterized the area since time immemorial.

## 27.9. REFERENCES

Acevedo C. (1982) Mitos de la Meseta Tarasca: un analisis estructural, Eds. Universidad Autonoma de Aguascalientes, Mexico.

Aguirre G. (1952) Problemas de la poblacion indigena de la cuenca del Tepalcatepec, Instituto Indigenista, vol. III, Mexico.

Alcala J. (1977) Relacion de las ceremonias y ritos y poblacion y gobierno de los indios de la provincia de Michoacan (1541), Eds. Balsal, Mexico.

Avila P. (1996) Escasez de agua en una region indigena de Michoacan: el caso de la Meseta Purepecha, El Colegio de Michoacan, Mexico.

Corona J. (1986) Mitologia tarasca, SEP-Mich., Coleccion Cultural no. 4, Mexico.

Sanchez A. (1987) Conceptos elementales de hidrologia forestal: agua, cuenca y vegetacion, Universidad Autonoma de Chapingo, Mexico.

# 28
# THE HISTORICAL DEVELOPMENT OF WATER AND SANITATION IN BRAZIL AND ARGENTINA

*José Esteban Castro & Leo Heller*

Figure 28.1 Location of Argentina and Brasil.

## 28.1. GENERAL BACKGROUND

The central hypothesis of this chapter is that understanding the historical trajectory of water and sanitation services (WSS) requires an analysis of how the development of these services is interwoven with, and often even determined by, wider processes of socio-economic, political and cultural development. In this regard, the history of WSS across Ibero America has important commonalities, which can be detected even in cases with such distinctive features as Argentina and Brazil. This interrelation becomes clearer during the nineteenth century, when the region became incorporated into the international world market under the influence of the leading capitalist countries, Britain in particular. This incorporation was significantly accelerated since the 1870s, which in the case of the two countries examined here prompted a number of interrelated processes: a) massive immigration, mainly of European origin, b) rapid population and urban growth concentrated in the main centres of production and trade, c) rising demand for basic infrastructure works (ports, railways, energy, communications, WSS, etc.) as a result of the expanding economies led by export industries, and d) the consolidation of the respective nation states in terms of territorial control, constitution of domestic markets and, in relation to our topic, important developments in sanitary policies including essential WSS.

In this connection, it is important to highlight the close interrelation between these processes at the national level and similar developments taking place in the leading industrial countries, particularly England, France, and the United States, since the mid-nineteenth century. There are several important developments that can be identified: a) the influence of the English system of water supply services inspired by the principles of free-market liberalism (privatism) and based on unregulated private water monopolies; b) the emergence of the sanitary movement among the educated elites in Europe and the United States, which developed a critical stance towards the privatist model and became instrumental, first in the introduction of stricter regulation of private water operators and later in the takeover of WSS by the public sector in most countries; c) the popular struggles led by the labour movements for improvements in working and living conditions; and d) the expanding role of the public sector in regulating private operators or directly providing basic infrastructure, particularly WSS; among the most important.[1]

The chapter is organized in chronological sequence, with a brief reference first to the colonial period, and the most important sections are dedicated to the nineteenth century and the twentieth century up to the 1970s. The ending briefly reviews the most recent developments since the 1980s, which cannot be treated in depth here for reasons of space but a reference to relevant literature for the contemporary period is provided.

---

[1] For a brief review with reference to the abundant literature on these topics, see Castro et. al., 2003; Swyngedouw et. al., 2002.

## 28.2. EARLY DEVELOPMENT (1500S-1850S)

There are important similarities in the development of WSS in Brazil and the Spanish American colonies, where the building of basic infrastructure was often the result of a mix of, broadly speaking, private and public initiatives.[2] In some cities, for instance, the authorities left the construction of water conduits, fountains, and other infrastructure to private individuals and to the influential religious orders, normally in exchange for some privileges. Later on, the authorities would sanction open access to these works in the name of the public interest, as actually happened, for instance, in the Mexican city of Puebla (Lipsett-Rivera 1993, 28-9). In other cases, however, like in Santiago de Guatemala (Webre 1990) or Mexico City (Musset 1991), the provision of essential water infrastructure was always under strict public control.

In the case of Argentina, during most of the colonial period water services in Buenos Aires were mainly a private endeavour, helped by the abundance of fresh water and the relative low density of the population. Shallow wells, complemented later by the services of street water vendors (*aguateros*) that tapped river water, remained the main sources until the second half of the eighteenth century, when the wealthiest families started to build rainwater collectors (*aljibes*) (Herz 1979). Although after independence from Spain (1810-16) the authorities carried out an assessment of the country's water services and developed a number of projects to upgrade and expand the systems, actual improvements were not introduced until the second half of the nineteenth century.

In Brazil, most sanitary interventions until the mid-eighteenth century were mainly carried out by individual initiatives. Among other factors contributing to this situation was also the low demographic density of most cities and villages, which had not yet developed a degree of economic, social and political significance that would merit the attention of the colonial authorities. The adoption of new water technologies was normally associated with economic activities, such as the introduction of water wheels in the sugar mills to boost the development of the sugar industry, which was the basic productive activity during the colony. The seventeenth-century gold rush fuelled the first migratory flows to Brazil, which led to rapid population growth in the towns and cities and triggered a demand for the provision of public services, particularly WSS. However, progress became concentrated in the most dynamic economic centres, particularly in the coastal areas and in the rich cities of the mining state of Minas Gerais. Even then, the improvements were mostly limited to water supply infrastructures and soon became insufficient in the face of the growing insalubrity of the cities, which became a fertile ground for the frequent outbreak of epidemic diseases.

The relocation of the Portuguese Court to Brazil as a result of the Napoleonic invasion of Portugal in the early nineteenth century awakened the interest of the central authorities in sanitary policies for the first time. Under the new conditions, the Portuguese monopoly

---

[2] We have explored the case of Spanish America in more depth in Castro (2006), especially Chapters 2 and 3.

Figure 28.2 Installation of the very first water mains in Brasilia, the new Brazilian capital, in the decade of 1950. (Source: http://www.caesb.df.gov.br/historia.asp, accessed 23/12/2005)

over the colony's trade was broken and Brazil developed open trade relationships with other countries, particularly with England. This created a demand for hygienic improvements in the port cities, which led to the centralization of these services in order to make commercial relationships secure. It was in this context that the first official post in charge of sanitary policy was created in the capital, the Health Inspector of Rio de Janeiro. However, the responsibilities of this officer were mostly limited to serving the needs of the Portuguese Court. It was only towards the mid-nineteenth century that the first public health institutions were created with the purpose of unifying and reforming the sanitary services of the Empire, but even then most actions continued to be restricted to Rio de Janeiro (Barreto, 1945). In the rest of the country, sanitary policies were left to the discretion of the local authorities, which in most cases lacked the resources and technical support needed for sanitary improvements. (Figure 28.2)

## 28.3. THE NINETEENTH-CENTURY MODERNIZATION (1850S-1910S)

In both countries, the public health consequences of the rapid urban growth fuelled by the promotion of foreign immigration in the second half of the nineteenth century prompted an increased awareness among the elites of the interdependence between social and sanitary problems. This awareness was helped by the succession of epidemics of yellow fever, cholera, and smallpox, which were especially acute in the most populated cities and did not respect class boundaries (Bordi de Ragucci 1992).

In Brazil, each new epidemic helped to expose the vulnerability of the population to these diseases (Hochman 1996). Additionally, the national economy, which was largely based on slave labour, was subject to recurrent instability in the production cycles owing to the impact of these epidemics. It was the comprehension of this interdependence, and the attempt by the authorities to improve the country's image in Europe in the context of the growing nationalist spirit characterizing the Brazilian elite of the time, that led to the implementation of sanitary initiatives at the federal level.

In Buenos Aires, it was only after increasing public pressure that in 1854 the Municipality finally decreed that the provision of water supply services was a public duty, which was entrusted to the Public Works Commission. This was followed in 1856 by the creation of the Commissions of Hygiene, Education and Public Works, though concrete actions were delayed until the 1860s. Then, the impact of cholera and yellow fever epidemics in the capital prompted the creation of the Water Supply Commission in 1867, which was entrusted to the British Director of the Western Railway Company, Engineer John Coghland. This interlink between water supply and the railway industry became a pattern in many Argentinian cities, as the railway companies had an interest in developing quality water resources for the operation of their locomotives (Catenazzi and Kullock 1997).

As already mentioned, the development of WSS in South America during this period was influenced by the institutional and technological models emanating from England, France, and, later, the United States and Canada, with varying degrees and forms of public (mainly municipal) and private participation. The main forms of private involvement, which became widespread since the 1880s in most countries of the region, were the concession of the services or the granting of building contracts to foreign private companies – mainly British, French, US, and Canadian – for developing the systems under state control and with public funding. However, this model of private water companies was short-lived because it was oriented to providing WSS mainly to the wealthy neighbourhoods of the most important cities. In the context of rapidly expanding urban centres affected by recurrent water-related epidemics, this privatist model of WSS prompted massive popular unrest and the reaction of the enlightened elites, particularly through the sanitary movement, which eventually would lead to the takeover of these companies by the public sector from the 1890s onwards.[3] Argentina and Brazil experienced a very similar development in this regard.

In Buenos Aires, the water system was inaugurated in 1869, but financial difficulties delayed the expansion of the services during the 1870s and 1880s, and in 1888 the works were given in concession to a private operator, the British-owned Buenos Aires Water Supply and Co. However, the private concession did not last and the water utility was renationalized in 1891 after the failure by the private operators to comply with contractual duties.

In the case of Brazil, after the creation of the Republic in 1889, the new constitution confirmed the autonomy of the provincial states in relation to health and sanitation policies. The federal government was only responsible for the situation in the federal capital, although it also contributed with technical and financial support for the states. On the negative side, this institutional setting proved to be an obstacle for the homogenization of sanitary policy in the country, and for the consolidation of the Brazilian state in general. Also, it was only at the end of the nineteenth century that the federal government declared the provision of WSS to be a public sector duty. In this stage, these services were granted in concession to foreign companies, mainly British, which had at the time hegemonic control of the Brazilian economy. Unsurprisingly, the private WSS companies targeted primarily the largest and wealthiest cities such as São Paulo, Campinas, Santos, Rio de Janeiro, Recife, Belém and São Luis, which were associated with the production and export of primary products like coffee, rubber and cotton. In these cities they almost exclusively limited their action to the wealthiest neighbourhoods, where the provision of water services was commercially viable. Nevertheless, like in the rest of Ibero America, most private water companies had a very short life and were swiftly taken over by the public sector. The only exceptions in Brazil were Rio de Janeiro's City Sewerage Company, which existed until 1947, and Santos City Company, that was in operation until 1953.

An important development of this period of Brazilian history was the liberation of slaves in 1888, which had strong repercussions in the context of rapid urbanization and the absence of sanitary policies. Liberated black slaves were left unprotected, but the immigrants that were brought to the country to substitute for them in the labour force received similar treatment, which substantially increased the mass of the excluded in the expanding urban centres. The modernization of the country and the ensuing improvement of sanitary conditions largely bypassed the popular classes, which triggered recurring social unrest and revolts. In relation to WSS, popular reactions often targeted the private water companies owing to the perceived elitist character of their services, which were limited mostly to the wealthy.

In this regard, both in Brazil and Argentina the adoption of sanitary policies and the expansion of WSS were intimately linked to the popular struggles for the improvement of living conditions and to the influential impact of the sanitary movement within the governing elites.[4] In this regard, the situation in Brazil at the beginning of the twentieth century was marked by the so-called "rediscovery of the *sertões* [poor rural areas]", through the systematic monitoring of the

---

[3] This process resembled very closely the situation in Europe. For instance in London, a similar development led in 1902 to the takeover by a public body of the eight private water monopolies that served the city (Castro et. al., 2003).

health conditions of the rural population, which was plagued with infections like malaria and Chagas disease. The reports of the health commissions in charge of these studies had a huge impact on the political and scientific elites, and triggered the creation of the Pro Sanitation League of Brazil, which brought together political leaders and scientists in the search for sanitary improvements. It is worth emphasising that these actions were prompted by the understanding among the elite members that the poor health and sanitary conditions were responsible for the underdevelopment of the country, while they failed to see the link between the situation of the poor and the reigning socio-economic inequalities. Nevertheless, these policies became a fundamental instrument in the consolidation of the state's control over the national territory, motivated by the awareness that the solution to the country's sanitary problem would not be possible without the concerted action of the public sector. This diagnosis was further confirmed in the case of essential WSS by the failure of the model based on private water companies, which led to the takeover of these services by the public sector and the expansion of the role of sanitary engineering in this process.

Similarly, the sanitary movement in Argentina played a key role in the expansion of WSS. Prompted by the appalling conditions in the capital, which according to some commentators resembled the situation in eighteenth-century Europe with annual mortality rates of between 30 and 35 per cent in the 1860s, and by the impact or recurrent epidemics in 1869, 1871, 1874, and 1879, the authorities launched a vigorous campaign in the late 1870s under the leadership of the mayor Torcuato de Alvear (1879-1887). Alvear formed a team of health experts from the educated elite that included leading members of the sanitary movement like Guillermo Rawson and José María Ramos Mejía, who had been politically active in matters of public policy and sanitation since their time as university students in the 1860s. These experts designed a master plan for the improvement of hospitals and sanitary infrastructure and created the National Department of Hygiene (1880) and the municipal office of Public Assistance (1883) to take the lead in its implementation (Álvarez 1999; Paiva 2000). In this context, after the renationalization of the water company in the capital in 1891, WSS became the exclusive preserve of the public sector and the state-owned utility expanded the share of households with water supply from 27 per cent in 1891 to 72.5 per cent in 1905. Moreover, in 1902 the responsibility of the public company was extended to provide WSS to the provincial capitals and other important cities, and was further consolidated nationwide in 1912 with the creation of National Sanitary Works (OSN). However, in smaller cities and rural areas, including the suburbs of the capital, WSS continued to be provided by small private companies, which would eventually become nationalized in the 1940s (Catenazzi and Kullock 1997).

---

[4] The import of social struggles in the development of WSS policies has been largely overlooked in the literature. See Castro (2006) for a discussion related to the Mexican experience but with insights into the situation of Ibero America at large; see also Castro (1997) for a brief reference to the Argentinean experience.

## 28.4. THE CONSOLIDATION AND CRISIS OF STATE-LED WSS (1910S-1980S)

The development of WSS in Argentina and Brazil during the early stages of this period was influenced by the general trend in the world economy whereby national states assumed a leading role in the expansion of basic infrastructure, including WSS, a situation that was further accentuated by the economic crisis of the 1930s. This state-led model was largely based on the role of the water engineers and sanitary experts working with the backing of the state apparatus. In some Ibero-American countries, in addition to this overwhelming role of the state, the expansion of WSS was also helped by the role played by co-operatives and other not-for-profit organizations, especially in rural areas and small towns, while commercial private water companies disappeared almost completely from the region by the 1950s. However, while in Europe and the United States the public sector succeeded in universalising WSS at some point after World War II, in Ibero-American countries the situation was different and in many cases actual progress was not forthcoming until the second half of the twentieth century. A notable exception to the rule was the case of Buenos Aires, where the public utility succeeded in providing full coverage for water supply in the federal capital by 1938 and almost universal coverage (94 per cent) of 400 litres per capita per day in the whole metropolitan area by 1947 (Pirez, quoted in Catenazzi and Kullock 1997).[5] Another important feature of the state-led model of WSS was the absence, in general, of citizen or user involvement, as public policies in the sector adopted a paternalistic character, the characteristic "leave it to the experts" approach that dominated public policy during much of the century in western countries (Dryzek 1997) whereby citizens and users were not expected to have an active role.

In Brazil, the period 1910-30 brought about the first important advances in the development of WSS, and for this reason it was termed "the Sanitation Era" (Hochman 1988). The ensuing Vargas Era, named after President Getúlio Vargas (1930-45), and the period of the Populist Republic (1946-50) gave continuity to these policies. During these years, the divide between the health and WSS sectors was widened, and health policies were progressively taken outside the Federal sphere, which favoured the expansion of private participation in this area. In the case of WSS, the overall trend was one of recurrent institutional rearrangements oriented towards the consolidation of the sector's autonomy, which would be achieved in the 1960s and 1970s. In this regard, local authorities and citizens in general were left outside the policy process, increasingly alienated by the new trajectories followed by the WSS sector. Some important landmarks of this long period were the introduction of postgraduate courses in Sanitary Engineering in the

---

[5] Without providing exact figures, this author argues that the coverage for sewerage was also very high, slightly lower than the figure for water supply.

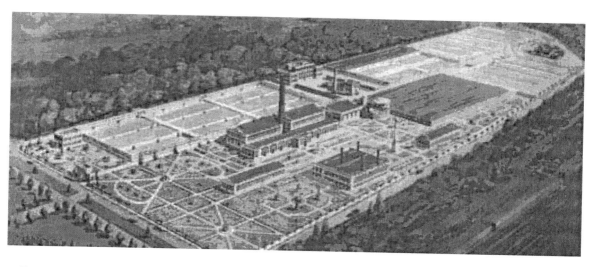

Figure 28.3 General San Martín Water Treatment Plant (aerial view), inaugurated in 1913 to serve Buenos Aires. The plant has currently a treating capacity of 3,100,000 m³/day (downloaded in 2005 from Aguas Argentinas's web site).

1950s, with the support of the United States, a country that had a significant influence in the development of WSS technologies. This influence can be best illustrated by the widespread implementation of US technology in the construction of water treatment plants across the country in this period (Iyda 1994). Also, in 1953 and after long debates, the first Ministry of Health was created, but the lack of resources severely restricted the impact of the new body in a context of appalling public health conditions, which was reflected in a poor impact on the reduction of the morbi-mortality rates during the 1950s.

Regarding WSS, the process of autonomization with respect to the health sector was further accentuated in the 1950s and 1960s, which laid the groundwork for significant transformations that would be introduced in the 1970s. In 1965, with the support of the United States International Development Agency (USAID), the government set up the Executive Group of the National Fund for Financing Water Supply, aimed at supporting the creation of WSS organized on the basis of autonomous public or mixed entities. The arrangement included a provision for the training of technical experts in the US or by US experts in Brazil. This was followed in 1966 with the adoption of the Ten-year Plan for Economic and Social Development, which included a detailed diagnosis of the WSS, a long-term financial model for the sector, and was aimed at extending coverage of water supply and sewerage respectively to 66 per cent and 61 per cent of the population by 1976. Another important development was the creation of the Fund for Sanitation Works (FISANE), which introduced the principle of financial self-sufficiency by transferring responsibility for the financial aspects of WSS to the National Housing Bank (BNH). This decision de facto positioned the BNH as the central agency for WSS, thus transforming the policy environment in a way that strengthened the monopoly control of these services by the provincial water utilities (companhias estaduais) vis á vis the municipalities. This process would be reinforced during the 1970s, which had a long-term impact on the institutional development of WSS in the country.

In Argentina, the period 1914-1950 saw the consolidation of the centralized system embodied in the National Sanitary Works (OSN). This was a crucial period in the country's history, with the radical transformations suffered with the demise of the economic model based on agricultural exports and the process of industrialization based on import substitution that characterized the 1940s and 1950s. In this period the capital Buenos Aires, the country's industrial core, underwent a process of rapid urban growth fuelled by internal and later international (from neighbouring countries) migratory flows. The ensuing urbanization process followed an anarchic pattern that posed enormous obstacles for maintaining the level of almost universal coverage achieved in the late 1940s, to the point that by the 1970s the coverage for water supply had fallen to an estimated 55-60 per cent of the population in the metropolitan area. The deterioration was compounded by an economic crisis in the early 1950s, which led to significant changes in OSN's management model and severely restricted investments in network expansion and system improvements. By the 1960s, problems such as the incapacity of OSN to recover costs had become chronic, and this prompted new institutional reforms including the transfer of responsibility for running the distribution networks to the provinces, municipalities, and cooperatives.[6]

The period from the mid 1960s to the early 1980s constituted a turbulent stage and both Argentina and Brazil were subject to oppressive military dictatorships, with a brief interruption in the case of Argentina during the period 1973-1976 when the country returned briefly to democratic elections. In this country, as another response to the ongoing crisis, in 1967 OSN was transformed into an autarchy, autonomous from the federal government, a process completed in the 1970s with successive changes that ended in the conversion of OSN into an independent state company in 1976. OSN's crisis worsened with the country nosediving into a process of deindustrialization and widespread economic decline in the following period. One of the last measures of the dictatorship in relation to WSS was the decentralization of OSN by transferring responsibility for these services to the provinces in 1980, when the company served 80 per cent of the country's population (Catenazzi and Kullock 1997).

In the case of Brazil, as mentioned earlier, the reforms of the 1960s paved the way for the consolidation of a WSS model based mainly on powerful state (provincial) water utilities. The

---

[6] It should be noted that during the rapid urban expansion of the metropolitan area from the 1940s to the 1960s, a large part of the population resorted to the self-organization of WSS largely helped by the abundance of fresh water and the availability of affordable technologies for drilling wells and septic pits. Self-reliance was the popular answer to the crisis of the state-led model of WSS. In some provinces, the role of the cooperatives in expanding the services during this period was paramount, particularly in La Pampa and Chubut (Catenazzi, 2002; Gouvello, 1999).

overwhelming political control of the country by the military allowed a process of delocalization of WSS, which were transferred from the municipalities to the state water utilities (Doria, 1992). The process was swiftly consolidated as a result of a fiscal reform carried out in 1965, which significantly weakened the financial autonomy of municipalities, and through the sanction in 1971 of the National Sanitation Plan (PLANASA). Under these conditions, most of the country's 4000 municipalities at the time accepted to delegate control over WSS to the state companies, in many cases even without a proper concession contract, as actually happened in major cities like Rio de Janeiro, Sao Paulo and Recife. An interesting development was the rebellion of around 1300 municipalities, which refused to give up control over WSS in their jurisdictions and would later (in 1984) create the National Association of Municipal Water and Sewerage Services (ASSEMAE) to defend their interests. ASSEMAE would become a major player in the struggles for the democratization of WSS policies in the country. On another count, the PLANASA introduced a significant change in the orientation of WSS development, as it prioritised water supply over sanitary actions including sewerage. Among its stated goals, the PLANASA aimed at expanding coverage for water supply to 80 per cent of the population by 1980 and to 90 per cent by 1990, while extending sewerage services to cover 65 per cent of the urban population by 1990. However, the PLANASA failed to achieve its goals, not least because of the serious economic problems that affected the country as a result of the 1979 oil crisis, which led the government to adopt severe emergency measures that substantially reduced investments in WSS (IPEA-UNDP 1996).

## 28.5. ELECTORAL DEMOCRACY AND CHANGING PARADIGMS IN WSS

Brazil and Argentina returned to electoral democracy in the 1980s, but in many respects the democratic experiences have been severely curtailed and the impact on the most vulnerable sectors of the population has been far from satisfactory. Like in previous historical periods, much of what happened with the WSS of these countries in this period has been influenced by processes that are, to a large extent, outside the field of influence of the individual countries and are often even unrelated to the specific problems of these services. In this regard, the main drivers of the policy and institutional changes implemented in the WSS sector worldwide since the 1980s are the result of the transformations undergone by the global economy during this period. For instance, since the 1970s there has been a significant increase of world trade, with the internationalization of capital flows and the global expansion of transnational companies, including a handful of global water companies now operating worldwide. Among other issues, these deep transformations were accompanied by changes in the policy environment at the international level, particularly public-sector reforms aimed at dismantling the "welfare state" in developed countries and creating the conditions for private-sector expansion in the provision of public services through the policies of liberalization, de-regulation and privatization, including those implemented in relation to WSS. At the time of writing this chapter, we are in a particular moment of this later stage, which is characterized by unresolved conflicts between rival models of WSS ownership and management worldwide. Argentina and Brazil have been at the forefront of this development, which provides important lessons that can be learnt from their recent experiences.

In Brazil, the political situation improved significantly with the first civil president appointed in 1985 after the military dictatorship,[7] and one of the first measures of President Sarney (1985-1990) was the creation of the Ministry for Urban Development and Environment, which took responsibility for WSS. Another landmark of this period was the 1988 constitution, which re-established a balance between provincial states and municipalities, particularly with regard to the provision of essential services including WSS. Simultaneously, however, starting with the policies of President Collor de Melo (1990-92), the WSS sector underwent a series of changes informed by the neo-liberal programmes of public sector reforms promoted by the international financial institutions worldwide. In particular, it is worth highlighting the implementation of the Programme for the Modernization of the Sanitation Sector (PMSS) and the passing of Law 8987, known as the "Concessions Law", by President Cardoso in 1995. These reforms, supported and funded by the World Bank, were aimed at transferring WSS to the private sector via concessions and other instruments.

In Argentina, President Alfonsín (1983-89) started a process of public-sector reform that included plans for the privatization of public services, but these measures were rejected in Congress by the political opposition. An important development in this period was the creation of the Federal Council for Potable Water and Sanitation (COFAPYS) in 1988, which aimed at fostering a new relationship between the central authorities and the provinces through the joint implementation of projects for expanding and enhancing the quality of WSS throughout the country. However, the country was subject to severe financial and economic problems that included a process of hyperinflation that eventually precipitated the end of President Alfonsín's administration and the call for early elections in 1989. The newly elected President, Carlos Menem, would introduce a bundle of radical reforms inspired by the international financial institutions based on the widespread privatization of public services.

The experience of WSS in Brazil and Argentina since the 1990s provides important lessons. Argentina under President Menem became the pilot case of the privatization policies promoted in the sector by the World Bank, to the point that the measures adopted by the federal government often went well beyond what was deemed acceptable for the international advisers. As a result, between 1991 and 1999 the proportion of the population served by privatized WSS companies in Argentina jumped from 0 to 70 per cent (Azpiazu et al. 2003). Contrariwise, despite the aggressive privatization policies developed by the Cardoso administration in Brazil since 1995, the strong mobilisation against the privatization of WSS organized by workers' unions,

---

[7] The first free elections in Brazil were held in 1989.

social movements, municipal federations, political parties, and prominent members of the Catholic Church, among others, in a context of higher institutional complexity than in the Argentinian case, significantly slowed down the implementation of these policies in Brazil. In 2006 only about 5-7 per cent of the Brazilian population was served by private water supply companies, and only around 50 municipalities out of a total of over 5600 had given their WSS in concessions to private operators.[8]

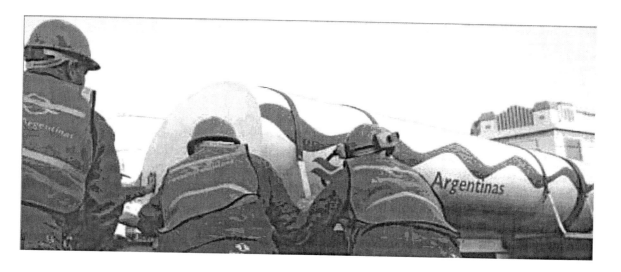

Figure 28.4 Aguas Argentinas' workers in 2003 (downloaded in 2005 from the company's web site). In 1993 WSS in Buenos Aires were granted in a 30-year concession to Aguas Argentinas, a private consortium headed by the French company Suez (later Ondeo). The concession, considered to be the world's largest, was cancelled by the Argentinean government in March 2006.

---

[8] We cannot discuss here this process in more detail. We have analysed these policies in historical perspective in Castro (2004).

## 28.6. CONCLUSIONS

We have explored here the history of WSS in Argentina and Brazil by highlighting the close intertwining between this development and the wider transformations in the global economy driven by the leading developed countries, particularly since the mid-nineteenth century. As our brief review shows, after the short-lived influence of the privatist model of WSS imported from England since the 1870s, WSS in Europe and the United States were placed in public hands or directly organized by the public sector. This was part of an overall trend whereby the public sector assumed increasing responsibility for the provision of basic infrastructure, an activity that became a state monopoly during much of the twentieth century. Despite the distinctive characteristics of the two countries examined here, it has been possible to identify important similarities in the broad patterns of development followed by the state-led model of WSS, which was particularly successful until the 1960s. Ever since, chronic political and economic crises have marred the further development of these services or, in the case of Argentina, have marked the relative decline of the country from the privileged position it had achieved by the 1940s in terms of WSS coverage. A balance of WSS in the late twentieth century in Brazil and Argentina casts an image of recurrent failure and regression from the ideals adopted in the late nineteenth century after the influence of the sanitarists and in the context of the consolidation of the nation state and the institutions of social citizenship. While in developed countries, the state-led model of WSS was able to achieve full coverage of these services at some point after World War II, in developing countries like Argentina and Brazil the history took an unfortunate turn away from the principle of universal access to essential WSS. By the 1990s, Latin America had become the most unequal region of the world, though certainly not the poorest, and this was reflected in the appalling situation of WSS.

Unfortunately, the policies that have been implemented since the 1980s to tackle the crisis have been based on an uncritical revival of the privatist principles that inspired the short-lived experience of private water monopolies in the late nineteenth century. As a matter of fact, these policies of privatization of WSS since the 1980s have completely ignored the lessons of history that we have briefly reviewed here. As a result, the institutional situation of WSS in our two countries at the beginning of the twentieth-first century is characterized by a lack of direction and, particularly, a vacuum in the financial and institutional aspects that is best expressed in the continued lack of articulation between the areas of public health, water resources management, urban planning, and WSS. Following our main hypothesis in this paper, finding potential solutions to the crisis will require an analysis of how the evolution of WSS is interwoven with the economic, social, political and cultural processes underpinning the protracted inequalities characterizing Brazil's and Argentina's development. This chapter aims at making a contribution in that direction.

Figure 28.5 Timeline of Brazil and Argentina water supply and sanitation events, 1500-2000.

## 28.7. REFERENCES

Alvarez A. (1999) Resignificando los conceptos de la higiene: el surgimiento de una autoridad sanitaria en el Buenos Aires de los años 80. In: História, Ciencias, Saúde — Manguinhos, Vol. 6, no. 2, 1999, pp. 293-314.

Azpiazu, D., A. Catenazzi, E. A. Crenzel, N. da Representaçao, G. Forte, K. Forcinito, and J. C. Marín (2003) Buenos Aires - Argentina Case Study Report, PRINWASS Research Project, Oxford: University of Oxford (http://users.ox.ac.uk/~prinwass/).

Barreto J. B. (1945) Finalidades, legislação, estrutura e posição hierárquica. In: O DNS em 1944. Rio de Janeiro: Arquivos de Higiene.

Bordi de Ragucci O. (1997) El Agua Privada en Buenos Aires. 1856 - 1892. Negocio y Fracaso. Buenos Aires: Government of Buenos Aires City, Historical Institute of Buenos Aires City and Vinciguerra.

Bordi de Ragucci O. (1992) Cólera e Inmigración. Buenos Aires: Leviatán.

Castro J. E. (2006) Water, Power, and Citizenship. Social Struggle in the Basin of Mexico. Houndmills, Palgrave-Macmillan.

Castro J. E. (Coord.) (2004) PRINWASS project, Oxford: University of Oxford (http://users.ox.ac.uk/~prinwass).

Castro J. E. (1997) Apuntes sobre el movimiento anarquista en la Argentina. In: Relaciones, Journal of the Universidad Autónoma Metropolitana-Xochimilco, México City, no. 15-16, 1997, pp. 211-230.

Castro J. E., Swyngedouw E. A., and Kaïka M. (2003) London: structural continuities and institutional change in water management. In: European Planning Studies, Special issue on "Water for the city: trends, policy issues and the challenge of sustainability", Vol. 11, no. 3, 2003, 283-298.

Catenazzi A. (2002) Universalidad y privatización de los servicios de saneamiento. El caso de la concesión de Obras Sanitarias de la Nación en la Región Metropolitana de Buenos Aires. 1993 – 2003. Research Report. Buenos Aires: National University of General Sarmiento.

Catenazzi A. and N. da Representaçao (2004) Analysis of the technical, infrastructural, and environmental dimensions of the Buenos Aires case study, in J. E. Castro (coord.), PRINWASS Project (http://users.ox.ac.uk/~prinwass/), Oxford: University of Oxford.

Catenazzi A. and Kullock D. (1997) Política de agua y saneamiento en el Área Metropolitana de Buenos Aires. Estrategias de acceso de los sectores de bajos recursos, antes y después de la privatización," Final Project Report. Buenos Aires, University of Buenos Aires, Secretariat of Science and Technology (UBACyT).

Clement J. P. (1983) El nacimiento de la higiene urbana en la América Española del

siglo XVIII. In: Revista de Indias, 49, 1983, 77-94.

Dória O. G. (1992) Município: o Poder Local. São Paulo: Editora Página Aberta Ltda.

Dryzek J. (1997) The Politics of the Earth. Environmental Discourses. Oxford: Oxford University Press.

Gouvello B. de. (1999) La recomposition du secteur de l´eau et de l´assainissment en Argentine à l´ heure néo- libérale. Lecture au travers du phènoméne coopératif. Thesis submitted to obtain the degree of Doctor. Paris: Ecole Nationale des Ponts et Chaussées, and Université París Val-de-Marne, Université Marne-la-Vallée-CNRS.

Hockman G. (1996) A Era do Saneamento: As Bases da Política de Saúde Pública no Brasil. Thesis submitted for the Degree of Doctor in Social Sciences. Rio de Janeiro: Rio de Janeiro's University Research Institute (IUPERJ).

Institute of Applied Economic Research (IPEA) – United Nations Development Programme (UNDP). (1996) Relatório Sobre o Desenvolvimento Humano no Brasil 1996. Rio de Janeiro and Brasilia: IPEA-UNDP, 1996.

Iyda M. (1994) Cem Anos de Saúde Pública: A Cidadania Negada. Sao Paulo: State University of Sao Paulo (UNESP) Editora.

Lipsett-Rivera S. (1993) Water and bureaucracy in colonial Puebla de los Angeles," Journal of Latin American Studies, 1993, no. 25, pp. 25-44.

Musset A. (1991) De l'Eau Vive à l'Eau Morte. Enjeux Techniques et Culturels dans la Vallée de Mexico (XVIe-XIXe Siècles). Paris: Éditions Recherche sur les Civilisations (ERC).

Paiva V. Medio ambiente urbano: una mirada desde la historia de las ideas científicas y las profesiones de la ciudad. Buenos Aires 1850-1915. In: Revista de Urbanismo (electronic journal), Santiago de Chile, University of Chile, no. 3, 2000(http://revistaurbanismo.uchile.cl/n3/paiva.html), .

Pedro A. (1969) Água e esgotos para o Brasil: a experiência do GEF. In: Brazilian Institute for Municipal Administration (IBAM), Seminário sobre Política de Financiamento para Serviços de Abastecimento de Agua. Curitiba: Brazilian Institute for Municipal Administration (IBAM), Municipal Consultancy and Planning (CONTAP), United States Agency for International Development (USAID).

Quevedo E. (2000) El Tránsito desde la Higiene hacia la Salud Pública en América Latina. In: Tierra Firme, no. 18, 2000, pp. 611-662.

Rezende S. C. and Heller L. (2002) O Saneamento no Brasil: Políticas e Interfaces. Belo Horizonte: Federal University of Minas Gerais (UFMG) Editora.

Swyngedouw E. A., Kaïka M. and Castro J. E. (2002) Urban water: a political-ecology perspective. In: Built Environment, Special Issue on Water Management in Urban Areas, Vol. 28, no. 2, 2002, pp. 124-137.

Webre S. (1990) Water and society in a Spanish American city: Santiago de Guatemala, 1555-1773, The Hispanic American Historical Review, 1990, Vol. 70, no. 1, pp. 57-84.

# 29
## THE GEOPOLITICS OF THIRST IN CHILE – NEW WATER CODE IN OPPOSITION TO OLD INDIAN WAYS

*Isabel Maria Madaleno*

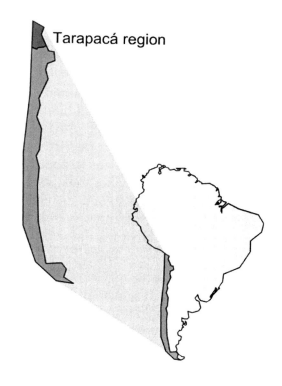

Figure 29.1 Location of Tarapacá region, Chile.

## 29.1. GENERAL BACKGROUND

This chapter seeks to detail work under progress intended to find out how for millennia Andean populations such as the Aymara Indians survived in remote and harsh environments, where water management had to be carefully designed both to permit daily needs and to prevent future shortages, whereas in a couple of decades modern water legislation and current natural resource regulations in Chile have given way to river exhaustion and ecological depredation.

The fundamental concept is the persistence of geopolitics of thirst in extreme northern Chilean regions, perceived as the suppression of indigenous land and water rights in Andean environments where rainfall is relevant and river sources lay together with copper and nitrate-rich deposits. Restrictive land and water regulations were further instated during military dictatorship (1973-1990), intended to transfer water concessions and mineral wealth to public and private enterprises. The geopolitics under discussion is a long way from Ratzel and Ritter's nineteenth century studies, though (Ritter, 1974; Mercier, 2000).

Political and strategic significance of geography is the main aim of geopolitics, geography being defined in terms of the location, size, and resources of places. In general, geopolitics analyses politics, history and social science with reference to geography. The word *geopolitik* was introduced by Swedish Rudolf Kjellen in 1905 and was used (misused) by Karl Haushofer to justify German expansionism, states and people being conceived as living organisms in a continuous struggle for survival (Mendonza, Jiménez and Cantero, 1988).

Karl Ritter (1779-1859) considered the spatial organisation of peoples and continents establishing European supremacy over other continental masses to be in close relationship with Europe's limited area and extensive coastal length, providing good ports and beneficial contacts between cultures as an easy overseas domain. Friedrich Ratzel (1844-1904) defined the state as a political organisation that in conformity with the political consciousness of a people controlled the territory where vital economic activities took place. States were unconceivable without a piece of land and without borders and so nations tended to disappear at the same pace as their land basis was taken away. There are, however, examples of not one but several peoples or ethnic groups keen on the same territory, not far at instances from the German *Geopolitik* of the interwar period.

Domination over territory was the basic principle in geopolitics and justified interceptive actions in order to achieve historical progress. Domination was political intervention sanctioned by state "legitimacy". Current innovation ranges from the increasingly noticeable replacement of ethnic groups or nations by economic groups and corporate interests about target spaces.

Coexisting with the Incas (1470-1572), not as peacefully with the Spaniards (1572-1820) and having been disputed by Bolivia, Peru and Chile in the advent of nineteenth century republican organisation, the Aymara Indians have dealt with war and disarray in their ancestral territory, Tarapacá. (Figure 29.2) Chile won Tarapacá, located between the Lluta and Loa rivers, during the War of the Pacific (1879-1884). For the political survival of the state, it was assumed that the option for "chilenisation", following the "one nation one territory" objective, was essential. More recently, economic development dictated that fast-return activities should be priorities to the detriment of self-consumption farming. Progress dictated that reticent indigenous populations should be outcast. And the solution was achieved with the 1981 Water Code that allowed the concession of river sources to wealthy businesses, the vital good becoming scarce as never before.

A joint Portuguese–Chilean research project, under Portuguese Tropical Research Institute coordination, started off comprehensive analysis and description of Andean populations' ways of life in austere ecosystems, such as the extreme northern Chilean territories. Millenary survival strategies were investigated within Aymara Indian communities, both via literature reviews and sample research in pueblos and riverine settlements so far covering an extension of 6,362 kilometres inside the Tarapacá Region in Chile and in neighbouring Peruvian and Bolivian villages and border cities. Field research dates from 2003 and 2004.

A total number of 30 rural and urban settlements were explored in Aymara territory, intended to typify major Indian ways of life and unique water management practices. The team interviewed 37 community representatives, most of them herders or farmers; teachers were also targeted, not to forget artisans and traders, in a quest for traditional Indian practices, ancestral customs and old Aymara lifestyles. Furthering the search, the team also surveyed politicians, in Chile and Bolivia, as well as regional and urban planning officers, indigenous population commissions, and enterprises involved in water extraction so as to evaluate both water management conceptions.

Results were then compared with recent water history and extensive legislation compiled, integrating and balancing current natural resources management in Chile. Being the source of life, and having been shared in an organized collective fashion for so long, the contending question was whether water privatisation could be considered a better solution for Aymara territories?

## 29.2. SPATIAL ORGANIZATION AT TARAPACÁ

Tarapacá is the northernmost region in Chile. It occupies about 7.8% of the whole country. A severe climate and scarce natural resources define the geography of this territory, namely insufficient water and limited arable soil availability in mountainous milieus. The region comprises the heart of the Aymara land, and the prevailing short water supply has dictated vertical economic systems associated with multiple ecological niches, for millennia. Field research has resulted in the following typified Indian habitats:

1. The *Puna*, a towering Andean plateau, located an average of 4,000 meters above sea level, where about 400mm of precipitation is registered annually. Animal husbandry has been exclusively practiced by indigenous communities called *Ayllus*, using dwarf wet pastures or *bofedales* of *Oxychloe andina*, *Distichia muscoides*, *Festuca rigescens*, and genera *Deyeuxia*, *Werneria*, *Lemma*, *Carex*, all of them plant species having freezing tolerance mechanisms to the cold (Villagrán et. al., 1999). Llamas, alpacas and sheep are

the few domestic animals able to cope with the low temperatures (minus 20° in July), and are usually brought downhill during the winter. Sheep were imported by Spaniards in the 17th century being less numerous than camelids that account for about two thirds of all flocks in the region (INE, 1997).

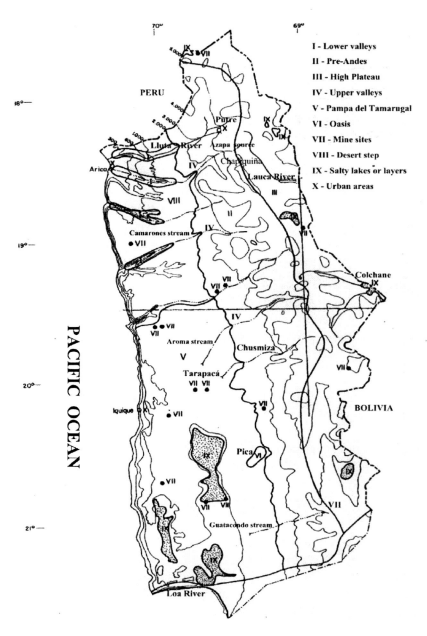

Figure 29.2 The first region of Chile, Tarapacá.

2. Sloped terraces follow, so-called Pre-Andes, for the mountains descend in gigantic steps into a series of interior plane deserts, here and there river meandered, fed by icy water from the heights, taking less than 200 km to fall westwards, toward the Pacific Ocean (Oyarzún, 2000). Irrigation channels are millenary techniques, intended to create microclimates high above the Andes, as to avoid soil erosion and deal with drought, a masterwork dating from ancient Tiwanaku supremacy (1.500 B.D. till 1.200 A.D.). Mixed vegetable and animal farming dominate on the irrigated terraces, built up in the Andes with careful and hard handwork, and displaying a diversity of crops. Potatoes were first harvested and eaten in the Central Andes, cultivated together with quinoa (*Chenopodium quinoa*) and oca (*Oxalis tuberosa*). These days quinoa is scarcer and oca has disappeared, replaced by maize, oreganos, and alfalfa that is used to feed camelids and sheep brought from the heights, as for domineering herding dairy cattle.

3. Upper valleys go after, strings of water that irrigate fig trees, sweet oranges, cherimoyas, guavas, avocados, tunas, *Prosopis chilensis* and sweet grapes. Fruit culture outweighs from the beginning of Christian era, it continued during the Inca domain, was even increased during Spanish colonization and is present up to today. Subsistence micro-farming is the rule in association with goats and sheep, the Indians farming the bottom of the deep valleys and not the slopes. Terracing is quite uncommon below 3,000 metres, whereas dazzling greenery along the streams unkindly contrasts the brownish tops where only cactus candelabra survive. Temperatures are higher but annual rainfall totals about 10 mm.

4. As altitude lowers from 2,000 meters, the meagre streams sometimes disappear underground, particularly in the southern parts of Tarapacá. In the old days Aymara Indians widely practiced *canchones* techniques, making good use of superficial waters in rare oases (Pica) as the upper salty soil layers were extracted during the hottest months, permitting the Indians to farm melons and watermelons there. Spaniards went much further, whilst they built long and large water tunnels bringing water from the springs located high in the Andes. Those remarkable underground architecture facilities were called *socavones*, intended to increase available water in the small disputed lower valleys and oases where they settled their haciendas (Figueroa 2003). Commercial fruit culture and horticulture are now sparse due to increasing demographic pressure. As observed in Table 1, an oasis like Pica had a 146.2% population growth in the last decade, against a 6.1% loss at Andean Colchane, minus 22.3% in the Pre-Andes and a 10.8% decrease at Camiña, an upper-valley municipality.

5. In between the oases and streams, around 1,000 meters high, only tamarugo (*Prosopis tamarugo*) could be found, an endemic tree adapted to water scarcity, able to seek deep table waters and to get humidity out of *camanchaca* (sea fog). For about nine millennia this salty soil-adapted species has dominated from a 19° 30' latitude southwards till the Loa River, in the so-called *Pampa del Tamarugal*, surviving the small endemic mammals, and the importation of goats and sheep from Europe. By the early 20th century the

tamarugo was destroyed by means of intensive mineral exploitation during the peak of nitrates exploitation cycle (1860-1925), because of its local usage as fuel. Nitrate deposits were located at the foothills of the Andes and mineral exploration gave way to valuable exports to Europe and North America, via Iquique port city, due to its intensive usage in vegetable farming and in warfare. The Tamarugo tree has undeniable ecological value, namely water retention, and its loss further damaged water reserves (Habit, 1985).

6. By seaside, dominated by sandy deserts, no farming is possible without generous irrigation. That's where port cities are settled, namely Arica (184,134 inhabitants) and Iquique (215,233). Arica possesses meagre water resources, the nearby Lluta river and Azapa creek being used for drop irrigated fresh vegetables, while Iquique, has no watercourse on hand or underneath (Sánchez & Morales, 2004). Arica's water supply comes from the Lauca River, captured at Chapiquiña dam, and Iquique uses both the Tarapacá stream and underground sources.

The accelerated demographic growth (Table 29.1) registered in littoral areas further pressured Tarapacá's scant hydrous capital, turning water transfer from Andean environments into an imperative with dramatic consequences for pastures, wildlife and irrigated vegetable farming.

Table 29.1 Recent census population data at Tarapacá.

| Altitudinal steps | Districts | 1992 | 2002 | % Change |
|---|---|---|---|---|
| High Andean Plateau | Colchane | 1,555 | 1,460 | -6.1 |
| | General Lagos | 1,012 | 1,295 | 28.0 |
| Pre-Andes | Putre | 2,803 | 2,179 | -22.3 |
| Upper Valleys | Camiña | 1,422 | 1,268 | -10.8 |
| | Camarones | 848 | 1,203 | 41.9 |
| Pampa, Oases and Lower Valleys | Huara | 1,972 | 2,593 | 31.5 |
| | Pica | 2,512 | 6,185 | 146.2 |
| | Pozo Almonte | 6,322 | 10,801 | 70.8 |
| Port Cities | Iquique | 151,677 | 215,233 | 41.9 |
| | Arica | 169,456 | 184,134 | 8.7 |

INE (2002)

## 29.3. OLD FARMING SYSTEMS

Climate is the most restrictive factor in the Tarapacá Region. Annual precipitation varies from 400 mm in the border areas with Peru and Bolivia to nothing at all as the latitude increases and the altitude diminishes. Causes of the water scarcity described are the proximity of the Humboldt Sea current, which brings cold and dry air to the Pacific Ocean littoral areas, in conjunction with the shelter and isolation the Andes Mountains provide the region, preventing Atlantic moisture from getting there. Singular exceptions are the Andes crests, where both glaciers and rivers have their origin, fed by an overwhelming easterly wind humidity (Aceituno, 1993, Mendez, 1993, Sánchez & Morales, 2004).

Complementary soil livelihood was the only way to survive, using subsistence farming systems, with Indians shifting from exclusive animal husbandry to irrigated vegetable farming, and adapting animals and food crops to water availability, differentiating not only the species they cultivated from step to step in the Andean slopes, but also giving them various degrees of attention, according to the limits of the ecosystem. The biodiversity of food crops was higher in the old days than it is today for European colonisation resulted in a break from traditional food habits and led to monocultivation. Nowadays maize dominates in Chile, sometimes at the neglect of potatoes, the most ancient Aymara staple food, whilst oca and quinoa are rarely consumed in the Andean regions at present. Environmental sustainability explains the spatial traditions described before, for it brought mutual benefits to all indigenous communities involved, commodities being traded according to family needs or surplus, wool, meat, milk, cereals, tubers, fruits even fish and sea shells being the basis of barter. Land and water were collective goods, used in shifts for the usufruct of all. Every family and every Indian had an assigned task, performed in turns (*mit'a*). Terracing was seen as the beautification of Mother Earth and the slopes were never excavated, terraces having been built-up with care by generations of Tiwanakotas and Aymaras, after them. Agriculture was never simple sow-crop repetitiveness but an act of love, in return for *Pachamama's* role in daily sustenance. That was made possible by the belief that Earth is the goddess of creation. Plant species were seen as human equals, for they have and award life, and therefore one ought be careful, take good care of them, and at the utmost understand the caprices of nature (Aguilar & Vilches, 2002). Human geography fitted physical geography in the old days.

## 29.4. AYMARA INDIANS – RECENT POLITICAL HISTORY

In order to fully comprehend changes in settlement, in food habits, and intensive twentieth-century Indian migrations to the coast, the team further researched the region's recent history. Rich as it was, it would not be advisable to describe the long political and military turmoil, and so I'll point out 7 main cycles corresponding to Tarapacá's social and economic development phases, regional demographic dynamics and education endeavours:

1. Chilean annexation of Arica and upland Putre, after victory in the War of the Pacific over Peru and Bolivia (1879-1884), was hard and bloody. The main reason for the havoc was the presence of several mineral deposits at Tarapacá and Antofagasta, presently named the First and Second Regions of Chile. About half of national exports are currently extracted in the conquered territories (Melcher, 2004);

2. Internationalisation of the Extreme Northern Territory, with intensive nitrate exploitation by foreign companies (Chilean saltpeter), was a dominant feature from 1891 to 1907, British presence having been the most remarkable;

3. Chilean repression of differences, known as *Chilenisation*, intended to preserve national cohesion and pacify long-lasting border disputes with Peru and Bolivia, followed. The policy launched the disintegration of traditional ethnic societies by means of political, social and cultural exclusion, even persecution, having been active from 1907 onwards but found quite hard and severe over the Aymara Indians in the 1930s (González, 2002). Indians' integration in Chile was further achieved, making them Spanish speakers, the Aymara language having been forbidden (1930-1945). Obligatory military service was imposed on adult male Aymaras, and they were forced to register landownership. The inability to meet deadlines and agree on terms led to informal land tenure that benefited the State of Chile, via broad "Fiscal Property" registration. Nitrate exploitation declined in 1925, replaced by sulphur (1900-1950);

4. Tarapacá's urbanization and cultural modernization continued after the Second World War, with increasing rural-urban migration as a consequence. One should point out that whereas Indian herders' mobility, community trade and reciprocity relations were always widespread, it was only when housing subsidies targeted the urban realm and colonization efforts privileged the peri-urban Azapa and Lluta valleys that Indians descended to the coast. Regarding mineral exploitation, the cycle of gold was then starting (1960-1990);

5. Submission to neo-liberal models came into action throughout General Pinochet's dictatorship (1973-1990). Agriculture sector modernization was a priority policy, much more effective than previous agrarian reform initiated during the Frei presidency (1964-1970) but quite troublesome during Allende's government (1970-1973). Private landownership was imposed in 1979, with a total neglect of collective practices, yet the most destructive act was the 1981 Water Code, which separated water from land rights and gave away to mine companies water usufruct (Diario Oficial, 1981). Traditional communal loyalties were then broken and family remained the last bond;

6. Democracy was reinstated in the 1990s, adherent to actual globalising patterns that ignored subsistence economies in the region. National programmes directed to indigenous pre-Hispanic Amerindians have been drawn up, and regional identities accredited via CONADI (National Commission for Indian Affairs). Aymara families accepted that shift as a good opportunity to provide educated young Indians job opportunities in the cities.

7. The socialist party has been running the country since the early 2000s. So far that didn't change Pinochet's water or land legislation under examination, nor committed

the government to comply with Indian demands. These demands are confined to a few high plateau and Andean villages with a long tradition of discussion of serious issues by a selected number of mostly male Indian representatives (Bengoa, 2001, Vega, 2002). Finally ethnicity was recognised in Chile.

## 29.5. WATER POSSESSION AND MANAGEMENT – RECENT LEGISLATION IN CHILE

Water is a scarce and limited natural resource. The productive usages of water in Chile are mainly four: agriculture, mining, hydroelectric power and fresh water supply (Donoso, 1999). As soon as population grew throughout the 20$^{th}$ century, particularly in Northern and Central regions, the gap between demand and the provision of water was augmented, the shortage being distressing in the First Region (Tarapacá) and alarming in the Second Region (Antofagasta). No wonder that several water codes were legislated and implemented in the country: in 1951, 1967 (connected with an Agrarian Reform bill) and 1981 (following Pinochet's coup d'État).

The 1951 legislation, subsequent to a population boom, clearly distinguished public from private uses and decentralized water management. Concessions of water rights to private sector businesses were regulated and transactions to other beneficiaries were permitted, providing they developed the same economic activity. In 1967, a new code deeply modified the right to exploit and use water resources stipulating those to be exclusively State affairs. Land titles could give the owners privilege over the precious resource, but that permission wasn't inherited or traded in the open market, the Chilean government having the unappealing entitlement to give or to take away water rights at any time, according to central planning parameters and determinations (Gentes, 2000; Lewin, 2003).

Such restrictive and centralized rulings, dictated by elected Christian-Democrat president Frei (1964-1970) and continued by Socialist Allende (1970-1973), affronted the neo-liberal economic system the military regime adopted after 1973 and consequently, not only landownership laws were soon reviewed (1979), but also the water code (1981). Water was henceforth considered a national good available for public usufruct. As a transacting commodity, though, surface as groundwater were supposed to be freely traded in the market, subject to demand from any individual or organization and dissociated from soil property. Water concession was supposedly free but wherever there's short supply the good has to be legally subject to the highest offer, better informed demand or quicker player moves, providing there is enough water in the source or river. Water was meant to have the same constitutional guaranties as soil property, which obviously turned land and water into separate entities, in detriment of the ecosystem approach. In practical terms, it meant that one might own a slice of soil but no water from underneath or from any creek that came through one's land, if by misfortune it had been granted as a permanent concession to any other enterprise or individual, upstream.

Water rights may be given permanently or for a period of time in Chile, continuously or discontinuously, to anyone and also in turns to several actors (1981 Water Code, Article 12).

It is measured in volume per time unit. In theory, the request for and possession of water concessions is free, non-taxed by the State of Chile. Nevertheless, claims are only necessary when and where supply isn't enough, such as in Extreme Northern areas, namely along the steep Andean slopes, lower pampas such as Tamarugal, oasis valleys with high demographic pressure (see Pica, at Table 1), and mostly in dry port city environments, where rainfall is 0 mm. Wherever and whenever requests are numerous, authorities must impose, by force of law, a public contest and give the rights of usage to the best offer. That's why within Tarapacá, Aymara ancestral territory in Chile, water concessions are in the hands of powerful mining companies, wealthy power plants, and affluent water supply enterprises. Concessions admit free and total usage of the precious water for any possible aim, change being admissible at any time or via transfer to another proprietor, in any way identical to landownership rights. Above and beyond, the 1981 Water Code easily leads to speculation, for any individual or business can ask for a permanent concession for free without necessarily using the water or even planning to use it. Legislation permits any concession owner to keep the water for future industrial or entrepreneurial businesses of any sort, turning the commodity into a value reserve for future transactions, while preventing others from usufruct of this vital natural resource.

Consequences of water legislation were serious limitations imposed on traditional economic sectors, and a clear privilege to the fast-return activities and lucrative segments, particularly mineral extraction. In matter of fact, 1981 Water Code article 110 gives mine enterprises free access to all water sources, either superficial or subterranean, comprised within their territorial concession, the rights being inalienable, inseparable from mining concessions and only extinguished with the business inactivity. Article 110 gives extractive companies of any sort an exceptional status because any other private individual or business has to ask for water concessions in a given property, all over Chile (Gentes, 2000).

It's interesting to note that very few scientific discussions have been conducted over the contending issue of water management in Chile. The available ones lay emphasis on the conclusion that the market is unable to resolve the situation by allocating the inestimable resource to its best-valued use, either consumptive or non-consumptive (Rios Brehm and Quiroz, 1995, Massud, 2002).

## CASE STUDIES FROM NORTHERN CHILE
### Industrial exploitation at Chusmiza:

A good example of old Indian ways in opposition to modern water businesses was recorded at Chusmiza, a remote Andean pueblo, very rich in sulphur springs of water, quite warm and recommended for bone therapy. The small village is located en route to Colchane, the high plateau border pueblo with Bolivia, in Iquique province. Interviews with local Aymara cacique and several other Indians resulted into a long list of complaints (February 2004). The Aymaras had been in a judicial dispute with the owners of a mineral water bottling enterprise, claiming the land and mineral water sources had been illegally taken from them, a litigation based on consuetudinary rights just reinstated. The Indians had recently won the right to suspend the bottling industry but so far had not gained the water concession itself. Enterprise representatives, by contrast, argued the production had just been suspended because of water scarcity, being not an Indian achievement at all.

Miracle cures of paralysis were attributed to Chusmiza thermal waters, quite hot and clear. Having been exploited as back as 1927, the sources have experienced quick depredation, and in the advent of the third millennium, the Indian pueblo is less known as a healing sanctuary and irrigated farming oasis, and better recognized for ongoing conflicts between the old ways natural resources were managed and the new devastative mono-exploitation formulas (Van Kessel, 1985).

Indigenous Law dates from 1993, brought up during the so-called re-democratisation process in Chile. As far as water is concerned, the 1981 Water Code reinstates consuetudinary practices but only in court cases between members of the same ethnic group. Furthering the water issue, the new code was of course dramatic for Indians. The Extreme Northern territories are mineral rich; nitrate, copper and other mining companies continuously put pressure on natural resources and private businesses easily get usufruct over the precious liquid, as suggested by Indian eviction from mine locations and mineral water fountains (see Table 1). The only way to see the Supreme Court of Justice ratify Indian claims is to search out a renowned lawyer firm interested in taking the litigation, giving proof the plaintiff community has been using that same stream or water source for ages, thus qualifying to get control of a certain portion. That was the case with Atacameño Indians to whom 100 litres per second were attributed in March 2004, in an Antofagasta lawsuit, under consuetudinary practices. Frequent problems stem from the possibility, either real or forged, that the opposite litigant attests the area under judgment has been acquired to any individual Aymara sometime in the past.

**River Loa desiccation:**

Debates over the governance of public services revolve around the triple involvement of governments, citizens and the private sector. Issues of good governance are extremely important in the case of natural monopoly services, the provision of which has serious social, economic and ecological consequences (Bakker, 2004).

The Loa River possesses the widest basin of all Chilean rivers (33.082$Km^2$) hitherto constituting the worst example of mismanagement and representing the most unfortunate consequence of the 1981 Water Code in northern Chile (see map in Fig. 1). Formerly running about 2 $m^3$ of water per second (Sanchez and Morales, 2004), the demographic occupation of Loa's riverine areas dates from 900 to 1,300 A.D., dispersed above 3,000 metres in the Pre-Andes or along the Upper Valley areas, where summer rains fed cactus (*Heliantocerus atacamensis*) and small bushes (*tolar*) as well as fodder, horticulture farming having been possible using ancestral *canchones* techniques, as described before. Carlos Aldunate studied Turi Vega, located at Salado River, one of Loa's tributaries, in the year 1985. A water spring had just been granted to the public copper mine group – *Codelco* Chile – causing direct damage to an irrigated alfalfa area of 1,500 hectares, used to feed about 2,000 animals, namely llamas, goats, horses and sheep. Local indigenous populations tended about 10 hectares of highly productive maize, wheat, potatoes and horticulture land (Aldunate, 1985). That resulted in the accelerated migration of Aymaras and Atacameño Indians to the city of Calama, located downwards at the middle flow of Loa river basin, on the fringes of the Atacama desert, and to the capital city of the Second Region of Chile, Antofagasta. Migrations further increased pressure over meagre Loa hydrous resources, Calama accounting for nearly 140.000 inhabitants nowadays. Finally, when Salado and upper Loa waters became scarcer, *Codelco* was permitted to drill underwater to the sources. Once in Northern Chile rainfall is only reasonable (400 to 300 mm per year) on the high plateau, soon the river wasn't meeting minimum water demands.

Less than two decades afterwards, exactly in the year 2000, Loa and its tributaries were formally decreed exhausted and new water rights solicitations were suspended, following Legal Resolution Number 197 (MOP, 2003). A recent view over the described scenario can be observed in a 2004 photo (Figure 29.3) taken on the verge of the Pacific Ocean. It depicts a humid month, February.

Figure 29.3 Loa River meeting Pacific Ocean. (Source: Isabel Madaleno 2004)

## 29.6. CONCLUSIONS

A mere 0.5% of the surfaces of the Tarapacá Region are currently classified as farmed soils; a scarce 0.8% are wet pastures used as fodder; around 31.5% are steppes and bushes; one per cent is urban fabric; a total percentage of 66.2 of the whole region is desert, either natural or induced (INE, 2000). Linkage to water code contents, land tenure regulations, and documentation of the Supreme Court jurisprudence over specific Indian communities' demands have given cause for reflection and nurtured concerns.

The forced accumulation of people in coastal drylands, as is the case with Pacific Ocean cities, lower pampas and bottom Andes oasis, whenever associated with the legal appropriation of the scarce superficial waters and groundwater located high above the mountains, easily leads to the advance of the desert and river dissection.

The water appropriation process described is connected with what I name the geopolitics of thirst. Geopolitics is acknowledged as the political domination of one nation and corporate interests over a certain territory, historically inhabited by another nation or ethnic group. Geopolitics has recently acquired a dynamic character, focusing a diversity of phenomena beyond the classical political evolution of states and borderlines. The issue here is water concession to given economic groups in Chile versus a deficient supply and shortage affecting both minority indigenous populations and the fragile local ecosystems. Water concessions to selected number of economic activities were politically driven choices detrimental to indigenous sustainable farming. The aim was to provoke human desertification and the abandonment of rich mineral deposit sites. It has been part of a des-aymarisation or cultural disintegration process.

The eviction of Aymara Indians is not new, first by Spanish colonization and later via Chilenisation procedures. Most historical and political periods converged into conflicting issues such as landownership and water availability control at Tarapacá. As far as the suppression of freedom and indigenous rights are concerned, the critical decades were the 1930s and the 1980s. Following the 1993 Indigenous Law, 1.1% of the 6,000 Tarapacá landowners regained collective land titles. Still, in the last Census to Agriculture, less than half the registered property was used as pasture in the high plateau; about one third was farmed in upper valley areas; a mere 24% was productive in the Pre-Andes for there's not enough water running through the channels in mountain slopes (INE, 1997). There was never a water shortage and ecological depredation as was reported after the 1981 Code.

Desertification is difficult to measure, because different cultures judge the problem differently. The Chilean government argues that management in waterless areas is obviously difficult and mine fields are more viable than subsistence farming. Indians see the *bofedales* dry above the Andes, sloped terraces are abandoned and rural exodus is widespread. People cannot eat or drink minerals, the elders argue. Indeed! Never so much was lost in such a short period of time, appropriated by so few.

## 29.7. REFERENCES

Aceituno P. (1993). Aspectos Generales del Clima en el Altiplano Sudamericano. El Altiplano. Ciencia y Conciencia en los Andes, 63-70. Santiago: University of Chile.

Aguilar M. & Vilches R. (2002). Terrazas Agrícolas, una estrategia cultural y tecnológica de desarrollo rural andino. La Paz: Isalp.

Aldunate C. (1985). Desecación de las Vegas de Turi. Chungará 14, 135-139.

Bakker K. (2004). Good Governance in restructuring water supply: a handbook. Ottawa: Federation of Canadian Municipalities.

Bengoa J. et al. (2004). La Memoria Olvidada. Historia de los Pueblos Indígenas de Chile. Santiago: Publicaciones del Bicentenario.

Diario Oficial (1981). Código de Aguas y Normas Complementarias. Santiago: Gobierno de Chile.

Donoso G. (1999). Análisis del mercado de los derechos de aprovechamiento de agua e identificación de sus problemas. Revista de Derecho Administrativo Económico, 295-314.

Figueroa C. (2003). Galerías Filtrantes en el Oasis de Pica: tecnología y conflicto social, siglos XVII-XVIII. Valparaiso: Congreso Americanista.

Gentes I. (2000). Culturas étnicas en conflicto – El Código de Aguas y las comunidades indígenas de agua en el Norte Grande (Chile). Revista Américas 16 (4), 7-50.

González S. (2002). Chilenizando a Tunupa. La escuela pública en el Tarapacá andino (1880-1990). Santiago: Dirección de Bibliotecas, Archivos y Museos.

Habit M. (1985). Estado Actual del Conocimiento sobre Prosopis tamarugo. Santiago: FAO.

INE (1997). Censo Nacional Agropecuario. Instituto Nacional de Estadísticas, Santiago.

INE (2000). Estadísticas del Medio Ambiente. Santiago: Instituto Nacional de Estadísticas.

Lewin P. (2003). Análisis de la eficiencia del mercado de derechos de aprovechamiento de aguas en Chile. 21 pp. www.rlc.fao.org/prior/recnat/pdf/lewin.pdf

Massud S. (2002). La Exportación de aguas del sudoeste de Bolivia a Chile. La Paz: Agualtiplano. www.aguabolivia.org/ExportacionAguas.htm

Melcher G. (2004). El Norte de Chile: su gente, desiertos y volcanes. Santiago: Ed. Universitaria.

Mendez C.S. (1993). Hidrologia del Sector Altiplanico Chileno. El Altiplano. Ciencia y Conciencia en los Andes, 71-77. Santiago: University of Chile.

Mendonza J.G., Jiménez J.M. & Cantero N.O. (1988) El Pensamiento Geográfico. Madrid: Allianza.

Mercier G., 2000. The Geography of Friedrich Ratzel and Paul Vidal de la Blache: A Comparative Analysis. 17 pp. Available at http://www.siue.edu/GEOGRAPHY/ONLINE/mercier.htm

MOP (2003) Determinación de los derechos de aprovechamiento de agua subterránea factibles de constituir en los sectores de Calama y Llalqui, cuenca del río Loa, II región. In Informe técnico. Antofagasta: Departamento Administración Recursos Hídricos. 153.

Oyarzún G. (2000). Andes, Chile. Santiago: Kactus.

Ríos Brehm M. & Quiroz J. (1995). The market for water rights in Chile. Washington: World Bank.

Ritter K. (1975). Introduction à la géographie générale comparée. Les Belles Lettres, Paris (1st edition in 1852).

Sánchez A. & Morales R. (2004). Las Regiones de Chile. Santiago: Editora Universitaria.

Van Kessel J. (1985). La lucha por el agua de Tarapacá; la visión andina. Chungará, 14, 141-155.

Vega V. G. (2002). Seeking for Life: Towards a theory on Aymara gender labor division. Chungará 34 (1), 101-117.

Villagran C., Castro V., Sánchez G., Hinojosa F. & Latorre C. (1999). La Tradición Altiplánica: Estudio Etnobotánico en los Andes de Iquique, Primera Región, Chile. Chungará 31 (1), 81-186.

# 30
## CASE OF TOKYO, JAPAN
*Yurina Otaki*

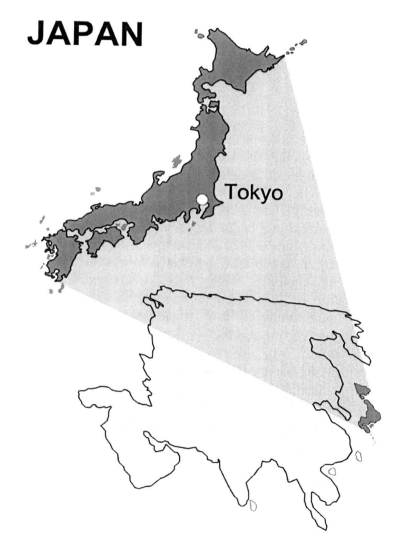

Figure 30.1 Location of Tokyo, Japan.

## 30.1. GENERAL BACKGROUND

Tokyo, called Edo until 1868, was a very small city in the beginning of the 17th century, but experienced unprecedented rapid growth after the Tokugawa shogunate brought the capital function to it in 1603. As there was not enough drinking water in Tokyo, a great deal of effort was made to establish the Edo water system (EWS). The EWS worked well until the 19th century, when cholera prevailed and the sanitary conditions became serious. Since people did not understand the causes of poor sanitary conditions, all of the EWS was thought to be a bad system needing replacement by a modernized one. As a result, in the 20th century Tokyo dismantled the EWS and completely modernized it regardless of the advantages or disadvantages. This was not a wise way to implement a system change. The best way would have been to make good use of the merits of the historical system and to adopt new technology to improve the problem areas.

## 30.2. BRIEF HISTORY OF WATER IN TOKYO

The consistent acquisition of water has been one of the fundamental tasks for urbanized areas. In the first stage of urbanization, people used water from springs, rivers, lakes, or wells near the residential areas. Later, when the cities experienced a population explosion by rapid urbanization, this growth caused an insufficiency of water and thus began the continuous efforts to secure enough water. An aqueduct was the first technology introduced to satisfy increasing demand. The water was taken from a water source of good quality and flowed down for a long distance to the cities, propelled only by the incline of the aqueduct. For example, the large-scaled aqueduct constructed in the 17$^{th}$ century in London brought water from a spring 60 km away. Most of the cities were perpetually looking for new water sources and constructing new aqueducts to satisfy the increasing demand. The second big change was brought by the innovation of the steam engine. The steam engine's ability to lift water allowed the supply of water to wide areas regardless of the altitude. However, the continuous expansion of the population and water use caused a large amount of wastewater beyond the natural capacity and this discharge threatened sanitary conditions. The drinking water was polluted and water-borne diseases, such as cholera and typhoid fever, spread by infected excrement, were prevalent. To improve such poor sanitary conditions, the modern water supply and sewerage system were developed.

Similarly, there were many efforts to obtain enough good quality water and to provide good sanitation in Japan. As stated above, Tokyo, called Edo until 1868, was a very small city in the beginning of the 17$^{th}$ century, but experienced unprecedented rapid growth after the Tokugawa shogunate brought the capital function to it in 1603. The population reached a high density; more than one million people lived in the area of 100km². As there was not enough drinking water in Tokyo, especially for those in the lowlands and reclaimed lands who could not use well water because of the high salt content, a great deal of effort was made to establish the Edo water system (EWS). The EWS worked well until the 19$^{th}$ century, when cholera prevailed and the sanitary conditions became serious. The epidemic became the driving force to transform the EWS into a modernized system.

## 30.3. OUTLINE OF THE EWS

The EWS was used during the 17$^{th}$ to 19$^{th}$ centuries in Tokyo. As there was not enough water in Tokyo, various methods were developed for a sufficient water supply.

Drinking water was supplied in three ways: the Tamagawa and Kanda aqueducts, deep wells, and water vendors (Figure 30.2). In Tokyo, there were many lowland and reclaimed land areas near the seashore where deep wells could not be used for drinking because of salt. To supply water to such areas, the aqueducts were constructed. The water was taken from rivers or ponds and run down a long distance only by taking advantage of the incline, then stored in a well (Figure 30.3) as terminal. As stored wells had a bottom, impurities in the water settled to the bottom. There was no water treatment technology such as filtration or disinfection. Mostly both

Figure 30.2 Water Vendor. A man who had a pair of buckets filled with water on his shoulders visited each house, and sold drinking water drawn from a well or aqueduct. (Source: Horikoshi 1995)

Figure 30.3 Schema of Stored Well. (Source: Horikoshi 1981)

deep wells and stored wells were used by multiple families. Water vendors were also important in areas where aqueducts and wells did not supply water.

Wastewater, which, unlike today, did not include excrement and urine, flowed down through a drainpipe into the rivers without any treatment. Excrement was stored in the toilet and brought to fields as an excellent fertilizer. Multiple families usually shared the same toilet, too. Because of this unique excrement collection system, the rivers and streets of Tokyo were cleaner than those of Paris and London.

## 30.4. SANITARY CONDITION OF THE EWS

Tokyo did not suffer any serious problem with sanitation until the end of the 19$^{th}$ century. The first outbreak of cholera occurred in 1858 and the second happened in 1882, following a world pandemic. No data exists on the number of patients and the total deaths in the 1858 outbreak. Figure 30.4 shows the cholera infection ratio in Tokyo from 1880 to 1990. The peak prevalence was in 1886. A typical water-born disease, the level of cholera infections can properly represent the sanitary conditions of a particular locale.

The city of Tokyo was divided into 15 districts in the end of 19$^{th}$ century (Figure 30.5). Some archives report that each area had a different sanitary condition and that Nihonbashi was particularly the worst. Figure 30.6 shows the difference of infection ratio at the peak in 1886.

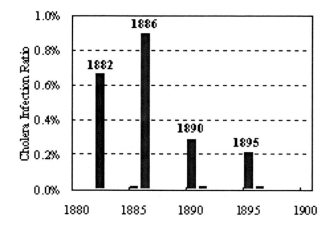

Figure 30.4 Cholera Infection Ratio in Tokyo (1880-1900). (Source: Statistic of Tokyo prefecture 1880-1990)

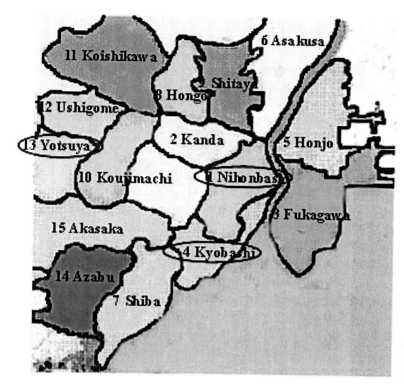

Figure 30.5 Map of Tokyo. (Source: ©APP company 2001, EDO Tokyo layer map (CD-ROM))

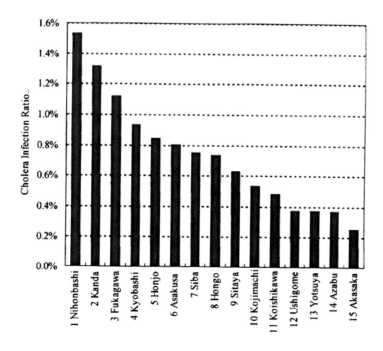

Figure 30.6 Difference of cholera infection ratio in 1886. (Source: Statistic of Tokyo prefecture 1880-1990)

Studying the reason for this difference reveals the key issues for constructing a water system with good sanitary conditions without expensive technology.

## 30.5. WHAT DETERMINED THE SANITARY CONDITIONS?

The potential factors determining the sanitary conditions were examined: population density, water stagnation, and the disposition of wells and toilets.

Generally, a high population density is apt to cause the prevalence of infectious diseases. The relationship between population density and the cholera infection ratio is examined in Figure 30.7. As expected, the districts with a higher population density tend to show the highest cholera infection ratios. However, a few districts with a similar population density, e.g., Fukagawa (3) and Yotsuya (13), display a different ratio of cholera infection. This indicates that population density alone does not fully explain the differences in sanitary conditions.

The water-flow tended to remain on the flat land, because the aqueducts flowed due to the incline. The stagnation of the water, which has the possibility to worsen the sanitary conditions, can be explained by the slant. The gentle incline means that the water tends to stagnate. As

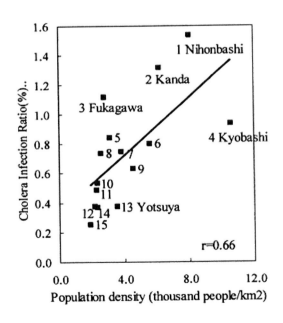
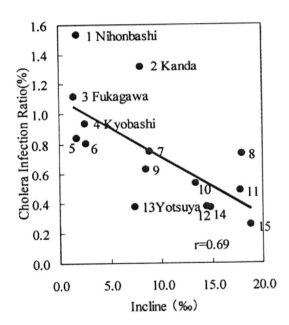

Figure 30.7 Population density and cholera infection ratio in 1886. (Source: Statistic of Tokyo prefecture 1886)

Figure 30.8 Incline and cholera infection ratio in 1886. (Source: Statistic of Tokyo prefecture 1886, Japan Map Center, Survey map of Tokyo 1887)

expected, the district with the most gentle incline had the highest cholera infection ratio (Fig. 30.8). The infection ratio, however, in some districts, such as Kanda (2) and Yotsuya (13), differed despite a nearly equal degree of incline. This difference can be explained by considering the population density.

In the Edo era, most of the people lived in "Nagaya," small row houses, with a high population density. Many shared the same well and toilet. Usually, the well was located near the toilet. Figure 30.9 provides an example of a map of a residential block. A detailed analysis on the disposition of the wells and toilets was done in three districts: Nihonbashi (1), Kyobashi (4), and Yotsuya (13). The sanitary conditions in Nihonbashi were the worst among these three districts, with a 1.5% cholera infection ratio in 1886. In Kyobashi and Yotsuya, the cholera infection ratios in 1886 were 0.9% and 0.4%, respectively.

If pathogenic bacteria existed in the toilets, the close proximity of the well and toilet could cause the higher risk of pollution of the well by underground or surface means. Especially, the close disposition of wells and toilets became a subject of discussion in the end of the 19$^{th}$ century. In 1878, the municipality of Tokyo enacted a law calling for the separation of wells and toilets by a distance of 5m (Horikoshi 1995). Figure 30.10 shows the histogram of the distance between the wells and toilets (Mitsui Research Institute for Social and Economic History; National Institute of Japanese Literature). As a result of the Kolmogorov-Smirnoff-test, the distribution

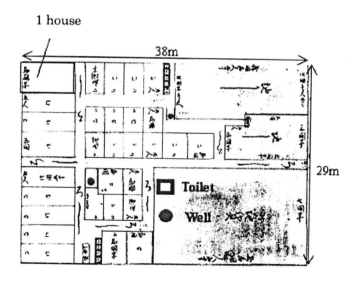

Figure 30.9 Map of a residential block. (Source: Mitsui Research Institute for Social and Economic History, Map of residential block)

demonstrates a significant difference with 95% confidence. The distance in Nihonbashi, the area with the worst sanitary conditions, was the shortest. The distance in Yotsuya was the longest. Consequently, it is clear that the distance between the wells and toilets is one of the important indicators of sanitary conditions. Keeping a reasonable distance between wells and toilets can be assumed to be the most useful way to avoid bad sanitary conditions.

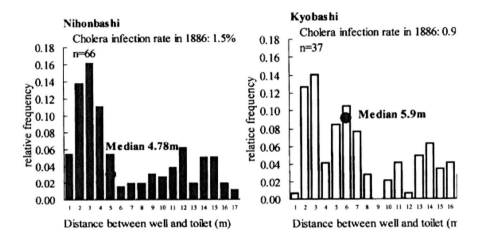

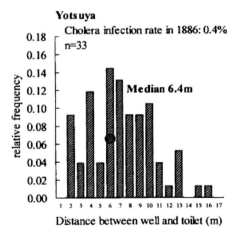

Figure 30.10 Histogram of the distance between wells and toilets.

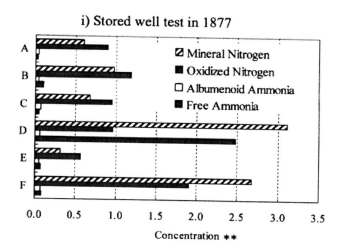

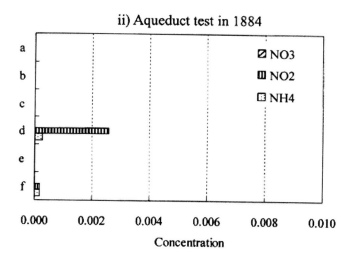

Figure 30.11 Water quality test.

There is evidence that the water was polluted when it was stored in the wells. In 1877 and 1884, the water quality was tested (Atkinson 1878; Municipality of Tokyo 1923, 923-932). Both tests inspected nitrogen-related substances ($NH_4^+$-N, $NO_2^-$- N, $NO_3^-$- N, etc.), which indicate contamination by excrement. A bacteriological check was not done in either test. In 1877, the quality of the water in the stored wells (6 wells: A-F) was measured and in 1884, the water quality in the aqueduct (6 points: a-f) was measured (Figure 30.11). From these figures, we find that the wells were contaminated and the distance between the wells and toilets is an important factor in urban sanitation.

## 30.6. CONCLUSIONS

The EWS had been operated without electric power or water treatment technology for 300 years in the area of highest population density. Although this water system had some problems in its sanitary conditions, these problems were thought to be manageable if the key points listed below were taken care of:

- Maintain a reasonable population density
- Prevent water stagnation
- Maintain a reasonable distance between wells and toilets.

However, since people did not understand the causes of poor sanitary conditions, all of the EWS was thought to be a bad system needing replacement by a modernized one. As a result, in the 20th century Tokyo dismantled the EWS and completely modernized it by imitating the systems of Western countries, regardless of the advantages or disadvantages. This was not a wise way to implement a system change. The best way would have been to make good use of the merits of the historical system and to adopt new technology to improve the problem areas.

Tokyo's experience can be a good example for developing countries facing the same situation as Tokyo in the 19th and 20th centuries. These countries should not blindly change their systems entirely in order to adopt Western ways, but instead harmonize their historical ways based on regional characteristics with new technology.

## 30.7. REFERENCES

Atkinson R.W. (1878) Water Supply of Tokio. Transaction of the Asiatic Society of Japan 6, 87-98.

Horikoshi M. (1981) Story of Well and Waterworks, Ronsosha (in Japanese).

Horikoshi M. (1995) Aqueducts in Japan, Sinjinbutuouraisha (in Japanese).

Mitsui Research Institute for Social and Economic History, National Diet Library.

Municipality of Tokyo (1923) History of Tokyo – Water Supply (2), 923-932 (in Japanese).

National Institute of Japanese Literature, Department of Historical Documents.

Statistic of Tokyo prefecture 1880-1990 (in Japanese).

# CASE OF TOKYO, JAPAN

# 31
## HEALTHY WATER FROM AN INDIGENOUS MAORI PERSPECTIVE

*Ngāhuia Dixon*

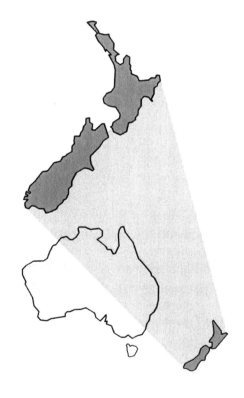

Figure 31.1 Location of New Zealand.

## 31.1. GENERAL BACKGROUND

This chapter considers waterways as important to health, and as sites of traditional food resources for the indigenous people of New Zealand. The waterway locations chosen for this study are the coastal mudflats of Tauranga and the inland Waikato River on the North Island of New Zealand.

Maori lived a subsistence traditional life in the 1800's. From the 1900's to the present day, in terms of health, they progressed with the assistance of sanitation introduced during colonisation. Irony played its hand though, when, through such settlement and pollution traditional Maori foods such as periwinkles and eels found in waterways have gradually depleted. The concern is that pollution and over-harvesting have affected habitats and supply of these traditional food resources.

In essence, then, the article takes two views of healthy waters: one supporting sanitation and good health, the other noting that seafood habitats and stocks have, and are being affected by ignorance of basic Maori water protection and management roles.

Waterways and their protection linked to health is a global issue. The resource is crucial for community and personal use and its protection has been a matter of concern for all, including Maori, the indigenous people of New Zealand. Water for communities in nearly all instances comes from natural waterways, and the quality and state of these waters is a constant focus/topic of discussion in water debates. These waters also hold traditional food resources such as certain fish and seafoods consumed by Maori tribes. The issue of healthy water is therefore of prime concern at the international as well as the local community level for some of the reasons mentioned.

This chapter aims to highlight links between water, health and food resources and although a Maori cultural stance is taken, other factors are not excluded, therefore where appropriate such references will be noted. The meaning of water within the chapter will include waterways, streams, rivers and coastline waters.

The study considers the situation of two tribal locations, Waikato and Tauranga local tribes who live respectively by the Waikato River, and the eastern coastal waters of the North Island of New Zealand (Figure 31.2). These two waterways were considered as they offered contrasts in terms of food resources, health issues, and they are also locations where a) the author was raised, and 2) where she now lives. These waterways are also identity markers for the tribal groups within the two locations and as such are respected physically and symbolically. Ironically these water bodies are also important to the New Zealand economy, since an export wharf is situated within the Tauranga coastal location, and the Waikato River is the main source of hydro-electricity for many of New Zealand's North Island communities

The period under examination is from the 1900's to the end of the century. This timeline takes into consideration settlement, population growth, new developments in the country regarding health, food productivity and other initiatives. Pre-1900 views will be given only if required to highlight the issues of waterways linked to this chapter's focus.

Figure 31.2 Map of New Zealand's North Island. (Photo: courtesy of Waitangi Tribunal, N.Z.)

The Maori population in New Zealand as at 2001 comprised 15% of the total population according to statistics information. (Statistics New Zealand 2005:1) Tauranga shows Maori tribal members made up 16.1% of the region's resident population of 97,900 as at June 2003. (Statistics Overview) Information is sparse on overall Maori population figures within the region. The Waikato area on the other hand, has just over 10,000 Maori according to Environment Waikato figures. This population figure has been taken as it is linked more closely to the area considered here. A larger Waikato provincial population count would provide a more general picture than that of Maori only.

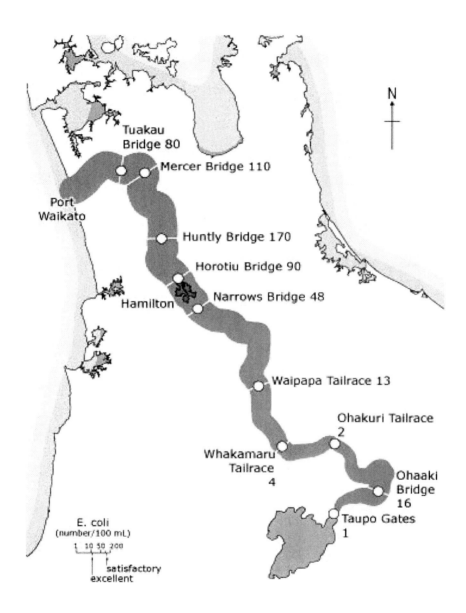

Figure 31.3 North Island's Waikato River with E. coli illustrated. (Source: Map courtesy of Environment Waikato, Hamilton, New Zealand.)

## 31.2. WATER DISCOURSE

Strang (2004, 3) notes that water transcends many boundaries in terms of its significance. Historians, theologians, and anthropologists have all written about water related to their disciplines. Other aspects of water such as hydrology and its biology have also been the central focus for many other researchers. While all these are important, water's cultural meanings for the Maori of New Zealand can also add to the debate regarding safe and healthy water.

Within New Zealand, two dominant communities have lived side by side since the 1800's but with different ideologies about waterways, their use, management, and states. Maori, acknowledged as the indigenous people of the land, settled the country during the fourteenth and fifteenth centuries, living and surviving on the bounty of the waterways and land. Through ancestral legacies, a cultural relationship with Nature emerged illustrating its importance together with its physical and symbolic sustenance of tribal life through the land and its waters. In return, tribal responsibilities were to protect these natural resources from misuse.

The other dominant group comprising settlers called "Pākehā" (white man) began settling New Zealand during the 1800's onward and introduced many new initiatives into the country.

Traditional Maori views of waterways especially with regard to their cultural importance, food resources and pollution were clearly defined for Maori by ancestors. In time these views contrasted with those of the Pākehā world, institutionally and community-wise became a central point of debate. The Foreshore Debate which legislated the foreshore as belonging to the Government for use by all the people of New Zealand, is one issue which comes to mind. Water for housing and development would also be affected, albeit for different reasons, i.e. the disposal of wastes from various sources including households into rivers, and/or the sea, thereby affecting the water in a negative way. Modern groups such as water companies and some local authorities who set community regulations have also been noted by Maori as only acknowledging the economic worth of waterways, thus providing opposing views to those of Maori tribes and many environmentalists.

## 31.3. MAORI WATER CATEGORIES

According to Maori traditions, water has many states and qualities. *WaiMaori*, or pure water, is water untouched by humans and comes directly from the mountains. This is the purest form of water and, as articulated by Durie (1994, 1, 13), it originates from *Ranginui* the Sky Father for the wellbeing of life. As it is commodified, and linked to human usage, its cultural and spiritual quality is affected through connections with hazardous substances, pollutants and forms such as human waste. *Waiora* literally means 'living waters', or symbolically and physically 'waters of life' and is recognized by Maori as the purest form of this natural resource. Its physical importance for drinking was matched by its symbolic properties for 'ritualistic purposes' according to Durie (1994). Best (1952, 39) states that *waiora* meant life and welfare. Another name for pure water is *waipuna*, or spring water. *Waitapu* or sacred water is a state which has spiritual connotations and is used for cultural purposes such as the opening of a meeting house or rituals related to birth.

Water in negative states is known by various names. *Waimate*, dead water, and *waikino*, or spoilt water, according to Maori have had their cultural and physical value affected and are unfit for human consumption, due to human or other forms of pollution. The question of human waste affecting the waterways is reviled by Maori and no doubt many other community members as well. Although waste is discharged into coastal waterways, protection from pollution here presents a more complex issue because of its vastness. Non-pollution of such waters has had to be addressed by worldwide organizations such as Greenpeace or local marine and environmental institutions.

Apart from these aspects the positive role of water in Maori traditions is as significant today as it was in the past although negative aspects, which we have touched on briefly above, are still a hotbed of tension for Maori. These philosophies regarding safe and healthy water provide in brief some fundamental differences between Maori and other groups such as farmers, industrialists and groups who have an economic interest in water. The Maori perspectives outlined are consistent throughout tribal groups, and support to protect the water's ability to provide, to cleanse, and to protect it for the common good is paramount in the Maori mind. Water and aspects of health linked to Maori will now be outlined.

## 31.4. PRE-1900 MAORI

Prior to the arrival of European explorers and settlers in New Zealand, some health issues were in place. (Durie 1994, 14, 15) Water was used for birth and healing purposes through rituals that were normally conducted in running water. For newborn babies, the mother held her offspring while a priestly expert knowledgeable in rituals would touch the baby with a leafy branchlet, then dip it in the stream and with it sprinkle the infant - this was done at a certain point in the ritual. Durie's table (1994, 14, 15) shows various links between Maori rituals, activities and health precautions undertaken prior and during colonization. (Table 31.1).The table shows that Maori were aware of the health issues linked to living in groups, and the precautions needed to be taken regarding food and its storage, rubbish collection, human waste, childbirth and eventual death. At the centre of all the issues as noted earlier was the role of water, and waterways in healing, physical and cultural cleansing and in rituals. (Durie 1994, 15)

The touching and handling of the dead was a form of ritual where washing with water and immersion in it would culturally cleanse those involved. From conception to death, then, according to Maori traditions, people are permanently linked to water in its various contexts.

The ancient Maori had taught his people that water is life and represents the welfare of all things. It was also noted by Durie (1994) that ancient Maori acknowledged flowing water from the earth was the purest form of water, and with this, humans would be safe from health hazards, culturally, spiritually or physically. The presence of water from springs, or streams, was far more important than that contained in 'man-made vessels' (ibid ) as water held in such

Table 31.1 The pā and public health precautions

| Amenity | Function | Health Advantages |
|---|---|---|
| Hilltop pā site | Community protection and comfort | Effective drainage<br>Minimal dampness<br>Military advantage |
| Rukenga kai<br>Waste disposal location | Refuse collection | Cleanliness<br>Reduced chances of contamination |
| Heketua<br>A Maori name for latrine or toilet | Community latrine | Safe disposal of human waste<br>Hygiene (measures) |
| Pātaka<br>Food storage house built above the ground to avoid rats and flies. | Food storage | Rodent-free<br>Year-round nutritional supply |
| Wharekōhanga<br>A special house for mothers giving birth | Childbirth | Reduced infection<br>Bonding<br>Recovery of mother and baby |
| Wharemate<br>A house especially erected for the dead being mourned. | Shelter for deceased and mourners | Avoidance of contamination<br>Mourning process eased<br>Moral and psychological protection of community as a whole |

(Source: Durie, M. Whaiora, 1994)

containers was seen to have lost its healing properties 'and could in fact be seen to add to the risk of contamination'. Here running water, as opposed to still or stagnant waters, was the key to these healing or cleansing practices.

With colonization, more stringent regulations began to emerge regarding the health benefits of clean and safe water, and the hazards of polluted water. Rules were put in place as local and central authorities were established.

## 31.5. POST-1900 DEVELOPMENT

Maori houses pre-1900 in the main used rainwater or water from streams and springs for household tasks. Rainwater, which ran down the roofs of houses, was caught in tanks positioned with pipes to channel it in the right direction. This was a common practice from the 1930's onward and caused concern among the health authorities of the time due to the possible corrosion of corrugated iron, used for roofing. Latrines during the same period were located outside and were still in use until health authorities introduced various sanitation and health regulations. For kitchen usage, running spring water was a preferred option during those years as it was regarded as clean, flowing and safe. This supplemented the tank water and vice versa, although carrying it was a burden for members of many Maori households, who had to go some distance in many instances to fetch it. This practice is known from the author's personal experience.

Housing settlements grew and the basic need of water for purposes such as bathing, washing clothes and toilets were of paramount importance to these new communities. Moreover, as Maori were part of these newer settlements, with many moving to town for work, they were able with their rural kin to benefit from these initiatives. Water issues therefore were being addressed for both Maori and Pākehā.

New technologies, industrial, housing and farming developments were introduced into the country during colonization and later. Today, for example, products such as fruit and livestock are exported by New Zealand farmers, industries in which Maori are also players. Many of these developments were beneficial for Maori and allowed for new industrial and technological advances such as machinery for land initiatives, manure to support land productivity, and other new products to assist with everyday living. Farms, meat abattoirs and building industries, to name a few, used these advancements to develop their economic interests. Healthwise, sanitation together with medicines addressed health issues linked to the burgeoning population. As housing developments grew, refuse accumulated and required disposal. This raised the issue of public rubbish dumps, in many instances near or close to waterways with traditional food resources of Maori and other community members. The effects, therefore, of these new initiatives would not be realized until later years, when the health of these waterways would be monitored by environmentalists. Today healthy water issues, and Maori values of waterways, have been a constant source of discussion between local authorities such as Environment institutions, city and town councils, environmentalists, central government and Maori tribal groups providing collaborative efforts to protect watereways.

With new settlements, Maori also benefited, but they still saw the need to assert their views regarding waste. They stated that waste water, human waste and other forms of toxins should be released onto the land, where they would be soaked up through the most natural means and then recycled back into the earth. Sewer connections to waterways were seen as abhorrent to Maori who, as mentioned earlier, class waterways as the 'waters of life' physically and symbolically.

In relation to health, medicines were introduced and assisted in treating people's health problems including Maori. The waterways were also undergoing changes of a different kind during the 1960's onward, as foodstocks were slowly being affected within rivers and coastal waters by waste and other discharges noted earlier. Farmers were also using chemicals such as pesticides, and meatworks were established where animal waste run-off such as urine, blood and animal fat created a waste cycle which eventually filtered into rivers like the Waikato, and coastlines like Tauranga. Timber industries using oils or preservatives all contributed to the slow demise of traditional Maori foods, such as the eel within rivers, and periwinkle stocks on coastal mudflats. The depletion of these traditional foods caused alarm not only because of species dying away, but also because the periwinkle was considered a delicacy, its absence on food tables considered culturally inappropriate for visitors. The significance of these two food delicacies will be presented after a brief outline of food resource sustainability and healthy communities is addressed.

Safe and healthy waterways allow sustainability of food resources. This concept can be described as the ability to collect food from a fishing ground or shellfish bed through practices designed to maintain the resource for future generations. In modern terms, this would be referred to as good practice, in maintaining the equilibrium of the environment to produce. For humans, this means food is available, fresh and there when required. For this scenario to be maintained, a suitable habitat for these foodstocks is crucial.

A habitat is an environment in which an organism can thrive. Optimal growth conditions allow for optimal productivity, etc. In terms of the food resource, the habitat is very important, as it nurtures the food/animal/invertebrate/plant resource, etc. Without the habitat, the food resource dies, is eliminated, expires, declines, disappears, fades, vanishes, wanes, wilts and withers. The aftermath of habitat destruction is change or the demise of the resource. To initiate change, some situations must be changed to create the desired outcome. The same rule applies to the environment. Change creates change (Ormsby 2004). This illustrates the effects that could occur in the case of water and its food resources such as the eel or *tuna*, and periwinkle or *titiko*.

Figure 31.4 Long-fin eel – National Waterways Project New Zealand. (Source: Photo Permission granted by Environment Waikato, Hamilton, for Peter Hamill, Marlborough District Council, New Zealand.)

Maori tribal groups of the Waikato River have been linked to eels since early tribal settlement. The long-fin eels or Maori *tuna* is found only in New Zealand's lakes and rivers such as the Waikato. They are caught in *hinaki* (eel nets), and were either dried, smoked or boiled for eating. For Waikato Maori and other tribal groups, it is considered part of the staple diet and a traditional delicacy, but it has been noted that a steady decline in its stocks has been taking place. This, environmentalists and Maori tribes claim, is due to effects of waste and pollution along their natural habitats, the Waikato river, and other such waterways. As the economic value of the eel was realized, it was eventually harvested commercially. Environment Waikato is an environmental local authority assigned the task of protecting the Waikato River and other environments within its catchment. They noted (2005) eel numbers in decline when the commercial harvest halved during the early 1990's. This outcome forced a quota management system in the North Island to protect the eels from further depletion. Eel habitats such as rivers were also being affected by pollution and waste products, thereby adding concern for all, including Maori of the Waikato. Today the eel is reduced in numbers and initiatives have been put in place by concerned authorities and Maori to arrest this decline. The question of declining traditional food resources has been an ongoing issue in Waitangi Tribunal Claims by Maori. Environment Waikato has studied the effects of waste discharges and hazardous substances, together with rural run-off and leaching going into this river, the longest in New Zealand and it is interesting in relation to eels.

The local authority Environment Waikato noted within the region three major point source discharges of waste which have affected the river (see Figure 31.3). These are the Hamilton sewage outlet, the Horotiu meatworks and Ngāruawāhia sewage, even though the output only generates 5% of the bacterial effects along the river. Further downstream towards the river outlet into the Tasman Sea, signs of more bacterial effects are noted. This is shown on the river map of bacterial buildup. The effects of bacteria are heightened probably due to the waste pumped into its waters as it flows to the Tasman Sea. Environmental scientists believe traditional Maori food sources have been depleted through such practices, in addition to rural and urban run-off, storm water disposal, farm run-off and dredging, to name a few.

## 31.6. RESOURCES AND NATURE GOVERNANCE

The Waitangi Tribunal, a formal group set up by, and under the control of, central government, addresses Maori claims on the despoliation of waters and lands under tribal guardianship. One such claim by Waikato stated that 'a serious and continuing deterioration in the quality and quantity of seafoods available to the Waikato-Tainui people' is a result of hazardous discharges into the river. Supporting claims by a Tainui tribal member also asked the Tribunal to investigate the despoliation of the Manukau Harbour adjacent to the Waikato River, a coastline also under the guardianship of Tainui tribes, recognized as indigenous guardians of the Waikato River. The Manukau Claim finalized in July 1985 stated that Maori values will not impede progress, and they are 'no more inimical to progress than Western values'. Maori, the Waitangi Tribunal said, had not sought to be entrenched in the past if values and actions would help progress the tribes as a whole and where a positive future is a desired outcome. Also included in the same Tribunal Report regarding the Waikato tribal groups were the following comments:

> [...] that those tribes have used and enjoyed the lands and waters of the Manukau and lower Waikato from early times to the present day. The river and harbour are as much their gardens as their cultivations on land

> [...] that the use and enjoyment of the waters have been severely limited by pollution from farm run-off sewage and industrial discharges (Waitangi Tribunal Report – Manukau Claim)

Support for these types of recommendations have also come from an unusual source, the Resource Management Act 1991, which provides legislation to protect the waterways and lands from various activities considered detrimental to the natural environment.

Shellfish and fish vary from the coastal waters to the inland rivers. The Tauranga coastal periwinkle is also a delicacy like the Waikato eel, and it has encountered similar problems regarding its habitat, to the long-finned eel of the Waikato River. It too has been affected by waste discharges from further housing developments and pollution in the area. This seashell food, called titiko in Maori with the appearance of a snail, is found on the tidal mudflats of the Tauranga inlet called Rangataua. The scientific name for titiko or mud snail is Amphibola crenata. (Ormsby 2004) Environmental information gathered by Bioresearches Ltd in 1974 stated,

> Amphibola is widespread in the protected high intertidal bays of the Harbour (Tauranga) and populations are of high density and excellent condition.

Tangihaere Ormsby undertook an investigation into this seafood in 2004 and provided a report for Ngā Pae o Te Māramatanga, a research group formally linked to tertiary institutions including the University of Waikato. Her methodology involved interviewing Maori families still living within the region who could recall the food resource being plentiful. A review of reports and other relevant literature on waters in the region and Tauranga environmental issues was undertaken by her.

Maori families have settled the [inlet] tidal waters of the inlet in question since the 1800's. Although they have now developed a lifestyle similar to the Pākehā, they have combined their eating habits, but still lean towards traditional seafoods such as the periwinkle for gastric sustenance. This, like the inland Waikato tribes, meant acquiring food from the waterways, lands and forests. It has been noted that the *titiko* has been depleted from the 1970's onward. According to Ormsby's findings, this slow demise can be attributed to different factors. Local authorities such as the city councils have built sewage ponds in close proximity to the habitat of the periwinkle and affected its environment. Water from household development around the inlet where the periwinkle enjoyed its habitat has also contributed to its demise with stormwater runoff and household waste pumped into the Rangataua inlet. She also points out that discharging effluent into the realm of Tangaroa, the Maori sea deity, constitutes a fundamental transgression evoking a culturally embedded abhorrence (Ormsby, 2004).

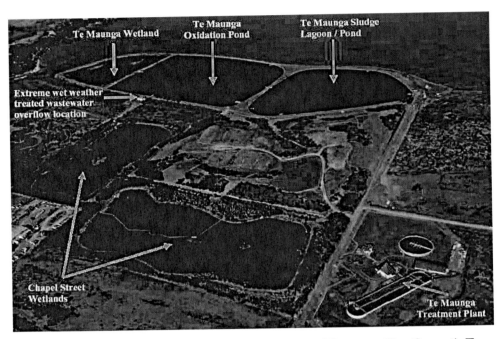

Figure 31.5 Oxidation ponds in Tauranga. (Photo: courtesy of Tauranga City Council, Tauranga)

Scientists, according to Ormsby (2004), state the titiko requires marshes in which to grow, and according to local interviews marshes were prominent vegetation around the inlet during the 1940's and 1950's. They provided the habitat for the young periwinkle. Mangroves now dominate the shoreline of this inlet and the breeding ground for the periwinkle has been eroded. The irony here hinges on legislative politics in action. An alternative plan regarding sewage collection and disposal for the borough of Mount Maunganui included the siting of these ponds in a predominantly Maori settlement location with the cultural and economic effects for these tribes barely considered.

Intertidal mudflats in Rangataua were reclaimed for the sewage ponds, or as Maori would say, 'confiscated' to service the borough of Mount Maunganui, a regional community. Community development here, it appears, has affected the habitat of food resources like the titiko.

Building of the sewerage ponds in close proximity to Maori seafood resources, it seems, was not considered as being relevant when the question of their location was discussed. This is highlighted through Waitangi Tribunal claims lodged by Tauranga Maori tribes, who have said their cultural roles, foods and lifestyles were not considered when the siting and building of sewerage ponds took place. For local tribes, the claim was made that local authorities saw the present site of the ponds as a perfect answer to sanitation woes, and that was all that mattered, Maori views being fairly peripheral in the final decision. This was seen as a suppression of cultural rights, and the physical desecration of the tribal waters so long protected through the traditional practices of rahui, i.e. the restriction of water activities when foodstocks are low, or when drownings take place.

Environment councils such as Environment Waikato and Bay of Plenty seem to be much more Maori-friendly as their philosophies show the environment should be cared for. It has been noted that to date they have friendly working relationships with tribal groups because of similar concerns for the natural world such as waterways.

## 31.7. CONCLUSIONS

In assessing and summarizing these views, they highlight Maori cultural thinking, and institutional perspectives regarding waterways, and the effects therein of these differences. New communities have forced changes in lifestyles for both Maori and non-Maori. Tribes are not ignorant to new ways to better themselves, and have accepted many decisions linked to the world around them through necessity. Other groups with different philosophies regarding waterways have now settled among Maori communities and view waterways in a different light. These perspectives have become a source of concern and dialogue between the communities

Figure 31.6 The amphibola or tītiko family of sea snails. A member of the amphibola family.

is now more robust. Water and sanitation issues such as the siting of sewerage ponds, water and Maori guardianship roles, and water and its healthy state are all factors of difference. Maori have only asked that their cultural views and roles as guardians of rivers and coastlines be considered and taken into account when dealing with the state of these waterways, their use, and the impact on Maori, and indeed the wider community. The views of Maori are now being taken more seriously.

What then does this tell Maori communities regarding the stance taken by others (including some Maori), with the exception of environment-focused local bodies? Community views from all sectors should always be explored and considered. This allows different points of view to be aired where more logical ideas and answers may be found regarding waste disposal, sanitation, pollution and other water related issues. The effects of development and community activity such as those mentioned above linked to water, which have been gauged by environmentalists, Maori and local environmental authorities in many instances are positive for many, but are now causing greater concern than before because of the state of waterways. Local communities and central government are now more aware of water and the environment as places which need to be protected at all costs. There is a sense of urgency regarding the protection of the waterways, and of food resources. The *tuna* and the *titiko* are important in themselves as part of the food and ecological chain and their protection, together with other seafoods is crucial if we are to protect, nurture and sustain the waters, habitats and stocks of these gifts of Nature. Waterways where habitats and food stocks are found should be the primary focus of action for and by communities, Maori and non-Maori alike. It is up to each and everyone to play their part to sustain the health of waterways and sustainability of their food resources. Communities can then say with pride that they supported the welfare of the *"waiora/*waters of life" and that it was a team effort.

## ACKNOWLEDGEMENTS

The author wish to acknowledge the information and pictures from Environmental Waikito.

## 31.8. REFERENCES

Best E. (1952) The Maori as He Was. Government Printer, New Zealand.
Durie M. (1994) Whaiora. Maori Health Development, Auckland. Oxford University Press.
Ormsby T. (2004) Kei Hea Te Kai (Where is the Food) – Report prepared for Research Unit, Nga Pae o Te Māramatanga, March, 2004. Ngā Pāpaka o Rangataua: Kei hea ngā kai? (The Crabs of Rangataua – (a symbolic name for the people).
Statistics New Zealand (2005) Te Tari Tatau – Hui Taumata 2005:1.
Statistics Overview – Tauranga City Council website http://ourcity.tauranga.govt.nz/statistics/overview/
Strang V. (2004) The Meaning of Water. Berg, Oxford, New York.
Waitangi Tribunal Report: Kaituna River Report - Wai 4 Summary, Wai 4 Report contents. http://wai8155sl.verdi.2day.com/reports/nicentr/kaituna/wai4014.asp

Environment Waikato http://wai8155sl.verdi.2day.com/reports/nicentr/kaituna/wai4014.asp

# 32

# THE MEDICAL IDENTIFICATION OF NEW HEALTH HAZARDS TRANSMITTED BY WATER

*Heikki S. Vuorinen*

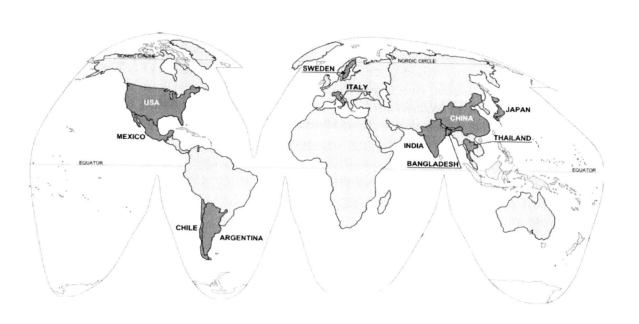

Figure 32.1 Key countries mentioned in the chapter.

## 32.1. GENERAL BACKGROUND

In this chapter the history of the identification of waterborne health hazards during the 20$^{th}$ century is presented. Mainly the international aspects are taken into account and especially the wide array of documents produced by WHO are used. The 21$^{st}$ century was opened by the third edition of *Guidelines for Drinking-water Quality* by WHO with the following definition of safe drinking water: *"Safe drinking-water, as defined by the Guidelines, does not represent any significant risk to health over a lifetime of consumption, including different sensitivities that may occur between life stages."* (WHO 2004, 1)

In the period from the late 19$^{th}$ to the early 20$^{th}$ century, a so-called bacteriological revolution occurred in medicine. The main health problems were identified to be caused by micro-organisms. Several diseases, such as cholera, typhoid fever and dysenteries, were found to be caused by the ingestion of water (or food) that was contaminated with human or animal excreta. In the early 20$^{th}$ century, the health problems associated with water pollution seemed to have been resolved in the industrialized countries when the chlorination of water and other water treatment techniques were developed and widely used.

In the 1960s, the water supply and its public health aspects could be presented in a quite straightforward and unproblematic way in a standard textbook of public health (Bruce 1969). It was even interpreted that waterborne infectious diseases had been conquered and the remaining serious risks to health transmitted by water were chemical and radioactive contaminants of water. When the optimism was high it could even have been said like George L. Waldbott did in the following: *"At the present time these diseases* (waterborne diseases) *occur only sporadically. They are being controlled by strict sanitary engineering. Only during floods, earthquakes, and other great disasters is there a threat of waterborne epidemics."* (Waldbott 1973, 255)

In the beginning of the 21$^{st}$ century, the attitude was totally different. *"Over 1 billion people lack access to safe water supplies; 2.6 billion people lack adequate sanitation. This has led to widespread microbial contamination of drinking water. Water-associated infectious diseases claim up to 3.2 million lives each year, approximately 6% of all deaths globally. The burden of disease from inadequate water, sanitation and hygiene totals 1.7 million deaths and the loss of more than 54 million healthy life years."* (WHO 2005, 2, 14)

## 32.2. INTERNATIONAL DRINKING WATER STANDARDS

Drinking water standards were developed nationally in different industrialized countries during the early 20$^{th}$ century. After the Second World War and especially in the 1950s, the requirements for safe and potable water became urgent with the increase in travel, particularly air travel. Safeguarding the public health in the international community was allocated to the World Health Organization (WHO), which took an active role in developing international standards for drinking water. There was a series of expert consultations and meetings in the 1950s sponsored by WHO. WHO published a special issue of water sanitation in the *Bulletin of the World Health Organization* in 1956. After several years of preparation, WHO published *International Standards for Drinking Water* in 1958, and revised editions were published in 1963 and 1971 (WHO 1958, WHO 2004). The Regional Office for Europe of WHO was active in the preparation of international standards and consequently *European Standards for Drinking-Water* were published in 1961, and a second edition in 1970 (WHO 1961; WHO 1970). WHO elaborated the international standards further in *Guidelines for Drinking-water Quality*, which appeared in three volumes: the first edition in 1984–1985, the second in 1993–1997 and the first volume of the third edition in 2004 (WHO 2004).

Anxiety about chemical and radiological environmental hazards to human health was mounting during the 1960s. WHO was organizing conferences, in which water pollution was among the topics, and a special issue on environmental health was published in the *Bulletin of the World Health Organization* in 1962. In 1971 the twenty-fourth World Health Assembly indicated several areas where action on the human environment was particularly needed, which included the provision of adequate quantities of potable water (WHA24.47). The United Nations held a conference on the human environment in July 1972 in Stockholm, Sweden. In the same year, WHO published a wide-ranging survey of environmental hazards to human health. Water pollution had a quite eminent place in this compilation work (WHO 1972, 47–71). In the survey the health hazards of water were grouped into: 1) biological hazards (pathogenic bacteria, viruses, parasites) and 2) chemical hazards, including radioactive substances.

In the 1980s WHO considered the International Drinking Water Supply and Sanitation Decade, 1981-1990, as an essential first stage in the global programme of health for all by the year 2000. (WHO 1981, 1) In this document, it was stated: "*The work facing those who aim to fulfil the Decade's targets is formidable. Approximately three out of five persons in the developing countries do not have access to safe drinking-water; only about one in four has any kind of sanitary facility, be it only a pit latrine ...*" (WHO 1981, 2)

From the 1990s onwards WHO started jointly with UNICEF to make assessments of how the water supply and sanitation situation in the world developed. There are obvious difficulties in obtaining reliable and comparable data from different countries and time periods. These reports are quite gloomy reading. At the turn of the millennia it was reported: "*Approximately 4 billion cases of diarrhoea each year ... cause 2.2. million deaths, mostly among children under the age of five... This is equivalent of one child dying every 15 seconds ... These deaths represent approximately 15% of all child deaths under the age of five in developing countries. ... Diarrhoea is the most important public health problem affected by water and sanitation and can be both waterborne and water-washed.*" (WHO/UNICEF 2000, 2–3) The newest assessment appeared in 2005 and it has practically the same story to tell, although some improvement in the worldwide water supply and sanitation situation could be observed from 1990 to 2002 (WHO/UNICEF 2005, 5, 24, 26, 36–38). However, although a lot of work seems to have been done to improve the quality of the data, the comparability of countries and years remains far from ideal (WHO/UNICEF 2000, 4–5, 77–79; WHO/UNICEF 2005, 6–7).

## 32.3. MODERN WATER TESTING AND TREATMENT

During the 20th century the health criteria for potable water were developed and bacteriological, virological and chemical analyses of water more and more replaced the ancient criteria based on the senses. The time-honoured ways of examining water based on the senses were, however, still there in the 20th century although in more standardized and technically refined ways. The taste, odour, temperature and appearance of water were included in the table of methods of examination for characteristics of water in *European Standards for Drinking-Water*. (WHO 1961, 38; WHO 1971, 42)

After the acceptance of the germ theory of infectious disease, the normal cycle of events in water microbiology has been: identification of a disease of unknown etiology, development of analytical techniques, and identification of the etiological agent (e.g. bacteria, protozoa, and virus). In spite of significant scientific advances, diseases of unknown etiology continue to comprise a high percentage of the total outbreaks of water-related disease (WHO 2003, 9).

During the 20th century new methods for water treatment were used: e.g. different methods of filtration were developed; for disinfection, water was ozonized, UV-irradiated or chlorinated. At the moment, a lot of hope for improvement in the quality of drinking water in developing countries seems to have been placed on the different household water treatment and safe storage practices (WHO/UNICEF 2005, 28–29). In spite of these new technologies, water still has to be boiled in many situations to make it safe for drinking, as recommended already during antiquity. Persistent health-related problems with piped water has kept the door open for the ancient occupation of water vendor to survive and even thrive in many areas of the world and helped to create an expanding industry of bottled water (WHO/UNICEF 2000, 78).

## 32.4. MULTITUDE OF IDENTIFIED BIOLOGICAL HEALTH HAZARDS OF WATER

The amount of different microbes, which were found to be able to be transmitted through contaminated water, escalated during the 20th century. To the bacterial diseases like typhoid fever, dysentery and cholera already known to be spread by water at the turn of the 20th century, a multitude of viral diseases like poliomyelitis were added (Tables 32.1 and 32.2.). Most of these bacterial and viral pathogens infect the gastrointestinal tract and are excreted in the faeces of infected humans (or other animals) and the transmission route is ingestion. In the microbial fact sheets of the most recent *Guidelines for Drinking-water Quality* a total amount of 17 groups of bacterial and 7 groups of viral pathogens were presented (WHO 2004, 222–259).

Table 32.1 Examples of main bacterial diseases capable of being transmitted through contaminated water or food prepared with such water.

| Disease: | Causative organism |
|---|---|
| Cholera | Vibrio cholerae |
| Bacillary dysentery | Shigella |
| Typhoid fever | Salmonella typhi |
| Paratyphoid fever | Salmonella paratyphi A, B & C |
| Gastroenteritis | Other Salmonella types, Shigella, Proteus spp., etc. |
| Infantile diarrhoea | Enteropathogenic types of Escherichia coli |

Table 32.2 Examples of virus diseases capable of being transmitted through contaminated water or food prepared with such water.

| Disease: | Causative organism: |
|---|---|
| Polio | Poliovirus |
| Hepatitis | Hepatitis virus A & E |
| Gastroenteritis | Rotaviruses, adenoviruses, caliciviruses |

Diseases caused by protozoan pathogens and helminthiases are among the most common diseases in humans and animals. Water plays an important role in the transmission of some protozoan pathogens (Table 32.3). The control of waterborne transmission of several protozoan pathogens has appeared to be difficult because they are resistant to processes generally used for disinfection or the filtration of water.

The role of water in the transmission of helminths is more complicated (Table 3). Important helminthiases are transmitted through water contact, like schistosomiasis, or the use of untreated wastewater in agriculture, like hookworm infections, or eating raw fresh water fish, like broad fish tapeworm infections, but are not usually transmitted through drinking water.

The history of cholera illustrates the difficulties in controlling waterborne infections. Cholera was already in the 19th century recognized to be caused by *Vibrio cholerae* and transmitted through contaminated food or drinking water, as well as by person-to-person contact through the faecal-oral route. Consequently efficient ways to keep it under control were introduced. Regardless of this, cholera continued to spread around the world in the 20th century. Again and again *V. cholerae* was observed to spread rapidly in crowded living conditions where water sources were unprotected and where the disposal of faeces was defective.

Table 32.3 Examples of main protozoa and helminths capable of being transmitted through contaminated water or food prepared with such water.

| Disease: | Causative organism: |
| --- | --- |
| Amoebic dysentery | Entamoeba histolytica |
| Giardiasis | Giardia lamblia |
| Anemia | Diphyllobothrium latum (broad fish tapeworm) |
| Schistosomiasis | Schistosoma mansoni, S. japonicum, S. haematobium |
| Hookworm disease | Ancylostoma duodenale, Necator americanus |

The sixth pandemic of cholera was fading away in the middle of the 20$^{th}$ century when a new biotype (El Tor) of *V. cholerae* started the current seventh pandemic in 1961. During this pandemic, cholera spread rapidly around the globe. This was assisted by the increase in international migration, wars, famine, recurrent refugee camps (especially in Africa) and the fact that the El Tor biotype was more likely to produce unobvious infection than the classical type of cholera. In 1991 it reached Latin America, which had been without cholera almost a century. Cholera spread widely in the Americas and has remained in Latin America after this new introduction. It is a recurring problem in many poor countries and an endemic disease in others in the beginning of the 21$^{st}$ century. (www.who.int/topics/cholera/en; Cholera 2004)

## 32.5. INDENTIFICATION OF CHEMICAL HAZARDS IN WATER

After the Second World War several chemical hazards connected with polluted water were identified. In the 1950s the etiology of the Minamata disease in Japan was identified to be the eating of fish and shellfish taken from Minamata Bay, which had received discharges of methyl mercury from a chemical plant. The problem of high mercury concentrations in aquatic food chains was also noticed in Sweden in the 1960s. In the early 1960s it was also recognized in Japan that cadmium poisoning (so called itai-itai disease) may result from eating rice irrigated by water containing suspended solids from a mining area. In the 1960s and 1970s growing attention was given to the organocholrine compounds and polynuclear aromatic hydrocarbons (PAH) in water and their accumulation in aquatic animals. (WHO 1972, 60–63; Waldbott 1973, 139–147, 161-165, 220–224)

The problems of water quality caused by the wastes from paper and pulp mills were internationally noticed already in the middle of the 1950s (Makkonen 1956). Although these problems were not recognized as important from a public health standpoint, their importance for those countries whose economy heavily depended on this industry was accepted. (WHO 1956)

The number of individual chemical substances or groups of chemical compounds dealt with in the international standards or guidelines for drinking water increased rapidly in the last decades of the 20th century (Figure 32.2.). In the first WHO *International Standards for Drinking-water* in 1958, there were only 15 chemicals listed, which were also included in the 125 fact sheets of chemicals or groups of chemicals in the third edition of the *Guidelines for Drinking-water Quality* in 2004 (WHO 2004, 296–460) That the amount of chemicals increased so much in 1971 was due to pesticides (Figure 32.2.). The phrase used in the fact sheets presenting the chemicals is *"(T)he 1971 International Standards suggested that pesticide residues that may occur in community water supplies make only minimal contribution to the total daily intake of pesticides for the population served".*

The year that a chemical or a group of chemicals was first mentioned in WHO *International Standards for Drinking-water* or *Guidelines for Drinking-water Quality* was normally not the year when a standard or guideline was established—most such standards and guidelines were established only years later.

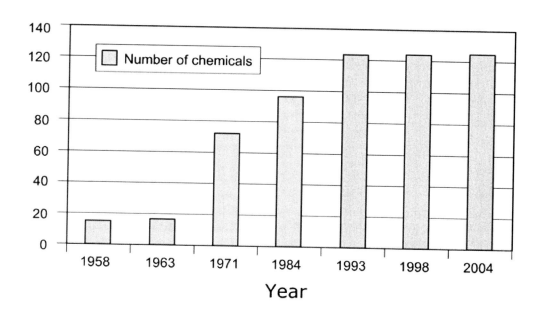

Figure 32.2 Number of chemicals in WHO International standards for drinking-water or in the Guidelines for Drinking-water Quality

In the third edition of the *Guidelines for Drinking-water Quality* published in 2004, a guideline value for 94 chemicals was presented; 25 chemicals were excluded because they are unlikely to occur in drinking water; and for 50 chemicals a guideline value was not established mostly because available data was inadequate or the chemical occurred in drinking water at concentrations well below those at which toxic effects may occur (WHO 2004, 488–493).

## 32.6. OBSTINATE CHEMICAL HEALTH HAZARD: LEAD IN WATER

Lead is a toxic substance and in drinking water it seldom if ever originates from natural sources. Lead was well known already in antiquity and because functional pipes can quite easily be made from lead, it was widely used in water supply systems. Lead was, however, considered to be hazardous to health already in antiquity and for that reason it was not a recommended material for water pipes, as clearly expressed, e.g., by Vitruvius (see Chapter five in this book).

Throughout centuries, evidence of toxic effects of lead accumulated but in spite of this many water supply distribution systems in the world were made of lead. After the Second World War it was recognized that the old lead pipes can expose people to elevated lead concentrations in water, but the major health threats of lead were identified to be occupational exposure and children's exposure to old plasters and paints containing lead. Replacement of the old lead pipes, which are still present in many old water supply systems in the world, has been considered expensive and economically unfeasible.

It is of no wonder that at the turn of the new millennia "*Owing to the decreasing use of lead-containing additives in petrol and lead-containing solder in the food processing industry, concentrations in air and food are declining, and intake from drinking-water constitutes a greater proportion of total intake. Lead is rarely present in the tap water as a result of its dissolution from natural sources; rather, its presence is primarily from household plumbing systems containing lead in pipes, solder, fittings or the service connections to homes. The amount of lead dissolved from the plumbing system depends on several factors, including pH, temperature, water hardness and standing time of water, with soft, acidic water being the most plumbosolvent.*" (WHO 2004, 392)

The health-based recommended maximum values for lead in drinking water changed upwards and downwards during the last half of the 20$^{th}$ century. The 1958 WHO *International Standards for Drinking-water* recommended a maximum allowable concentration of 0.1 mg/litre for lead, the 1963 International Standards lowered this to 0.05 mg/litre, the 1971 International Standards increased it to 0.1 mg/litre, the 1984 Guidelines lowered it again to 0.05 mg/litre and the 1991 Guidelines lowered the recommended value further to 0.01 mg/litre, which was also the guideline value in 2004. (WHO 2004, 392–394)

## 32.7. UNEXPECTED RESULT OF THE PROVISION OF SAFE DRINKING WATER: WIDESPREAD INTOXICATION BY ARSENIC

In some areas of the earth, drinking water has always contained toxic amounts of some elements because of the geology. Arsenic can occur in drinking water in high concentrations in some areas of the world, e.g. Argentina, Bangladesh, Chile, China, India, Mexico, Thailand and the United States of America. Besides acute poisoning, exposure to arsenic can cause serious long-term health effects, for instance, skin lesions, and skin and internal cancers. (www.who.int/topics/arsenic)

High amounts of arsenic have been found already in prehistoric mummies excavated on the north coast of Chile, and it caused widespread chronic arsenic poisoning in the same area still in the middle of the 20$^{th}$ century (Aufderheide & Rodríguez-Martín 1998, 321–322). By the 1970s, endemic arsenic poisoning was well recognized in Latin America and the Western Pacific, where drinking water could contain high concentrations of arsenic (WHO 1972, 59–60). Already in 1958, a maximum allowable concentration of 0.2 mg/litre was set for arsenic (WHO 1958, 27). This value was lowered to 0.05 mg/litre in 1963 and further to 0.01 mg/litre in 1993 (WHO 2004, 307).

The most serous public health problem arsenic has caused, however, is because modern technology was employed to get safe drinking water for people. People in the area of modern Bangladesh have used drinking water easily contaminated by pathogens probably from time immemorial. Tube wells have been installed since the 1940s and especially widely from the 1970s onwards *"to provide "pure water" to prevent morbidity and mortality from gastrointestinal disease"*. (Smith et al 2000). The clear and microbiologically safe water produced by the tube wells was not chemically analysed.

Arsenic-induced skin lesions began to be identified in patients coming from Bangladesh in the 1980s and arsenic contamination of water in tube wells was confirmed in 1993 (Smith et al 2000). The scale of the disaster was realized during the 1990s. *"It is estimated that of the 125 million inhabitants of Bangladesh between 35 million and 77 million are at risk of drinking (arsenic-) contaminated water."* (Smith et al 2000; www.who.int/topics/arsenic)

## 32.8. EMERGING HEALTH HAZARDS TRANSMITTED BY WATER

Towards the end of the 20$^{th}$ century, it was widely realized that health problems connected to water were not solved: the old biological hazards of water had not disappeared and many of them were re-emerging together with new ones that were emerging. A review of scientific literature identified 175 species of infectious organisms associated with diseases considered to be emerging (Taylor, Latham & Woolhouse 2001). The reasons for the worrisome situation can be grouped under four headings: new environments, new technologies, scientific advances, and changes in human behaviour and vulnerability (WHO 2003, 13).

The current climatic change (increasing temperatures, changing rainfall levels and winds) will influence various infections due to contamination of drinking water. Such infectious diseases as giardiasis and cryptosporidiosis would spread via contaminated drinking water particularly if extremes of rainfall and drought were to intensify (McMichael 2001, 303). There is also some evidence that the spread of cholera is facilitated by warmer coastal and estuarine waters associated with algal blooms (McMichael 2001, 303). A new biotype of the cholera vibrio was identified in 1961 and again in 1992, and both were connected to the spread of cholera around the globe.

Today there is a global shortage of potable water especially in the poor countries. *"From 5% to possibly 25% of global fresh water use exceeds long-term accessible supplies and is now met either through engineered water transfers or overdraft of groundwater supplies... Much of this water is used for irrigation, with irretrievable losses in water-scarce regions. All continents record such withdrawals. In the relatively dry region of North Africa and the Middle East up to 30% of all water use is unsustainable."* (WHO 2005, 17)

However, the problems with water are not limited to the developing world; new types of biological health hazards transmitted by water are also emerging in the highly industrialized Western countries. One of the problems is a protozoan genus called *Cryptosporidium parvum*. Cryptosporidia are not eliminated by the chlorination of water, and consequently large outbreaks have been reported from different parts of the Western world (Ford 2004). Cryptosporidia are not the only problem faced by modern societies; for instance, in recent years in Finland there have been several quite large waterborne epidemics caused by *Campylobacter jejuni* (*Tartuntataudit Suomessa 2001*, 11–12; *Tartuntataudit Suomessa 2002*, 12).

Many waterborne threats to human health are transmitted by parasites that need a host living in water or needing water in some part of its lifecycle. Mosquitoes spreading different types of malaria are highly dependent on water. Malaria has been a human health hazard from time immemorial and continues to be one of the most important diseases in the world, leading to some 1.3 million deaths each year (www.who.int/water_sanitation_health/publications/facts2004). This is despite all the efforts made to curb it during the 20$^{th}$ century. There were some at least temporary triumphs mostly in Europe, where DDT was widely used as an insecticide after the Second World War and, e.g., the Pontine Marshes in Italy were successfully drained already before the Second World War and thus the millennia-old breeding grounds of malaria-carrying mosquitoes destroyed.

Schistosomiasis has been, like malaria, a long-established water-related plague of humans at least from the beginning of irrigation in agriculture. It continues to wreak havoc among people. WHO has estimated that 160 million people are infected with schistosomiasis, which causes tens of thousands of deaths every year, mainly in Africa (www.who.int/water_sanitation_health/publications/facts2004).

Anthropogenic changes in ecosystems escalated during the 20$^{th}$ century. Many of the health hazards caused by the rapidly changing ecosystems in the late 20$^{th}$ and early 21$^{st}$ century world were conveyed by water. Malaria, Japanese encephalitis, schistosomiasis and cryptosporidiosis were among the infectious diseases likely to be highly sensitive to ecological change related to water (WHO 2005, 24).

## 32.9. CONCLUSIONS

At the beginning of the 20th century, it seemed that the health hazards caused by contaminated water could be avoided. However, during the century it was realized that the old waterborne health problems remained and new ones emerged. In an economically, politically and socially more and more incorporated world, international standards for an ever-growing number of microbiological and chemical elements in drinking water were introduced in the second half of the 20th century.

Poverty is intimately connected to the health problems caused by contaminated water. Most of the over one billion people who lack access to safe water supplies live in developing countries. However, waterborne health problems are not a monopoly of the developing world. Many health hazards are truly global. Most probably widespread anthropogenic ecological changes will cause serious waterborne health risks for a significant portion of the world population in the near future.

## 32.10. REFERENCES

Abbreviations: WHO = World Health Organization
Many of the documents of WHO are available in www.who.int/

Aufderheide AC & Rodríguez-Martin C. (1998) The Cambridge Encyclopedia of Human Paleopathology. Cambridge University Press, Cambridge.
Bruce F. E. (1969) Water supply, sanitation, and disposal of waste matter. In Hobson W. (Edited by) The Theory and Practice of Public Health. pp. 91-103, Oxford University Press, London.
Bulletin of World Health Organization 1962: 26: No. 4.
Cholera 2003 (2004). Weekly epidemiological record. 79, 281–288. (see also www.who.int/wer).
Ford B. J. (2004) Human behaviour and the changing pattern of disease. In Mascie-Taylor, N., Peters, J. and McGarvey, S. T. (Edited by). The Changing Face of Disease: Implications for Society. pp. 182–204, CRC Press, Boca Raton.
Makkonen O. (1956) Wastes from paper and pulp mills: a problem of the northern European countries, with special reference to Finland. Bulletin of World Health Organization 1956: 14: No. 5–6: 1079–1088.
McMichael T. (2001) Human frontiers, environments and disease. Past patterns, uncertain futures. Cambridge University Press, Cambridge.
Smith A. H., Lingas E. O. & Rahman M. (2000) Contamination of drinking-water by arsenic in Bangladesh: a public health emergency. Bulletin of World Health Organization 2000: 78: 1093–1103.
Tartuntataudit Suomessa 2001. Kansanterveyslaitoksen julkaisuja KTL B7/2002: 11–12.

Tartuntataudit Suomessa 2002. Kansanterveyslaitoksen julkaisuja KTL B8/2003: 12.

Taylor L. H., Latham S. M., Woolhouse M. E. J. (2001) Risk factors for human disease emergence. Phil. Trans. R. Soc. Lond. B 356: 983–989.

Waldbott G. L. (1973) Health effects of environmental pollutants. The C. V. Mosby Company, Saint Louis.

WHA24.47 Problems of the human environment. www.who.int/governance/

WHO (1956) Introduction. Bulletin of World Health Organization 1956: 14: No. 5–6: 839–843.

WHO (1958) International Standards for Drinking Water. World Health Organization, Geneva.

WHO (1961) European Standards for Drinking-Water. World Health Organization, Geneva.

WHO (1970) European Standards for Drinking-Water. Second Edition. World Health Organization, Geneva.

WHO (1972) Health hazards of the human environment. prepared by 100 specialists in 15 countries. World Health Organization, Geneva.

WHO (1981) Drinking-Water and Sanitation, 1981–1990: A Way to Health. World Health Organization, Geneva.

WHO (2003) Emerging Issues in Water and Infectious Disease. World Health Organization, Geneva.

WHO (2004) Guidelines for Drinking-water Quality. Third Edition. Volume 1, Recommendations. World Health Organization, Geneva.

WHO (2005) Ecosystems and human well-being: health synthesis: a report of the Millennium Ecosystem Assessment. Core writing team: Carlos Corvalan, Simon Hales, Anthony McMichael; extended writing team: Collin Butler et al. World Health Organization, Geneva.

WHO/UNICEF (2000) Global Water Supply and Sanitation Assessment 2000 Report. www.who.int/water_sanitation_health/monitoring/globalassess

WHO/UNICEF (2005) Water for Life: making it happen. World Health Organization, Geneva.

# 33
# CONCLUSIONS
*Petri S. Juuti, Tapio S. Katko & Heikki S. Vuorinen*

Figure 33.1 View from Dar es Salaam, Tanzania early 2000's. (Photo: Osmo Seppälä)

# PART III – CONCLUSIONS

The water and sanitation development in Kenya over the years has been multi-sectoral. While the state has played a leading role in policy formulation and implementation, initially water use and supply policy initiatives were *ad hoc* and later based on clear development plans. The dominance of the state in development led water to be regarded as a social good as opposed to an economic commodity. The political and administrative portfolio under which the water and sanitation development existed has constantly changed over the years.

In Tanzania, during the "German period" (1888-1916), all water development in the Buhungukira area, Tanzania was organised by the chief and headmen and focussed on man-made dams, while the colonial administration dreamed about pumping water from Lake Victoria to the dry areas in the Shinyanga region. Through the British indirect rule, the chiefs were supported in organizing huge labour-intensive interventions, for instance tsetse clearings in Buhungukira to provide land for the expanding human and cattle populations.

In the early 1930s, the drilling of boreholes was tried, but due to a series of failures, a return was made to man-made small dams. After WW II the administration introduced machinery and payment for labour, while towards the end of the colonial period (1888-1961), large-scale interventions faded away.

Following the villagisation programme of the independent country, donors supported large piped schemes to provide the new villages with water. Again, planners anticipated a grand solution with a grid of boreholes from which the new villages were to receive water. The failure rates were high and the cost for diesel to pump the water went up drastically due to the world oil crises. In the late 1970s, the strategy was changed into shallow wells dug by villagers and donor support for cement rings and hand pumps. Maintaining the pumps turned out to be a major problem, and soon the pumps were taken off and villagers drew water by buckets from the wells, although there was also an interesting development of village-level operated and maintained (VLOM) hand pumps. The century was concluded by villagers themselves being largely responsible for developing water sources and improving access.

In 1915 the first waterworks in Nigeria was completed at Iju on the outskirts of the Lagos city. For several decades supply was limited to the Island of Lagos, including the European settlements at Ikoyi. It was only from the 1950s that supply was extended to the mainland communities. Along with the growth of Lagos city, the increasing incapacity of the state has created a yawning gap in the water supply sector. This has been filled to some extent by private operators of various descriptions—mainly retailers of water: the "Hausa" porter, the water tanker driver and the sachet water producer and hawker.

Older neighbourhoods in the original municipality of Lagos are generally covered by the network of pipelines though most of them have become aged, leaky or damaged. The public water supply even in highbrow areas such as Ikoyi and Victoria Island is generally erratic, compelling residents to depend on water vendors for their supplies. The retail business is dominated by retired or retrenched workers and other persons who eke out a living, as water tanker drivers or by selling water to individual homes and for construction purposes. This is an indication of residents' potential ability and willingness to pay while the actual problem seems to be the lack of an operational institutional framework – the basis for the proper management of any water company.

The earliest sources of water supply throughout the world were springs and wells, while surface waters also had to be used in times of drought. In South Africa and Finland, the evolution of rural water supply was and still is largely based on cattle farming. The Finnish government started to support rural water supply through the first financial support act in 1951. This was preceded by the work of the Parliamentary Committee for the Rationalizing of Households, involving mainly female parliamentary members and other influential women. Women and children have usually been the ones to carry water in both case countries, South Africa and Finland, and also elsewhere. Especially in developing economies — such as Nepal (see Chapter 37 by Rautanen) — the role of women is of high importance.

Although the political development of Finland and Hungary are different, the development of environmental services had many similar elements.The history of Tampere and Miskolc shows how water-related problems emerged slowly and remained unrecognized for long by the public. Despite the different political, economic and social development of Finland and Hungary, roughly similar environmental problems appeared at the same time. When these problems started to affect everyday life, decision-makers adopted legislation in order to solve the emerging problems. (See chapter 25 by Juuti & Pál )

The first water pumping device in Riga, Latvia run by horses, was built in 1663 to take water from the River Daugava. In 1863 a new pumping station with steam engine plunger pumps was put in operation. In 1904 a groundwater intake station, located 20 km from Riga was taken into use. By 1912 installation of the sewer system within the centre of Riga Old Town was almost completed and in 1913 a sewer installation project was started on the western bank of the river Daugava. After World War II the industrial production of Riga grew along with the size of population. To meet the increasing water demand, artificial groundwater recharge was started in 1953.

By 1991, the year of regaining independence after the Soviet period, most of sewage in Riga was discharged untreated or partially treated into the river Daugava and further the Baltic Sea. The same year the first wastewater treatment plant of the city was put into operation. By 2005 Riga used a number of sources for its water supply, including surface water, natural groundwater and artificial recharge. During the past decade the water quality of inland waters in Latvia has begun to improve while other environmental problems still remain - one of them being the eutrophication of raw water sources.

The case study of a water history in an indigenous region in central-western Mexico reveals some common elements throughout the study period. The scarcity of water in the area has very long history and over the centuries the indigenous population have developed many strategies for water use and management that have allowed it to confront the problem. Their strategies are based on ecological principles and they include: social, community-based control of water that assures access to the entire population and the conservation of this resource; efficient water use and management that has given rise to rational forms of exploitation; diversified water use and management that has optimized the exploitation of all available sources of supply, including rainwater; and, multivalent water use and management that has permitted the development of a range of productive and domestic activities and maximized exploitation through recycling. Water has a sacred and divine character in the area and that has helped to cope with scarcity problem.

There are important similarities in the development of water supply and sanitation in Brazil and the Spanish American colonies, where the building of basic infrastructure was often the result of a mix of private and public initiatives. In both Brazil and Argentina, the public health consequences of rapid urban growth fuelled by the promotion of foreign immigration in the second half of the nineteenth century prompted an increased awareness among the elites of the interdependence between social and sanitary problems. In Brazil, each new epidemic helped to expose the vulnerability of the population to these diseases. In Buenos Aires, it was only after increasing public pressure that in 1854 the Municipality finally agreed that the provision of water supply services was a public duty.

The development of water and sewerage services in Argentina and Brazil during the early stages of this period was influenced by the general trend in the world economy whereby national states assumed a leading role in the expansion of basic infrastructure, including water supply and sanitation, a situation that was further accentuated by the economic crisis of the 1930s. The period from the mid-1960s to the early 1980s constituted a turbulent stage when both Argentina and Brazil were subject to oppressive military dictatorships.

In the 1980s Brazil and Argentina returned to electoral democracy, but in many respects the democratic experiences have been severely curtailed and the impact on the most vulnerable sectors of the population has been far from satisfactory.

The experience of water services in Brazil and Argentina since the 1990s provides important lessons. Argentina under President Menem became the pilot case of the privatisation policies promoted by the World Bank, to the point that the measures adopted by the federal government often went well beyond what was deemed acceptable to the international advisers. As a result, between 1991 and 1999 the proportion of the population served by privatised water and sewerage companies in Argentina jumped from 0 to 70 per cent. In contrast, in Brazil the strong mobilisation against privatisation significantly slowed down the implementation of these policies.

Unfortunately, the policies that have been implemented since the 1980s to tackle the crisis have been based on an uncritical revival of the privatist principles that inspired the short-lived experience of private water monopolies in the late nineteenth century. These policies of water service privatisation since the 1980s have completely ignored the lessons of history.

In Tokyo, Japan the Edo Water System (EWS), was used from the 17th to 19th centuries. As there was not enough water in the city, various methods were developed for a sufficient water supply. Drinking water was supplied in three ways: the Tamagawa and Kanda aqueducts, deep wells, and water vendors. In lowland and reclaimed land areas near the seashore, aqueducts were constructed. Water was taken from rivers or ponds and run down a long distance only by using the incline, then stored in a well with a bottom, where impurities were settled out. Excrement was stored in the toilet and brought to the fields as an excellent fertilizer.

Until the end of the 19th century, Tokyo did not suffer any serious problem with sanitation. The first outbreak of cholera occurred in 1858 and the second in 1882, following a world pandemic.

The EWS had been operated without electric power or water treatment technology for 300 years in the area of highest population density. Although this water system had some problems in its sanitary conditions, they were thought to be manageable if a reasonable population density was maintained, water stagnation prevented, and a reasonable distance between wells and toilets guaranteed.

Water has an important place in the traditional religion of Maori in New Zeeland and indigenous people in Mexico, which supports the sustainable use of water. The ethnographic case of Mâori shows how the traditional perceptions of water must be considered when utilising water resources. The case of Northern Chile shows how different cultures judge the problems of water management differently and how the Aymara Indians have been in repeated conflict with the Chilean State. Consequently the rivers have been depleted, the sloped terraces are abandoned and rural exodus is widespread. These cases imply that community and native views must be taken into consideration in environmental and water management.

In the middle of 20th century, it was considered that the remaining serious waterborne public health risk was chemical and radioactive contamination of water. However, during the late 20th century the biological hazards transmitted by water were again recognized. The overall amount of different biological and chemical health hazards transmitted by water increased manifold during the last half of the 20th century.

Although the greatest problems in water supply and sanitation are possibly linked to urban population growth, the majority of the world's population in 2006 still live in the countryside and rural centres

PART III – CONCLUSIONS

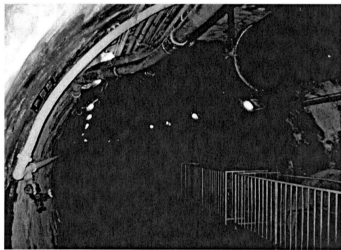

Figure 33.2 View from Paris sewers. Note oval shape in smaller sewer. (Photo: Petri Juuti)

# PART IV

# FUTURE CHALLENGES IN WATER SUPPLY AND SANITATION SERVICES AND ENVIRONMENTAL HEALTH

When the water of a place is bad it is safest to drink none that has not been filtered through either the berry of a grape, or else a tub of malt. These are the most reliable filters yet invented.
- Butler, Samuel

A cup of cold Adam from the next purling stream.
- Tom Brown, Works (vol. IV, p. 11)

Photo: Petri Juuti

# 34
## INTRODUCTION
*Petri S. Juuti, Tapio S. Katko & Heikki S. Vuorinen*

Figure 34.1 Old iron handpump in Tampere, Finland. (Photo: Petri Juuti)

# INTRODUCTION

Now, in the beginning of the 3rd millennium, there are already six billion humans on the planet. It has been estimated that some 1.5 billion of them lack safe water and over 2.5 billion adequate sanitation. It is estimated that by 2025 the global population is going to be around 8 billion and by the year 2050 the maximum population of around 9 billion will be achieved. This growth means a vast challenge to expand and improve water and sanitation systems. It is much easier to estimate and analyse the situation of water and sanitation services in slums than to provide these services to their dwellers. Suggested means include improved accountability, decentralizing assets, private sector participation, improving regulatory systems, reforming the water and sanitation sector, charging for services, and supporting client provision.

In addition to new investments, many cities in the world need to repair and replace outdated infrastructure. The most critical issue is that drinking water and sewer mains as well as many treatment plants are now far beyond their estimated design life and need replacement. Chapter 35 by Laurel Phoenix first gives examples of the effects of breaking mains or outdated systems on local citizens and economies in the United States. The economic repercussions and public and environmental health consequences of these events are discussed. Next, barriers to updating water infrastructure found at local, national, and international levels are reviewed and the importance of this problem for public health, the economy, and the future are discussed.

For many centuries, communities in southern Africa have had to learn how to cope with drought conditions. Since the 1870s, bustling urban centres of industrial activity, commerce and trade have mushroomed. The 1980s brought severe drought conditions that affected the whole of southern Africa. In the final decade of the twentieth century, the conventional approach to water restrictions changed when the introduction of water demand management strategies produced remarkable results. Chapter 36 by Johann Tempellhoff describes how urban societies in South Africa have managed with water shortages in the late 20th century.

Sanna-Leena Rautanen describes in Chapter 37 how water supply and sanitation services have developed in Nepal. The Kingdom of Nepal is a remote Himalayan country with immense diversity in both geographical features and culture and until the 1950s, Nepal was practically isolated from the outside world. Even if the country has been catching up by many human development indicators, it still is one of the poorest countries in the world.

It is estimated that one-third of the world's urban population lives in slums. Regardless of the characterization of slums, their dwellers face higher challenges such as higher morbidity and infant mortality rates than either non-slum dwellers or the rural population.

While definitions of slums and squatter areas may vary from region to region and from city to city, the term is used in this context to describe a wide range of low-income settlements and/or poor human living conditions. Thus slums include both deteriorated and informal housing areas. The water supply and sanitation services of slums are described in Chapter 38 by Hukka and Seppälä.

# 35
# AGING AMERICAN URBAN WATER INFRASTRUCTURE

*Laurel E. Phoenix*

Figure 35.1 Key places mentioned in the chapter.

## 35.1. GENERAL BACKGROUND

Many cities need to repair and replace aging infrastructure, which is difficult and costly in an already built environment. Aging water and sewer infrastructure needs upgrading to fix leaky systems, replace outdated plants, increase capacity to address higher demand, respond to increasing source water pollution requiring more treatment, comply with more stringent 1972 Federal Water Pollution Control Act (otherwise known as the Clean Water Act (CWA)) and 1974 Safe Drinking Water Act (SDWA) regulations, and increase water security. The age of the infrastructure is the most important issue. Many large cities, including New York, Washington, D.C., Baltimore, Philadelphia, New Orleans, St. Louis, Chicago, and Los Angeles, still use drinking water mains that are 90 to 150 years old.

There are now over 700,000 miles of drinking water pipes in the United States (AWWA), many of which need replacement over the next 20 years. The 600,000 – 800,000 miles (USGAO 2004) of sewer pipes face the same problem. Moreover, many treatment plants are now beyond their estimated design life and need replacement. Infrastructure has aged as the country's population has grown. In 1900, the U.S. population was 76 million, it is now 296 million (as of May, 2005), and is predicted by the U.S. Census to grow to over 363.5 million by 2030. This chapter first gives examples of the effects of breaking mains or outdated systems in urban areas on citizens, the economy, and the greater environment. Barriers to updating water infrastructure at the local, national, and international levels are then reviewed. Finally, what some cities are doing to confront this problem, and its import for public health, the economy, and the future are discussed.

## 35.2. BRIEF BACKGROUND

Cities first built their own public water supply systems, and later added sewers. First attempts to provide drinking water in cities were initiated by private businesses to serve only the wealthier portion of the population, and were eventually replaced by public systems obligated to provide water to every city resident. Wastewater disposal was originally a citizen's responsibility, evolving from outhouse to septic pond before cities laid sewer pipes to move the effluent away from buildings and into the nearest waterbody. Where possible, cities drew water from upstream and dumped sewage downstream. Each city had no feeling of responsibility for the sewage it was dumping that a downstream city would have to use for drinking water. Eventually, some cities started treating their effluent, and later the CWA provided massive funding in the 1970s for cities with no sewage treatment to finally build treatment plants, or cities with primary treatment to upgrade to secondary treatment. The SDWA provided revolving funds distributed as grants or low-interest loans.

Recently, problems with water and wastewater infrastructure are increasing as more pipes leak each year and increased demand for water causes cities to pump water at a higher pressure through the pipes, stressing them further. Raw water is continually being degraded by too many discharges into waterways (legal and illegal) as well as inputs from polluted runoff, making it more difficult for water treatment plants to provide safe drinking water. Compliance with new stormwater regulations requires significant investment in laying stormwater pipes separated from sanitary sewer pipes and then treating the stormwater. New federal requirements to improve water security to prevent terrorist attacks are another financial demand on cities. Cities in the United States now need to replace their aging water and wastewater infrastructure over the next 20 years, and will be hard-pressed to afford it amongst so many competing needs.

## 35.3. CURRENT EVENTS HIGHLIGHTING AGING INFRASTRUCTURE

### When pipes break

Pipes break because of age, composition, the substrate in which they are placed, and the stresses they endure. Water infrastructure shows its age in several ways. Because cities are still using some of the original pipes laid during their systems' initial construction, it is common for drinking water or sewage mains to burst under the streets. Composition in conjunction with age results in pipes with different estimated life spans. Pipe vulnerability to cracks is also influenced by the substrate in which the pipes are laid (e.g., thixotropic soils that "liquefy" in response to earthquakes, areas with high water tables) and local weather (e.g., some sediments are more prone to expansion and contraction with freeze/thaw events). Vibrations from railroad tracks or roads can also stress pipes. Cross contamination can result when a crack in a drinking water pipe results in lower pressure and then sucks in contamination from leaking sewer pipes or other underground pollutants such as a chemical plume from a factory spill or leaking underground tanks.

The propensity of pipes to break is compounded by not always knowing where pipes are laid, as no records or maps were initially produced. This makes it harder for construction or utility crews to avoid accidentally puncturing pipes, much less find and fix leaks. For example, a Washington, D.C. crew digging to fix a drinking water pipe unknowingly punctured a sewer pipe that then spewed raw sewage into a residential basement on two occasions. Lacking maps, crews spent several days digging up landscaping and the street searching for the leak. (Fisher 2004)

**Breaking mains**

Like sewer mains, bursting water mains often cause a spectacular mess. Sinkholes in roads can appear, like the 43-foot wide sinkhole on Chicago's Lake Shore Drive that engulfed several cars and flooded the road with millions of gallons of water in 2002. (ICIJ 2003) A New York City sinkhole that opened in January of 1998 on 5th Avenue was roughly 35 feet deep, swallowing a car and rupturing a gas line that sent flames two stories high. (Associated Press 1998) Several streets were under a foot of water for days, electricity was cut off, street and subway transportation was halted, and business suffered. Basements were flooded, threatening circuit-breaker boxes and boilers; forcing 50 residents to evacuate. This extraordinary damage was due to a 128-year-old, 48-inch water main breaking. That size of pipe usually breaks about 6 times a year in the city, but New York City has an average of 550 line breaks per year along its 6,048 miles of pipes. (Kifner 1999) Another 48-inch main bursting on the Upper East Side first looked like river rapids coming down the street, and then waves moved down store aisles over sinking floors. (Lipton 1999) More recently this same neighborhood had a 15-inch sewer pipe and 12-inch water main break together, creating a sinkhole, leaving 400 residents without water, and flooding basements, elevator shafts and nearby businesses. (Urbina 2004) A 119-year-old, 10-inch cast iron pipe burst in 1999 near City Hall, flooding and closing streets and threatening to undermine and topple a 15-story-tall construction crane. (Kifner 1999) In Baltimore, the drinking water diverted from a broken 36-inch main then burst the 54-inch main to which it was transferred. (Vozzella 2004) These examples typify the suspension of economic activity, transportation, energy, and water access caused by deteriorating infrastructure and illuminate the threat to public welfare.

**Drinking water problems**

Cities are not always sure of the composition of long-buried pipes, making water treatment all the more difficult. They may be hollowed-out logs, cast-iron, ductile iron, lead, copper, concrete, ceramic impregnated with asbestos, and polyvinylchloride (PVC). Pipe composition can affect tap water quality. For instance, lead poisoning from drinking water pipes has long been a concern. From the early 1900s through the Depression, lead pipes were commonly used for service lines (this is the name of the pipe connecting the water main under the street to the plumbing of each building). Lead was again used during WWII when copper was scarce. (Cohn 2004) The internal plumbing in many older homes also is made of lead. Even pipes not made of lead can contaminate water if soldered with lead solder, and half-tin and half-lead solder was used until it was banned in 1987. Valves and water meters can be made of brass, lead, and other metals. Brass fittings, even when labeled "lead free", can legally have up to 7% lead. This means that even a new house with no lead pipes or solder can still have lead leaching into the tap water.

Treatment plants try to prevent lead leaching by adjusting water chemistry. Often, they will add something like orthophosphate, which reduces the water's acidity, but later adds extra phosphate to water entering wastewater treatment plants. This forces wastewater plants to either spend more money to remove the phosphate, or exceed their discharge permits[1] for the amount of phosphorus allowed in their effluent. Too much of this biological nutrient promotes eutrophication and subsequent degradation of the receiving water.

A 1991 regulation to switch from using chlorine to chloramines is thought to explain why lead levels in Washington, D.C.'s drinking water jumped sharply in the last few years, affecting thousands of homes. This has renewed municipal interest in old lead pipes and solder in the many cities that still have lead mains or service lines. Several cities such as Boston, (Ebbert & Massey 2004) Seattle, and Los Angeles are now testing their water for lead levels to see if the Washington, D.C. problem may be theirs as well. Washington, D.C. has added orthophosphate to the water supply to reduce lead level readings in home tap water, although it is expected to take a full year for the orthophosphate to coat the interior of the pipes and reduce the lead levels. Cities with lead pipes have always manipulated water chemistry to mitigate the presence of lead pipes, since they do not have the money to replace them. However, since the source waters for every city have their own unique chemistry, and each city has a different mix of pipe materials, solutions for one city may not work for another. The lead pipe problem is of special significance as a public health threat because of the neurological damage it can cause to fetuses, babies, and small children. Despite the well-known dangers of lead for these cohorts, the cost of replacing lead service lines is so great that many cities have covered up high lead levels in reports so the Environmental Protection Agency (EPA) would not force them to replace lines. (Leonnig & Nakamura 2004) Washington, D.C. had tried to cover up as well, but was exposed by the local press. It is chilling to know that local public agencies charged with providing safe drinking water are the very ones who knowingly deliver toxins.

Aside from pipe composition problems and unethical water treatment plant administrators, a relatively mundane problem for cities is spending money to treat water only to lose a good deal of it in transit because of leaks. For example, Buffalo, New York loses close to 25% of its treated water through its 150-year old cracked pipes, but cannot afford to search for and repair them. (*Rusty, leaky, localities...* 2003)

Although lead pipes and leaky pipes are ubiquitous, problems stemming from these are far less dangerous than the more typical and frequent contamination events around the country. The most infamous event was the Cryptosporidium contamination of the water supply in Milwaukee, Wisconsin in 1993. A filter failure caused 400,000 to be sickened and roughly 100 people died. Other contamination events happen sporadically, usually after rain or snowmelt events. The increased turbidity and associated first flush of fertilizers or sewage sludge (spread on fields) contaminates groundwater and surface water with E. coli and associated bacteria.

---

[1] The National Pollutant Discharge Elimination System (NPDES) is a program set up to fulfill the Clean Water Act goals of eliminating water pollution by requiring every effluent pipe into a lake or river to obtain a permit limiting the type and quantity of pollutants discharged.

**Wastewater problems**

Older sewers may be made of brick, concrete, or clay. Sewers typically have a 50–75 year life span, which is now up for most sewers and far overdue for some. These sewer channels or pipes are now failing under a variety of stresses. Cracks can form in pipes brittle with age or cheap materials, or from added pressure of traffic, buildings, or freeze/thaw events. New construction as well as illegal connections from existing foundation tiling, sump pumps, roof drains, and manhole covers, called inflow, add a heavier sewage load that increases pressure in the pipes and use up capacity that could otherwise service new growth.

Sewer pipes can also corrode. Higher-strength wastewater has a variety of chemicals that can corrode pipes, and even hydrogen sulfide gas given off human waste eats away at concrete pipes and manhole access points. Iron sewer pipes in very large systems, often oversized to accommodate stormwater flows, undergo even more corrosion when flows are low and more oxygen contacts interior surfaces. In coastal areas, pipes can be corroded from the outside if in contact with salt water.

Roots can then push through cracks and corrosion points and thus reduce system capacity by partially blocking pipes. This type of groundwater intrusion, called infiltration, can also enter the sewage pipes and add to their sewage load.

Besides sewer mains leaking or bursting and causing problems on city streets, sewer overflows are increasingly polluting local waterways. Sanitary sewer overflows (SSOs) and combined sewer overflows (CSOs) are a significant problem for older cities that built storm sewers and sanitary sewers as one system. During periods of heavy rain, the additional water load overwhelms a sewage treatment plant's capacity, necessitating the release of raw sewage uphill from the plant in order to avoid backups into the city. This direct release into a waterbody is the combined sewer overflow. The pathogens in raw sewage combined with the "first flush" of urban nonpoint source pollutants (e.g., oil, grease, metals, animal droppings, sediment, etc.) pollute the waters into which they are dumped. Compliance with federal regulations to address both CSOs and the treatment of storm sewer runoff is very expensive, and cities still struggle to find funding.

To avoid the costs of updating wastewater systems to eliminate CSOs, municipal wastewater plants have lobbied the EPA for sewage "blending." Blending allows the diversion of untreated sewage around the secondary treatment systems, and then the untreated sewage can be "blended" with fully treated sewage and released into waterways. This would allow viruses, parasites and helminthes to be discharged into the waters. EPA is crafting new rules for blending. It is considering this despite its knowledge of the extent and the consequences, as evidenced by its 2004 report "Report to Congress: Impacts of Control of CSOs and SSOs." (United States Environmental Protection Agency 2004) The EPA has already noted that 850 billion gallons of untreated stormwater and wastewater are released each year from CSOs, and between three to ten billion gallons of untreated wastewater are released each year from SSOs.

Another capacity problem is generated when municipalities grow too fast. Many areas do not have concurrency ordinances, which require that the necessary public facilities and services are available before new buildings can be built. Without concurrency ordinances, far too many permits can be given to build without ensuring that the local water or wastewater plants have the capacity to serve this new development. Growth moratoriums can result. In Soledad, California, the city council declared a building moratorium when they learned that the sewage treatment plant was already near capacity. (Times Wire Reports 2004) But for Centreville, it was Maryland's Department of the Environment that declared a building moratorium as Centreville kept approving building permits when its current wastewater plant was often exceeding its capacity and when new permits being considered far exceeded the capacity of the new wastewater plant that was not even operational yet. (Guy 2004) If Centreville had had a concurrency ordinance, new building permits would not have been issued without the local utility agreeing there was enough wastewater capacity to serve them.

An interesting problem related to municipal growth is that as new pipes and new plants may get installed to attract or service new growth, aging pipes in the original parts of town remain ignored. Depending on the layout of the system, this could strain the older pipes to bursting and burden the original residents with the "costs" of subsidizing new growth.

In some areas, the public is responding to aging infrastructure by suing utilities. Because cities avoid expenditures on repair and updating and the EPA has not enforced the law, infuriated citizens and downstream/downcurrent cities are turning to lawsuits to force municipalities to fix their systems. A recent example is a successful lawsuit against the City of Los Angeles initiated by Santa Monica Baykeeper. The city must now spend $1.7 billion repairing or replacing sewer lines (at least 488 miles of lines must be replaced) and $300 million on other control and maintenance issues. (McGreevy 2004)

## 35.4. CONSEQUENCES OF AGING INFRASTRUCTURE

The current state of water infrastructure has serious negative impacts. Public health is threatened. Pipe materials can create serious health threats, either because their composition (e.g., lead) leaches dangerous substances into the water supply, or because their age and composition result in numerous cracks and breaks that can either contaminate treated drinking water during delivery or nearby waterbodies. Cross contamination is a good example of how cracks in subsurface pipes allow treated drinking water to be contaminated before it gets to the tap. Since there is no way to test for this type of contamination, it is not discovered until an unusual number of people enter emergency rooms with gastrointestinal problems. Decreased pressure from cracked pipes can also threaten buildings if pressure is not high enough for firefighting purposes. Aside from pipes breaking, antiquated sewer systems are not treating all of the wastewater. Every sewer overflow event contaminates nearby waterbodies and endangers the people who frequent them. Raw sewage contains numerous pathogens, including bacteria (e.g., *E. coli, Salmonella typhi, Vibrio cholera, Shigella,* and *Yersinia*), viruses (e.g., Adenovirus, Astrovirus, Noraviruses, Enterovirus, Rotavirus, Poliovirus, and Hepatitis A), protozoa (e.g., *Cryptosporidium parvum, Giardia lambia, and Entamoeba*) and helminthes (flukes, tapeworms and roundworms). There are more beach closings and shellfish warnings each year in the United States because of these contamination events, and more people sickened from swimming in the water or eating shellfish. (Dorfman 2004)

Deteriorating infrastructure also harms the local and national economy. The suspension of economic activity and vital services with every water or sewer main break that floods streets lowers productivity and income. The potential economic value of properties can be lost if properties become devalued because they are near combined sewer overflows. Future economic activity may not be realized if combined sewer overflows prevent the revitalization of urban waterfronts.

Pipes made of dangerous materials or older pipes breaking more often results in health problems or economic disruptions. This draws cities inexorably closer to a needed infrastructure overhaul before the cost of repairing breaks or replacing lead pipes starts exceeding the cost of planned replacement at the end of a pipe's expected lifetime.

If these costs grow at a faster pace than public funding can be obtained, some cities may consider consolidating and/or privatizing their water or sewer systems. Consolidation offers obvious per capita savings but may have political disadvantages for communities that view neighboring communities as their economic competitors. Private companies claim to have the financial might to invest millions for the necessary upgrade, which may offer savings. However, if the choice of privatizing is forced upon cities because of crisis, this places cities in a weak position for negotiating long-term contract specifics that influence quality or cost of service. In consequence, this also weakens the local economy.

## 35.5. WHAT WILL IT COST TO UPDATE WATER INFRASTRUCTURE?

As of 2001, the combined annual shortfall for capital investments and operations for water and wastewater systems was predicted to be $32-$34 billion. (Water Infrastructure Network) The United States General Accounting Office (GAO) in their water infrastructure report compiled investment predictions from the Congressional Budget Office (CBO), EPA and Water Infrastructure Network (WIN). (USGAO 2004) These figures are estimates in billions of dollars for costs to update and expand water infrastructure between 2000 and 2019 (Table 35.1)

Thus, up to $1 trillion is needed over the next 20 years for water and wastewater upgrade and expansion. This would include water and wastewater treatment systems, pipelines, storage, and expansion for growing populations.

Table 35.1 The estimates in billions of dollars for costs to update and expand water infrastructure between 2000 and 2019. (Source: USGAO 2004, 13)

| Prediction | CBO | EPA | WIN |
|---|---|---|---|
| Low estimate | $ 492 | $ 580 | |
| High estimate | $ 820 | $1194 | |
| Single estimate | | | $ 940 |

The CBO had predicted that the costs to utility customers over this same time period would raise their rates from 0.5% of their household income in the late 1990s to between 0.6% and 0.9% of their household income. (Congressional Budget Office 2002) This relatively small increase could be easily accommodated by the vast majority of customers, but since they could not pay it up front, utilities would still need to receive revolving funds or other large loans to update infrastructure first and then recoup their costs slowly from increased consumer rates over this 20-year period. These predicted rate increases were compared to rates that utility customers pay in other industrialized countries, and were easily within those ranges. In fact, households in Belgium, Sweden, Spain, and Denmark pay from 144% to 160% more than US households, and Finland, France, England and the Netherlands pay over 200% more than US households.

From a local utility perspective, the costs to replace aging pipes of various compositions and ages tend to increase at a somewhat even rate over time, but the expenditures required for replacing treatment plant components tend to rise sharply every 25-40 years. (AWWA 2001, 16) Consequently, the overall local costs for upgrading subsurface mains and treatments plants rise and fall sharply over time. The American Water Works Association (AWWA) produced a graph for the projected total replacement expenditures due to wear-out for the Bridgeport, Connecticut water facility (Figure 35.2). Projection graphs for other facilities around the country will look different depending on the age of their pipes and plants. The graph illustrates not only the increase in costs over 50 years, but the extreme differences in costs for certain years. These graphs have been called "Nessie Curves," since the sharp waves of treatment plant expenditures look like the dorsal fins of the mythical Loch Ness monster in Scotland. The value of a graph like this is to prepare utility managers and their cities to know how much money they will need by which years, so they can begin to look at financing options.

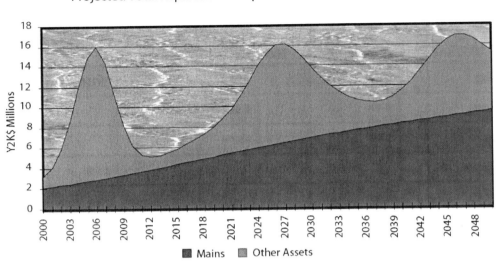

Figure 35.2 A graph for the projected total replacement expenditures due to wear-out for the Bridgeport, Connecticut water facility, 2000-2048. (Source: Reprinted from Dawn of the Replacement Era, by permission. Copyright © 2001, American Water Works Association.)

## 35.6. BARRIERS TO UPDATING WATER INFRASTRUCTURE

Examining how different levels of government view the problem, and how industries avoid or take advantage of the problem, can highlight the tensions between initiative, responsibility, technology, and funding. This section will first survey the many local level factors that delay updating infrastructure and then discuss the federal perspective. The international influence of nongovernmental organizations and their support of transnational water corporations will then be covered to explain why those same corporations then have increasing influence over federal and local governments.

### Local influences

"Water is by far the most capital intensive of all utility services, mostly due to the costs of these pipes". (AWWA 2001, 10) Massive capital projects face an uphill battle to obtain funding for a multitude of reasons. First, not only do citizens never think about water infrastructure, they have no personal memory of waterborne diseases since those were common many generations ago. Thus they assume that the waterborne diseases of the distant past will remain so. They also have a lifetime history of water and sewer bills being their cheapest utility bills. If no one remembers paying the initial costs of laying the pipe network, paying high utility bills for water, or having to suffer from waterborne illnesses, this combines with invisible subsurface infrastructure to result in no public saliency. Researchers have discussed (Levin et al. 2002) the difficulty for public water systems to set rates high enough to maintain their systems. Besides middle class flight from big cities, they mention the downward pressure on rates exerted by bill payers, who may also be voters. They add that limited political terms influence local politicians to avoid tackling projects whose benefits will not become obvious until after the politician is out of office.

City officials have ignored these "out of sight, out of mind" problems for decades to deal instead with other problems more obvious and salient to the public. Local officials must always prioritize expenditures to allocate funds amongst competing needs. Because of this, maintenance and rehabilitation of sewer and drinking water pipes is the first thing to go during tight fiscal times. No officials want to make hard decisions to repair or update infrastructure until they have public support for those expenditures. Although one may argue that a politician's relatively short term in office, typically 2-4 years, prevents them from understanding the seriousness of aging infrastructure so they will act to replace it, the hidden nature of subsurface infrastructure coupled with the massive investment required would make it an unpopular issue to broach to the public even from the most seasoned politician.

The public will not support large capital investments until they are convinced of the need by suffering themselves from main breaks, boil orders, or other problems. The public agitates for aboveground issues of which they are more aware and from which they will receive immediate perceived benefit, such as road repair, a new school, or another police officer hired. Politicians do not mind this, as they like their accomplishments simple, visible and perhaps, memorialized.

A city may have used historic water revenues to pay for other, more visible public services. Before the SDWA and its attendant costs, it was common for cities to use their water supply as a "cash cow" to help fund other city services. For example, San Francisco used its extra revenue from Hetch Hetchy (the dam built in the Sierra Nevada mountains to create their water supply) water to pay for other services, and now is facing up to eight billion dollars in seismic improvements to the century-old Hetch Hetchy delivery system, which now serves 2.3 million people. (Epstein 2000) Many cities in the country still combine their revenue from many sources into a common "pool" and then disburse funds from that pool to whichever projects or services most demand their attention. This prevents an accurate accounting of water infrastructure income relative to necessary expenditures and makes it easier to shift water-related revenue away from needed repairs.

City revenues can decrease for many reasons. Cities can lose population and its associated revenue base. Suburban and exurban growth outside of the city lure some citizens to suburbs, so the higher tax base leaves the city. This growth also lures businesses to suburbs, so the business tax base moves to suburbs as well. This results in "stranded capacity," or not enough income from current customers to cover capital facilities costs. (AWWA 2001) This creates a domino effect and leaves the city short of the critical number and variety of businesses sufficient to support the downtown—starting the city's slide downward.

In an effort to create jobs to replace those exiting the city with businesses, some cities offer subsidies to corporations to encourage business to locate there. These subsidies are large, must be paid for with local taxes, and may not recoup the jobs lost. Some corporations may leave only a few years after having accepted the subsidies and moved in, because they can now make more money moving away to some other city or country. Consequently, money is spent luring new businesses to the area instead of maintaining local infrastructure for current users. This problem is exacerbated by global economic competition, which makes it all the harder for cities to spend funds on infrastructure repair when they see their own economic development as so critical. Another popular economic development gamble cities have recently invested in the building of new stadia to attract sports teams and their related revenue streams. Although Washington, D.C. is facing a $300 million cost to replace lead pipes that deliver drinking water, (Nakamura 2004a) it has voted on building a stadium to bring in the Expos baseball team at a cost of $584 – $614 million in public funds. (Nakamura 2004b)

Revenue can also be lost from the industrial sector. Cities used to receive significantly more revenues from local industries than they do now. Transnational corporations have no incentive to invest in local infrastructure, since they have investments scattered across the world and have the flexibility to leave if infrastructure declines. Industries my also insist that the city upgrade infrastructure or the corporation will leave the city and/or state and take needed jobs with it. (Ordonez) Thus, although industry benefits from municipal infrastructure, it no longer wants to help pay for it.

Sufficient revenue can also be elusive due to new state laws. In several states, city and state governments are constrained by state law that limits the revenue the cities or state may receive. California and Colorado have initiated these taxpayer bill of rights laws, and now are suffering from too few revenues to support state and local government services. Under these laws, the only way for cities or the state to increase revenue is by putting it to a vote. Citizens tend to vote down requests for increased taxes or fees unless they are already feeling the pain of the problem and have personally recognized the need for the investment.

## Federal influences

Public water and wastewater and associated public health issues have traditionally been considered a local government responsibility. This is evidenced by the trend in federal spending. Congress (Figure 35.3) provided little money for cities to upgrade their infrastructure before the SDWA or the CWA, when Congress then funded about 75% of the cost. This unusually high funding was in response to public outcry over polluted waters. Despite Congress having

---

[2] The Water and Sewer Authority (WASA) of Washington, D.C. was forced to take on a new employee with a public health background to serve as an authority on the public health ramifications of any WASA decisions.

Figure 35.3 Water tower - a visible part of the otherwise hidden water infrastructure (Photo: T. Katko)

current information on the extent of the problem, the necessity for federal funding, (USGAO 2004; USGAO 2002) and the shortfalls on infrastructure investment, (AWWA 2004) federal funding is now down to only 5%. Part of this reduction can be attributed to the devolutionary movement of the last 25 years, where the federal government has turned over more social programs to the states while leaving them under-funded.

Perspective is important. Federal government officials tend to see the problem as a local infrastructure problem, and some water treatment administrators[2] see it as merely an engineering problem, without relating it to the larger issues of public health and urban economies. These issues pit civil engineers against health officials against politicians, and the public suffers.

Lobbying by various interest groups also weakens Congress' will to act. To reduce costs, municipal and private wastewater systems lobby Congress to weaken regulations, such as asking for the "blended sewage" provision. In contrast, to weaken municipalities and force them toward privatizing, the major international water and wastewater companies contribute to campaigns and lobby to reduce federal aid to cities to fund infrastructure improvements, among other things (see next section for more details). Even when some funding is given,

Congress is weighed down by considerations of how small private utilities should be treated relative to public utilities. For example, should they both get equal subsidies?

Greater corporate influence in government makes it difficult for the public interest to be realized. It is now harder to get Congress to regulate water polluters or utilities, and harder still to get agencies in the Executive branch such as the EPA to rigorously implement or enforce the existing regulations. Plaintiffs expecting higher courts to always uphold the law are surprised by conservative judges taking the side of business and stressing the costs to business.

**International influences**

Transnational corporations (TNCs) that build and manage water and wastewater systems are growing in influence as they push for privatization at international, national, and local levels. At the international level, the World Bank, the International Monetary Fund (IMF), and the Inter-American Development Bank (IADB) aid TNCs. These organizations can pressure poor countries into privatizing their public utilities as a condition for getting loans or receiving debt relief. (Grusky 2004) Water corporations then receive contracts with guaranteed high profits, and have the option of suing a country if they fail to make enough money or something else goes wrong with the privatization agreement in any city. This profitable expansion into many countries further empowers corporations to look for more business, especially in the United States, where infrastructure is failing and the public can afford to pay more for water.

At the national level, TNCs and their trade association, the National Association of Water Companies, are increasingly sending campaign contributions to key officials on congressional committees that pertain to water utilities, tax reform, or tort reform. These groups lobby Congress to limit federal aid to cities for infrastructure improvement, so cities will be forced to turn to privatization to access enough capital funding. (Public Citizen) They also ask for limitations on a local government's ability to take over a private company through eminent domain. The water industry wants reduced liability from lawsuits if people get sick from their drinking water. They also lobby for the same federal funding or low-interest loans as public utilities can get so the corporations do not have to risk as much of their own assets. Lobbyists ask for a change in the federal tax code so that customer connection fees would not be considered taxable income to the corporation. All this lobbying helps these companies gain a stronger foothold at the local level, avoid responsibility, or increase their net profits.

TNCs buy up their American competition to reduce their operating costs through "efficiency" and also to eliminate the competition. In some cases, buying a new water company can then make the foreign transnational look like an American company. For example, German RWE bought the British company Thames, which then bought American Waterworks. This gives a local community in the United States the naïve perception that if they are going to privatize, they will feel safer if foreigners do not own the company.

Local-level strategies of TNCs involve going into a community that looks promising for privatization and developing a "champion" for their cause. (ICIJ 2003) This local ally will either be a local official or well-known civic personality. They teach this person how to help sell the idea of privatizing the local water or wastewater utility, and shower them with gifts and money. Others are recruited to go to local planning meetings to speak in favor of privatization or to write favorable letters to the editor. Newspapers are approached to write articles on the benefits of privatization. Collectively, all these strategies help to convince locals that turning to privatization will be cheaper and produce higher quality than their public utility could ever achieve.

**A Plethora of Barriers**
Therefore, among competing needs, water and wastewater infrastructure is given low priority at the local level by citizens for whom the problem does not exist; and by politicians for whom the solution is not an easy, popular, or quick fix. Local revenue is reduced when local industries ask for lower, wholesale utility rates to prevent them from leaving town and taking their jobs with them. The federal government views water and wastewater infrastructure as a local issue and consequently is not interested in funding repair or replacement. Powerful interest groups lobby Congress to attend to short-term private interests instead of the long-term public interest. Multinational water corporations gain power through their lucrative contracts granted by or because of international nongovernmental entities that wield great power over the world's poorer countries. Such power allows them to have inordinate influence over federal officials in the legislative and executive branch to further pit local and federal interests against each other. This weakens local utilities further, making more of them vulnerable to the tactics water corporations use at the local level.

There is little initiative at any government level to attack this problem of aging infrastructure, but much initiative on the part of multinational corporations to help drive a wedge between cities with infrastructure problems and the potential federal funding they could receive. The history of who is responsible for providing local public water and wastewater services obscures the relevance of adequate infrastructure to public health and economies outside of each service area. Without considering these broader impacts, Congress can continue to consider infrastructure merely a local problem and not feel responsible for funding it. Lack of technology is not a problem, but merely obtaining the funding to purchase it. Indeed, much of the infrastructure needing replacement is not highly technological (e.g., subsurface pipes) but merely expensive to dig up and replace.

## 35.7. HOW ARE CITIES TRYING TO ADDRESS THEIR INFRASTRUCTURE PROBLEMS?

Some cities are taking advantage of new technology. A relatively low-tech solution is manipulating water chemistry in drinking water to avoid leaching lead and copper from pipes. For subsurface repairs of cracked sewer pipes, they are now using *in situ* repairs that do not require ripping up large portions of streets. These trenchless technologies allow for lining pipes to fill cracks or pipe bursting to replace pipes without ripping up the landscape. To replace old treatment plants, some coastal cities are considering combining the upgrade of their water supply with building desalting plants to address increased water demand. In California, where receiving additional water from the Colorado River is no longer an option, over 20 desalting facilities are being considered, despite significant public concern over private companies (German RWE or Connecticut's Poseidon) building and running these operations. (Boxall 2004) Years ago during a California drought, the wealthy community of Santa Barbara built its own desalting plant and only runs it when water from other sources is unavailable. In Massachusetts, several cities are studying the possibilities of desalting plants, but are concerned about the pros and cons of possible privatization, through contracting with Aquaria or other private developers. (Daley 2004)

Upgrading water treatment plants also gives some communities the chance to consider regional collaboration to build one large plant rather than several small ones. Regionalization through utility consolidation can save money on customer service, billing costs, and water testing costs. The Tri-Town Board previously mentioned found consolidation to offer savings on several fronts. (Redd 2004) Motivated by poor source water, seven small towns in Montana are collaborating to find a cleaner and more abundant source of water. (Associated Press State and Local Wire 2004)

Cities must also consider various ownership, management, and funding solutions. Cities faced with infrastructure problems have the option of privatizing their public utility, accessing enough funds through loans or bonds to do the infrastructure upgrades themselves, or taking public control of a private utility if the private utility has not been doing a good job. There are numerous examples of cities that have privatized their water, wastewater, or both. Some have just wanted their systems upgraded and managed, while others have wanted to increase their water supplies through finding additional water sources or increasing plant capacity. A few cities that have recently privatized their systems have later cancelled their contracts, as Atlanta did with United Water (owned by the French conglomerate Suez). New Orleans was considering a water and sewer contract with US Filter (owned by French Veolia, formerly Vivendi) for several years but then reversed its position.

For a city to afford to update its own utilities without turning to privatization is hard, since various means of generating revenue are all difficult. Because Americans are habituated to expect low water and sewer charges, any move to increase rates is met with intense opposition. A local public will often go into debt willingly to build a sports stadium, but will rarely support additional debt for hidden infrastructure. However, in cities where either water supply or quality is becoming a public concern, they are able to move away from flat fees (e.g., you pay one rate no matter how much water you used) or fees per gallon (e.g., you pay the same amount for each increment of 1000 gallons you use) to increasing block rates (e.g., each subsequent block of 1000 gallons of water will cost you more than the preceding block – block rates also known as "inverted cost structure"). Making people pay for what they actually use either increases revenue for the city or increases conservation by the water users.

There are examples of cities that have successfully maintained control over their own systems. In reassessing what they could do to become more efficient, several cities have improved their public utilities and saved millions of dollars. (Public Citizen 2002) These programs in comprehensive asset management can typically realize a 20–25% savings for a utility. Phoenix, Arizona saved $77 million over 5 years in their water and wastewater treatment systems. San Diego, California will save $78 million over six years in their wastewater department. Miami-Dade Water and Sewer in Florida saved $52 million in four years.

Cities have difficult decisions to make on how to spend limited revenue. Certainly, spot repairs and water chemistry are merely short-term fixes for old infrastructure. Long-term fixes start with saving/collecting money through better management, more efficient rate structures, and consideration of regionalizing services. Using hi-tech solutions for water supply problems like desalting is a long-term solution that can continue to grow with a population. Privatization can be a long-term solution if a contract can be crafted that ensures both parties a fair return.

## 35.8. CONCLUSIONS: DESIGNING A FUTURE

More than just a commodity, water is an economic and social good. This places responsibility for its management and oversight in the public sphere. However, water management and water policy have always been piecemeal affairs in the United States. National water policy analysts have long charged that American water policy is fragmented. (Davies & Mazurek 1998; Deyle 1995) This fragmentation is endemic to a policy system of incremental decision-making and numerous agencies at all levels of government having some water-related responsibilities, whether they concern quality, quantity, distribution, or treatment. Combining this fragmentation with the current weakening of the Clean Water Act and reduced EPA enforcement under the Bush administration, it is no wonder water quality is declining. In light of this, one realizes that for Congress to assume that cities should be primarily responsible for infrastructure costs ignores two major points. First, most cities cannot control the quality of the raw water their treatment plants must purify. The quality of this raw water is influenced by federal and state water quality regulations and land use regulations (or lack thereof) of other local governments. When sufficient regulation is lacking, resulting in poor quality source water, water treatment plants must either raise their treatment costs and bill their customers more, or serve their public substandard water. Second, urban infrastructure across the country supports the entire economy, making this a national and not just a local issue.

The initiative to attack this problem must come from the public if they want their elected officials to fulfill their responsibility. Who, indeed, does the governing? Clearly, it is the responsibility of citizens to oversee their government and demand that it act in the public interest.

The public interest not only includes the current US population, but its future residents as well. The population is expected to grow from its current 296 million to 420 million by 2050, (U.S. Census Bureau 2004) with the greater portion of this growth due to immigration. Adding approximately 120 million people to the US population in only 45 years is unprecedented. What are the ramifications of this for urban areas? First, cities will need new water sources. Purchasing and accessing these additional sources will increase the costs of water service. Second, supplying this greater demand for water either means that bigger mains are needed or more pressure is needed to deliver through the smaller existing mains, ultimately stressing pipes further and causing more pipeline breaks. Third, there will be a greater need for efficiency. These higher costs mean cities will not be able to afford line leaks of treated water, so further investment in replacing pipes will also mean water bills will rise. Rising water bills would not be detrimental for most residents, but a portion of any city will consist of residents too poor to pay higher water and sewer bills. Newly arrived immigrants will likely swell these ranks further. As cities will need increased revenue to pay for rising costs, setting rates to recoup their expenditures must include a human right-to-water-minimum, the cost of which will not rise. Otherwise this turns an infrastructure problem into a social equity and subsequent public health problem. And finally, more water demand in any urban area subsequently produces more sewage. Adding more of this degraded water to local waters means the concentration of pollutants rises in these waters. This would then burden nearby or downstream water treatment plants, *unless* money was invested into mandatory tertiary treatment for all wastewater plants. Otherwise, either water treatment costs increase even more, or public health declines due to substandard drinking water.

Undoubtedly, the provision of safe drinking water and removal and treatment of wastewater is critical for maintaining public health. The infrastructure problems cities currently confront will only get more serious as infrastructure ages and populations grow. The infrastructure problem is tightly linked with regulations intended to protect ambient water quality or drinking water. Updating infrastructure and water quality regulation are far more important issues than the public or Congress have recognized. Once rate payers start acting like citizens rather than just consumers, they will force federal government officials to provide funding while local utilities do their part by managing assets efficiently and charging utility customers in ways that encourage water conservation. Citizens becoming more involved and incensed over the threat to public health can then make sure that federal government officials on down to local treatment plant officials do their jobs with the primary goal of protecting public health rather than their budgets. Because water infrastructure is in its "replacement era" (AWWA 2001) phase, cities can no longer patch up systems with short-term fixes. It is time to make long-term investments because one way or another, the public will pay dearly.

## 35.9. REFERENCES

Associated Press (1998) Water Main Break Floods Parts of Manhattan. [Online at CNN] The Associated Press, January 2, 1998.

Associated Press State and Local Wire (2004) Towns in six Montana counties seek regional water system. Billings Gazette, October 17, 2004.

AWWA. Straight Talk On Drinking Water Infrastructure. http://www.awwa.org/Advocacy/StraightTalk/docs/Straight%20Talk%20on%20Infrastructure.pdf

AWWA (2001) Dawn of the Replacement Era: Reinvesting in Drinking Water Infrastructure. May. Washington, D.C.: The AWWA Water Industry Technical Action Fund.

AWWA (2004) to US House Subcommittee: Funding gap for drinking water infrastructure is real and is big. April 29, 2004. Water Quality & Environment News, www.waterchat.com.

Boxall B. (2004) A Wave of Desalination Proposals. Los Angeles Times, March 14, 2004.

Cohn D. (2004) District residents seek tests for lead. The Washington Post, February 4, 2004, B05.

Congressional Budget Office (2002) Future Investment in Drinking Water and Wastewater Infrastructure, November 2002.

Daley B. (2004) Thirsty Towns Consider Desalting Plants, Boston Globe, April 10, 2004. A1.

Davies J. C. & Mazurek J. (1998) Pollution Control in the United States: Evaluating the System. Washington, D.C.: Resources for the Future;

Deyle R.E. (1995) Integrated Water Management: Contending with Garbage Can Decisionmaking in Organized Anarchies. Water Resources Bulletin 31(3), 387-398.

Dorfman M. (2004) Swimming in Sewage. Natural Resources Defense Council.

Ebbert S. & Massey J. (2004) Lead levels in water high in 10 locales. The Boston Globe, April 29, 2004, A01.

Epstein E. (2000) Price estimate doubles for fixing Hetch Hetchy. S.F. mayor seeks ways to cover $8 billion in repairs, San Francisco Chronicle, November 15, 2000.

Fisher M. (2004) Somebody Needs to Draw WASA a Map. Washington Post, May 20, 2004, B01.

Grusky S. (2004) The IMF, the World Bank and the Global Water Companies: A Shared Agenda. In Will the World Bank Back Down? Water Privatization in a Climate of Global Protest. International Water Working Group.

Guy G. (2004) Town's sewage problems mount, Baltimore Sun, May 7, 2004, B01.

ICIJ, International Consortium of Investigative Journalists (2003) Low rates, needed repairs lure 'Big Water' to Uncle Sam's plumbing. http://www.enn.com/news/2003-02-14/s_2665.asp.

Kifner J. (1999) Water Main Bursts Near City Hall, Complicating Commute Home. New York Times, November 11, 1999.

Leonnig C.D. & Nakamura D. (2004) Several U.S. Utilities Being Investigated for Lead, Washington Post, October 14, 2004, B01.

Levin R.B., Epstein P.R., Ford T.E., Harrington W., Olson E. & Reichard E.G. (2002) U.S. Drinking Water Challenges in the Twenty-First Century, Environmental Health Perspectives 110, Supplement 1, February.

Lipton E. (1999) Ground Opens to Flood Upper East Side. New York Times, December 30, 1999.

McGreevy P. (2004) LA to replace sewage pipes, Los Angeles Times, August 6, 2004, B01.

Nakamura D. (2004a) WASA to Replace 2,800 Lead Pipes over Next Year, Washington Post, November 13, 2004, B01.

Nakamura D. (2004b) Stadium Back on Calendar for Vote, Washington Post, November 18, 2004, B01.

Ordonez F. Town Presses State for Sewer Funds, Cites Threat by Firm to Relocate, Boston Globe, Globe West, 1.

Public Citizen (2002) Public-public partnerships. Public Citizen's Critical Mass Energy and Environment Program, Nov. 21, 2002. http://www.citizen.org/documents/sparklingppps.pdf

Public Citizen. Turning Up the Tap: How the Private Water Industry Wants to Boost Profits – At the Expense of Taxpayers. http://www.citizen.org/documents/Fact%20Sheet%20-%20corporate%20wish%20list%20-%20PDF.pdf , accessed on May 28, 2004.

Reckless Abandon: How the Bush Administration is Exposing America's Waters to Harm (2004) Earth Justice, National Wildlife Federation, Natural Resources Defense Council, and Sierra Club. August 2004.

Redd C.K. (2004) Water board moves toward regional plant, Boston Globe, April 4, 2004.

Rusty, leaky, localities deal with old water systems (2003) http://www.win-water.org/wine_news/040103article.html

Times Wire Reports (2004) Building Moratorium Likely to be Extended, Los Angeles Times, November 9, 2004, B07.

United States Environmental Protection Agency (2004) Report to Congress: Impacts and Control of CSOs and SSOs. August, 2004. EPA 833-R-04-001.

Urbina I. (2004) Upper East Side Block Closed by Main Breaks, New York Times, April 9, 2004.

U.S. Census Bureau (2004) U.S. Interim Projections by Age, Sex, Race, and Hispanic Origin. http://www.census.gov/ipc/www/usinterimproj/natprojtab01a.pdf

USGAO, United States General Accounting Office (2002) Water Infrastructure: Information on Financing, Capital Planning, and Privatization. Report to Congressional Requesters. August 2002. GAO-02-764.

USGAO (2004) Water Infrastructure: Comprehensive Asset Management Has Potential to Help Utilities Better Identify Needs and Plan Future Investments. Report to the Ranking Minority Member, Committee on Environment and Public Worlds, U.S. Senate. March 2004. GAO-04-461.

Vozzella L. (2004) A Flood of Broken Mains. Baltimore Sun, April 8, 2004.

Water Infrastructure Network. http://www.win-water.org/

# 36

# FROM WATER RESTRICTIONS TO WATER DEMAND MANAGEMENT: RAND WATER AND WATER SHORTAGES ON THE SOUTH AFRICAN URBAN LANDSCAPE (1983-2003)[1]

*Johann W. N. Tempelhoff*

Figure 36.1 Locations of Pretoria and Johannesburg, South Africa.

---

[1] Aspects of this study were first dealt with in Tempelhoff 2003b.

## 36.1. GENERAL BACKGROUND

In southern Africa droughts are a frequent occurrence. The region stands out on the global rainfall map for the fact that its annual rainfall is well below the world average. Furthermore, the South African landscape varies from an arid to semi-arid landscape in the west, to a fairly well-watered region along the eastern and southern seaboard. For many centuries communities resident in southern Africa have had to learn how to cope with drought conditions. Sparse populations scattered over large tracts of land were, up to a century ago, a common feature of the southern African landscape. Since the 1870s bustling urban centres of industrial activity, commerce and trade have mushroomed. This was a direct result of the discoveries of diamonds (1871) and gold (1886) in the interior of South Africa. Increasing population and rapid population growth had the effect that the problem of drought had a less direct impact on people resident in the urban centres – providing the water supply systems were able to cope with demand. If water was in short supply, stringent measures were introduced by local authorities.

By the second half of the twentieth century, South Africans in the urban areas of the Highveld, a grassland region spanning across large parts of the Gauteng Province and Free State, became familiar with intermittent water restrictions in times of drought. The 1980s was noted for the severe drought conditions that affected the whole of southern Africa. (Figure 36.2) The neighbouring states of Angola, Botswana, Lesotho, Zambia and Zimbabwe had lost almost one third of their agricultural output, as result of drought conditions. (Vogel 1994, 18)

Also the economically active nucleus of South Africa, which is today known as the Gauteng Province, was affected by the drought conditions. This region is primarily an urban landscape, with scatterings of rural areas. Cities such as Johannesburg, Pretoria, Germiston, Springs, Roodepoort, Vereeniging, Sandton, Midrand and Randburg, were at the time some of the most populous (primarily white) cities in this region.

When the effects of the drought conditions finally had an impact on the urban areas, water restrictions had to be introduced. Particularly suburban gardeners, nurseries, car wash concerns, hotels and industries that consumed large amounts of water, were hard hit by the measures. The drought conditions were also harsh for people of colour who, until 1994, had to contend with discrimination in terms of the apartheid policies of the former government. These policies had, since 1948, permeated the structures of governance that discriminatory practices in respect of black water consumers, were the order of the day in the integrated management of water supplies, from the local municipalities, to the black townships. Simultaneously, urbanisation was taking place at a rapid rate and it was difficult for the authorities to keep pace with growth in demand on many service levels.

In the final decade of the twentieth century, the conventional approach to water restrictions changed. The introduction of specific water demand management strategies, which had been put to the test internationally (Anon. 2001a, 12-3; Abderrahman 2000, 465-473), produced remarkable results. Instead of merely promoting awareness, direct actions were taken by the authorities to manage water scarcity. Moreover, consumers irrespective of race and creed

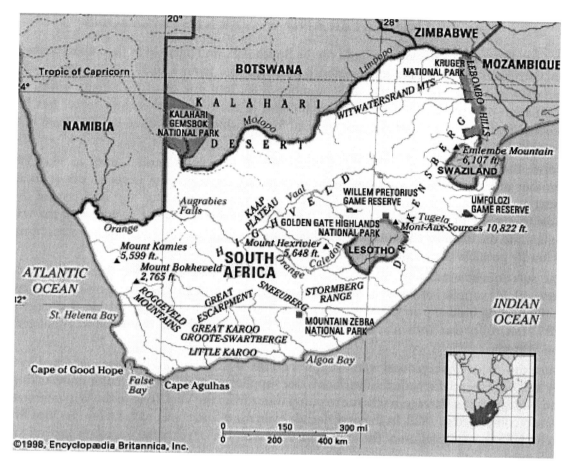

Figure 36.2 Southern Africa. (Source: Encyclopaedia Brittanica 2002).

were made familiar with basic principles of conservation that acted as an incentive for water conservation.

In the discussion to follow we will consider the manner in which Rand Water, one of the major bulk water suppliers in South Africa, in the final quarter of the twentieth century changed from basic strategies of water restrictions to water demand management in the Gauteng region.

In particular the focus will be on the drought periods 1983-7 and 1995 by explaining how water restrictions were introduced in the supply area of Rand Water. Then attention will be given to public responses to the awareness campaign to save water in both instances. Finally, attention will be given to the long-term strategies aimed at making consumers more careful of water wastage.

## 36.2. RAND WATER AND WATER RESTRICTIONS

Ever since the major discoveries of gold on the South African Highveld in the 1880s, water has been a scarce commodity. As Johannesburg mushroomed from a small mining town, with a population of about 250 residents in September 1886, into one of Africa's major cities, with a population of more than two million by the end of the twentieth century, water has been an endearing theme. Its availability was crucial for securing the sustained development of the economically active region.

Before the establishment of the Rand Water Board (currently Rand Water) in May 1903, a number of private companies had been responsible for supplying water to domestic and industrial consumers on the Witwatersrand. The private sector water suppliers proved to be unable to cope with the growing demand. Moreover, as a result of the pollution of the water supplies and the inability to contend with the problem outside formal governance structures, it was hardly possible to procure sufficient supplies of potable water.

The subsequent establishment of a public water utility had one major objective: it had to secure reliable supplies of sound quality water in large quantities for use by domestic consumers and the mining industry on the Witwatersrand. This proved to be a formidable challenge (up to 1993 the utility was known as the Rand Water Board. In that year the name was changed to Rand Water.).

Once the major underground water supplies on the Witwatersrand had been depleted by the early 1900, the search for sufficient surface water supplies started. The first major undertaking of Rand Water was the Barrage, a scheme to pump water from the Vaal River to the Witwatersrand. This was completed in 1923. In the next decade even more water was added from the Vaal River system, with the completion of the Vaal dam. For a number of years the measures proved to be satisfactory.

At the start of the 1950s it was possible for board chairman, A.S. van Lingen, to note with a sense pride that since 1927 the Rand Water Board had never needed to introduce water restrictions. (Minutes 1950-1951, 165-6) This apparent state of euphoria was short-lived. By the early 1960s there simply were not sufficient water supplies for the Witwatersrand. A number of factors contributed to this state of affairs. (For more on this see, Tempelhoff & Haarhoff 2003)

Firstly, large parts of South Africa were enveloped in what was considered, at the time, as the worst drought conditions since the Great Depression in the 1930s. Secondly, the South African economy was in a phase of unsurpassed growth (the boom conditions were similar to trends in other parts of the world where capitalist societies were quick to recover. *See* Hobsbawm 1996, Ch. 9. South Africa's economic growth was affected by the Sharpeville (1960) uprising, but conditions improved after 1963. See Natrass 1981, 25-26, 28-29; and Scher 1993, 414-418; Also see Andrews et al. 1965, 121) Thirdly, in the wake of the economic growth, the process of urbanisation had taken place at such a rapid rate that it was not possible to meet the growing demand for more water supplies. The consequence was that Rand Water, between 1960 and 1974, simply had to contend with rapidly diminishing water supplies. (Gardiner [s.d], 7; 31/S. See, for example, a number of press releases issued by the board between 1966-70) As a bulk supplier of water to local authorities and industries on the Witwatersrand (Figure 36.3), elementary strategies were introduced in collaboration with the central government's department of water affairs to cope with the crisis of insufficient water supplies.

The early measures aimed at conserving the available supply were fairly crude. Consumers, particularly households, were asked to limit their consumption by abstaining from filling or topping up, for example, swimming pools (Gardiner [s.d], 7) and watering gardens. (TA TRB2/1/691 File 124/11/4 Vol. 1.) However, industrial complexes were allowed to continue with their usual water consumption. Their operations were considered to be in the interest of the economy. (31/S. Notice to all consumers; Anon. 1966) Sometimes tariffs were increased to pay for more expensive water schemes. But again the authorities were careful not to stand in the way of the large industrial consumers and their production processes.

By the 1970s water restrictions had become a way of life as cyclical drought conditions and rapid growth continuously put the available water supplies of Rand Water to the test. The state had been instrumental in locating sufficient water supplies and undertook the imaginative Tukhela-Vaal water supply scheme in the 1970s. In the 1980s construction work started on the Lesotho Highland Water Project. These projects were intended to meet the demand for more water in the Witewatersrand region until well into the 21$^{st}$ century.

Figure 36.3 Rand Water area of supply. (Source: Rand Water)

## 36.3. THE DROUGHT OF 1983

In the 1980s drought conditions forced Rand Water to introduce restrictions – this time more comprehensively than ever before. (See Department of water affairs) At the start of 1983 the Vaal dam was only 44 per cent full. (Anon. 1983a) The drought appeared to be widespread. Reports from the Northern Transvaal and the homelands in the Eastern Cape suggested that the drought was affecting the whole country. (Anon. 1983j; Anon. 1983k) However, it was particularly in the industrialised and most densely populated urban region of South Africa – the Witwatersrand – that the drought was most serious. It was for all intents and purposes a drought that affected both the agricultural economy of South Africa as well as the hydrological infrastructure serving the urban areas.

A number of factors shaped the macro environment under which the Witwatersrand, the larger part of what is today Gauteng, had to contend with the drought conditions.

Firstly there had been a considerable shift in consumption patterns. For example, after the Soweto uprisings (1976-7), black South Africans resident in the townships of the Witwatersrand had started using more water. The townships were being developed. Massive projects were undertaken to erase the housing backlog. Many of the new houses were provided with sophisticated sanitary equipment such as taps and toilets were standard facilities. Previously most of these facilities had been located outside black residential structures. As the black South Africans in the urban centres nurtured a new sense of civic pride, they started with the beautification of their environment. Gardening – a high-consumption domestic activity – was one of the pastimes to which they resorted. (Anon. 1978; See Tempelhoff 2003a, ch. 9 &10)

Internationally South Africa had been politically isolated. The defence force was involved in a bush war on the country's borders. Local industries had the task of manufacturing most of the arms and ammunition necessary to conduct extensive military campaigns. Water was a prerequisite for the industrial manufacturing processes. What's more was that South Africa's economy, at the time, was in a phase of spiralling inflation. It was thus not possible to summarily push up water tariffs in a punitive attempt at keeping consumption low. In the political arena the political dimensions of water restrictions were also exploited to the full as a new racially conservative parliamentary opposition grouping, the Conservative Party (CP), came into existence in 1982. For the first time since 1948, the Afrikaner nationalist government had to deal with an official rightwing opposition. (Giliomee 2003, 603-9) Also the Progressive Reform Party (primarily comprising English speaking and liberal-minded South Africans) (Giliomee 2003, 610-1) frequently, especially on the level of local politics, joined the fray on excessive water tariff increases. Making consumers pay more for water thus proved to be a potential political minefield for local authorities as well as Rand Water, as the bulk water-supplying utility.

## 36.4. THE CRISIS OF THE THREE DAY COLLAPSE

The significance of a potential water shortage was brought home in a very direct manner in mid-January 1983 when a total of 17 suburbs in Johannesburg were left high and dry after two major power failures, lasting three days (Anon. 1983e; Municipal reporter, *The Star* 1983a; Anon. 1983g), disrupted the Rand Water's distribution system. (McNamara 1983b; Anon. 1983d; Minnie 1983a; Anon. 1983b; Anon. 1983c) Overnight the available water supply to the Witwatersrand dropped by 30 per cent. (McNamara 1983b) People were angry at the authorities for not informing them of the causes of the breakdown. (*Rand Daily Mail* 1983)

Information on precisely what was responsible for the breakdown was withheld from the media. The national electricity supply utility, Eskom, later indicated that one of its power supply lines had overheated. Another contributing factor was Rand Water's pumping operations, which were running at capacity because of the hot and dry weather conditions. (Minnie 1983b) Then, when the power break came, the emergency power supply was insufficient to keep the Zuikerbosch pumping station of Rand Water in operation. (Anon. 1983f) After the power had been restored, a spokesperson for the board told the journalists that although Zuikerbosch was in operation again, it could take a considerable period of time before the reservoirs would be filled properly. Three reservoirs of Johannesburg, Hursthill, Brixton and Northcliff, had dropped to well below their minimum levels. In an effort to contend with the crisis, members of the South African defence force, civil defence groups and workers of the council used 11 tankers to supply water to domestic consumers. (McNamara 1983b; Minutes 1982-3, 581)

In the months to follow, local authorities on the Witwatersrand introduced water restrictions. (McNamara 1983b; Anon. 1982a) The first measures were not severe. It was assumed that the drought would be broken in due course. (Krüger 1983) The drought conditions did not end. As the crisis deepened, the government stepped in with the minister of the environment, Sarel Hayward, stating in parliament that the whole region served by the Vaal River would be subject to water restrictions. (Anon. 1983h; Anon. 1983i; Minutes 1982-3, 581: 1) Water consumption had to be cut by 20 percent over the corresponding period for 1982. Taken in conjunction with the average population growth of 5.2 per cent per annum on the Witwatersrand at the time, it implied that savings had to be made in excess of 27 per cent on the average consumption. (79th annual report, 2)

## 36.5. PUBLIC RESPONSES

At first the public participated in open debate in the media. There were those who felt the proposed fines to be imposed were insignificant. Officials acknowledged it and agreed that fines should be increased substantially. (Anon. 1982b) There were also those who felt themselves to be 'victims' of the restrictions. They alleged that the local authorities were wasting too much water as a result of rusted water pipelines. (*Die Valderland* 1983; *The Star* 1983b)

By mid-March 1983, the first wide-ranging measures were introduced by all the local authorities that relied on Rand Water for their bulk water supplies. (Anon. 1983l; Anon. 1983m) From the outset, it was clear that not all consumers would accept the measures aimed at reducing water consumption. Wealthy property owners in the residential suburbs of Sandton, north of Johannesburg, simply did not care to pay more for water. Councillor J.F. Oberholzer, chairman of Johannesburg city council's management committee, who also served on the board of Rand Water, at the outset warned against this state of affairs. He was in favour of strict measures against transgressors. More expensive water, at the time, appeared to be some sort of a solution to the problem. (Anon. 1983o; *Beeld* 1983a) At one stage he even propagated jail sentences for water misuse. (Anon. 1983p; Minnie 1983c) Shortly afterwards Rand Water's chairman, Dale Hobbs, explained in the media that strict measures were to be introduced. He gave consumers a brief period to prepare for the proposed measures. (McQuillan 1983a) The media supported the savings drive and dedicated a lot of editorial space to water-saving strategies. (Anon. 1983r.; Anon. 1983t; Anon. 1982c; Raine 1983)

With the level of the Vaal dam at the lowest since 1966, the management of Rand Water became concerned that the crisis could deepen. The levels of most dams of the department of water affairs, close to the Witwatersrand, were low. The Hartebeestpoort dam, west of Pretoria, was only 36 per cent full. The Loskop dam, northeast of Pretoria, had 21 per cent of its capacity. The Sterkfontein dam southeast of Johannesburg was 69 per cent full. (McQuillan 1983a) Many popular local holiday resorts, *inter alia* the Jim Fouché resort on the Free State side of the Vaal dam, were closed to visitors. (Anon. 1983n)

Towards the end of 1983 the first steps were taken to introduce sliding scale tariffs. It implied that consumers would be charged higher monthly tariffs when they exceeded limits laid down by the local authorities. (Bath 2003) The percentage savings in consumption were determined by the department of water affairs. Then Rand Water passed the information on to the local authorities. They, in turn, could decide how they wished to achieve the proposed savings. Not all resorted to water restrictions. Pretoria city council, for example, informed its consumers that they had to use water sparingly (Anon. 1983s. The politically dangerous situation created with the sliding-scale tariffs later became evident when a number of office and apartment blocks in Pretoria were reported to be facing municipal water accounts of tens of thousands of Rands. See *The Star* 1984a). Other local authorities, such as Sandton (*The Star* 1983c), and Vanderbijlpark (Anon. 1983v), conformed by maintaining water restrictions.

However, by the time a sliding-scale system of water tariffs came into effect, local authorities were geared for it. (Minutes 1983-4, 212: Extraordinary meeting) In the public, as well as the private sector, there were meanwhile numerous projects aimed at finding alternative water supplies (at the beginning of 1983 the board had also started pumping water from the Betty Shaft of Cornelia Mine. Minutes 1982-3, 613:1). Increasingly the trend was to start making use of underground sources. Domestic consumers in some towns and cities started hiring contractors to drill boreholes in private suburban gardens. It became fashionable, and an economically viable proposition, for suburban households to start making use of boreholes for gardening purposes. The borehole sinking and fitting industry soon flourished. (Carlisle 1989)

## 36.6. TIDE TURNS AGAINST CONTINUED RESTRICTIONS

By 1984 the tide started turning against continued water restrictions. Despite fairly good summer rains, consumers were told that water restrictions had to be kept in place. (Anon. 1983u; Anon. 1983w; Anon. 1983x; Van Heerden 1983; Minnie 1983d; Wessels 1983) As the summer of 1984 approached, consumers became critical of the clampdown on the use of hosepipes in gardens. (Clarke 1984b) One critic was *The Star* newspaper's James Clarke, who was of the opinion that for many months people had carried around buckets instead of using their hosepipes. In the process they had saved merely a small amount of the Vaal dam's water. (Clarke 1984a) Clarke spoke for many consumers. In what was by that time described as the 'decriminalisation' of water consumption, the Sandton Town Council in September 1984 unanimously agreed to scrap all water restrictions. However, residents were warned not to squander water. They were to pay severe penalties for abusing the privilege. Residents using 20 kilolitres (kℓ) would pay 38 cents per kℓ. If they used more than 300 kℓ per month they could pay as much as R2 per kℓ. (Howell 1984. For the purposes of this discussion 1Euro= R8.00. Typically R2 would then amount to Euro 0.25. In 2005 the average exchange rate was 1 Euro = R8, or $1=R6. Ed.)

The unilateral action by Sandton drew an angry response. Rand Water decided to clamp down on domestic wastage and warned consumers that if they used water for more than two hours per week, an additional levy over and above the basic charge would be levied. (Commitee of the whole board meeting 1984, 96) J.F. Oberholzer of Johannesburg's city council was perturbed by the arrogance of the 'mink and manure set' that were under the impression that 'they can buy anything – they think money is the only thing that counts'. (Gault 1984; *See* also later criticism in Bond 2002, 127-128)

At the time the residents of Sandton were daily using on average 618 ℓ *per capita*, whereas other residents in other areas consumed about 245 ℓ. (Gault 1984) Following the example set by Sandton, Bedfordview, Boksburg and also Pretoria lifted bans on the use of hosepipes. (Unwin 1984a) Others, such as Randburg, chose to follow a conservative route by complying with the call for a ban on hoses. (Stein 1984b; Anon. 1984j)

The war of words soon turned into a party political conflict. (*The Star* 1984b; Ryan 1984; Anon. 1954; Spoormaker 1984) Ultimately it led to the resignation of one of the town councillors of Sandton, who had driven the campaign of resistance against what he considered to be unreasonable water savings demands. (Anon. 1984e) The government at first appeared to be intent on maintaining water savings. (Clarke 1984c; Unwin 1984b; Anon. 1984f) Then the stringent measures were gradually eased and some local authorities, such as the city council of Johannesburg, towards the end of 1984, announced that residents could start making limited use of hosepipes. (Stein 1984a; Anon. 1984g; Goosen 1984) They were, however, reminded that they would have to be prepared to pay for excessive water consumption. At the time some municipalities were already facing severe penalties for over-consumption. (Anon. 1984h; Anon. 1984i) There were indications, by December 1984, that after many months of restrictions the board's consumers were beginning to comprehend what it meant to save water. (Anon. 1984l)

As the drought entered its third year in 1985, the department of water affairs required consumption to be confined to 70 per cent of the average at the time the drought originally started. The board's customers – 47 local authorities and more than 800 other large consumers of water – were then subjected to even more severe restriction. (80th annual report, 3) The levels of the Vaal dam had meanwhile dropped further. (Anon. 1985; Clarke 1984e; Anon. 1984k; Beattie 1984; Wessels 1985)

Consumer groups were up in arms as local authorities pushed up water rates. (*Beeld* 1983b; Anon., 1984d; *The Citizen* 1984; Anon. 1984a) Then the experts started criticising the government. Dr Bob Laburn, a former chief engineer of Rand Water, on receiving a prestigious award for service to the engineering profession, stated that shortages of cash and bad government planning made a substantial contribution to the prevailing water crisis. In terms of the size of the national budget surprisingly little money was being spent on the strategic resource of water. Laburn was of the opinion that the state engineers and those engineers in the employ of the country's water boards had to join forces by pooling their skills and knowledge. It then had to be their responsibility to try and find some solutions. (McQuillan 1984)

On 16 February 1985 the government responded with the prime minister, P.W. Botha, making a call in parliament for all South Africans to participate in a national day of prayer for rain and relief. The date was scheduled for 22 February. In total 16 church denominations in South Africa responded. (*Die Vaderland* 1984) According to some newspaper reports, the rain started falling in some places as people gathered in churches to pray. (Anon. 1984b; Anon. 1984c)

## 36.7. THE RESPONSES OF INDUSTRY & PUBLIC AWARENESS

Since the imposition of water restrictions in March 1983, there had been frequent reports of large industrial concerns using innovative measures to save water. Not only were they showing a sense of social responsibility, they were in fact working on measures aimed at saving substantial money.

In the Western Transvaal (currently Northwest Province), industries in the vicinity of Stilfontein managed to cut consumption of water by as much as 26 per cent. (Anon. 1983q) In the Vaal Triangle several of the large industrial concerns reported that they had started recycling water. (Herselman 1983) The electricity utility, Eskom, embarked on a R250 million scheme at its Kendall power station in the Eastern Transvaal (currently Mpumalanga province) to conserve water. Similar innovative technologies were introduced at many of its other power stations. (Jeans 1983; Clarke 2002, 121) This was to pave the way for considerable successful efforts by various industries to resort to dry-cooling methods instead of using valuable water. On occasion the minister of environmental affairs, Sarel Hayward, in promoting the government's decentralisation policy, called on industries to start contemplating moving away from the hubs of the South African economy to ease the strain on the over-extended water supplies. (McQuillan 1983b) That appeared to be a sensitive matter and many captains of industry on the Witwatersrand did not like the sound of the proposal.

In the public sphere the drought phenomenon became a popular topic of discussion. It was clear that, as a result of the process of urbanisation, many residents of the Witwatersrand had lost touch with nature. It was simply taken for granted that water should come from a tap and they apparently had grown ignorant of the real implications of drought conditions. Moreover, the reality of the complex set of processes that raw water had to go through before it ran through the mains of a factory or a domestic environment had become insignificant to the average domestic consumer. The public campaign of Rand Water and local authorities on the Witwatersrand, at the time, was aimed at making ordinary people aware of the importance of water. As soon as conditions became critical, the 'issue' seemed to interest most people. The drought was the 'hot' news of the day.

Experts were of the opinion that the dry weather conditions were the worst in more than a century. It was even predicted that the next severe drought would only again manifest itself in about 150 years' time. (Anon. 1986c) Prof. Peter Tyson, a climatologist of the University of the Witwatersrand, explained that droughts occurred in cycles of nine years. He was of the opinion that as the country approached the 1990s, there would be a wet spell. (Martin 1988) Dr Henk van Vliet, director of South Africa's Hydrological Research Institute, was of the opinion that expensive new technology would have to be introduced to find alternative sources of water. It was becoming an increasingly expensive commodity that could have a counterproductive effect on the economic processes of the country. (Leeman 1986a)

Despite all the knowledge that had been accumulated, there was still uncertainty as to how drought conditions had to be dealt with. Suppliers simply had to contend with a basic planning matrix of water consumption. The problem could be addressed, it was explained, but it would be expensive to build dams and storage areas in advance to cope with potential droughts. The funds were simply not available to invest in spare dams over the long term. (Du Toit 1986)

While the experts expressed their opinions, the lives of ordinary people were directly affected by the drought. In October 1986, residents of smallholdings in Vanderbijlpark, close to the Vaal River, used wheelbarrows to cart their domestic water supplies home. Others collected water in drums in town. Residents then called on the minister of agricultural economics and water affairs to help. Schools started closing down after the drought conditions forced local residents to move away. (Jensen 1986) The boreholes in the region were beginning to run dry. Many had been sunk at the onset of the drought, but they kept drying up. (Van Aardt & Kotzé 1986)

## 36.8. RELIEF AT LAST?

Towards the end of October 1986 there were outstanding rains. In the media commentators were confident that the worst drought conditions were something of the past. (Anon. 1986d; Boshoff 1986; Wigget 1986) The rains kept improving and in December it was reported that all parts of the country were receiving good rains. (Koch & Venter 1986) But restrictions were still not lifted. Only on 30 October 1987 did Rand Water lift the restrictions that had been in place since 2 March 1983. (Minutes 1987-8, 352: 1) This followed in the wake of an announcement by the government that restrictions were to be lifted. (Government notice 2264, 1987)

The effect of the water restrictions was significant. When the first measures were introduced in 1983 it was decided that water consumption should be restricted to 70 per cent of the consumption in the corresponding months in 1982. By October 1987 the allocated quota of consumption stood at 1824 Mℓ/d. The aggregate saving that had been achieved since 1 April 1984 had been 23.55 per cent. (Minutes 1987-8, 361: 1)

The restrictions – even after being lifted in 1987 – still had a sting in the tail. When water tariffs were increased by 12 per cent in 1988 (*The Star* 1988a), consumers did not hesitate to criticise the authorities. The Afrikaans daily newspaper, *Beeld,* commented in an editorial article that it was unfair to the consumer to pay more for water. The newspaper felt Rand Water had to investigate its cost-efficiency. (*Beeld,* 1988a; *Beeld* 1988b) The fact of the matter was that in the management structures of the organisation it was argued that the increased rainfall allowed water consumption to drop. Consequently the board had to secure revenue by means of raising water tariffs. (RWA, Newspaper clippings 1988-9; Rumney 1988, 32) Management stressed in the media that the levels of the Vaal dam were an entirely unrelated matter. The board was interested in coping with the increasing cost of processing water. (*The Star* 1988a) Some local authorities, such as Edenvale and Randburg, tried to accommodate domestic consumers either by not increasing the price of water, or increasing it by a fraction only (Van Buuren 1988; Anon. 1988b. In the case of Randburg, the tariff was increased by 18 per cent in 1990, thereby indirectly recovering the potential losses incurred as a result of the lower tariffs of 1989. *See* Anon. 1990).

The campaign of discontent, however, continued to gain momentum. In April 1988 Prof. J.A. Dockel, the head of the department of Economics at the University of South Africa (Unisa), explained that Rand Water had a monopoly of the water supply on the Witwatersrand. That was why it was possible for the utility to push up the tariffs as it pleased. (Anon. 1988a) Board chairman Dale Hobbs defended the increase by stating that the board had to pay 30 per cent more for the water it purchased from the department of water affairs. (Anon. 1988a) In May 1988 the government stepped in when the minister of environmental affairs and water affairs, G.J. (Gert) Kotzé, announced that the water tariffs of the department were to be reduced. Making the announcement in parliament, the minister expressed the hope that water providers would also pass on the benefit of a lower price for water to the consumers. (Vermaak 1988; Anon. 1988d) Rand Water did not respond immediately. Hobbs told the media that he first wanted to get the details from the department of water affairs before making an announcement on the situation of the board's tariffs. (Anon. 1988c) The media seemed undecided about the issue. In its response, *The Star* thought that Rand Water's increase in the price of water to 62 cents per kℓ was moderate. It was, after all, providing water at a low rate in a semi-arid region, the newspaper claimed. (*The Star* 1988b)

Even when the board ultimately reduced its increase by half of the original hike, there were still queries. The media now gave it a political spin with arguments that the opposition Conservative Party was taking the government to task for, in effect, promoting inflation by allowing increases in services such as water. (*Die Vaderland* 1988) For a while there was confusion. Local authorities indicated that it was difficult to make adjustments to the tariffs. The new rate that the board had conceded to was an increase of 6.25 per cent. Some local authorities had already started charging the original increase of 12 per cent. Other councils had not yet introduced it. (Fray 1988; Anon. 1988e) As the row intensified, there were indications that the government and local authorities were beginning to respond to attempts to bring the consumer price index down. (*The Star* 1988c; Delmar 1988; Anon. 1988f) Ultimately the promised cuts did not materialise to their full potential. (Anon. 1988g; Anon. 1988h)

The tendency was for the price of water simply to increase. The management of Rand Water persistently pointed out that the days of cheap water were something of the past. In an exclusive interview with an Afrikaans daily newspaper, it was explained that the cost of pumping water had not declined. It was still the same, despite a drop in the price paid for water. It was also anticipated that upon the completion of the Lesotho Highland Water Project, which was scheduled to provide the Witwatersrand with 75 per cent more water, water could cost the consumer as much as 300 per cent more. In the process of preparing for the new supply, Rand Water incurred great costs in developing the water mains needed to transfer the water. According to Rand Water's board secretary, Tony de Witt, the pipes used for transferring the water cost in the vicinity of R2000 per metre (*about EUR 250 or $334. Ed.)* This was about the same price as the construction of one metre length of a double highway. (Du Toit 1988) *Beeld* was still not satisfied with the reasons that had been given by Rand Water and warned that it was due to the ineptness on the side of the authorities that consumers became angry. It brought governing bodies and boards into discredit with the people. (*Beeld* 1988c)

The public debate on the rising price of water did not bring down tariffs. In fact, Rand Water's tariff was increased in March 1989 by 16 per cent. Of that, four per cent was meant to cover a rebate offered to consumers in the previous year. It now had to be recovered. (Bath 2003) In the announcement, the board also indicated that it was scrapping preferential tariffs for some of its consumers. This implied that municipal authorities, the transport services and the mines were to be most affected by the new measures. The reason for the step was that the board did not want to reduce its reserve funds. (Anon. 1989a) To make matters worse, there were indications that the department of water affairs intended adding a levy to the tariff. (Melville 1989) The Consumer Council protested against the hike (Anon. 1989b), and it was pointed out in the media (Figure 36.4) that consumers were now beginning to feel the pinch of a scarce commodity. (*The Star* 1989)

Figure 36.4 Andy, the cartoonist of The Star, satirised the rate increases in March 1989, following the appointment of a new chief executive officer, Vincent Bath. (Source: Rand Water.)

## 36.9. LESSONS LEARNT THE HARD WAY

A few lessons were learnt from the period of 57 months between 1983 and 1987 in which restrictions had been in force. It was found that the most effective way of persuading consumers to limit water consumption was by means of keeping them informed on the availability and use of water. (83nd annual report, 14) Consumers, it appeared, would cut down on water consumption if and when the tariffs rose significantly. Then, a period of exhaustion would eventually set in. Under these circumstances, the price was of lesser importance than the need for the commodity. This is evident in the following data (Figure 36.5):

Experience also taught that after periods of restricted use, the return to the previous order of consumption takes a considerable time. (84nd annual report, 13)

Not all consumers of water were affected in the same way by the measures to reduce water consumption. In 1986 the Bureau for Market Research conducted research for the Water Research Commission on the socio-economic effects of water restrictions. (Collins 1986) Because of the sensitive nature of the findings, the report was kept under wraps. According to informed sources, the report could give an indication of how the restrictions were depriving some sectors of the population of a certain quality of living. (Collins 1986) The irrigation farming activities, mining, electricity supply, central government, industries, local authorities and households were among the consumers investigated. (Anon. 1986a, 17) More important was the fact that the suppliers of water were becoming aware that black South African consumers could also be affected by rising water tariffs.

Between the severe drought conditions of the 1960s and 1983, the consumers of water from Rand Water were subjected to water restrictions on no fewer than six occasions. In percentage time frames it constituted a combined period of 34 per cent of the time over 17 years. (80th annual report, 3) For the planners at Rand Water, there was one truth that held: the Witwatersrand, like most parts of South Africa, was a water-stressed region.

The 1980s were decisive years in Rand Water's history as far as water restrictions were concerned. Perceived as the result of the drought conditions that had started in the 1970s (79th annual report, 2), the restrictions were to shape the way in which officials of Rand Water, as well as the department of water affairs, would plan in future. (Woodgate 1988) One example of creative thinking was made public in May 1986 when Prof. D.C. Midgley, a well-known hydrological engineer of the University of the Witwatersrand, told a conference in Pretoria that bad distribution systems played an important role in the water crisis that prevailed. He explained that the demand for water in the country, as a whole, had far exceeded the yield.

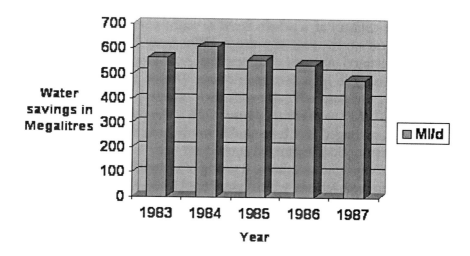

Figure 36.5 Water savings 1983-1987. (Source: 83nd annual report, 14)

One method of addressing the need was to transfer water by means of inter-basin supplies in order to meet expected shortfalls. He foresaw a probable solution in better relations between Southern African states. It could lead to the Okavango/Zambesi water system being brought in to help South Africa. (Leeman 1986b; Anon. 1986b)

Within seven years after the longest period of water restrictions in the history of Rand Water, the region served by the water utility was once again faced with drought conditions. Conditions had changed substantially. South Africa had become a multi-racial democracy in 1994 and a new government now had the responsibility of working in the interest of all the country's people. The racial divide of the former dispensation no longer applied. It was consequently necessary for the authorities to bear in mind that water restrictions, if they were to be introduced, had to be effective from the lowliest squatters' camp to the most upmarket residential area serviced by Rand Water.

Apart from the prevailing drought, Rand Water's deteriorating water supply was partly caused by the construction and improvements that were underway at the time at the Vaal dam. The dam's capacity had to be increased to cope with the increased supply of water that was scheduled to flow into the dam once the Lesotho Highland's Water Project (LHWP) had been completed. Perhaps more important from a humanitarian perspective, was the fact that the residents of the black townships in Rand Water's supply area now required the authorities to provide a more consistent and profuse supply of water. Their living conditions were beginning to change and they had a constitutional right to more water.

In January 1995 a journalist visiting the southeastern Gauteng region where the Vaal dam was situated noted that the dam was so empty that it appeared like a bathtub of which the plug had been pulled. (Du Plessis 1995) By February there were indications that restrictions could be introduced as the country as a whole faced drought conditions. Countrywide the rainfall was down 44 per cent on the annual average. (Woodgate 1995) Towards the end of March 1995, Rand Water's general manager of operations, Attie van Rensburg, told the media that it would soon be necessary to introduce restrictions. (Anon. 1995) The representatives of Rand Water's engineering division held talks with their counterparts at the department of water affairs and forestry (DWAF) to reach consensus with a view to introducing restrictions as from 1 May 1995. It was envisaged that households would have to effect a reduction of 30 per cent in their regular consumption. Industrial consumers were to cut back 10 per cent and agricultural consumers were expected to achieve a saving of as much as 50 per cent. (Minutes 1994-5, 417:1) The subsequent restrictions, which lasted until December 1995, were an attempt to cope with the crisis of total dam storage dropping to 25 per cent of capacity. (Rand Water Board: annual report 1996, 2)

## 36.10. A NEW APPROACH TO RESTRICTIONS

What was interesting about the drought period was that Rand Water started with new and innovative measures to try and get the message across that all consumers in a new multi-racial society had to save water. (Davis 1995, 13-17) At the April 1995 board meeting the matter was discussed. The department of water affairs and forestry required an overall saving of 20 per cent in consumption. In the process of communicating the information with the consumers, Rand Water formed a Rand Water Users' Conservation Forum. This body was to meet routinely to discuss and coordinate issues in respect of water restrictions. It was also agreed that bulk consumers would be given savings requirement quotas for the month. Subsequent monthly targets also had to be passed on to the consumers. (Minutes 1995-6, 48:1) The forum, which also went by the name of the Rand Water Users' Council, made a number of creative proposals for ways in which water could be conserved. (Rand Water Board: annual report 1996, 2)

What was interesting about the new system of restrictions was that consumers were not to be penalised for their inability to reach target consumption figures in the first month. Instead, an attempt was made to first get public support for the measures. If that arrangement did not work, it was argued, Rand Water would be in a better moral position to introduce penalties. It was considered essential to see to what extent the public would respond to the moral request for limited water consumption. (Minutes 1995-6, 48:1) In an effort to get support for the proposed water restrictions, the marketing division had worked out an extensive plan in the media for community awareness programmes and potential steps to discipline consumers. (Minutes 1995-6, 48-9) The advertising campaign to bring the restrictions to the attention of consumers, on all levels of society, was estimated to cost R7.5 million (*About Euro 1 million, or $1.25 million*) over a period of eight months. (Minutes 1995-6, 84:1)

Following the publication of an official notice by the department of water affairs (Notice 653 of 5 May 1995), Rand Water was informed in June 1995 that irrigation farmers had to save 40 per cent of their consumption, local authorities 20 per cent, mines and industries 10 per cent and intensive livestock feeding programmes five per cent. The petroleum manufacturing industry, Sasol, and Eskom, had to save five per cent respectively. (Minutes 1995-6, 126)

In implementing the restrictions, the board of Rand Water approved an additional rate, if the limits were exceeded, of initially R1 per kℓ, with an increase of R1 per kℓ per month to a total of R5 per kilolitre if bulk consumers did not meet the required savings targets. (Minutes 1995-6, 127:1) Provision was also made for a ban on the use of all garden sprinkler systems and hosepipes. (Minutes 1995-6, 127:1)

At a board meeting on the measures, the focus tended to fall on *per capita* water consumption. There was awareness that it was important for local authorities to be consistent in respect of policies on punitive measures. The board was also informed that the broader public had not responded favourably to the extensive advertising campaign. It was necessary to take effective measures to ensure that consumers maintain the limits on supply. (Minutes 1995-6, 128:1) The restrictions were not having the desired effect by July 1995. Several members of the board expressed concern about the state of affairs. It was acknowledged that consumption in the less affluent areas of Rand Water's supply areas constituted a fraction of the total amount of water consumed. On the surface it appeared as if consumers had grown insensitive to water restrictions. What made matters worse was the fact that the media, along with local authorities, appeared to be angry about the introduction of the new tariffs. (Minutes 1995-6, 157-8:1)

## 36.11. AWARENESS CAMPAIGN AUGUST 1995

By August 1995, some headway was made with the conservation awareness campaign when a plan was introduced that had been strongly propagated by the department of water affairs and forestry. (Clarke 2002, 121) The board was informed that the strategy was multifaceted and was constantly monitored to secure the effectiveness of the campaign. The main points of the strategy were: to build an awareness of the problem; gain participation and ownership of solutions; allow behaviour patterns to change; and finally entrench and reinforce behaviour changes. (Minutes 1995-6, 179:1) The marketing division busied itself with an extended advertising campaign. There were contributions in terms of editorial space in publications and pamphlets on effective gardening and educational videos to get the message across to consumers. The marketing people were constantly making available press kits and doing extended public relations. Posters and leaflets were distributed and direct contact forums were conducted and brought into service. These included a Rand Water Users Committee, a horticultural forum and the plumbers' forum. (Minutes 1995-6, 179-80:1)

Extensive research was conducted. With a profile of the audience in mind, an intensive advertising campaign was introduced. Public opinion was also tested beforehand. By mid-July 1995 it was reported that public awareness had increased by 7 per cent. This was substantially higher than in similar campaigns, where a maximum of 2 per cent of response had been recorded. (Minutes 1995-6, 179:1) The rate of saving water had gone up from 3.8 per cent in June 1995 to 7.1 per cent in July 1995. (Minutes 1995-6, 179:1)

The discussion forums also stepped up their support for the campaign. In municipal circles local byelaws were reviewed, codes of practice, standards and minimum plumbing qualifications reconsidered. Moreover, there was a review of the equipment standards pertaining to water restrictions and greater attention was paid to promoting awareness of the most suitable plants (preferably indigenous) for specific regions. The services of water loss management firms were used and on the planning front a long-term water management strategy was undertaken in conjunction with all Rand Water's stakeholders. (Minutes 1995-6, 179:1)

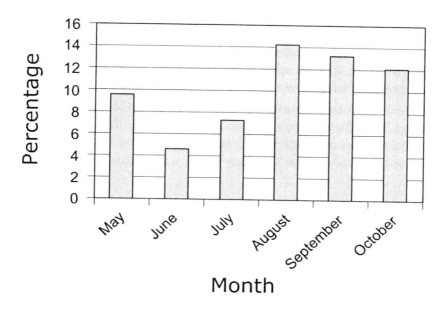

Figure 36.6 Savings in water consumption May to October 1995. (Source: Rand Water)

## 36.12. RESPONSES IN THE PUBLIC SPHERE

Perhaps more important was the fact that South Africa's transition to a multi-racial democracy was noted for its tolerance of free speech and protest. Activists increasingly tended to speak their mind on numerous issues. Environmental concerns were among the popular public debates of the day. It had been accepted for a long time that the Lesotho Highlands Water Project (LHWP) would provide valuable water to the Gauteng Province, formerly roughly the area of the Witwatersrand-Pretoria-Vaal Triangle region. Activists were now, however, beginning to insist that water should be used sparingly. Mega-sized dam construction projects were no longer seen as a solution to water shortages. Instead more constructive alternatives had to be devised. In the years to come the pressure on water planning strategists to find sensible solutions, only mounted. (For more on this trend, see Bond 2002, 127-131) This trend proved to be an incentive for Rand Water to work with even greater determination at conservation strategies.

It was a daunting task. By September 1995, complacency amongst consumers had once again set in. They were then made aware of the implications of penalties. There were fears that public responses were on the decline. Public responses on the telephonic information line of Rand Water in particular had dwindled. (Minutes 1995-6, 225:1) This was an indication of the campaign's losing its effect. In November, a month before the restrictions ended, statistics were available on the campaign trends since May 1995. (Minutes 1995-6, 314:1) (See Figure 36.6)

It was now evident that local authorities were implementing the restrictions in different ways. The overall savings, after the media campaign and also after the introduction of additional rates had been introduced, were disappointing. It was reported that the mines and metered areas in the local authorities' areas of the 'formerly white' townships had responded well to water conservation.

In the black township the savings had been limited. There was one exception. For one month there was a saving of 10 per cent in Soweto. (Minutes 1995-6, 314:1) At the same time it was pointed out that the *per capita* consumption in non-metered areas was about 100 ℓ/d. In metered areas the consumption was 300 ℓ/d. The planners were of the opinion that it was much easier to apply more stringent measures in the metered areas. It was also recommended that it was important to continue with operations to place meters in the formerly black townships. (Minutes 1995-6, 314:1)

Restrictions were not having the desired effect. Consequently the board gave permission to the chairman and the chief executive officer to hold talks with the minister of water affairs and forestry with a view to increasing the rate for consumption over the limit to as much as R20 per kℓ. (Minutes 1995-6, 315:1) Ultimately the proposed plans to substantially increase the price were based on comprehensive investigations. Research conducted on patterns of savings in different urban centres brought interesting facts to the fore. Small samples were used in the research programme. For the researchers it was important to determine to what extent the advertising campaign had produced the desired effect. (Table 36.1)

Table 36.1 Savings in water consumption in 10 local authority areas in September-October 1995 (Minutes 1995-6, 316:1)

| Area | Percentage saving |
|---|---|
| Lenasia | 33 |
| Pretoria | 25 |
| Springs | 20 |
| Roodepoort | 15 |
| Embalenhle | 17 |
| Johannesburg | 11 |
| Tembisa | 11 |
| Soweto | 9 |
| Katlehong | -2,5 |
| Shoshanguve | -6 |

## 36.13. A NEW PHILOSOPHY ON WATER USE: WATER DEMAND MANAGEMENT (WDM)

By December one of the learning processes in dealing with water restrictions in the new South Africa had come to an end. What remained to be done was for the experts to sit down and plan for future restrictions by looking at the phenomenon from a more indigenous perspective. All South Africans had to be made aware of the value of that scarce ubiquitous commodity called water.

From the outset it was clear that awareness campaigns aimed at promoting water saving with consumers had to be extended well beyond times of drought. In the months of the water restrictions campaign of 1995, Rand Water earned R70 million through penal tariffs. In view of the need for water conservation and more skills, management decided to propose to the board that the revenue earned by means of penalties on excess quotas be used for water conservation projects. (Minutes 1996-7, 63:1) Of this sum, R30 million was to be spent during the financial year 1996/7. R10 million was allocated for the eradication of alien plant species in the upper Tukhela river area. It formed part of the Working for Water project of the department of water affairs and forestry (the basic point of departure was that alien plants consumed excessive amounts of water. In order to conserve water it was necessary to eliminate these plants within the reaches of the river). The remainder of R30 million was to be spent on projects aimed at attending to leaks and water wastage in residential homes (Minutes 1996-7, 63:1), and concentrating on water demand management (WDM). This important principle, based on strategies aimed at conserving water, was subsequently ensconced in the *National Water Act* 35 of 1998. (Glazewski 2000, 532)

The primary objective of WDM was simply to discourage water wastage. In motivating the introduction of the plan, Rand Water's CEO Vincent Bath remarked:

> Allocation of water resources in the water scarce region of the interior of our country is likely to become an increasingly controversial issue as available resources become fully utilised. (Rand Water Board: annual report 1995, 10)

He went on to explain that in future, the existing apportionment of the Vaal river's water to Rand Water would grow increasingly until the allocation for agriculture could become static or even go into decline. It was also not unlikely that the 55 per cent of the Vaal River's water being used by Rand Water could literally increase to two thirds. For these reasons it was necessary, he explained, for WDM to be introduced for the first time in Rand Water's area of supply. (Rand Water Board: annual report 1995, 10) Rand Water announced its intention of introducing a sliding scale tariff for water that had been designed to ensure that the essential water required for sustaining life was provided at an affordable cost to all people, especially those with a lower income. The water destined for other uses was to be provided at discretionary rates, but the objective was to discourage waste. (Rand Water Board: annual report 1995, 10) By 1996 the philosophical tenets had been anchored that water tariffs should – after making provision for the essential human needs of survival and hygiene – be focused on reflecting economic realities. (Rand Water Board: annual report 1996, 6) This made provision for a free allocation of 60 kℓ of water that was subsequently introduced in July 2001 throughout South Africa by the ministry of water affairs and forestry. (Van Burick 2001)

The management of Rand Water conceded that many problems were experienced with implementing the water restrictions in 1995 and it was considered essential to introduce a response and control framework in which water demand could be managed as a permanent feature of Rand Water's activities. (Minutes 1995-6, 448:1) In a report to the board, it was explained that there was severe stress on the water supply because of overuse and population increase in the service area of Rand Water. Already in the 1980s, this situation had created the need for the development of the Lesotho Highlands Water Project (LHWP). The agreement for the scheme was signed between the governments of Lesotho and South Africa in 1986. The construction project started shortly afterwards. Unless the prevailing demand was contained, more costly schemes such as the LHWP would have to be developed, Bath explained. It was envisaged that the price of water was destined to rise because of the need to transfer it over large areas. Local authorities were slow to respond to the lowered supplies. The smaller consumers simply did not believe that there was a water supply problem. There was general ignorance about the prices paid for water, and widespread resentment about having to pay for water. This state of affairs affected the credibility of local authorities and prevented them from doing their job properly. Similarly local authorities had a poor performance record when it came to dealing with their consumers' queries and problems, Bath explained. (Minutes 1995-6, 448:1)

In the report it was stressed that water conservation was not merely an esoteric issue. It was a set of principles. These had to be perceived in terms of financial impact. Water conservation principles could affect all levels of capital expenditure in the country. (Minutes 1995-6, 448:1) It was argued that if sensible water savings could be induced, and the 1992 additional water scheme could be halted, Rand Water could save as much as R85 million *per annum* based on deferred implementation of capital work. (Minutes 1995-6, 448:1)

Overall WDM, however, required substantially greater inputs from Rand Water. In a number of investigations conducted by officials, it was determined that there were great losses in revenues for local authorities in the former black municipal areas. It was estimated that in Tembisa there was a loss of 68 per cent of the available water supply. In the neighbouring Kempton Park water losses amounted to a mere nine per cent. (Minutes 1996-7, 63:1)

There were many leaks in the existing macro-supply system. More important were the leaks that existed in the domestic water system of toilets, standpipes and yard taps. (Minutes 1996-7, 63:1) There was a marked lack of awareness amongst consumers of the nature of the problem. People did not seem to care about the cost of water and the prevailing scarcity of the resource. Tenants tended to pay a flat rate for water. Those who paid did not care much for the value and cost of water. (Minutes 1996-7, 63:1)

For Rand Water, overseeing the WDM programme consisted of four components. First there were the necessary repairs that had to be done to pipe work and/or domestic fittings and plumbing. This was a responsibility that ordinarily rested with municipalities and individual householders. This strategy was considered desirable in an attempt at cutting down on wastage. (Bath 2003) Secondly it was necessary to start with a campaign of making people aware of the value of water. In the third phase there was the need to create jobs and train local plumbers. Finally there was the phase of capacity building for local authorities which could provide some support in reaching small consumers to promote water conservation and put an end to water leaks. (Minutes 1996-7, 63:1)

*Repairing water leaks:* It was recommended that four different areas of operations be approved. An overall amount of R15 million had to be made available for the implementation of the project. In Odi 1 (a rural development project) there were an unknown number of water units that needed repairs estimated to cost in the vicinity of R2.4 million. In Tembisa, 12,000 units were to be overhauled at an estimated cost of R4.2 million, while in Soweto there were 15,000 units that would cost R5.7 million to repair, and in the inner city of Johannesburg 1,500 apartments were in need of repairs to the value of R2.7 million. (Minutes 1996-7, 64:1)

The board was at first not keen to give its approval (Minutes 1995-6, 450:1), but as the project gained momentum it became apparent that a number of benefits could be derived from the initiatives. In 1999 management reported on progress and indicated that it was in search of partnerships to make the project even more comprehensive. (Minutes 1999-2000, 688-90:1)

Soon the department of water affairs and forestry as well as the Gauteng provincial government responded and strengthened the initiative. (Minutes 1999-2000, 689:1; For the outcome see Anon. 2001b, 15) The authorities knew that the ordinary consumer had to be reached if the project was to become successful.

*Teaching children:* Ever since the start of the 1990s, community outreach programmes of Rand Water were focused on children as representatives of the community that would ultimately be active in promoting the objectives of the organisation. Special attention was also given to the elderly. (Rand Water Board: annual report 1994, 11) Initiatives launched under the auspices of the "Water Wise" campaign in 1993 were further enhanced. Educational tours and creative classes were presented at the Vereeniging pumping station where children were informed on the significance of water. The Water Wise Educational Team (WWET) was the result of a partnership between Rand Water and the Delta Environmental Centre. (Rand Water annual report 1998, Chief executive's report, 24)

After five years of operations, more than 17,000 learners had been reached by means of educational activities. (Rand Water annual report 1998, 25) In the course of 1998, the team started with an edutainment programme at the Delta Environmental Centre to promote water awareness by means of community theatre productions. The programme led to the appointment of members of the Soweto community as environmental educators. (Rand Water annual report 1999, Chief executive's report, 18) Within the context of outcomes-based education, principles of science and technology were conveyed to learners of various communities within the area Rand Water served. (Rand Water annual report 1999, 19) By 2000 a sensory garden at Delta, which formed part of the Water Wise project, made it possible for disabled people to experience the beauty of a garden environment. (Anon. 2000, 19-21)

*Community forums:* A major point of interaction between Rand Water and the organised public was the trend of community forums. These were used particularly to promote a series of environmental conservation objectives. By 1998, when a policy statement on the mission of Rand Water was outlined, it was evident that the organisation had excelled in this area. In the process it had managed to win many friends and established valuable partnerships in the interest of water and environmental conservation. At the time Rand Water was involved in the catchment forums of Blesbokspruit, Klip river (Upper Klip, Rietspruit and Lower Klip), Taaibosspruit and Sasolburg, Vaal River and Hartebeestpoort dam, as well as the Barrage. In the case of the Barrage, Rand Water had a special interest and was an active participant in the River Property Owners' Association (founded in the 1980s), the Vaal River Safety Association, the Vaal Barrage Conservancy and the forum for Riparian Development on the river. (Minutes 1998-9, 224-5:1: Anon. 1998, 27)

## 36.14. CONCLUSIONS

Up to the 1980s, water restrictions to a large extent focused on only one or two segments of society in the Rand Water service area. These measures proved to be partially ineffective. It was fragmented and many choices remained open to consumers. Those who could pay summarily continued to consume large amounts of water. Otherwise they resorted to limited savings measures. Communications strategies proved to be incomplete and it frequently led to the awareness campaigns creating the wrong type of reception with consumers. Rand Water was also to a large extent still essentially an 'invisible' organisation. Dale Hobbs, as chairman, started with innovative strategies of making the public aware of Rand Water and its task. His successor, Vincent Bath (1989-2002), as chief executive officer of the water utility carried on with the task.

The transition to a new dispensation in 1994 saw the principle of the democratisation of water being applied. From the outset consumers, also in the previously disadvantaged parts of Rand Water's supply area, were allowed to become part of attempts at saving water. The effect of the first phase of restrictions was that the authorities had to give more dedicated attention to awareness campaigns. Subsequently the profits generated with the higher water tariffs in the period of restrictions were passed on to consumers at the lower end of the scale. By, for example, retrofitting existing plumbing facilities in township houses, Rand Water made it possible for unemployed people to be trained as plumbing contractors. They were also enabled to earn an income for themselves. At the same time, valuable savings were realised on the ground level. Young people were also educated in using water sparingly.

In conjunction with the DWAF, the principle of a consumer-friendly approach to water was implemented on all levels of society throughout South Africa. Also Rand Water played a role in this campaign. It had a lasting effect.

Drought conditions in South Africa most definitely are not something of the past. In the years to come South Africans, especially in urban areas, could well be subjected to severe limitations in terms of available water supplies. They will then hopefully become more sensitive to the fact that water is a scarce commodity that has to be used with the utmost care if we want to secure the sustained development of society in a water-scarce region of the African continent.

## 36.15. REFERENCES

**Principal sources**

A. Rand Water Archives (RWA):

31/S.
   Notice to all consumers of Rand Water Board water, Secretary to the board, Johannesburg, 13 January 1966.

79th annual report, balance sheet and accounts of Rand Water. Financial year to 31st March 1984, p. 2.

80th annual report, balance sheet and accounts of Rand Water. Financial year to 31st March 1985, p. 3.

83nd annual report, balance sheet and accounts of Rand Water. Financial year to 31st March 1988, p. 14.
   84nd annual report, balance sheet and accounts of Rand Water. Financial year to 31st March 1989, p. 13.

Commitee of the whole board meeting, 28 September 1984, p. 96.

Gardiner, J.L., 'Some historical aspects of the water supply to the PWVS complex', (undated manuscript), p. 7. Africana collection.

Minutes 1950-1951:
   pp. 165-6: 622nd meeting of the Rand Water Board, Headquarters, Johannesburg 22 December 1950.

Minutes 1982-1983:
   pp. 581: 1 008e gewone vergadering, hoofkwartier, Johannesburg, 25 February 1983. Hoofingenieur se verslag, nr. 6844. L.H. James.

Minutes 1983-1984:
   p. 212: Extraordinary meeting, headquarters, Johannesburg, 1984.03.02.

Minutes 1987-1988:
   p. 352: 1 062e gewone vergadering, hoofkwartier, Johannesburg, 30 October 1987. Sekretaris se verslag, nr 8045. A.J. de Witt
   p. 361: 1 063rd ordinary meeting, headquarters, Johannesburg, 27 November 1987. Report MO 7900. A.J. de Witt.

Minutes 1994-1995:
   p. 417: 1 141st ordinary meeting, headquarters, Rietvlei, 26 January 1995. General.

Minutes 1995-1996:
   p. 48: 1 144th ordinary meeting, headquarters, Rietvlei, 26 April 1995. Report V.J. Bath: Corporate services division.
   p. 84: 1 145th ordinary meeting, headquarters, Rietvlei, 25 May 1995. Report A.J.J. van Rensburg: Corporate services division.
   p. 126: 1 146th ordinary meeting, headquarters, Rietvlei, 29 June 1995. Report V.J. Bath: Chief executive.
   p. 127: 1 146th ordinary meeting
   p. 128: 1 146th ordinary meeting
   pp. 157-8: 1 147th ordinary meeting, headquarters, Rietvlei, 27 July 1995.
   p. 179: 1 148th ordinary meeting, headquarters, Rietvlei, 31 August 1995. Report V.J. Bath: Corporate services division.
   pp. 179-80: 1 148th ordinary meeting.
   p. 225: 1 149th ordinary meeting, headquarters, Rietvlei, 28 September 1995. Report V.J. Bath: Corporate services division.

p. 314: 1 151st ordinary meeting, headquarters, Rietvlei, 30 November 1995. Report V.J. Bath: Chief executive.

p. 450: 1 154th ordinary meeting, headquarters, Rietvlei, 20 March 1996. Report V.J. Bath: Corporate services division. Board decision

p. 448: 1 154th ordinary meeting, headquarters, Rietvlei, 28 March 1996. Report V.J. Bath: Corporate services division.

Minutes 1996-1997:
    p. 63: 1 156th ordinary meeting, headquarters, Rietvlei, 30 May 1996. Report V.J. Bath: Corporate services division.

Minutes 1998-1999:
    pp. 224-5. 1 183rd ordinary meeting, headquarters, Rietvlei, 1998.10.29:

Minutes 1999-2000:
    pp. 688-90: 1 192nd ordinary meeting, headquarters, Rietvlei, 26 August 1999.

Newspaper clippings 1988-9. Margin note by the manager operations

Rand Water Board (1994) annual report 1994, p. 11.

Rand Water Board (1995) annual report 1995, p. 10.

Rand Water Board (1996) annual report 1996, p. 2.

Rand Water (1998) annual report
    - Chief executive's report

Rand Water (1999) annual report
    - Chief executive's report

## B.

Government notice 2264 of 9 October 1987

TA TRB2/1/691 File 124/11/4 Vol. 1. N. Johannesburg region. Water supply scheme – restrictions. R.W. Hedding, Anglo American Corporation, Johannesburg – Circular to all consumers from the Bryanston mains, 1961.08.29.

Department of water affairs, Management of the water resources of the Republic of South Africa, (Government Printer, Pretoria, 1986), pp. 1.3, 1.5 and 1.13.

Notice 653 of 5 May 1995, issued by the Minister of Water Affairs.

## Literature

Abderrahman W.A. (2000) 'Water demand management and Islamic water management principles: a case study'. in International Journal of Water Resources Development, 16(4), Dec 2000, pp. 465-473.

Andrews H.T., Berrill F.A., de Duingand F., Holloway J.E, Meyer F. and van Eck H.J. (eds) (1965) South Africa in the sixties: a socio-economic survey (Second edition), p. 121.

Anon. (1954) 'Verslap perke – inwoners' in Beeld, 27 September 1954

Anon. (1966) 'Nywerhede nie geraak: waterbeperking nou 30 p.s.' in Die Transvaler, 8 July 1966.

Anon. (1978) 'Soweto becomes more water conscious' in The Star, 13 June 1978

Anon. (1982a), Waterperke tref wyd' and 'Water' in Beeld, 14 January 1982

Anon. (1982b) 'Onwettige natlei gaan nie die sak seermaak' in Die Vaderland, 17 February 1982.

Anon. (1982c) 'Pyp help water benut' in Vaalweekblad, 9 September 1982

Anon. (1983a) 'Waterperke straks gou hier' in Vaalweekblad, 4 January 1983.

Anon. (1983b) 'Krisis met water' in Die Transvaler, 13 January 1983

Anon. (1983c) 'Water hit by power failure' in The Citizen, 13 January 1983

Anon. (1983d) ''n Waterlose Gouydstad toe 'krag' ophou om te vloei' in Beeld, 14 January 1983.

Anon. (1983e) 'Water normal, but probe result secret'. Rand Daily Mail, 17 January 1983

Anon. (1983f) 'Eskom meets on power cut'. Rand Daily Mail, 18 January 1983.

Anon. (1983g) 'Fout by pompstasie is reeds herstel'. Beeld, 19 January 1983.

Anon. (1983h) 'Govt steps in as crisis at Vaal dam deepens'. Rand Daily Mail, 11 February 1983

Anon. (1983i) Rand Water Board to apply restrictions'. The Citizen, 11 February 1983

Anon. (1983j) 'Droogte se greep nie gou gebreek' in Beeld, 24 February 1983.

Anon. (1983k) 'More water curbs if drought goes on' in Rand Daily Mail, 1 March 1983

Anon. (1983l) 'Vaal dam may be empty in 1984' in The Citizen, 16 March 1983

Anon. (1983m) 'Waterperke: só raak dit u' in Die Vaderland, 17 March 1983.

Anon. (1983n) 'Min water: gewilde oorde toe' in Vaalweekblad, 29 March 1983.

Anon. (1983o) 'Duurder water vir die Goudstad' in Die Vaderland, 31 March 1983.

Anon. (1983p) 'Jail term wanted for water misuse' in The Star, 31 March 1983

Anon. (1983q) 'Myne spaar baie water' in Die Vaderland, 26 May 1983.

Anon. (1983r) 'Wen 'n lekker prys met 'n waterwenk' in Die Valderland, 26 May 1983.

Anon. (1983s) 'Pretoria rejects hose ban' in Rand Daily Mail, 8 September 1983.

Anon. (1983t) Water watch: 'Water wastage warnings' in The Star, 15 September 1983.

Anon. (1983u) 'Rains and high winds lash Witwatersrand' in The Star, 22 September 1983.

Anon. (1983v) 'Waterbeperking: nuwe rekord vir Vdb na nuwe beperkings' in Vaalweekblad, 23 September 1983.

Anon. (1983w) 'Rantsoenering kom dalk nog' in Die Vaderland, 19 October 1983.

Anon. (1983x) 'Nog nie genoeg besparing van water nie' in Beeld, 25 October 1983

Anon. (1984a) 'Feeling drained – women who complained of leak' in The Citizen, 23 February 1984.

Anon. (1984b) 'Rain falls as people pray: prayers draw thousands' in The Citizen, 23 February 1984.

Anon. (1984c) 'Reën val toe biduur begin' in Beeld, 23 February 1984.

Anon. (1984d) 'RWB urged to review new water tariffs' in The Citizen, 15 March 1984.

Anon. (1984e) 'Valente sal nou bedank' in Beeld, 27 September 1984.

Anon. (1984f) 'Kwotas mag nie oorskry word nie: tuinslange is nie die grootste kommer' in Beeld, 29 September 1984.

Anon. (1984g) 'Oberholzer-waarskuwing oor hoë verbruik: Goudstad kry vergunning oor tuin natmaak' in Beeld, 9 October 1984

Anon. (1984h) 'Water board pulls in R3-m in fines since April' in The Citizen, 12 October 1984

Anon. (1984i) 'R3m. Uit waterboetes geïn' in Die Vaderland, 12 October 1984.

Anon. (1984j) 'Tuinslange sommige dae toegelaat: Randburg verslap sy waterperke' in Beeld, 18 October 1984.

Anon. (1984k) 'Natste Oktober in 5 jaar' in Vaalweekblad, 6 November 1984.

Anon. (1984l) 'Kwota word nog elke week oorskry: waterperke langste nog in werking' in Beeld, 12 December 1984.

Anon. (1985) 'Vaaldam sak nog: damme is 8 p.s. voller' in Beeld, 6 November 1985.

Anon. (1986a) 'Waterbeperkings se invloed ondersoek' in Vaalweekblad, 23 May 1986.

Anon. (1986b) 'SA not short of water – expert' in The Star, 27 May 1986.

Anon. (1986c) 'Eers weer oor 150 j. só droog' in Beeld, 8 August 1986.

Anon. (1986d) 'Pretoria gets 107 mm and Jo'burg 80 mm – and more to come' in The Star, 29 October 1986.

Anon. (1988a) 'Verhoogde waterprys omnlogies, sê kenners' in Beeld, 19 April 1988.

Anon. (1988b) 'Randburg holds down water rate' in The Star, 26 April 1988.

Anon. (1988c) 'Minister haal geld uit Hobbs se water' in Beeld, 3 May 1988.

Anon. (1988d) 'Prysverhoging van water teruggetrek' in Die Vaderland, 3 May 1988.

Anon. (1988e) 'Watertarief 'n dilemma' in Die Vaderland, 5 May 1988.

Anon. (1988f) 'Jo'burg water may not cost much more' in The Citizen, 14 May 1988.

Anon. (1988g) 'Water duurder' in Die Vaderland, 25 May 1988.

Anon. (1988h) 'Edenvale se water duurder' in Beeld, 26 May 1988.

Anon. (1989a) 'Water se prys styg met 16 persent' in Beeld, 1 March 1989.

Anon. (1989b) 'Bulk water price hikes slammed' in The Citizen, 3 March 1989.

Anon. (1990) 'Increase in water tariffs from 20 April 1990' in Randburger, May 1990.

Anon. (1995) 'Water cuts are on way despite downpours' in The Citizen, 31 March 1995.

Anon. (1998) 'Residents launch river conservation initiative' in Water, Sewage & Effluent, June 1998.

Anon. (2000) 'Rand Water's sensory garden uses Water Wise planting' in Parks and Grounds, No. 114, April-May 2000

Anon. (2001a) 'Sustainable water use' in Water & Environment International, Jun/Jul 2001.

Anon. (2001b) 'Rand Water partnership launches Water Wise campaign' in Imiesa 26(8), August 2001.

Bath V.J. (2003) Personal disclosure, former chief executive official, Westcliff, 7 March 2003.

Beattie A., 'Hope fades as rainy season draws to close' in The Star, 28 February 1984

Beeld (1983a) 'Water-vasvat in Goudstad' (Municipal reporter), 31 March 1983.

Beeld (1983b) 'Water word duurder' (Municipal reporter), 30 November 1983.

Beeld (1988a) 'Waterprys' (Editorial comment), 5.4.1988.

Beeld (1988b) 'Mors of betaal!' (Editorial comment), 19.4.1988.

Beeld (1988c) 'Drupsgewys' (Editorial comment) 4.5.1988.

Bond P. (2002) Unsustainable South Africa: environment, development and social protest. University of Natal Press, Pietermaritzburg.

Boshoff C. (1986) 'Dit stroom na damme: uitkoms!'. Die Vaderland, 29 October 1986.

Carlisle L. (1989) 'Groundwater can offer better value'. Business Day, 15 August 1989.

Clarke J. (1984a) 'Spring and the hosepipe dilemma'. The Star, 7 September 1984.

Clarke J. (1984b) 'Water level in Vaaldam is well above average'. The Star, 22 September 1984.

Clarke J. (1984c) 'No, Sandton can't use the hose yet'. The Star, 28 September 1984

Clarke J. (1984d) 'As Vaal dam fills up, the waters let out'. The Star, 14 November 1984.

Clarke J. , (2002), Coming back to earth: South Africa's changing environment. Jacana, Houghton.

Collins M. (1986) 'Results of survey on water curbs to stay secret for now' in Business Day, 15 May 1986.

Davis A. (1995) 'Rand Water's response to the drought'. Imiesa, 20(9), September 1995, pp. 13-17.

Delmar P. (1988) 'Lower PWV water tariffs now likely' in The Citizen, 5 May 1988.

Die Valderland (1983) 'Water word so vermors, kla die publiek' (Municipal reporter), 22 March 1983.

Die Valderland (1984) 'Nasionale biddag vir reën en nood', (Parliamentary reporters), 16 February 1984.

Die Vaderland (1988) 'Standpunt: waterprys', editorial comment, 4 April 1988.

Du Plessis A. (1995) 'Vaaldam se 'prop' is uitgetrek'. Beeld, 1 February 1995.

Du Toit A. (1986) 'Water and the Vaal system'. The Star, 11 September 1986.

Du Toit A. (1988) 'Meer water, maar baie duurder'. Die Vaderland, 1 June 1988.

Clarke J. (2002) Coming back to earth: South Africa's changing environment. Jacana, Houghton.

Fray P. (1988) 'Municipalities will pay for water hike'. The Star, 4 May 1988

Gault R. (1984) 'Obie threatens strong action: outcry over Sandton water vote'. The Star, 25 September 1984

Giliomee H. (2003) The Afrikaners: biography of a people. University of Virginia Press.

Glazewski J. (2000) Environmental law in South Africa. Butterworths, Durban.

Goosen M. (1984) 'Jo'burg eases ban on hoses'. Rand Daily Mail, 9 Ocober 1984.

Herselman M. (1983) 'Water word nie gemors: nywerhede dra 'n groot deel by'. Vaalweekblad, 22 April 1983.

Hobsbawm E. (1996) The age of extremes: a history of the world, 1914-1991. Vintage Books, New York.

Howell C. (1984) 'Sandton turns on taps'. Rand Daily Mail, 25 September 1984.

Jeans F. (1983) 'R250-m plan to save water'. The Star, 16 September 1983.

Jensen D. (198

6) 'Waternood vlak by die Vaalrivier: Heelwat huise al leeg'. Die Vaderland, 14 October 1986.

Koch I. and Venter A. (1986) 'Riviere en spruite bruis en die damme word vol: Wnderlike reën!'. Beeld, 1 December 1986.

Krüger D. (1983) 'Waterbeperkings taamlik verslap'. Die Transvaler, 18 January 1983.

Leeman S. (1986a) 'Shortage of pure water may hit consumer'. The Star, 14 May 1986.

Leeman S. (1986b) 'Water crisis caused by bad distribution – expert'. The Star, 22 May 1986.

Martin S. (1988) 'Cycle of wet years may be coming'. The Star, 27 February 1988.

McNamara L. (1983a) 'Reef faces drastic water rationing' in The Star, 13 January 1983.

McNamara L. (1983b) 'No relief in sight for 17 dry suburbs'. The Star, 14 January 1983.

McQuillan S. (1983a) 'Water chief warns of tough new curbs'. The Star, 25 March 1983.

McQuillan S. (1983b) 'PWV area system over-strained – Minister'. The Star, 5 May 1983.

McQuillan S. (1984) 'State is lambasted over the water crisis'. The Star, 14 February 1984.

Melville B. (1989) 'PWV water costs up by 16% from April 1'. Business Day, 1 March 1989.

Minnie J. (1983a) 'Jo'burg battles to keep the water flowing' in Rand Daily Mail, 14 January 1983

Minnie J. (1983b) 'Eskom makes a promise'. Rand Daily Mail, 19 January 1983.

Minnie J. (1983c) 'Jo'burg clamps down on water'. Rand Daily Mail, 31 March 1983.

Minnie J. (1983d) 'Rationing of water still on the cards'. Rand Daily Mail, 22 November 1983.

Natrass J. (1981) The South African economy: its growth and change. Oxford University Press, Cape Town.

Raine S. (1983) 'How they save water the good ol' US way'. The Star, 21 September 1983.

Rand Daily Mail (1983) 'Official insolence' (Editorial comment), 18 January 1983.

Rumney R. (1988) 'Flowing uphill: gopod rains and a drop in water consumption pose pricing problems'. Finance Week, 12-18 May 1988.

Ryan C. (1984) 'Ratepayers petition mayor for easing of water curbs'. The Star, 26 September 1984.

Scher D.M. (1993) 'The first five-year plan'. In Liebenberg B.J. and S.B. Spies, South Africa in the 20th century. JL van Schaik Academic, Pretoria.

Spoormaker M. (1984) 'Laat ons natlei, versoek tuiniers'. Die Vaderland, 27 September 1984.

Stein J. (1984a) 'Weekend watering times announced by Council'. The Citizen, 10 October 1984.

Stein J. (1984b) 'Randburg hose clampdown to be eased'. The Citizen, 18 October 1984.

Tempelhoff J.W.N. (2003a) The substance of ubiquity: Rand Water 1903-2000.

Tempelhoff J.W.N. (2003b) 'Whatever happened to good old water restrictions?' Rand Water and water demand management (1983-2003)' presented at the biennial conference of the South African Historical Society at the University of the Free State, Bloemfontein on 1 July 2003.

Tempelhoff JWN & Haarhoff, J. (2003) 'The development of an urban awareness of water shortfall: Rand Water and the drought conditions of the 1960s'. First draft presented to the South African Cultural History Conference, Heidelberg 29 August 2003. The research is scheduled for publication.

The Citizen (1984) Correspondence: Thayes, Greenside – Editor, 'Stop threatening water users', 15 February 1984.

The Star (1983a) Municipal reporter, 'End of 'drought' in suburbs' in The Star, 17 January 1983.

The Star (1983b) Correspondence: Under Pressure – Editor, 'No need for bullying over water' in The Star, 12 April 1983.

The Star (1983c) Municipal reporter, 'Sandtonians lead the way in water saving' in The Star, 19 September 1983.

The Star (1984a) 'R10 000 water bills fuel looming Pretoria clashes' (Pretoria correspondent), 8 February 1984.

The Star (1984b) 'Sadton and the drips' (Editorial comment), 26 September 1984.

The Star (1988a) 'PWV water price up 12 pc' (Municipal reporter), 13 April 1988.

The Star (1988b) 'Bargain basement water' (Editorial comment), 18 April 1988.

The Star (1988c) 'Play it again' (Editorial comment), 4 May 1988.

The Star (1989) 'Vaal's dirty washing' (Editorial comment), 2 March 1989.

Unwin J. (1984a) 'Pressure mounts for easing of hosepipe curbs'. The Star, 22 September 1984.

Unwin J. (1984b) 'Water Board chief warns municipalities'. The Star, 29 September 1984.

Van Aardt S. & Kotzé A. (1986) 'Droë boorgate: saak nou by Wentzel'. Vaalweekblad, 10 October 1986.

Van Buuren C. (1988) 'Raadsake Edenvale: busdiens sal behou word'. Die Vaderland, 28 April 1988.

Van Burick N. (2001) 'Gratis water binne 'n jaar in alle gebiede'. Beeld, 7 July 2001.

Van Heerden D. (1983) 'Droogte se rug gebreek, maar...'. Die Vaderland, 17 November 1983.

Vermaak M. (1988) 'Water tariffs are to be reduced'. The Citizen, 3 May 1988

Vogel C. H. (1994) Consequences of droughts in southern Africa, 1960-1992. PhD Thesis, UW.

Wessels N. (1983) 'Voller damme keer rantsoene ... maar waterperke bly' in Die Vaderland, 6 November 1983.

Wessels N. (1985) 'Drastiese plan om waterkrisis af te weer' in Die Valderland, 8 January 1985.

Wigget R. (1986) 'Strate word strome' in Vaalweekblad, 31 October 1986.

Woodgate S. (1988) 'Renewed water curbs unlikely in near future' in The Star, 13 January 1988.

Woodgate S. (1995) 'Harsh water cuts' in The Star, 16 February 1995.

# 37
## WATER AND NEPAL – AN IMPRESSION
*Sanna-Leena Rautanen*

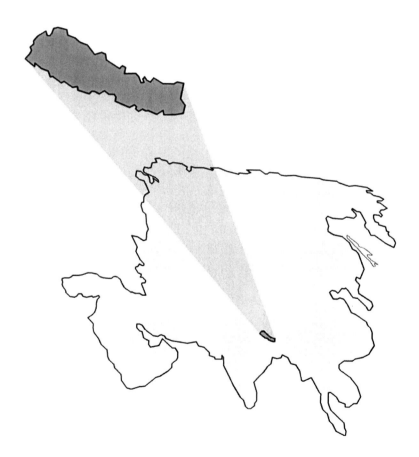

Figure 37.1 Location of Nepal.

## 37.1. GENERAL BACKGROUND

The Kingdom of Nepal is a remote Himalayan country with immense diversity in both geographical features and culture. Until the 1950s, Nepal was practically isolated from the outside world, with the majority of the people living in medieval conditions. Even if Nepal has been catching up by many human development indicators, it still remains behind many of its South Asian neighbours and is one of the poorest countries in the world. Nepal has a young, unstable democracy, which at this writing has experienced radical changes over the past year. Instead of His Majesty's Government of Nepal, there now is the Government of Nepal working out a new future.

Traditional technologies are still strongly present in most locations, even in the quickly changing urban areas. Traditional technologies are used daily in many rural practices, with several communities returning to their old *Kuwa* or *Dhara* (traditional water sources, springs which are sometimes protected) for drinking water after the more modern systems have failed.

The Kingdom of Nepal is a landlocked Himalayan country bordering China in the North and India in the South. Ancient Indian cultures from the South and Tibeto-Burmese cultures from the North have influenced Nepal, and the impact is seen in its cultural, ethnic and linguistic diversity. This mountainous country of 23.5 million people has more than 100 caste/ethnic groups and some 60 distinct languages or dialects. The climate ranges from tropical Terai in the Southern belt to the Himalayan mountain range in the Northern Region, the lowest point being 70 m at Kanchan Kalan and the highest at 8,848 m on Mount Everest. Extreme slopes, high snow-blocked passes, flooding rivers and land slides, as well as tropical lowlands with tropical hazards have historically led to the evolution of strongly localised cultural and religious patterns, and persistent poverty. Nepal is predominantly a rural society with 80 per cent living in the rural areas.

Geographically and economically, Nepal belongs to the Indo-Gangetic plain. The Himalaya-Ganga represents a complex highland-lowland interactive system with diverse biophysical and human-built environments having significant symbolic cultural and religious meanings. Water shapes and affects all aspects of life, from worship to the location of settlements, from agricultural practices to logistics. Many sacred rivers of the Hindu religion have their origin in the Himalayan range. In addition to the Hindu population, Nepal also has Buddhist, Muslim, and several localised animistic and combined religious traditions, with water having many deep ritualistic meanings in various traditions — Hindu, Buddhist or animistic alike. (Sharma 2003)

There has been some encouraging progress since Nepal started to modernize its public administration and infrastructure over 50 years ago. For example, access to basic services, including primary education, health care and water supply, have increased significantly, as have the number of households with access to electricity and sanitation. The regional disparities are evident. Nepal is one of the poorest countries in the world, with a per capita income of USD 230 per year. According to the World Development Report 2005, 42% of the total population and 44% of the rural population live below the national poverty line. Assessed against the international poverty line, 37.7% of population live below one USD per day, and 82.5% below two USD. (World Development Report 2005)

## 37.2. A SHORT POLITICAL HISTORY OF NEPAL

Nepal was controlled by a hereditary autocratic and highly centralised Rana prime ministership between 1846 and 1951. Until the 1950s, the majority of the people lived in medieval conditions. The first political party in Nepal was established only in 1946, and the first constitution of Nepal was adopted in 1948. Nepal joined the United Nations in 1955, and during the late 1950s its diplomatic missions were established, marking the beginning of foreign aid to and development projects in Nepal.

Decentralisation is one of the issues which have been on the official development agenda since the First Five Year Development Plan (1956-61). On the other hand, *democratic* local governance and development have always been rocky. The first general election was held in 1959, but in 1960, King Mahendra dismissed the cabinet, dissolved parliament, and banned political parties. The 1962 Constitution, the third constitution since 1951, created a non-party *Panchayat* (council) system of government. Dahal et.al. note that under the "*Panchayat Democracy*" no political parties were allowed, human rights were denied, corruption was rampant, state money was misused, economic development did not take place, and "nationalist" slogans were given, yet, there were no improvement in the day-to-day life of the people. (Dahal et al. 2002).

A modified version of the *Panchayat* system was approved after a 1980 referendum. Direct parliamentary elections were held in 1981. In the late 1980s there was a serious economic crisis, which followed a dispute with India and the subsequent closing of most border crossings. There were strong demands for political reform, and after months of violence, King Birendra dissolved parliament. The opposition formed an interim government, and a new constitution creating a constitutional monarchy and a bicameral legislature became effective in late 1990. Multiparty legislative elections were held in 1991, marking the beginning of democracy in Nepal. Mid-term elections in late 1994 resulted in a hung parliament and the communists emerged as the single largest party, forming a minority government. Yet, the communist government was dissolved already the next year, in 1995, provoking a radical leftist group, the Nepal Communist Party (Maoist), to begin an insurgency in rural areas aimed at abolishing the monarchy and establishing a people's republic. (Dahal et al. 2002)

Despite decades of foreign aid and projects, Nepal's economic growth has not improved noticeably over the years and by many indicators, Nepal is still amongst the poorest countries in the world. One reason why the gain achieved by developmental activities is concealed is the estimated population growth. (Central Bureau of Statistics 2001). Yet, there are a number of other reasons for poverty too. Sustainable development and all other efforts for human development and poverty reduction are seriously undermined by the continued political instability, human rights violations and such governance problems as corruption, lack of transparency and the absence of democratically elected local governments.

Many donors have laid increasing emphasis on policy reforms. Economic liberalisation, decentralisation, human rights, democracy, civil society and good governance are amongst the preconditions for development assistance today. As a matter of fact, the outside advice and resources have influenced the governance vision and objectives, effectively embedded in the Constitution of 1990 which made strong provisions for decentralisation.(Dahal et al. 2002)

The practical aspect of decentralization is stipulated in the Local Self-Governance Act of 1999, which seeks to delegate authority and responsibility to local bodies, empower local authorities to collect taxes and develop local plans. These initiatives, including social mobilization, should have been implemented and strengthened in a gradual manner by strengthening local government institutions. According to the Tenth Plan, adopted in 2003, *"Decentralization is an important mechanism for improving service delivery to local communities and enhancing effectiveness of public spending."* (The Tenth Five Year Development Plan 2003, para 146 & 147). Should the situation return to normal, this act will have very practical implications for local water resource management by stipulating the legal framework for users' groups, i.e. water users' committees.

## 37.3. HEALTH AND POPULATION

The first population census in Nepal was made in 1911-12. It reported 5,638,749 persons, but was also realistic about this figure: it was by no means the correct figure due to very difficult logistics. The next census for 1950 counted 8.5 million people, the 1960 census 9.4 million. In 1980 the population count was up to 14.6 million. The latest figure is 24.2 million. Yet, this is a figure that varies, depending on who is reporting it based on what. See Figure 37.2 for the total population 1950-2025, with the expected growth of the urban population. Population growth has remained around two percent annually, and is now estimated to be 2.2 per cent in rural areas and 5.5 per cent in urban areas, the overall total population growth rate being 2.5 per cent. This has had many impacts. For instance, population growth has led to fragmented land holdings and the depletion of forest products, the ratio of the population to arable land being one of the highest in the world. (Population Division of the Department of Economic and Social Affairs of the United Nations Secretariat 2003).

The average total fertility rate in Nepal has changed from six (1975-1980) to four (2000-2005), being still higher than in Asian countries on average. Life expectancy in Nepal has gradually increased faster than in the world in general. It has increased, but at 60 years, it is still lower in Nepal than its neighbouring South Asian countries. The reduction in infant mortality rate has been a major contributory factor (Figure 37.3). Still, infant mortality rates are among the highest in the region even if it dropped from 100 (per 1,000 live births) in 1990, to about 64 in 2002. (UNICEF 2004). Due to high maternal mortality, life expectancy for women is lower than for men. Note that there are no reliable statistics as many births and deaths go unreported.

Health is about human well-being and in many ways is closely linked to water and sanitation. The health problems of poor water supply have been recognized since the First Five Year Development Plan. A typical water- and sanitation-related health indicator is diarrhoea, which in Nepal accounted for 16-25% of childhood deaths in the early 1990s. (Nepal Human Development Report 1998, 59: Table 4.3) Skin diseases, worms, diarrhoea, dysentery, and gastritis are all rampant in rural areas, and can all be linked to poor water supply and sanitation. Diarrhoea is listed in official hospital records as one of the five leading causes of both mortality and morbidity. (WHO 2003)

## 37.4. WATER USE AND WATER RESOURCES

Nepal is rich in water resources, with mean annual precipitation ranging from more than 6000 mm along the southern slopes of the Annapurna Himalayan range to less than 250 mm in the northern border near the Tibetan Plateau. Water is considered one of the principal natural resources supporting the economy of Nepal. Up to 84 per cent of the electricity in Nepal is produced by hydroelectric generation with an estimated potential of 400 megawatts still being available. (Water Resources Strategy Nepal 2002)

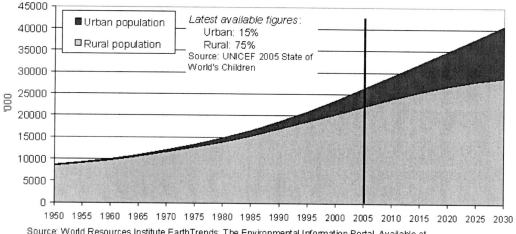

Figure 37.2 Total population of Nepal 1950-2025.

The importance of irrigation development in Nepal was recognized a long time ago. There are numerous small canals (*raj kulos*) from the seventeenth and eighteenth centuries, constructed by the government sector in and around Kathmandu valley. The first large public sector irrigation canal system (the Chandra Canal System) was constructed in 1922 and is still in operation. (FAO 2002) Irrigation remains an important issue as it contributed 40 per cent of GDP in 1996. Of the total estimated irrigated land, 25 per cent was served by a public system and the rest comprised farmer-managed systems. Of the public systems, 91 per cent of total water withdrawal was from surface water. Of all water withdrawal, it is estimated that 99 per cent is for agriculture, that is, for irrigation. The figures are estimated as no irrigation system as such actually measures the volume of water. (FAO 2002)

Non-formal associations have existed for a long time in almost all farmer-managed irrigation systems. Water Users Associations received legal status after the promulgation of the 1992 Water Resources Act, and have since become a prerequisite for the transfer of public schemes to users, similar to drinking water users groups as defined in the Local Self-Governance Act 1990. (FAO 2002)

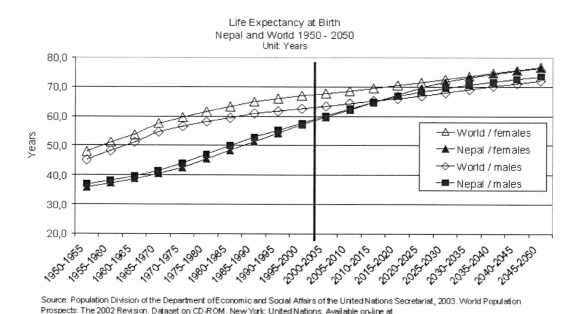

Figure 37.3 Life expectancy at birth, Nepal and world 1950-2050.

There are many ways of estimating the drinking water supply coverage. The Millennium Development Goals for Nepal state the target for the year 2015 by aiming to halve the proportion of people without sustainable access to safe drinking water, from 54 per cent without drinking water in year 1990 to 21 per cent in 2015. (Nepal – Millennium Development Goals Progress Report (2002). The recent mid-term assessment of progress towards the drinking water and sanitation targets by UNICEF and WHO reported that the drinking water coverage in rural areas had improved from 67 per cent in 1990 to 82 per cent in 2002. The respective figures for improved sanitation coverage were 7 per cent and 20 per cent. (UNICEF & WHO 2005). However, estimating especially access to adequate sanitation facilities is very complicated: some count temporary pit latrines, some count only water-flush permanent toilets, some count only latrines constructed in certain programmes, leaving self-help latrines out of the count.

The Millennium Development Goals progress report for Nepal recognises that the local governments must continue to play a key role in responding to the community demands and in the institutional strengthening of the water users' groups. This target sets priorities for development assistance including development of new technologies to enhance rural water supply, such as rainwater harvesting technology, and investigation of the arsenic contamination of the groundwater in Terai districts. (Nepal – Millennium Development Goals Progress Report (2002)

Sanitation is an essential service closely linked to health and the overall quality of life. The sanitation policy of His Majesty's Government of Nepal (HMGN) adopted the key elements of guiding principles of international initiatives, statements and declarations relevant to the sanitation sub-sector. The policy objectives emphasized the links between sanitation and public health, the integration of investments in sanitation into wider awareness and behavioural change programmes, and the need to ensure that all water supply programmes include sanitation as an integral component and vice versa. The strategies comprise general elements as well as specific topics. The specific topics include: the involvement of women, appropriate technology, knowledge and awareness creation, community participation, resource mobilization, legislation, co-ordination and integration, and institutional arrangements.

National Sanitation Action Steering Committee, Nepal, recognises that in its broader context national sanitation policy should aim at improving and sustaining political commitment at all levels, appropriate sanitation-related legislation and enforcement, sufficient funding and resource mobilisation, and well-functioning sanitation technologies, together with further research and development. Participatory approaches to the problem identification, analysis, promotion, implementation, and monitoring and evaluation of the sanitation programmes have to be utilised. (Nepal State of Sanitation Report 1999/2000)

## 37.5. WATER SUPPLY AND SANITATION SERVICES

The systematic development of rural water supplies is very recent. The following outline of the history of water supply and sanitation in Nepal follows the division into phases introduced by Sharma (2004).

**The first phase (1880–1951)** refers to the period of highly centralised Rana rule. The few attempts to construct infrastructure dealt mainly with roads, bridges and some public buildings. According to Sharma (2003), the first comprehensive law enacted in Nepal was *The Mulki Ain* in 1854. It related to administrative and personal issues, paying attention to the caste hierarchies with even execution following from disobeying caste norms. The *"water line"* referred to the caste from which water was accepted. Another water issue included in the early legal text referred to irrigation and access to irrigation water. (Sharma 2003) Until the mid-1900s, there was a high mortality rate in Kathmandu Valley due to cholera, typhoid and smallpox. Poor sanitation, a lack of education, poor nutrition and the absence of health care services added to the high morbidity of the general public. The first piped water was brought to Kathmandu in 1891 in the form of the *Bir Dhara Works*. Direct connections were provided for the Rana palaces and homes of ruling elite. Access to the piped water supply was a status symbol and an indication of a close link to the ruling class. Some stand posts were provided for the general public as "a gift from the rulers," one cinema was served and many fountains constructed. New systems were built as the population grew, and *Bir Dhara Works* was followed by the *Tri Bhim Dhara* system in 1928. The first water works office was established during the following year, 1929. (Dixit 2002, 252-253)

**The second phase (1951-1970)** covers the periods of the First, Second and Third Five Year Development Plans, the beginning of the efforts to modernise Nepal. All of these plans mention drinking water, the focus remaining on agriculture and irrigation with hydro-power potential being highlighted. During these plans, piped water was introduced to urban areas. The Third Plan listed district and zonal headquarters as focus areas, and suggested that information should be gathered also from elsewhere. Also an *Experimental Village Sanitation Programme* was introduced, supported by UNICEF, in the village of Lalitpur.

**The third phase (1971-1980)** covers the periods of the Fourth and Fifth Five Year Development Plans. In the beginning of this period, the coverage of piped drinking water was six per cent, according to the Fourth Plan. It further stated that: *"Drinking water projects have been undertaken or are in progress in 12 zonal and 33 district headquarters out of a total of 14 and 75 respectively. During the Plan period, 37 drinking water projects in district and zonal headquarters and selected village Panchayats will be started."* (The Fourth Five Year Development Plan 1970-1975) The interest was extended from the main urban areas to the district headquarters. The fifth plan also makes a note of a rural drinking water programme, but does not specify the number of projects under this heading. (The Fifth Five Year Development Plan 1975-1980)

During this period, drinking water was included under the social services together with health and education, the sector receiving 20% of the budget allocated for the *Panchayats*. There was a new emphasis on basic needs and consequently also the national development plans allocated more budget money for the drinking water supply. The World Bank gave a loan to improve drinking water and wastewater services in 1972. The Department of Water Supply and Sewerage was established in 1972 and the Water Supply and Sanitation Board in 1974. The Department of Water Supply together with UNICEF were the key actors, the Department serving communities with more than 1500 persons and UNICEF acting through the Ministry of *Panchayat* and Local Development with its Community Water Supply (CWS) programme. CWS introduced the idea of water supply as the entry point into reduced child mortality and health improvements. In these plans "sanitation" refers to making plans and surveys in major urban areas already served by drinking water. (Dixit 2002, 253; Sharma 2004, 48-50)

**The fourth phase (1981-1990)** covers the periods of the Sixth and Seventh Five Year Development Plans. It extended the serious interest from the District headquarters to outlying areas along with the International Drinking Water Supply and Sanitation Decade. During this period, the drinking water supply became the third most important sector for investment in the national development plans, surpassing such other interest areas as transport, communications, agriculture and irrigation. Halfway through the decade, it was suggested that the goals set for universal coverage were too ambitious and that more serious effort was needed. At this point, a large number of donors, international non-governmental organisations (INGOs) and non-governmental organisations (NGOs) stepped in. Finland started to support the Rural Water Supply and Sanitation project in 1990, and become one of the main donors in the rural water supply sector at the time.

**The fifth phase from 1990** onwards brought water also to the more remote areas. The water supply and sanitation sector become one of the priority sectors for the Government investment in Nepal as a part of the poverty reduction strategy. The overall sector objectives of the government are to improve the health and productivity of the people by making safe water supply and sanitation facilities available to the entire population. The access to piped water supply in rural areas rose from 16.3 per cent in 1991 to 29.1 per cent in 1996. The objective for sanitation is more challenging. Less than one-fifth of the households have a latrine. Poor sanitation is clearly reflected in the health figures, as the leading cause of childhood deaths is diarrhoea, causing 16-25 per cent of the total deaths. (Nepal Human Development Report 1998, 69) The Ninth Five Year Development Plan (1997-2002) aimed at achieving full water supply coverage by the end of the plan period. In practice by the end of the plan period, the drinking water supply coverage had changed from 60 per cent to 71.6 per cent. Consequently the Tenth Plan (2003-2007) no longer aimed at full coverage but the target was set at 85 per cent of the population being served by basic water supply and 50 per cent by sanitation facilities by the year 2007. Note that the calculations for the service coverage vary, the figures being drastically different depending on who and how the figure was counted. To reach the targets the Tenth Plan aimed at the expansion of the sanitation facilities, upgrading water supply *service levels* and the rehabilitation and major repairs where needed. (Figure 37.4; Water Resources Strategy Nepal 2002, 12-14; The Tenth Plan 2003)

## 37.6. CONCLUSIONS: WATER FUTURES

Water and sanitation services should be placed into a wider context of sustainable livelihoods recognising water, health and sanitation as human rights. These are elements of human development, and have a role to play in environmental sustainability and the integrated management of water resources, as well as in the elimination of poverty. Nepal is geographically challenging in many ways: landslides, flooding and climatic conditions are some of the aspects that shape the options. Natural conditions alone set a locality-specific starting point for sustainable water management, livelihoods and governance. Adding cultural diversity, political instability, difficult logistics and communications, strong regional disparities on the level of human development and well-being, and many other human factors, the only certain conclusion

Figure 37.4 Women queuing by the traditional but now protected water source. (Photo: Sanna-Leena Rautanen)

is that there cannot be a one-fit-for-all solution. For instance, water sector action can do a lot in poverty reduction and as poverty is about everyday life and about personal experience, problems caused by poverty are very community- and household-specific. Consequently the solutions should be made available accordingly. An enabling environment should be truly dynamic as ultimately the decisions and related actions have to be strongly localised. Human resource development and overall capacity building at all levels is a must.

Poverty alleviation is about culture. A culturally sensitive, dynamic, adaptive and systemic approach to poverty alleviation through water action should make the most of the lessons learned in the past. Appropriate and locally feasible technological options respond to actual needs, can be constructed cost-effectively with the local skills and local materials, and thus can be replicated to increase the number of households with improved sanitation facilities. A revised approach should also make the most of the World Water Forum 2003 statement principles by providing options to build on people's energy and creativity at all levels, recognising the difference between the various actors and internal dynamics of local culture. A culturally poverty-sensitive approach is about empowerment and builds the capacity of people in households and communities to take action based on the situation and experiences in each unique location. It recognises that committed and compassionate leadership and good governance are needed for changing long-accustomed roles and practices, leading to new responsibilities of authorities and institutions to support households and communities in the management of their water resources. Formal and informal institutions have to be recognised, as these have to support the change. Traditional practices and cultural diversity should be seen as strengths, and local capacity, indigenous knowledge and creativity should be encouraged.

## 37.7. REFERENCES

Central Bureau of Statistics (2001) Nepal in Figures, Kathmandu, Nepal. http://www.cbs.gov.np

Dahal D.R., Uprety H. & Subba Ph. (2002). Good Governance & Decentralisation in Nepal. Centre for Governance & Development Studies in cooperation with Friedrich-Stifting (FES). Modern Printing Press, Kathmandu, Nepal.

Dixit A. (2002) Basic Water Science. Supported by World Health Organisation and Nepal Water for Health. Nepal Water Conservation Foundation. Format Printing Press, Kathmandu.

FAO, Food and Agriculture Organization of the United Nations (2002) Water Resources, Development and Management Service. AQUASTAT Information System on Water in Agriculture: Review of Water Resource Statistics by Country. Nepal Country Profile version 1999. Rome: FAO. http://www.fao.org/ag/agl/aglw/aquastat/main/index.stm

The Fifth Five Year Development Plan 1975-1980. Appendix 2 Allocation of the Fifth Plan Outlay, CD ROM.

Fourth Five Year Development Plan 1970-1975, Chapter XXIV Drinking water and Sanitation. CD ROM

Nepal Human Development Report (1998) Nepal South Asia Centre, submitted to United Nations Development Programme, Kathmandu, Nepal. 295 p. http://www.undp.org.np

Nepal – Millennium Development Goals Progress Report (2002) The Millennium Development Goal 7, Ensuring Environmental Sustainability, Target 10. United Nations Country Team of Nepal & HMG Nepal, Jagadamba Press, Kathmandu, Nepal. http://www.undp.org.np

Nepal State of Sanitation Report (1999/2000) National Sanitation Action Steering Committee, Kathmandu, Nepal.

Population Division of the Department of Economic and Social Affairs of the United Nations Secretariat, 2003. World Population Prospects: The 2002 Revision. Dataset on CD-ROM. New York: United Nations. Available on-line at http://www.un.org/esa/population/ordering.htm. Also at the World Resource Institute Earthtrends http://earthtrends.wri.org/searchable_db/

Sharma S. (2003) Water in Hinduism: Continuities and Disjunctures between Scriptural Canons and Local Traditions in Nepal. Water Nepal, Vol. 9/10, No. ½, 2003, Journal of Water Resources Development. Nepal Water Conservation FoundationFormat Printing Press, Kathmandu. p.215-247. ISSN 1027-0345.

Sharma S. (2004) Cost-Effectiveness of rural water programmes – Technology choice or malpractice? In Sharma, S., Koponen, J., Gyawali, D. & Dixit, A. Aid Under Stress – Water, Forests and Finnish Support in Nepal. Himal Books for Institute of Development Studies, University of Helsinki, and Interdisciplinary Analysts, Kathmandu. Institute of Development Studies, University of Helsinki. pp.44-57. ISBN 99933-43-48 X

The Tenth Plan - Poverty Reduction Strategy paper 2002-2007 (2003) Summary in English, National Planning Commission, His Majesty's Government of Nepal, Kathmandu, Nepal. May 2003.

United Nations Childrens' Fund (UNICEF) (2004) Monitoring the Situation of Children and Women. Available on-line at: http://www.childinfo.org/. New York: UNICEF. And: United Nations Children's Fund (UNICEF). 2004.State of the World's Children: Girls, Education, and Development. Available on-line at: http://www.unicef.org/sowc04/. New York: UNICEF. World Resources Institute web-source http://earthtrends.wri.org/

UNICEF and WHO (2005) Meeting the MDG Drinking Water and Sanitation Target – A Mid-Term Assessment of Progress. Country, regional and global estimates on water and sanitation. http://www.unicef.org/wes/mdgreport/who_unicef_WESestimate.pdf; http://www.unicef.org/wes/mdgreport/waterCoverage0.php; and http://www.unicef.org/wes/mdgreport/sanitation0.php

Water Resources Strategy Nepal (2002) His Majesty's Government of Nepal, Water and Energy Commission Secretariat, Singha Darbar, Kathmandu, Nepal. January 2002.

WHO (2003) Country Health Profile Nepal. WHO Regional Office for South-East Asia. Last modified 29 October, 2003. http://w3.whosea.org/cntryhealth/nepal/

World Development Report (2005) A Better Investment Climate for Everyone. A Co-publication of The World Bank and Oxford University Press, 2004, The International Bank for Reconstruction and Development / The World Bank, Washington www.worldbank.org. Oxford University Press, New York.

World Resources Institute (2005) Population, health and human well-being. Searchable database at http://earthtrends.wri.org/searchable_db/ Note: various dates for the latest up dates.

# 38
# IMPROVEMENT OF THE LIVES OF SLUM DWELLERS - SAFE WATER AND SANITATION FOR THE FUTURE CITIES OF THE FOURTH WORLD

*Jarmo J. Hukka & Osmo T. Seppälä*

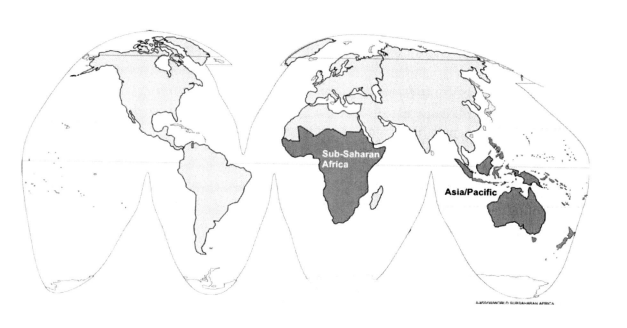

Figure 38.1 Key regions mentioned in the chapter.

## 38.1. GENERAL BACKGROUND

Slums are a physical and spatial manifestation of increasing urban poverty. *Poverty* is not defined only as a lack of income and economic resources, but it covers the concept of vulnerability and, i.a. safe drinking water, sanitation services and infrastructure. Regardless of the characterization of slums, slum dwellers face greater challenges such as higher morbidity and infant mortality rates than either non-slum dwellers or the rural population.

At the start of the third millennium, 47 per cent of the world's population lived in urban areas. Within the next two decades, this figure is expected to increase to 56 per cent. Even more challenging is the fact that 98 per cent of the projected global population growth during the next two decades will occur in developing countries. The vast bulk of this increase (86 per cent) will occur in urban areas. Of the total world's urban population increase, 94 per cent will occur in developing countries.

This population increase implies that about 39,000 new dwelling units will be required each and every day in developing countries during the next two decades – to cater for population growth alone. An increasing percentage of urban dwellers are earning their livelihood from the informal sector. One of the results of these trends has been a rapid growth of slums and informal settlements, where more than half of the population in many cities and towns of developing countries are currently living and working (UN-Habitat 2002).

Target 11 indicates directly that slums are a development issue that needs to be addressed. Slums can no longer simply be considered as an unfortunate result of urban poverty, but need to be envisioned as a major development issue.

Access to water and sanitation underpins every other issue of development in the world's poorest countries — from improving access to education to tackling disease, eradicating hunger and promoting gender equality. The significant economic benefits to the world, and particularly to developing countries can be expected, if the Millennium Development and World Summit on Sustainable Development goals are met.

It is estimated that up to one-third of the world's urban population lives in slums (Table 38.1). The comparatively more rapid growth in urban areas of developing countries suggests that the problems associated with slum dwelling will worsen in those areas that are already vulnerable. More than 70% of the least developed countries and of Sub-Saharan Africa's urban population lived in slums in 2001 (UN-Habitat 2003b). For example, more than one million people in the

Kenyan capital, Nairobi, are living in destitute situations that need improvement (Kithakye 2004). Regardless of the characterization of slums, slum dwellers face higher development challenges such as higher morbidity and infant mortality rates than either non-slum dwellers or the rural population (UN-Habitat 2003b).

As a part of the United Nations Millennium Declaration, the international community has made a commitment to the world's poor, the most vulnerable, in precise terms, established in quantitative targets. One of the three targets of the Millennium Development Goal (MDG) 7 "Ensure Environmental Sustainability" is Target 11: "By 2020, to have achieved a significant improvement in the lives of at least 100 million slum dwellers". This target comes in response to one of the most pressing challenges of the Millennium. The Expert Group Meeting on *Defining Slums and Secure Tenure* held in Nairobi in November 2002 recommended five key dimensions for improving slums: Access to safe water, Access to sanitation, Secure tenure, Durability of housing and Sufficient living area (UN-Habitat 2003b).

Table 38.1 Distribution of the world's urban slum dwellers, 2001 (UN-Habitat 2003a)

| Region | Urban population (000) | % in total population | % slum dwellers in total urban population |
|---|---|---|---|
| Sub-Saharan Africa | 231,052 | 34.6 | 71.9 |
| Asia Pacific | 1,211,540 | 35.4 | 43.2 |
| Latin America and Caribbean | 399,385 | 75.8 | 31.9 |
| Middle East and Northern Africa | 145,624 | 57.7 | 29.5 |
| Transition economies | 259,091 | 62.9 | 9.6 |
| Advanced economies | 676,492 | 78.9 | 5.8 |
| **World** | **2,923,184** | **47.7** | **31.6** |
| *Developing countries* | *2,021,665* | *40.9* | *43.0* |
| *Least developed countries* | *179,239* | *26.2* | *78.2* |

Access to water and sanitation underpins every other issue of development in the world's poorest countries – from improving access to education to tackling disease, eradicating hunger and promoting gender equality (Burrows, Acton, and Maunder 2004). Evaluating the health and the socio-economic benefits of safe water and adequate sanitation results in a compelling argument in support of further resource allocations to improving access. Based on the present analysis, achieving the water and sanitation MDG target would definitely bring economic benefits, ranging from USD 3 to USD 34 per USD invested, depending on the region. Additional improvement of drinking-water quality, such as point-of-use disinfection, in addition to access to improved water and sanitation, would lead to a benefit ranging from USD 5 to USD 60 per USD invested (Hutton and Huller 2004).

From a health point of view, achieving the water and sanitation MDG target, by using simple technologies, would lead to a global average reduction of 10% of episodes of diarrhoea. Choosing more advanced types of technologies such as provision of regulated in-house piped water would lead to massive overall health gains (Hutton and Huller 2004).

The International Drinking Water Supply and Sanitation Decade (IDWSSD 1981–90) provided a considerable extension of basic services. In fact, the IDWSSD was very successful: 1,000 million people gained access to safe water and 700 million got sanitation. Yet, presently 1.1 billion people lack access to improved water services and 2.4 billion to improved sanitation. Yet, of all the target-setting events of recent years, the UN Summit of 2000, which set the Millennium Development Goals for 2015, is the most influential. One of the goals is: *to halve the proportion of people without access to safe drinking water* (UN General Assembly 2000).

A Ministerial Conference at The Hague, the Netherlands in March 2000, agreed on a Ministerial Declaration on Water Security in the 21$^{st}$ Century. The Ministerial Declaration addressed seven major challenges, including *meeting basic needs*, i.e. it was recognized that access to safe and sufficient water and sanitation are basic human needs and are essential to health and well-being, and to empower people, especially women, through a participatory process of water management. Furthermore, the World Summit on Sustainable Development in Johannesburg, South Africa, from 26 August to 4 September 2002 also added the 2015 target of *reducing by half the proportion of people without sanitation* (UN 2002 and WSSD 2002). The UN General Assembly also adopted a resolution in December 2003, by which it proclaimed the period from 2005 to 2015 the International Decade for Action, *"Water for Life."* (UN General Assembly 2003).

The UN-Habitat (2004) report points out that it has been known for more than a decade that urban-based economic activities account for more than 50 per cent of GDP in all countries. This economic contribution of urban settlements is normally ignored or overlooked. Undervaluing urban areas can place the economic and social futures of countries at risk. Therefore new strategies should be identified for protecting and sustaining urban economies and subsequently, national economies, over the long term.

## 38.2. BASIC CONCEPTS AND DEFINITIONS

### Urban locality

Urban population size is defined as the number of persons residing in urban localities (UN 1989). Different countries use diverse criteria to define an *urban locality* reflecting a variety of social and geographical conditions. The most common criterion is a minimum number of persons residing in the locality, e.g. at least 200 persons in Denmark, Greenland, Iceland, Norway and Sweden, 20,000 persons or more in Mauritius and Nigeria, and 30,000 or more persons in Japan. Most countries choose a minimum between 2,000 and 5,000 persons. Some countries use quite different criteria to define an urban area, such as the number of dwelling units in a locality, population density, types of economic activity and living facilities.

The figures are based on a UN publication (1989) in which, instead of imposing uniform definitions on all countries, national definitions of urban localities are accepted.

- An urban agglomeration is defined as an area with a population concentration that usually includes a central city and surrounding urbanized localities.
- A large agglomeration may comprise several cities and/or towns and their suburban fringes. Although this concept is common in national statistics, some countries use the concepts of metropolitan area or city instead.
- A metropolitan area is similar to an urban agglomeration, but may be a specially designated administrative unit and may sometimes cover certain rural areas in terms of population characteristics.
- A city is not usually regarded as the same as an urban agglomeration. When a city is used to represent an agglomeration, the size of the agglomeration is generally underestimated, unless the areas of population concentration happen to be solely within the boundaries of the city.

### Urbanization and urban growth

Hardoy and Satterthwaite (1989) define *urbanization* as follows:

> The process by which an increasing proportion of the population comes to live in urban centres. A nation which is urbanizing has an increasing proportion of its population living in urban centres—but the term *urbanization* usually implies not only this change in the distribution of population but also processes which cause this change which are usually a combination of economic, social and political change.

*Urban growth is* defined by them in this way:

> The growth of the population living in urban centres. This is not the same as urbanization, because if the rural population and the urban population are both growing at the same rate, there is urban growth but not necessarily growth in the proportion of people living in urban centres.

The *level of urbanization* means the percentage of the population living in urban centres. Since the figures for the level of urbanization in a nation are based on that nation's own definition as to what is an urban centre, they are not comparable figures for other nations.

**Peri-urban areas, slums and the Fourth World**

The urban poor population is seldom homogeneous and the type of urban community affects the strategies and methods to be used in a community services development programme. In relation to the use of these different terms, Rossi-Espagnet (1984, cited by Harpman, Lusty and Vaughan 1988) suggests the following definitions:

- *Shanty towns:* once a commonly used term, but now considered pejorative, referring to the external view that the low-income settlements are only makeshift huts.
- *Slum:* usually referring to the old, deteriorating tenements in the city centre (originating from the word slump meaning 'wet mire' where working-class housing was built during the British industrial revolution in order to be near the canal-based factories).
- *Squatter settlements:* originally referring to the fact that the inhabitants squat on, or do not have legal tenure to, the land but now often referring to the new slums where the inhabitants sometimes do have legal title. *Squatments* is contrived from squatter settlements to include a broader range of the new slums and not simply to imply that all the inhabitants in such settlements are squatting. Besides this familiar term, many adjectives have been officially applied to specify further the settlements, among them marginal, transitional, uncontrolled, spontaneous, sub-integrated, non-planned, provisional, unconventional, and autonomous.

There are also many different local terms for low-income urban settlements (Harpman et al. 1988; Bairoch 1988; Peri-Urban Network... 1993 and UN-Habitat 2003c):

- Brazil: favelas, alagados, vilas de malocas, corticos, mocambos
- Peru: barriadas, pueblos jovenes, barrios marginales
- Venezuela: barrios, ranchos
- Mexico: colonias proletarias, colonias paracaidistas, jacales, ciudades, perdidas, asentamientos irregulares, colonias populares
- Panama: barnida de emergencia
- Chile: poblaciones, callampas, campamentos
- Ecuador: barrios, urbanizaciones, ranchos
- Argentina: villas miserias
- Colombia: barrios clandestinos, tugurios, invasion
- El Salvador: colonias ilegales, tugurios
- Morocco: bidonvilles (can or drum towns)
- Tunisia: gourbivilles, bidonvilles

- Ethiopia: chica
- Zimbabwe: periurban septic fringes
- Turkey: hisseli tapu, gecekondu, gecekodular (erected in a single night)
- Iran: halabi abad (canned-foods town), alatchir (peasant hut), and gode (hole, quarry)
- Iraq: serifas (hut)
- India and Pakistan: bustees, jompris, chawls, ahatas, cheris, katras, juggies
- Korea: panjachon
- The Philippines: barong-barong
- Indonesia: kampong (little village)
- Kenya: vijiji (village)

Definitions of slums and squatter areas may vary from region to region and from city to city. Slums usually are defined as run-down housing in older, established, legally built parts of the city. Squatter settlements are mainly uncontrolled low-income residential areas with an ambiguous legal status regarding land occupation. *Peri-urban areas* are commonly referred to as squatter settlements, marginal, transitional, and uncontrolled neighbourhoods, the informal sector, low-income areas, shantytowns, urban slums, or illegal settlements (Bairoch 1988, WASH 1992). A broad definition of peri-urban areas by the Peri-Urban Network (WASH 1992) also includes inner-city tenement buildings and low-cost boarding houses. Peri-urban areas, the neighbourhoods and parts of cities where most of the urban poor live, usually have extremely limited or no access to central water supplies, sewage or septic systems, garbage collection, and other services.

The term slum is used in this chapter to describe a wide range of low-income settlements and/or poor human living conditions. The term slum includes the traditional meaning, i.e., housing areas that were once respectable or even desirable, but which have since deteriorated, as the original dwellers have moved to new and better areas of cities. The condition of the old houses has then declined, and the units have been progressively subdivided and rented out to lower-income groups. A typical example is the inner-city slums of many historical towns and cities in both the industrial and the developing countries (UN-Habitat 2002).

The term slum has, however, come to include also the vast informal settlements that are quickly becoming the most visual expression of urban poverty. The quality of dwellings in such settlements varies from the simplest shack to permanent structures, while access to water, electricity, sanitation and other basic services and infrastructure tends to be limited. Such settlements are referred to by a wide range of names and include a variety of tenurial arrangements. UN-Habitat (2003b) defines urban slums as residential areas that lack adequate access to water and sanitation, security of tenure, with a poor structural quality of housing and insufficient living area (Table 38.2).

Table 38.2 Slums: five key definitions (UN-Habitat 2003b)

| Key slum indicators | Definition |
| --- | --- |
| 1 Access to improved water | A household is considered to have access to improved water supply if it has a sufficient amount of water for family use, at an affordable price, available to household members without being subject to extreme effort, especially to women and children. |
| 2 Access to improved sanitation | A household is considered to have adequate access to sanitation if an excreta disposal system, either in the form of a private toilet or a public toilet shared with a reasonable number of people, is available to household members. |
| 3 Security of tenure | Secure tenure is the right of all individuals and groups to effective protection by the state against forced evictions. People have secure tenure when there is evidence of documentation that can be used as proof of secure tenure status, and when there is either *de facto* or perceived protection from forced eviction. |
| 4 Structural quality/durability of dwellings | A house is considered 'durable' if it is built on a non-hazardous location and has a structure that is permanent and adequate enough to protect its inhabitants from the extremes of climatic conditions such as rain, heat, cold and humidity. |
| 5 Sufficient living area | A house is considered to provide a sufficient living area for the household members if not more than two people share the same room. |

The term "fourth world" has been used by UNESCO (1980, cited by Tabibzadeh, Rossi-Espagnet and Maxwell 1989) to describe a sub-proletariat whose housing, sanitation, clothing, and food are inadequate; whose cause is not championed by politicians and unions; who have limited information, education, and voice; and who, because of indifference or intolerance, and the way that they are affected by the law and by administrative practice, are systematically prevented from exercising the rights that other people take for granted.

Less than half of the population in urban centres of Africa, Asia and Latin America have water piped into their homes. Less than one third of them have adequate sanitation (Table 38.3; UN-Habitat 2003c).

Table 38.3 Developing regions: Urban population lacking access to improved water and sanitation (UN-Habitat 2003c, modified by the authors 2005)

| Sub-region | Urban population (%) | % Urban classified slum | Population lack of improved water (%) | Population lack of improved water | Population lack of improved sanitation (%) | Population lack of improved sanitation |
|---|---|---|---|---|---|---|
| Northern Africa | 52 | 28.2 | 3.8 | 2,876,330 | 19.12 | 14,472,500 |
| Sub-Saharan Africa | 34.6 | 71.9 | 18.1 | 41,820,410 | 56.7 | 131,006,480 |
| Latin America and the Caribbean (including Bermuda) | 75.8 | 31.9 | 7.2 | 20,875,572 | 19.7 | 78,678,840 |
| Eastern Asia | 39.1 | 36.4 | 5.6 | 29,858,190 | 32.7 | 174,350,510 |
| Eastern Asia excluding China (optional) | 77.1 | 25.4 | 2.6 | 1,592,630 | 23.8 | 14,578,690 |
| South-central Asia | 30 | 58 | 6.9 | 31,221,390 | 34.3 | 155,202,010 |
| South-eastern Asia | 38.3 | 28 | 10 | 20,285,400 | 14.6 | 29,616,680 |
| Western Asia | 64.9 | 33.1 | 9.1 | 11,369,810 | 18.2 | 22,739,620 |
| Oceania (excluding New Zealand and Australia) | 26.7 | 24.1 | 18 | 372,960 | 9.5 | 196,840 |
| Total | 40.9 | 43 | 8.3 | 168,152,850 | 30.7 | 620,842,190 |
| World | 47.7 | 31.6 | | | 21.2 | |

For many slum dwellers, the official national statistics belie the actual conditions, the actual living environment where up to 500 people have to share one toilet or communal tap. In Mahira, a section of the Haruma slum in Nairobi, there is one self-help toilet with ten units and two bathrooms for a settlement of 332 households with 1,500 inhabitants. Individual city studies

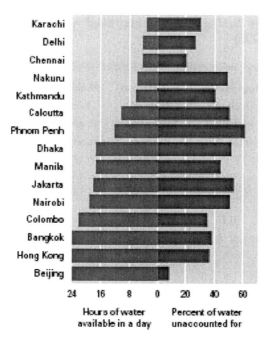

Figure 38.2 24-hour water: a pipe dream (Human Settlements Program 2003, cited by World Bank 2003).

show that the number of urban dwellers who are inadequately served is much higher than officially acknowledged (UN-Habitat 2003d).

Because a quarter to more than half of urban water supply is unaccounted for and the supply is limited, the access points must be widely shared, which dramatically increases waiting times and often simply overwhelms the system (Figure 38.2). The poor people bear the disproportionate share of the impact of inefficient water and sanitation services.

## 38.3. APPROACHES TO SERVE THE URBAN POOR

Biswas (2000) considers it most important to continue radically changing the mind-sets of the decision-makers and water managers in order to safeguard safe drinking water and sanitation systems for urban citizens within the next generation. Otherwise there would not be the adequate political will and sound system management needed to resolve this difficult and complex problem, and to change "business-as-usual" attitudes.

Since most infrastructure systems in the majority of countries rarely deliver a sustained level of service, or in economic terms, benefits, ways of improving the performance of existing systems must be found. Although there is pressure to create new services for the growing populations, the policy must be 'get more from what you have already' (UN-Habitat 2004). This implies shifting more resources to the rehabilitation and maintenance of existing water services systems to ensure that infrastructure continues to support the productivity of urban economies, as well as the environment and health in urban settlements.

World Development Report 2004 (World Bank 2003) recommends various approaches to address failing water and sanitation services. Some approaches try to make services work for poor people through targeted actions. Others seek to improve services overall on the premise that making services work for all is necessary for making them work for poor people. The approaches include, i.a. the following:

- Improving accountability
- Decentralizing assets
- Private sector participation
- Improving regulatory systems
- Reforming the water and sanitation sector
- Charging for services
- Supporting client provision

The lack of good governance principles is regarded as one of the root causes of all major constraints within our societies (UNESCAP 2002). Good governance is participatory, consensus-oriented, accountable, transparent, responsive, effective and efficient, equitable and inclusive, and follows the rule of law (Figure 38.3). It also assures that corruption is minimised, the views

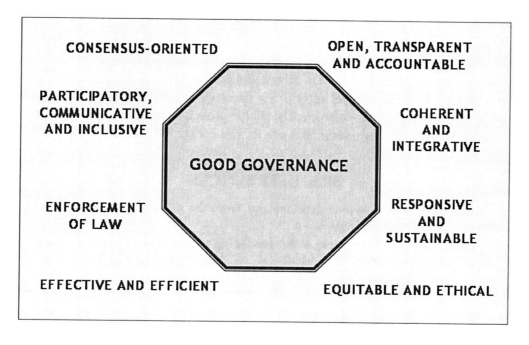

Figure 38.3 Characteristics of good governance (UNESCAP 2002).

of minorities are taken into account and that the most vulnerable in society are listened to in decision-making. It is also responsive to *the future needs of the society*.

Governance can be defined as a process of decision-making and a process by which decisions are implemented or not. Since governance is a process of decision-making and a process by which decisions are implemented, the analysis of governance should concentrate on multiple, complex and inter-related formal institutions. These are, e.g. laws, rules, regulations, and agreements; on informal institutions, such as customs and codes of behaviour; as well as on the stakeholders involved in the decision-making process who implement the decisions made.

Various estimates show that 15-60 per cent of the urban population in developing economies is served by water vendors (i.a., Briscoe 1985, Wegelin-Schuringa 1999, cited by Njiru 2004). Therefore, Njiru (2004) suggests that viable utilities — small water enterprise partnerships — have the potential to improve water services in peri-urban areas. The enabled business environment of small water enterprises (SWEs) would mean improved services especially for the urban poor living in informal settlements.

In South Africa, a quantity of 6,000 litres of free water per household per month was introduced by the government in 2000 (UN-Habitat 2003d). This kind of tariff structuring combined with better non-revenue water and wastage reduction strategies and practices would offer the ways and means to provide adequate services to the urban poor.

## 38.4. CONCLUSIONS

It is much easier to estimate and analyse the situation of water and sanitation services that slum dwellers are facing every day than to provide those services to them in practice. Yet, the various approaches suggested aforesaid should be tested and put into practice by international, national and most importantly by local-level actors. The foreseen future challenges - caused by rapid urbanization and urban growth – also clearly call for proactive futures-oriented approaches to make the essential infrastructure services work for the Fourth World.

## 38.5. REFERENCES

Bairoch P. (1988). Cities and economic development: From the dawn of history to the present. The University of Chigago Press, Chigago. 574 p.

Biswas A.K. (2000). Water for urban areas of the developing world in the twenty-first century. 23 p. In: Uitto J. I. and Biswas, A.K. (eds.). 2000.

Briscoe J. (1985). Defining a role for water supply and sanitation activities in the health sector in the Asian region: An issues paper for the Asia Bureau, USAID University of North Carolina, Chapel Hill, USA.

Burrows G., Acton J. & Maunder T. (2004). Water and sanitation: the education drain: WaterAid report. London, UK, WaterAid.

http://www.wateraid.org.uk/documents/Education%20Report.pdf.

Hardoy J.E. & Satterthwaite D. (1989). Squatter citizen. Life in the urban Third World. Earthscan Publications Ltd, London. 374 p.

Harpman T., Lusty T. and Vaughan P. (eds.) (1988). In the shadow of the city – Community health and the urban poor. Oxford University Press. New York. 237 p.

Human Settlements Program (2003). Local action for global water goals: Addressing inadequate water and sanitation in urban areas, (or the State of the World's Cities: Water and Sanitation). International Institute for Environment and Development for UN-Habitat: London,U.K. Processed.

Hutton G. & Huller L. (2004). Evaluation of the costs and benefits of water and sanitation improvements at the global level. World Health Organization.

Kithakye D. (2004). New housing policy targets Kenyan urban slum dwellers. <http://www.unchs.org/marathon.asp>.

Njiru C. (2004). Utility-small water enterprise partnership: serving informal urban settlements in Africa. Water Policy. Vol. 6, No. 5. pp. 443-452.

Peri-Urban Network on Water Supply & Environmental Sanitation. (1993). Voices from the City. Newsletter. Vol. 3, September. WASH Operations Center. Arlington, Virginia. 16 p.

Rossi-Espagnet A. (1984). Primary health care in urban areas: Reaching the urban poor in developing countries. A state of the art report by UNICEF and WHO. Report No. 2499M. World Health Organization Geneva.

Tabizbadeh I., Rossi-Espagnet A. & Maxwell R. (1989). Spotlight on the cities–improving urban health in developing countries. World Health Organization. Geneva. 174 p.

Uitto J. I. & Biswas A.K. (eds.) (2000). Water for urban areas. Challenges and perspectives. Water Resources Management and Policy. United Nations University Press. 245 p.

UNESCAP (2002). What is good governance? <http://www.unescap.org/huset/gg/governance.htm>. 4 p.

UNESCO (1980). Fourth world and human rights. Paris.

United Nations (1989). Prospects of world urbanization 1988. Department of International Economic and Social Affairs. Population Studies No. 112. New York. 204 p.

United Nations (2002). Report of the World Summit on Sustainable Development. Johannesburg, South Africa, 26 August – 4 September 2002. 167 p. http://www.johannesburgsummit.org/html/documents/summit_docs/131302_wssd_report_reissued.pdf.

UN General Assembly (2000). United Nations Millennium Declaration. 9 p. http://ods-dds- ny.un.org/doc/UNDOC/GEN/N00/559/51/PDF/N0055951.pdf?OpenElement.

UN General Assembly (2003). Press Release GA/10224 23/12/2003. http://www.un.org/News/Press/docs/2003/ga10224.doc.htm.

UN-Habitat (2002). Cities without slums. World Urban Forum. 15 p.

UN-Habitat (2003a). The challenge of slums: Global report on human settlements 2003. Earthscan/UN-Habitat. London.

UN-Habitat (2003b). Guide to monitoring Target 11: Improving the lives of 100 million slum dwellers. 15 p. http://www.unchs.org/mdg/.

UN-Habitat (2003c). Slums of the world: The face of urban poverty in the new millenium? Monitoring the Millennium Development Goal, Target 11-World-wide slum dweller estimation. Working Paper. 90 p.

UN-Habitat (2003d). Water and sanitation in the world's cities: Local action for global goals.

UN-Habitat (2004). The state of the world's cities 2004/2005: Globalization and urban culture. Earthscan. London. 198 p.

Water and Sanitation for Health Project (1992). WASH Update. Supplement No. 1 to Progress Report 14. 6 p.

Wegelin-Schuringa M. (1999). The SSIP model and community-based providers of water and sanitation services. UNDP Conference on Public-Private Partnerships for the Urban Environment. Bonn.

World Bank (2003). World Development Report 2004. Making services work for poor people. Oxford University Press. 271 p.

World Summit on Sustainable Development (2002). Plan of Implementation.

http://www.johannesburgsummit.org/html/documents/summit_docs/2309_planfinal.htm

# 39
# CONCLUSIONS: DOES HISTORY MATTER? PRESENT WATER GOVERNANCE CHALLENGES AND FUTURE IMPLICATIONS

*Petri S. Juuti, Tapio S. Katko & Heikki S. Vuorinen*

Figure 39.1 Local conditions are to be taken into account.

In historical context, the growth of urban centres is a continuous and even an escalating trend. Many of these centres are located in developing economies, while the ensuing problems are concentrated on the poorest people – as always. The most severe constraints include poor living conditions, a lack of democracy, poor hygiene, illiteracy, corruption and a lack of proper water and sanitation services. Especially women and children suffer most from these constraints.

At the same time, in the developed societies aging infrastructure is and will be a major problem. For instance, in the USA, many large cities including New York, Washington, D.C., Baltimore, Philadelphia, New Orleans, St. Louis, Chicago, and Los Angeles use drinking water mains that are 90 to 150 years old. The 600,000 – 800,000 miles of sewer pipes, and many treatment plants need replacement.

Because many cities all over the world are still using some of the original pipes laid during their systems' initial construction, it is common that drinking water or sewage mains tend to burst and leak. Cities with lead pipes have always manipulated water chemistry to mitigate the presence of lead pipes, since they do not have the money to replace them. The lead pipe problem is of special significance as a public health threat because of the neurological damage it can cause to foetuses, babies, and small children.

The current state of water infrastructure has serious negative impacts. Public health is threatened. Besides, deteriorating infrastructure also harms the local and national economy. Yet, the public will not support large capital investments until they are convinced of the need by suffering themselves from main breaks or other problems. The initiative to deal with this problem must come from the public if they want their elected officials to fulfil their responsibility. Clearly, it is the responsibility of citizens to oversee their government — central and local — and demand that it acts in the public interest.

Once rate payers start acting like citizens rather than just consumers, they will force government officials to provide funding while local utilities do their part by managing assets efficiently and charging utility customers in ways that encourage water conservation.

The main task in South Africa and in other dry or semi-dry areas of the world is how to solve the problem of the shortage of water. A few lessons were learnt from the drought period of 57 months between 1983 and 1987 in which restrictions in the water supply were in force in Johannesburg. It was found out that the most effective way of persuading consumers to limit water consumption was by means of keeping them informed on the availability and use of water.

In 1994, South Africa became a multi-racial democracy and, for instance, Rand Water in Johannesburg started with new and innovative measures to try and get the message across that all consumers in a new multi-racial society had to save water. This important principle, based on strategies aimed at conserving water, was subsequently ensconced in the *National Water Act* (35) of 1998. Up to the 1980s water restrictions to a large extent focused on only one or two segments of the society in the Rand Water service area.

Drought conditions in South Africa and in many other areas of the world most definitely are not something of the past. In the years to come, such areas especially with urban centres could well be subjected to severe limitations in terms of available water supplies.

The case of Nepal plainly shows that there cannot be a one-fit-for-all solution to water supply and sanitation services. Local institutions have to be able to respond to the needs expressed by the communities, while technical and financial support has to be available locally. Appropriate ways of organizing the water supply and sanitation can do a lot in poverty reduction. A culturally poverty-sensitive approach implies empowerment and highlights the capacity of people in households and communities to take action based on the setting and know-how in each unique location.

Globally, the past development and foreseen future challenges of rapid urbanization, and fundamental infrastructure services, call for proactive futures-oriented approaches to provide operational services for this fourth world.

It is important to take into account that if new water pipe networks are constructed, at the same time we should also build a sanitation system. When making the big decisions concerning water supply and sewerage, it is also necessary to be ready to make big investments, like in sparsely populated areas. Many such areas have water pipes but not an expensive sewer system. Clean water and wastewater are closely connected. But for the peri-urban settlements of the developing countries, these are still a distant reality even in the new millennium.

Although water and wastewater facilities are mainly hidden underground, all city dwellers come into contact daily with the products of water and sewage works: potable water, wastewater, cleaner water bodies as well as easier and safer everyday life – or too often their non-existence. If services are now at a high operational level, they were not achieved easily and without heavy inputs and efforts. This is something to keep in mind when assessing future options and considering required strategies.

PART IV – CONCLUSIONS

# EPILOGUE

# LOCAL SOLUTIONS BASED ON LOCAL CONDITIONS

*Petri S. Juuti, Tapio S. Katko & Heikki S. Vuorinen*

In rivers, the water that you touch is the last of what has passed and the first of that which comes; so with present time.
- Leonardo da Vinci

No publication is a staple of life. It's not bread and water. You have to make it noteworthy in people's minds and even in their hands as they're holding it."

- Timothy White

# EPILOGUE: LOCAL SOLUTIONS BASED ON LOCAL CONDITIONS

Local conditions -political, economic, social, technological, environmental and legislative dimensions - are all to be taken into in water services management (Photos: Tapio Katko, Tuusula Area Water )

In this book the main focus is on water and sanitation services and their evolution over time. We emphasize that the history of water and sanitation services is strongly linked to the current dialogue on water management and policy issues – and available future options.

The 29 cases from all the continents covering various historical phases indicate that the level of water supply and sanitation is not necessarily bound with time and place as much as the capability of society to take responsibility for developing the living environment of its citizens. In some cases, the situation was better in earlier times than nowadays. For their time, even wells and latrines were dramatic improvements compared to the more traditional options.

Below some key findings are presented according to the basic horizontal themes of the book: population growth, health, water consumption, technological choices and water governance.

## Population density & poverty

Throughout history major problems seem to be concentrated largely on the same people — the poor, if not the poorest of the poor. They suffer from poor sanitation, a lack of good water, ensuing health problems, poor education and often a lack of good governance and basic democracy. Women, children – and especially girls – are often the ones who fetch the water from distant sources. This daily task prevents the latter from going to school, learning to read etc. Eventually they become adults who might not be able to act as active citizens and all in all have an effect on their own lives, let alone the society as a whole.

From the point of view of our and the environment's wellbeing, it is essential that water is good and safe – regardless of whether it is from piped systems or point sources like wells. The same is the case with sanitation — it is a question of being connected either to a sewer or using proper on-site sanitation solutions. It is vitally important to operate and maintain these systems properly. A well and an eco-toilet, especially in areas with scattered settlements, will also provide in the future durable and ecological solutions. Investing in water supply and sewerage and thus also in the environment is always worthwhile. Studying alternative solutions for water and sanitation services – from on-site systems to rural villages, communities, cities and metropolies – needs more resources, so that we could find the best solutions and paths of action for different conditions – there is no such thing as cookie-cutter solutions for systems which have such direct interaction with the environment.

For economic but also for several other reasons, it is not feasible to have waterborne sewerage everywhere while obviously it is needed in densely-populated areas at the moment. In dispersed rural areas, at least, alternative on-site systems can be considered. These alternative systems seem to be particularly subject to local conditions.

## Health

Public health has always been a major factor influencing the ways in which the water supply problem has been solved by societies. The source of water supply was chosen according to its salubrity: clear, odourless water e.g. from springs or wells was preferred. Already starting in antiquity it was known that certain kinds of water caused health problems. Stagnant and marshy waters have been avoided throughout time. New waterborne health hazards were recognized from the 19th century onwards: microbes, chemical pollutants.

We know that people have used their senses to perceive the quality of drinking water at least from antiquity onwards and most probably a long time before. Related to technological development, new methods of studying the quality of water were introduced from the 19th century onwards including chemical and microbiological studies. Concerning these two factors in historical perspective, we can see a clear continuity and also strong changes that both are dependent on the scientific and technological level of the society.

Already in antiquity various methods like sieves, filtration and boiling were used to improve the quality of drinking water. However, it was only in the 1800s that filtration of drinking water in urban centres became a common practice. Disinfection of drinking water by different methods was introduced in the early 20th century. Through proper use of the water treatment technologies, the salubrity of drinking water could be guaranteed to an ever-growing population.

The importance of good quality drinking water for the urban population was also realized already in antiquity. Yet, the importance of proper sanitation for the health of town people was not discovered until the 19th century. The building of "modern" urban sewerage systems started in Britain and rapidly spread all over the globe.

**Water use**

Water used in large quantities has been deemed an essential part of a civilized way of life in different periods: Roman baths needed a lot of water as does our current way of life with water closets, showers and jacuzzis. Particularly high rates of water use are noticed when it is not properly charged for. The evidence indicates that as soon as water but also wastewater is charged for according to their real costs, wastage diminishes remarkably. Although on a global scale the great majority of water is used for irrigation, the highest priority for the use of water is for the community water supply.

Throughout history there have been different solutions to guarantee an ample amount of water for human settlements. Indigenous people have been very ingenious in drawing their water. They have considered water a very crucial and often a sacred element. In the long run, the availability of water has been one of the crucial factors for the development of a society – cities and communities.

**Technological choices**

In some cases, the technological choices may have been erroneous or less successful. For instance, lead pipes were considered hazardous for health already during antiquity while they continued to be used in house connections until recently. Sometimes latercomers have been able to avoid such choices while many of the modern countries face the lead problem still in 2006. Selection of separate sewers paved the way for proper wastewater treatment unlike those with combined sewers. The use of meter-based billing has a lot of advantages. Thus, such path dependencies may have clearly positive impacts – not only negative as is too often assumed.

Ultimately water supply and sanitation systems need continuous maintenance and adequate rehabilitation. This was already evident in the Roman aqueducts: calcium carbonate incrustation formed inside the conduits needed constant removal, otherwise it would have stopped the water flow. The same is true for the modern systems, the maintenance of which must be taken care of, otherwise they do not function properly. The bigger – if not the biggest – problem is related to the need for continuous replacement and renovation.

One long-term issue has been, and still is, whether to use ground or surface water as raw water for community purposes, or more generally, what sources to use. For small systems often ground water may be available but for bigger systems ground water at further locations or surface water may be needed. This is connected to the current question, how far is it economically feasible to expand such systems? Once the systems expand, other criteria such as vulnerability are also to be considered.

## Governance

The World Water Development Report 2003, produced practically by the whole UN family and thus almost all the sectors of human life and society, pointed out how *the water crisis is largely a crisis of governance.* The report also listed "many of the leading obstacles to sound and sustainable water management: sector fragmentation, poverty, corruption, stagnated budgets, declining levels of development assistance and investment in the water sector, inadequate institutions and limited stakeholder participation" [1]

The findings of our book are largely along the same lines. People should be allowed and encouraged to use their own experiences and abilities to solve their problems. Such empowerment will most probably have more sustainable results than any mere top-down approaches. In overall good and effective water governance we obviously need to balance the centralised requirements such as legislation and the decentralised requirements of water services management in local conditions.

The findings of this book also imply the importance of involving all the stakeholders in decision-making in their proper roles as well as the participation of people and citizens – the ultimate users, beneficiaries and payers of these services.

Interestingly enough, some of the basic principles of sustainable and viable water governance and services were formed more than 2000 years ago. By using these principles, many of the present problems could be avoided and solved. In spite of this, mankind does not by and large use these principles due to a lack of proper governance but also resistant attitudes among people. It seems to be difficult, if not impossible, to resist the advantages and profits gained in the short-run by some, instead thinking of long-term benefits potentially achieved by many.

---

[1] http://www.unesco.org/water/wwap/wwdr2/pdf/wwdr_2_brochure.pdf

It seems to be very challenging for mankind to adopt systems and issues of fundamental importance – such as water supply and sanitation. Unfortunately it is psychologically much easier to promote bottled water or handing-over water services to international private operators – both of which involve parties/actors/companies interested mainly in short-term profits. Indeed, any sustainable water services will require long-term actions and planning, which the current Western culture largely ignores. The time frame and related thinking seem to become shorter and shorter.

Yet, change itself should not be an end in itself. Some decisions made in antiquity and in the late 19th century had a minimum frame of a century and often even more. In the foreseeable future, such a time frame should also be used if any sustainable results are to be achieved. This is also shown by early and quite successful solutions, although other problems and future challenges were risen later as shown by this book. Therefore, more futures-oriented thinking should be practised.

More than just a commodity, water is an economic and social good. This places responsibility for its management and oversight in the public sphere. Balancing water use priorities, water quantity and water quality is of high importance for the future. While the water supply will continue to have the highest priority, water quality issues will be relatively even more important than quantity. At the same time, it is more and more important to use water wisely and avoid wastage of this important natural resource. In a global context, water pollution control and sanitation are probably the biggest challenges – removing wastewater loadings from communities, industries and agriculture in many parts of the world.

Finally, the historical cases of the book clearly reveal that there is probably a wider diversity of options and development paths – whole sets of institutional arrangements – than believed or recognised so far. The role of capital cities has not been as dominant as earlier assumed. In many cases, the remarkable networking of professionals has taken place in the early phases.

There is a huge variety of development paths and solutions in urban water supply and sanitation. Local conditions, traditions and people have to be at the core of decision-making when future solutions are considered. However, since water sources for every city have their own unique location and quality, and each city has its own unique physical, social, cultural and administrative morphology, solutions of one city may not work for another.

In the long historical perspective, it is evident that, regardless of the political system, good solutions can be found based on local conditions, needs and traditions. Although water – and particularly water services – are largely dependent on local conditions, it is useful to make comparative studies between cities and communities in various regions and cultures, and identify possibly applicable and replicable principles and practices.

# ABOUT AUTHORS

## EDITORS

### Petri S. Juuti

Head of the IEHG group (www.envhist.org), Dr. Juuti (petri.juuti@uta.fi) is a historian and Docent/Adjunct Professor in Environmental History (at University of Tampere) and in History of Technology (at University of Oulu). He is currently working as a Senior researcher at University of Tampere. Previously he has worked as a Senior researcher for the WaterTime project funded by the European Commission, as an Assistant Professor (2002-2004) and as a researcher at University of Tampere and the Ministry of the Interior as well as in the businessworld. His major area of interest is environmental history, especially the urban environment, city-service development, water supply and sanitation, urban technology, pollution, and public policy. His interests also cover development studies, social and economic history and political history. He is the author of over a dozen books, the two latest ones are *Brief History of Wells and Toilets* (2005) and *Water, Time and European Cities* (2005, with Tapio Katko).

### Tapio S. Katko

Dr. Tapio S. Katko (tapio.katko@tut.fi) is a Docent /Adjunct Professor in water services development at Tampere University of Technology where he heads the CADWES (Capacity Development in Water and Environmental Services) research group. He also holds a docentship in Environmental Policy at University of Tampere and in Environmental Sciences at University of Jyväskylä. He has several years of practical, teaching, and research experience in and for Finland; earlier in his career he also worked in Eastern and Central Africa. Dr. Katko's current research deals with institutional, management and policy issues and long-term development and strategies of water and sanitation services. He has written over a dozen books and monographs and some 50 peer review papers. In 1998 he received the Abel Wolman Award of the Public Works Historical Society, and in 2006 jointly with Petri Juuti the "Highly commended" Marketing and Communications award of IWA in the category "Best popular presentation of water science ". In 1998-99 he was an International scholar of the Society for the History of Technology and he has also received three national writers' awards.

### Heikki S. Vuorinen

MD Heikki S. Vuorinen (heikki.vuorinen@helsinki.fi) is 55 years old, married, and has three children and two grandchildren. He is an Adjunct Professor (docent) of History of Medicine at University of Tampere and University of Helsinki. He specialises in the history of public health and has written numerous articles on different aspects of the history of public health, a textbook about the history of diseases (2002) and a monograph about public health in Finland in the mid-19th century (2006). Dr. Vuorinen has for years lectured on different aspects of medical history, especially the history of diseases, at the universities of Tampere and Helsinki. Currently he is writing a textbook on medical history.

# AUTHORS

## CHAPTER 6 – WATER SUPPLY IN THE LATE ROMAN ARMY

**Ilkka Syvänne**

Ilkka Syvänne (ilkkasyvanne@yahoo.com) completed his Master's Degree in Finnish History in 1997 at the University of Tampere. In 2000, he began his postgraduate studies in General History and achieved a Doctoral Degree in 2004. The subject matter of the Doctoral Dissertation (The Age of Hippotoxotai, Roman Art of War in Military Revival and Disaster (491-636) was the late Roman art of war that also included short analyses of the water supply as well other usages of water in warfare. In late 2004 an article relating to the subject matter of the Master's Thesis in Finnish History was published in Kolmas Artturin Kirja. His other fields of interest besides academic ones include a number of international business ventures mainly in the field of real estate brokering.

## CHAPTER 11 – COLONIAL MANAGEMENT OF A SCARCE RESOURCE: ISSUES IN WATER ALLOTMENT IN 19TH CENTURY GIBRALTAR

**Lawrence A. Sawchuk**

Lawrence A. Sawchuk (sawchukl@utsc.utoronto.ca) is an Associate Professor at the University of Toronto, Scarborough, Canada. He has been studying the effects of environment and disease on the health of the Gibraltarian people for 30 years. In 2001, he published *Deadly Visitations in Dark Times*, which recounts the social implications of the 19th century cholera epidemics in Gibraltar.

**Janet Padiak**

Janet Padiak (padiakj@mcmaster.ca) is currently holding a Postdoctoral Fellowship at McMaster University, Hamilton, Canada. Her dissertation focused on the morbidity and mortality of the British soldiers stationed in Gibraltar during the 19th century. Other areas of interest include infant mortality, suicide, and marriage within the colonial military.

## CHAPTER 12 – COPING WITH DISEASE IN THE FRENCH EMPIRE: THE PROVISION OF WATERWORKS IN SAINT-LOUIS-DU-SENEGAL, 1860–1914

**Kalala Ngalamulume**

Kalala Ngalamulume (kngalamu@brynmawr.edu) is an Associate Professor of Africana Studies and History at Bryn Mawr College. He holds a Ph.D. from Michigan State University. His fields of specialisation include 1) Social History of Medicine in Africa, 2) Urban History in Africa, 3) and French Empire in West Africa.

## CHAPTER 13 – WATER SUPPLY IN THE CAPE SETTLEMENT FROM THE MID-17TH TO THE MID-19TH CENTURIES

Petri S. Juuti, Harri R.J. Mäki & Kevin Wall

**Harri Mäki**

Harri Mäki (harri.r.maki@uta.fi) has a Master's degree in history from Tampere University. He is now working on his doctoral thesis about the history of water supply in South African towns in 1850–1920. His key interests are environmental and urban history.

**Kevin Wall**

Dr Kevin Wall (kwall@csir.co.za ) is the Manager: Urban Management and Infrastructure, of CSIR Knowledge Services, in Pretoria, South Africa. A qualified civil engineer and town and regional planner, he is a former Assistant City Engineer of the City of Cape Town. His current area of specialisation is national policy and strategy formulation in water and sanitation and housing.

## CHAPTER 15 – WATER, LIFELINE OF THE CITY OF GHAYL BA WAZIR, YEMEN

**Ingrid Hehmeyer**

Ingrid Hehmeyer (ihehmeye@ryerson.ca) is an Assistant Professor of History of Science and Technology at Ryerson University (Toronto, Canada). With a doctorate in the history of agriculture, she specialises in human-environmental relationships in the arid regions of ancient and mediaeval Arabia. She currently conducts field research in the Republic of Yemen which focusses on questions of water management and the methods of astronomical timekeeping used for allocating water, during both day and night, in medieval Islamic times.

## CHAPTER 16 – HISTORY AND PRESENT CONDITION OF URBAN WATER SUPPLY SYSTEM OF TASHKENT CITY, UZBEKISTAN

Dilshod R. Bazarov, Jusipbek S. Kazbekov & Shavkat A. Rakhmatullaev

**Shavkat A. Rakhmatullaev**

Shavkat A. Rakhmatullaev (rakhmatullaev@rambler.ru) is PhD research fellow with Tashkent Institute of Irrigation and Melioration. He holds a MSc in civil engineering from TIIM and a BSc in watershed science from Colorado State University. His areas of scientific interest are sedimentation, reservoir operation and river basin planning.

## CHAPTER 17 – PHILADELPHIA WATER INFRASTRUCTURE, 1700-1910 ARTHUR HOLST

**Arthur Holst**

Arthur Holst (Arthur.Holst@phila.gov) is Legislative Manager for the Philadelphia Water Department. He earned his Ph.D. in Political Science at Temple University where he also earned a Master's in Public Administration and a Bachelor's in Business Administration. He has contributed to a number of reference works on various subjects related to political science, history and the environment.

## CHAPTER 18 – PRIVATISATION OF WATER SERVICES IN HISTORICAL CONTEXT, MID-1800S TO 2004

Petri S. Juuti, Tapio S. Katko & Jarmo J. Hukka

### Jarmo Hukka

Dr. Jarmo J. Hukka (jarmo.hukka@tut.fi) holds an adjunct professorship in futures research in the water sector at Tampere University of Technology, Finland. He has 31 years of professional experience from the water services sector, including 10 years of long-term overseas experience in the Cayman Islands, Sri Lanka, Kenya and Kosovo. He has also been a coordinator for Finland Futures Academy and an acting director at Finland Futures Research Centre, Turku School of Economics and Business Administration, Finland. His current research interests include water sector reforms, governance, management, leadership, assets management, vulnerability and strategic development of water services. He has authored 80 publications.

## CHAPTER 21 – HISTORY OF WATER SUPPLY AND SANITATION IN KENYA, 1895 – 2002

### Ezekiel Nyangeri Nyanchaga

Dr. Ezekiel Nyangeri Nyanchaga (samez@wananchi.com), University of Nairobi, Department of Civil Engineering, Kenya, has more than 24 years of experience from planning, design and implementation of both rural and urban water supply and wastewater, irrigation and drainage, infrastructure engineering, urban water demand management, preparation of contract documentation and contract implementation, environmental impact assessment and audit, and preparation of operation and maintenance manuals for water supply and sewerage works. In particular, Dr. Nyangeri has been involved in the development of monitoring and evaluation systems and procedures for water supply and sewerage works. He has experience from performance evaluation for water supply projects. He is a registered engineer and licensed (panel I) water and wastewater engineer and EIA lead expert registered by the National Environmental Management Authority. He is a senior lecturer, Department of Civil Engineering, University of Nairobi. He has a network of scholarly peers in Kenya and Finland who have been involved in similar assignments

### Kenneth S. Ombongi

Dr Kenneth S. Ombongi (kombongi@hotmail.com) is an experienced historian, holding a PhD from Cambridge University, based in the Department of History University of Nairobi, Kenya. He has been a key participant in a wide-ranging forum of historical discourse and has been involved in many historical projects.

## CHAPTER 22 – THE HISTORY OF WATER CONSERVATION AND DEVELOPMENT IN THE MWAMASHIMBA AREA OF THE BUHUNGUKIRA CHIEFDOM AND IN RUNERE VILLAGE, TANZANIA

### Jan-Olof Drangert

Dr. Jan-Olof Drangert (jandr@tema.liu.se) is an Associate Professor at Linköping University, Sweden, Dept. of Water and Environmental Studies. He has been the team leader for large research projects in Sweden and Kenya. After a period of research work on rural water and sanitation, he has accumulated experience from interdisciplinary research and consultancy work related to urban water and sanitation issues in developing and developed countries during the last ten years. He has a special interest in the historical evolution of sanitation in urban areas. He is also a team leader for international training activities in ecological sanitation.

## CHAPTER 23 – PROVISION AND MANAGEMENT OF WATER SERVICES IN LAGOS, NIGERIA, 1915-2000

### Ayodeji Olukoju

Ayodeji Olukoju (aolukoju2002@yahoo.com), B.A. First Class Honours, University of Nigeria, Nsukka, M.A. and PhD (Ibadan), is Professor and Dean of Arts, University of Lagos. He is a specialist in African and comparative economic and social history, and has published books and articles on maritime, urban, business and culture history.

## CHAPTER 24 – EXPANDING RURAL WATER SUPPLIES IN HISTORICAL PERSPECTIVE: SIX CASES FROM FINLAND AND SOUTH AFRICA

Petri S. Juuti, Tapio S. Katko, Harri R. Mäki & Hilja K. Toivio

### Hilja Toivio

Hilja Toivio (hilja.toivio@uta.fi) is an undergraduate student of philosophy at University of Tampere, Finland. Her main subject at the Department of History is Finnish history. Her main areas of interest are agricultural history and the history of the horse. She has written articles on the history of horse traffic in Tampere. She also has a special interest in genealogy.

## CHAPTER 25 – SISTER TOWNS OF INDUSTRY: WATER SUPPLY AND SANITATION IN MISKOLC AND TAMPERE FROM THE LATE 1800S TO THE 2000S

Petri S. Juuti & Viktor Pál

**Viktor Pál**

Viktor Pál (viktor.pal@uta.fi) is a post-graduate student, researcher at the University of Tampere, Finland. His research has focussed on the history of Central and Eastern Europe, the environmental and technological history of urban areas and the politicisation of water. He also lectures at the Aleksanteri Institute at the University of Helsinki and has written scientific articles in English, Finnish and Hungarian.

## CHAPTER 26 – WATER SUPPLY AND SANITATION IN RIGA: DEVELOPMENT, PRESENT, AND FUTURE

**Gunta Springe**

Dr.biol. Gunta Springe (gspringe@email.lubi.edu.lv) is an Assistant Professor at the Faculty of Geography and Earth Sciences of the University of Latvia and the Head of the Laboratory of Hydrobiology, Institute of Biology, University of Latvia. Her research interests are related to the assessment of water quality, its protection and management.

**Talis Juhna**

Dr. sc.ing. Talis Juhna (talisj@bf.rtu.lv) is an Assistant Professor at the Department of Water Technology, Riga Technical University (Latvia). His research interests involve removal of natural organic matter from surface water, biological stability, biofilm formation and molecular methods for detection of pathogens in drinking water.

## CHAPTER 27 – THE HISTORY OF WATER IN ONE WESTERN MEXICAN CITY: MORELIA, 1541-2000

**Patricia Avila García**

Patricia Avila (pavila@oikos.unam.mx) holds a Ph.D. in Social Anthropology and an M.A. in Urban Development and Civil Engineering. Winner of the National Award for Social Sciences (2003). Author of two books and more than 50 papers published in various books and journals. The focus of her research are problems associated with water in Mexico, from the social and political perspectives. Currently Dr. Avila is a researcher at the Ecosystems research centre of the National University of Mexico (UNAM) involved in two projects: urban sustainability and quality of life; and water culture in the indigenous regions of Mexico.

## CHAPTER 28 – THE HISTORICAL DEVELOPMENT OF WATER AND SANITATION IN BRAZIL AND ARGENTINA

### José Esteban Castro

Dr José Esteban Castro ( J.E.Castro@newcastle.ac.uk) is a Senior Lecturer in Sociology at the University of Newcastle upon Tyne, United Kingdom. A sociologist by training, Dr Castro has an interdisciplinary background, and his main research interests in relation to water concern issues of water politics and the interwovenness of social and ecological change in historical perspective. His work has focussed on a number of countries, including Argentina, Brazil, Mexico, the United Kingdom, and Portugal. Recent publications include the book *Water, Power, and Citizenship, Social Struggles in the Basin of Mexico*, Palgrave-Macmillan, 2006.

### Leo Heller

Dr Léo Heller (heller@desa.ufmg.br) is a Professor at the Department of Sanitary and Environmental Engineering, Universidade Federal de Minas Gerais, Brazil. He is a civil engineer and has an MSc. in Sanitary Engineering and a PhD. in Epidemiology. He spent a sabbatical at the University of Oxford in 2005-2006. His current main research areas are environmental health and public policies of water supply and sanitation services. Among his publications related to the subject of this book is: Rezende, SC and Heller, L. *Basic sanitation in Brazil: policies and interfaces* [Saneamento no Brasil: políticas e interfaces], Belo Horizonte, Ed. UFMG, 2002.

## CHAPTER 29 – THE GEOPOLITICS OF THIRST IN CHILE – NEW WATER CODE IN OPPOSITION TO OLD INDIAN WAYS

### Isabel Maria Madaleno

The author (Portuguese Tropical Research Institute; Isabel-Madaleno@netcabo.pt) has been committed to socio-economic research within the Latin American geopolitical space. Scientific missions have targeted predominantly Brazil, Chile, Bolivia, Peru and Mexico. Scientific peer-reviewed papers, book chapters and a manual have been published as research results. The author with a PhD in Geography and History from the Spanish Salamanca University has studied subjects such as urban and regional planning, agriculture, and ethno-botany. For over a century, the Portuguese Tropical Research Institute has developed databases on ancestral ecological farming systems and water management, and uncovering lost or shattered cultures in Africa and Latin America.

## CHAPTER 30 – CASE OF TOKYO, JAPAN

### Yurina Otaki

Originally, Yurina Otaki (yurina@iii.u-tokyo.ac.jp) was an urban engineer involved with sewage treatment and related micro-organisms. She took a master's degree in engineering at the University of Tokyo, Japan in 1996. After that, she became interested in the history of water and sanitation, and took a master's degree in arts and sciences at the University of Tokyo, Japan in 2002.Presently, she is interested in people's attitude toward water and how they use water in their daily life, especially in Asian cities.

## CHAPTER 31 – HEALTHY WATER FROM AN INDIGENOUS MAORI PERSPECTIVE

**Ngāhuia Dixon**

Ngāhuia Dixon (ngahudix@waikato.ac.nz)is a Senior Lecturer at the School of Māori and Pacific Development, University of Waikato, Hamilton, New Zealand. His areas of interests are the contrasting cultural and scientific views of water, symbolic representations related to environment and water, and ideological differences within community groups.

## CHAPTER 35 – AGING AMERICAN URBAN WATER INFRASTRUCTURE

**Laurel Phoenix**

Laurel E. Phoenix (PHOENIXL@uwgb.edu) is Assistant Professor of Public and Environmental Affairs, University of Wisconsin, Green Bay, USA. Her research interests: Anti-environmentalism, watershed management, drinking water, environmental planning. She holds degrees in Geography and a Ph.D. in Watershed Management and Hydrology. An associate editor for Water Resources IMPACT, her recent publications are 'Vulnerable Groundwater Drinking Sources in Door County, Wisconsin.' Applied Environmental Science and Public Health. 2, 2004 (2):1-9; 'Forging New Rights to Water.' Water Resources IMPACT 5, 2003 (2):3-4. (With Clay J. Landry); and 'Rural Municipal Water Supply Problems: How do Rural Governments Cope?' Water Resources IMPACT 4, 2002 (2):20-26.

## CHAPTER 36 – FROM WATER RESTRICTIONS TO WATER DEMAND MANAGEMENT: RAND WATER AND WATER SHORTAGES ON THE SOUTH AFRICAN URBAN LANDSCAPE (1983-2003)

**Johann Tempelhoff**

Johann WN Tempelhoff (D. Litt et Phil [SA]; GSKJWNT@puknet.puk.ac.za) is an environmental historian in the School of Basic Sciences, at North-West University, South Africa. He is the author of *The substance of ubiquity: Rand Water 1903-2003*, (2003). He is also editor of *African water histories: transdisciplinary discourses*, (2005). He is currently researching the cultural history of water in southern Africa.

## CHAPTER 37 – WATER AND NEPAL – AN IMPRESSION

**Sanna-Leena Rautanen**

Ms. Sanna-Leena Rautanen (srautanen@yahoo.co.uk ) has an MSc in Civil Engineering from Tampere University of Technology, Finland, where she is a post-graduate student and researcher. In Nepal she was the Field Specialist of the Rural Water Supply and Sanitation

Support Programme. Her special areas of interest in the water and sanitation sector are poverty, gender and ethnic equity, good governance and democratic local development.

## CHAPTER 38 – IMPROVEMENT OF THE LIVES OF SLUM DWELLERS - SAFE WATER AND SANITATION FOR THE FUTURE CITIES OF THE FOURTH WORLD

Jarmo J. Hukka & Osmo T. Seppälä

### Osmo T. Seppälä

Osmo Seppälä (osmo.seppala@fcg.fi) is a Project Director at Plancenter Ltd / Helsinki Consulting Group Ltd (Helsinki, Finland). He has a doctoral degree (D.Sc. Tech. Civil Eng. Water and Environmental Engineering), and has over 20 years of professional water and environment sector experience, as a consultant and a researcher. He has worked in water and sanitation projects in Tanzania 1981-82, Sri Lanka 1987, in Zanzibar 1993-96 and in Kenya 1997-99. He has also worked in several positions at Tampere University of Technology 1989-93 and 1999-2004. His main research interests include water sector reforms and visionary management in water services. He has studied and applied various futures research methodologies in his doctoral dissertation and recent research projects in water management.

# INDEX OF PERSONS

Abd al-Rahim bin 'Umar Ba Wazir 203
Abegg W. A. 244, 246
Alcmaeon of Croton 48
Aldunate C. 457
Aetius 77
Alfonsín 440
Allende 454, 455
Alvear de T. 435
Ammianus Marcellinus 70
Anderson D. M. 148
Anderson J. 177
Anderson T. F. 298
Aseto O. 283
Aspar 77, 78
Audibert 150
Avebury 241
Badois M. 155, 156
Bahaaeddin Ibn Qaraqosh 24
Bakker K. 241, 247
Barnato B. 183
Barraqué B. 237, 242
Bath V. 544, 552, 553, 555
Batory S. 403
Beaussier 151
Belisarius 83, 85, 86
Bergström J. W. 244
Best E. 479
Birendra 565
Biswas A. K. 584
Blake N. 236
Blomfield 304
Borgnis-Desbordes 159
Botha P. W. 540
Boyle R. 182, 184, 189
Bramah J. 37, 100
Brandström P. 323
Brière de l'Isle 156
Brown J. C. 236
Bruun C. 55
Bryce J. 185
Budd W. 111
Calishoff S. 236
Cardoso 440
Chadwick E. 104, 105, 106, 108, 123, 138

Chamberlain J. 241
Clarke J. 539
Coghland J. 433
Colenso 180
Collor 440
Colot 157
Columella 50, 55
Cook J. 177
Cooley A. 228
Crapper T. 37, 100
Cuerauahperi 416, 427
Cuerauaperi 417
Cummings A. 37, 100
Dahal D. R. 565
Delamere 274
Desmars M. 249
Diocles of Carystus 50, 52, 54
Dockel J. A. 542
Don G. 137
Durie M. 479, 480
Eberth C. 111
Elizabeth I 36
Evans O. 222, 223, 226
Evenor 50
Eyoum G-P. 152
Faidherbe 151
Farr W. 110, 138
Fortie M. 323
Francis I 36
Franklin B. 222
Frederick I (Barbarossa) 35
Frei 454, 455
Frontinus 46, 55, 57, 64
Galen 50, 51, 52
Gamble J. G. 175, 176, 187
Gaspari K. C. 236
Gillman C. 323, 326
Gouin 155
Grey G. 179
Grmek M. D. 62
Hadrian 58, 72
Hardoy J. E. 579
Harrington J. 36
Harrison H. 334

Hassan J. 236
Hausen 124
Haushofer K. 448
Hayward S. 537, 540
Heraclius 72
Heron of Alexandria 75, 76
Heron of Byzantium 75, 77
Hietala M. 241
Hippocrates 50, 63
Hirsch A. 111
Hobbs D. 538, 542, 555
Hodge A. T. 47, 63
Huber R. 120, 244
Huggins P. M. 325, 327, 328, 329, 331
Isaac 18
Jacobsen 251
Jakande L. 346
Jaureguiberry 156
Joseph 24, 25, 26
Julius Africanus 76, 80, 81, 86, 87
Justinian 71, 72, 73
Jusuf 24
Kjellen R. 448
Koch R. 110, 111
Kollmann P. 323
Kotzé G. J. 542
Laburn B. 540
Lapis M. 418
Latrobe B. H. 222
Le Blanc J. 156, 157
Lewis 274, 279
Lillywhite 274
Lingen van A. S. 534
Lubbock J. 241
Mahendra 565
Malcolm D. 341
Malenfant 160
Mankurwane 370
Manzoni H. 275, 276, 280
Marks S. 2, 184, 185
Martin B. 251
Maurel 153
Maurice 70, 77, 82, 83
McGregor Ross 275, 280
Melosi M. V. 236, 237, 243, 244, 251
Menem C. 440, 504
Mfugaji B. 338, 339, 340, 341

Midgley D. C. 545
Mill J. S. 239
Millward R. 248
Mitro B. 39
Moffat 297
Montshiwa 370
Morris P. 240
Morrison J. A. 273
Moss J. 249
Mossweu 370
Moswete 370
Moule H. 37, 100
Murray R. W. 174
Murto M. 389, 393
Mäkijärvi E. 366, 367
Mäkijärvi J. 365, 366, 367
Ndalawa 334
Nelson B. 236
Nelson M. C. 118, 240
Njiru C. 586
Noble J. 368
Nottbeck von W. 244, 245
Oberholzer J. F. 538, 539
Ocupi Tiripeme 416, 427
Okello J. A. 283
Okun D. 247, 248
Ormsby T. 485, 486
Palladius 50, 55, 63
Palmberg A. 111, 112
Paulus Aeginata 55, 63
Perham M. 327, 332
Peter 77
Pettenkofer von M. 111
Philiscus 62
Phocas 72
Pinet-Laprade 152, 153, 154
Pinochet 454, 455
Plato 50, 65
Pliny the Elder 46, 50, 51, 54, 57
Polyaenus 70
Porter C. 186
Praxagoras 50
Priska T. 390
Procopius 70, 72, 73
Pryor J. H. 83
Python 62
Pälli V. 363

Ramos Mejía J. M. 435
Rathbone R. 148
Ratzel F. 448
Rawson G. 435
Rees R. 251
Rensburg van A. 546
Richard 157
Richter J. G. 244
Ritter K. 448
Robinson H. 370
Rogers J. 118, 240
Roosevelt N. I. 222
Rossi-Espagnet A. 580
Rounce R. V. 335
Russell 304
Salah el Din, Saladin 24, 25
Saldanha da 166
Salim 207
Sarney 440
Satterthwaite D. 579
Sayid O. 323
Schramm 250
Sendo M. 333
Sekhou A. 153
Sharma S. 570
Sheikh Ali bin Salim 274
Shiga K. 111
Sidonius Apollinaris 80
Sikes H. L. 279
Simpson W. J. 296, 297
Smith 273
Snow J. 109, 110, 111, 112, 113, 139, 265
Stanley H. 322
Stephens J. L. 25
Strang V. 479
Swanson M. 190
Swederus G. 109
Swynnerton 278, 279, 326
Syrianus the Magister 70, 77
Tarr J. A. 236, 251
Thales of Miletus 48
Thatcher M. 238, 247
Theophylact 70, 77
Tiberius 80
Tinubu B. 346, 348
Titus 27, 50
Trédos 154

Twain M. 26
Tyson P. 541
Valens' aqueduct 72
Valière 150, 154, 155, 156
Vargas G. 436
Vegetius 70
Velleius Paterculus 80
Vespasian 27
Vinci da L. 36, 593
Vitruvius 46, 50, 51, 53, 54, 55, 57, 71, 496
Vliet van H. 541
Wagenaar Z. 166
Waldbott G. L. 490
Webster G. 74
Williams G. B. 296
Witt de T. 543
Woolf A.G. 236
Wuolle B. 247, 248
Zonaras 71

# INDEX BY PLACES

Adderley Street 174
Adiyan 346, 348
Africa 13, 74, 102, 132, 136, 148, 251, 261, 301, 356, 494, 498, 510, 532-533, 577, 582-583
Agege 344, 345
Ahvenisto 124
Ajegunle 344, 345, 349, 352, 353
Ajeromi 345
Alexandra reservoir 176
al-Mukalla 208, 209
Altona 113
Amir Timur square 215
Andalusian 132, 143
Angola 532
Ankhor 214, 215
Annapurna 567
Antonine wall 58
Antwerp 242
Apennine Peninsula 47
Arabia 42, 198, 211
Arabian peninsula 101, 198
Arantepacua 414
Aranza 418
Argentina 268, 429-446, 497, 504, 580
Arizona 527
Asian 216, 564, 566, 586
Atlanta 115, 526
Atlit Yam 19
Aurajoki 32
Austria 111, 382
Auximus 86
Babylonia 13, 21
Bakel 153
Baltic countries 238
Baltic region 268
Bangladesh 497, 499
Barcelona 239, 242, 243
Barrage 185, 188, 191, 534, 554
Bechuanaland 357, 369, 378-380
Bedfordview 539
Belgian Congo 152
Belgium 111, 242, 291, 519
Belmont 231
Berea 181

Berea Road 180
Berlin 35, 244, 246
Beroe 77
Beshagoch 214
Betty Shaft of Cornelia Mine 538
Bir Yusuf 24
Bissau 152
Blesbokspruit 554
Bloemfontein 189
Boksburg 539
Bologna 55
Boston 89, 515, 529, 530
Botanic Gardens 180, 187
Botswana 1, 532
Bozsu 214, 217, 219
Braamfontein 183, 188
Brazil 268, 429-446, 504, 580
Britain 55, 58-59, 74, 100-101, 104-106, 108-109, 111, 236, 240, 246-247, 273, 430, 596
British East Africa Protectorate 267, 272, 273, 275, 315
Budapest 382, 387
Buffalo 515
Buhungukira 321, 322, 324-329, 331-332, 337-338, 341, 502
Bükk Mountains 387-388, 393, 397
Bupamwa dam 333
Buyogo 335
Cairo 24, 27, 101
California 211, 243, 517, 522, 526, 527
Calmar 31
Camperdown 182, 188
Camperdown Dam 182
Carthage 83, 85
Cape 19, 34, 41-42, 99, 101-102, 165-166, 168-170, 172, 174-180, 183, 187-195, 261-262, 342, 373, 379, 380, 536, 560
Cape of Good Hope 166, 175, 187
Castle Chenonceau 38
Castle of Good Hope 34
Chamborg Castle 36
Cape Peninsula 166, 177, 188
Cape Town 19, 34, 99, 101-102, 165-166, 168-170, 174-178, 180, 183, 187-192, 261-262

Carpathian Basin 382
Central Asia 101, 215, 583
Central Europe 13, 21, 124, 238, 382
Central Square 244
Centre Square 222-226
Charapan 414
Chasalawe 328, 330-331
Cheran 414, 421, 426
Cheranatzicurin 422, 426
Chestnut Street 222-223
Chicago 89, 236, 512, 514, 590
Chigatai Street 217
Chilchota 414
Chile 268, 445, 447-462, 497, 505, 580
China 4, 12, 152, 497, 564, 583
Chirchik 215
Cistern of Aetius 77
Cistern of Aspar 77, 78
Citadel of Cairo 24
Claremont 178
Cologne 59
Colorado 41, 522, 526
Colorado River 526
Congella 188
Connecticut 519, 520, 526
Constantinople 72
Corinthian Avenue 228, 229
Crete 3, 5, 10-11, 14, 24, 26, 47, 93, 95, 143, 215, 277, 306, 348, 351, 358, 433, 492, 514, 516, 576
Curianguro 416
Currie's Fountain 180, 187
Cyprus 19, 60
Czechoslovakia 391
Dakar 152-153
Danube 382
Delaware River 231
Denmark 35, 519, 579
Digo 298, 301, 303, 305, 317
Disa Gorge 178
Dodoma 329-332
Dougga 60
Douglas reservoir 179
Durban 102, 174, 180-182, 187-192, 261, 373
East Africa 1, 267, 272-275, 279, 283, 297-298
East Indies 166
Eastern Cape 168, 536
Eastern Cape Province 168

Eastern Transvaal 540
Egypt 4, 12, 13, 21-22, 47, 62, 64, 94, 98, 136
Elburgon 283, 284
Eldoret 272-273, 290, 292, 316
Embu 303
Enarosura 283
England 36-37, 66-67, 100-101, 115, 123-124, 139, 236-238, 240-241, 246, 248, 250, 253-257, 262, 329, 379, 390, 430, 432-433, 442, 519
Erongaricuaro 414, 421, 424
Esek 18
Europe 14, 19, 27, 37, 47, 94, 100-101, 104-106, 108-109, 114, 138, 236-240, 244, 251, 260, 268, 430, 433-436, 442, 451-452, 491, 498
Fairmount 224, 226, 228-230
Farobiy Street 217
Far East 269
Festac 344
Finland 1, 8, 20-21, 31-33, 37-40, 101, 105, 111, 114, 117-120, 122-128, 237, 239-240, 244, 253, 264, 268, 290, 355-359, 363, 365-367, 373, 377-378, 381-383, 392, 395, 397-399, 498, 503, 509, 519, 571
Fish River 168
Florida 527
Fort Hall 302, 316, 317
France 36, 38, 101, 104, 111, 125, 148, 152, 154-155, 237-240, 242, 248-249, 253, 261-262, 353, 430, 433, 519
Free State 532, 538, 561
Gadern Estate 278
Gambia 153
Garissa 278
Gauteng 532, 533, 536, 546, 549, 554
Gauteng Province 532, 549
Gbagada 345
Genoa 133
Germany 101, 104, 109, 111, 120, 125, 154, 238, 240-241, 250, 290, 306, 358, 390
German 89, 104, 111, 239, 240, 246, 250, 254, 290, 322, 341, 390, 448, 502, 524, 526
Germany 101, 104, 109, 111, 115, 120, 125, 154, 238, 240, 241, 250, 257, 290, 306, 358, 379, 390
Germiston 532
Ghayl Ba Wazir 197-211, 261
Gibraltar 1, 101, 131-144, 260-261
Gorun 156
Gothenburg 244

Government Gardens 174
Grahamstown 102, 174, 178-180, 187-189, 191-192, 261
Greece 3, 6, 10, 47, 63
Green Hills Estate 278
Greenmarket Square 167
Great Britain 74, 101, 104
Grey Reservoir 179, 187
Hadrian's aqueduct 72
Hadrian's wall 58
Hamburg 112-113
Hamina 38
Hanko 121, 122
Harappa 22
Hartebeestpoort dam 538, 554
Harts River 369
Harvard University 37
Hejőcsaba 384
Helsingör 35
Helsinki 114, 119-121, 124, 126-127, 244, 246-248, 357, 359, 383
Helsinki University 248, 357
Hely-Hutchinson reservoir 102, 176, 188
Hernád 386, 387, 388, 391
Highveld 532, 534
Himalayan 510, 564, 567
Hungary 37, 42, 268, 381-383, 387, 389-390, 392, 394, 397-400, 503
Huru-Huru 327
Hämeenlinna 121-122, 124-125
Ididi River 327
Iju waterworks 346
Ikeja 344, 345, 354
Ikorodu 345
Ikoyi 344-345, 352-353, 502
Ilangabafipa 330
Ilkko Water Cooperative 360, 362, 378
Ilmajoki 357
India 108-109, 152, 166, 497, 564-565, 581
Indus 4, 12, 22
Iraklion 3, 11
Iran 203, 581
Ishasi 346
Isiolo 278
Isolo 344
Israel 19
Italy 47, 57-58, 64, 96, 238-239, 242, 296, 498

Jameson dam 180
Japan 27, 269, 287, 290, 463-474, 494, 505, 579
Jericho 13, 22
Johannesburg 102, 174, 177, 183-188, 190-192, 261, 531-532, 534, 537-539, 551, 553, 578, 590
Joseph's Well 24-26
Juupajoki 21
Jyväskylä 121
Kabale 328, 341
Kadyrya 217
Kahama 323
Kahuga 331
Kakamega 298, 303-305, 315, 317
Kangasala 41, 357, 359-366, 374-376, 378-379, 389, 400
Kangundo 292
Karen Estate 278
Karoo 357, 372, 373, 379
Kathmandu 568, 570, 573-574
Kavirondo 303, 305, 317
Kechkuruk 214
Kelston 36
Kenya 267, 271-320, 342, 502, 581
Kenya Colony and Protectorate 267, 275
Kericho 290, 292
Kerugoya 303, 316
Ketu 344
Khor 154, 157, 160
Kibigare Estate 278
Kibrai water supply facility 217
Kijima 324
Kikuyu Estate 278
Kilimanjaro 289
Kiliwi 330
Kimberley 189, 370, 371
Kinangop 283
Kinja 283
Kirawa Water Co. Ltd 278
Kisembe Estate 278
Kisesa 336
Kissonerga-Mylouthkia 19
Kisumu 272-273, 292, 296-297, 304-305, 318-319
Kisumu City 273
Kitale 273, 290, 292, 305
Kitisuru Estate 278
Klip river 186, 554
Klipspruit 177, 186

615

Kloof Nek 178, 188
Knossos 10, 14-15, 23-24, 26, 47
Kokkola 121
Kosice 388, 391
Kosofe 344
Kotze's Spring 175
Kotka 121
Kukcha 214
Kunni farm 365-366
Kuopio 121
Kuramin district 215-216
Kuwinda Estate 278
Kwazulu-Natal 374
Kwimba 322, 327-328, 335, 341
Kwimba federation 327
Kyrönsalmi strait 33
Lagos 1, 268, 343-353, 452, 502
Lagos Island 344, 348, 352-353
Lahti 121-122, 124
Lake Kaukajärvi 362
Lake Kirkkojärvi 362-363, 375-376
Lake Näsijärvi 125, 363, 383-384, 389, 394-395, 397
Lake Pakkalanjärvi 365
Lake Pitkäjärvi 362
Lake Province 336, 341
Lake Roine 363, 383, 395
Lake Ukkijärvi 360-361
Lake Vesijärvi 360, 374
Lake Victoria 267, 272, 322-324, 337, 502
La Mojonera 414
Lampsar 149, 152, 154-157
La Pacanda 416
Latin America 251, 257, 262, 442, 494, 497, 577, 582-583
Latvia 268, 401-402, 406, 408-410, 503
Lebowa 372
Lee 390
Leghorn 133
Lerma-Chapala 412
Lesotho 532-535, 543, 546, 549, 552
Lielahti 389, 392-393
Lihasula farm 360
Lillafüred 388
Limpopo 371
Limpopo River 183
Limuru 278
Linköping 244, 255, 342

Lisbon 242
Lithuania 2
Lodwar 278
Loire Valley 36, 38
London 35, 100, 109, 114, 123, 133, 139, 240-241, 265, 434, 464, 466
Long Street 168
Los Angeles 211, 512, 515, 517, 529, 530, 590
Los Reyes 414
Lower Klip 554
Lübeck 27, 239
Maboko 329-330
Machakos 277, 291, 303-305, 316-318
Magadi 289
Madagascar 152
Makhana 154-155, 157, 159-160
Malampaka 324
Malindi 290, 292, 315
Mandera 278
Marsabit 278
Martintelep 384
Maryland 517
Massachusetts 67, 228, 526
Maswa District 327
Mereroni River 273
Meseta Purepecha 411-415, 419, 423, 427-428
Mesopotamia 1, 12-13, 22
Mexican city 431
Mexico 268, 411-428, 431, 444-445, 497, 503, 505, 580
Mhalo 329-331
Mhande 329-331
Mhande Hills 330
Mhunze Hill 331
Michoacan 411, 414-415, 419
Middle East 4, 203, 205, 498, 577
Midrand 532
Mikkeli 121
Mill Creek 231, 232
Milner dam 180
Milner Park 186
Milwaukee 515
Minamata Bay 494
Mirema Estate 278
Miskolc 268, 381- 400, 503
Mkula 336
Moame 322, 324, 328

Moame river  322, 324, 328
Mocke Reservoir 176, 188
Mohenjo-Daro 13, 22
Molteno reservoir 170, 187
Mombasa  272-273, 276, 289, 292, 296-298, 305, 316, 318
Montana  526, 528
Morelia 411
Mount Everest  564
Mount Kenya  272
Mount Maunganui 486
Mowbray 178
Moyale  278
Mozambique  152
Mphahlele Rural Area  357, 371
Mpumalanga province  540
Mukalla  208-209, 211
Muqaddim Salim from al-Suda  207
Muranga  303
Mushin  344-345, 352-353
Mwabayanda  335
Mwamashimba  321, 326, 328, 336-337, 341
Mwanza  322, 334, 336, 341-342
Mwitu Estate  278
Mältinranta  244-245, 385, 394
Nahuatzen  414-415, 421
Nairobi  271-274, 278, 282, 292, 296-297, 300-301, 305, 307, 311, 314-319, 577, 583
Naivasha  272, 284, 296, 297, 304, 317, 319
Nakuru  273, 284, 290, 292, 296-297, 304, 317, 319
Nandi  279, 301, 303, 305, 317
Nanyuki  289, 299, 301, 316
Natal  180, 188, 190
Ndabeni  177
Ndagaswa river  324, 337
Near East  13, 22, 64
Nepal  40, 503, 510, 563-574, 591
Nera  322, 341
Netherlands  42, 129, 166, 238, 249, 255, 519, 578
New Doornfontein 186
New Orleans  512, 526, 590
New York  42, 89, 90, 115, 130, 193, 211, 222, 236, 257, 342, 488, 512, 514-515, 529-530, 560, 574, 587, 590
New Zealand  12, 269, 475-480, 482-484, 488, 583
Ngong  289
Nigeria  267, 268, 343-354, 502, 579

Nile 13, 21, 26, 62
Niuwejaars river 180
Njoro River  273
Nol-Turesch  289
Nordic counties  237
Nordic countries  238, 244, 383
Northern Transvaal  372, 536
Northwest Province  540
North Africa  47, 74, 132, 498
North America  102, 236, 239, 452
North-West University  148
Nummela  127
Nyahonge  324, 328-329, 341
Nyahururu  292
Nyamiselya  330
Nyang'hanga  329, 330-332
Nyang'hanga Hill  329-330
Nyang'hanga Mbuga  329
Nyangalata river  324
Nyeri  272, 290, 292, 298, 299, 312, 315, 318
Näsijärvi  42, 125, 130, 363, 383-384, 389, 394-395, 397
Ocumicho  426
Ogba  344
Olabanaita  283
Olavinlinna castle  33
Old Dutch Road  180
Olobanaita  284
Oporto  242
Orange Kloof  178
Orange Street  168
Orile  344, 352
Oshodi  344, 345
Ostia  59
Ostrobothnia  38, 357-358, 375
Oulu  121, 124
Oworonsoki  345
Pakistan  13, 22, 581
Pakkala  365
Palestine  203
Paphos  60
Paracho  414, 418
Pareo  416
Paris  27, 35, 100-101, 104, 107, 154-156, 353, 466, 506
Parrish street  228
Patzcuaro  414, 416

Pennsylvania 102, 222
Pergamon 55
Philadelphia 102, 221-234, 262, 512, 590
Phoenix 1, 251, 510, 511, 527
Pichataro 414, 416, 419, 421, 424, 426
Pinetown 181
Platteklip Stream 170
Pointe de Barbarie 150
Pointe du Nord 154, 160
Pompeii 55, 57-59
Poplar 228
Portugal 238, 242, 257, 431
Portuguese Guinea 152
Port Elizabeth 177, 189
Port Natal 180
Port Victoria 273
Porvoo 121, 125-126
Pretoria 2, 185, 189, 531-532, 538-539, 545, 549, 551
Pyhäjärvi 383, 389, 396
Randburg 532, 539, 542, 558, 559, 561
Rangataua 485, 486
Rehoboth 18
Rhenish Prussia 236
Ridge Ways Estate 278
Rietspruit 554
Rift Valley 272, 304, 315
Riga 1, 268, 401-109, 503
Riku Eco Village 374
Rio Balsas 412
River Hernád 388, 391
Roman Britain 58
Rome 6, 14, 26-27, 35, 55-57, 59, 61, 76, 96
Rondebosch 178
Rongai 274, 283
Rongai River 274
Roodepoort 532, 551
Roselyn estate 278
Royal Ontario 198, 210
Roysambu estate 278
Ruaraka Mango Farm 278
Runere 324, 337, 342
Rutubiga 336
Ruutana 374
Sahalahti 357, 365-367, 374
Saint-Louis-du-Senegal 147-164, 260
Sajó 383, 385-388, 390-393
Samburu 278, 316

Sandton 532, 538-539, 559-560
Sanjo 330-331
Santa Barbara 526
San Diego 527
San Francisco 521, 529
Sapiwi 336
Saratoga Avenue 183, 187
Sasolburg 554
Savonlinna 33, 122
Scandinavia 21, 396
Schuylkill River 222, 224, 226
Sea Point Service Reservoir 188
Seattle 515
Sebzar 214
Segovia 57
Senegal 102, 147-164, 260
Senegal River 149, 150, 158-159
Sergiopolis 74
Shaikhantakhur 214
Shinyanga 322, 327-328, 334, 341, 502
Shomolu 345, 352-354
Singara 83
Simiyu River 327
Slovakia 388
Soli 335
Somolu 344
Sor 150, 154, 156-157, 160
Sortavala 121
Sosiani River 273
South Africa 1-2, 19, 34, 101-102, 148, 165, 167, 173 – 174, 177, 180, 183, 186, 189-192, 268, 279, 355-357, 368, 370-371, 373-377, 503, 510, 531- 534, 536-537, 540-542, 545-546, 549, 551-552, 555, 578, 586, 590-591
South America 269, 433
South Asian 564, 566
Soweto 536, 550-551, 553-554, 557
Spain 57, 132, 134-136, 139, 141-143, 237-239, 242, 431, 519
Springs 20, 53, 274, 532, 551
Spring Valley Estate 278
Spiral Well 24
Square of the revolution 215
Steenbras Valley 178
Stilfontein 540
Stockholm 31, 109, 119, 126, 240, 244, 491
Strand Street 168

St. Germain Palace 35
St. Peter's church 35
Stuhlmann Bay 324
Sukumaland 322-324, 326, 335-336, 338, 342
Sumiainen 358
Sundsvall 244
Surulere 344, 352
Sweden 31, 37, 39, 109, 111, 118, 120, 124, 126, 239-240, 244, 281, 289, 307, 491, 494, 519, 579
Syrdarya 215
Syria 74
Szinva spring 386, 388
Taaibosspruit 554
Table Bay 166
Table Mountain 99, 102, 166-167, 170, 175-178, 187, 191
Tabora 322-323, 326
Tala 292, 305
Tala Town 292
Tallinn 27, 41-42, 238, 239, 255
Tammerkoski Rapids 125, 244, 383, 396
Tampere 8, 39, 114, 120-127, 244-245, 268, 359, 362-363, 365-366, 374-376, 381-400, 503, 509
Tanaco 414
Tangancicuaro 414
Tanganyika 152, 323
Tangiers 133
Tanzania 259, 262, 264, 267, 32-323, 337, 341-342, 501-502
Tapolca 382, 385
Tashkent 101, 213- 220, 261
Tauranga 269, 476-477, 482, 485-486
Taveta-Lumi 290
Thailand 497
Thames 115, 390, 524
Thugga 60
Tiber 27
Tibetan Plateau 567
Tiflis 86
Tigoni Estate 278
Tingambato 414
Tisza 382, 390
Transvaal 186-188, 190, 373, 536, 540
Trier 55, 59
Tokyo 100, 269, 463-474, 505
Toronto 146, 198, 210
Torresdale 231
Tukhela-Vaal 535

Tukhela river 551
Tukuyu 279
Tulenheimo manor 360
Tunisia 52, 60, 580
Turkestan territory 215, 216
Turku 31-32, 119, 121-122, 124-126, 358, 383
Uganda 272, 273, 296, 318
UK 132, 241, 246, 250, 390, 586
Umbilo Rivers 181
Umbilo Waterworks 180, 187
Umgeni River 180, 191
Umhlatuzan 181
Umlaas River 181-182, 187-188, 191
United States 37, 358, 430, 433, 436-437, 442, 497, 510, 512-513, 516, 518, 524, 527
United States of America 497
Upper Klip 554
Ur 13, 22
Uruapan 414
US 236, 243, 251, 390, 433, 437, 519, 526-529, 561
USA 19, 101-102, 104, 221-222, 237, 243, 590
Usukuma 323
Uzbekistan 101, 213-220, 261
Vaal dam 534, 536, 538-540, 542, 546, 558-559
Vaal River 184,-185, 189, 191, 534, 537, 541, 552, 554
Vaal Triangle 540, 549
Vaasa 20, 120-121, 125, 128, 264
Valencia 242-243, 254
Valetta 133
Valkeakoski 389, 396
Vanderbijlpark 163, 195, 538, 541
Venice 239, 242
Vereeniging 532, 554
Vesuvius 58
Victoria Island 352-353, 502
Victoria reservoir 176
Vihti 127
Viipuri 121-122
Vissoi 283
Vuohenoja 120
Waai Vlei 178
Waikato 476-477, 482, 484-485, 487
Waikato River 269, 476, 478, 484-485
Wajir 278
Wales 132, 237-238, 246, 248, 250, 253, 262
Washington, D.C 512-513, 515, 522, 590
Waterhof Estate 175

Well of Saladin 24-25
Wemmershoek Valley 178
Wernher Beit 185
Westacre 283
West Africa 102, 152, 268, 344
Western Europe 13, 21, 27, 109, 390
Western Pacific 497
Western Transvaal 540
West Africa 102, 152, 268, 344, 354
West Germany 390
Wisconsin 515
Wissahickon Creek 222
Witwatersrand 184, 534-538, 541,-543, 545, 549
Wonderfontein 183, 185, 187
Woodhead Reservoir 102, 176, 188
Woodstock 178
Wynberg 176, 178, 187
Yaba 344
Yemen 101, 197-212, 261
York 59
Yorkshire 247
Zambia 532
Zamora 414
Zimbabwe 532, 581
Zinapecuaro 416-417
Zinziro 414
Zuurbekom 185-186, 188

# INDEX BY SUBJECT

accountability  294, 510, 585
activated-carbon  395, 406
activated sludge  392, 408
aeration  311, 392
African Land Development (ALDEV)  276, 280, 312
algae  395
alum  49, 53, 71, 73
American Water Works Association (AWWA)  255, 512, 519, 520, 521, 523, 528, 529
analogy  53
ancient civilization  4
animal waste  148, 159, 482
aquatic food  494
aquifer  200, 203, 209, 337
arsenic  497, 500, 569
artesian well  151, 155, 390
artificial recharge  126, 386, 395, 396, 408, 410, 503
asset  527
atmospheric alteration  108
attitude  52, 110, 219, 308, 384, 490, 606
bacteriological revolution  110, 123, 266, 490
bank filtration  361
barrel  28, 29, 151
basic need / basic needs  27, 421, 482, 571, 578
bathrooms  47, 112, 384, 583
biodiversity  453
biologically degradable  409
biological oxygen demand  393
biological wastewater treatment  362
booster station  337, 366
bottled water  351, 353, 492, 598
brick  22, 167, 177, 180, 214, 222, 223, 383, 387, 516
building standard  180
bulk supply  276
bursting  134, 514, 516, 517, 525
calcium carbonate  63, 64, 597
canal  109, 110, 215, 217, 219, 403, 568, 580
capacity building  309, 553, 573
capital intensive  520
castle  27, 30-34, 36, 38, 42-43, 100, 167, 613, 617
catastrophe  6, 255, 262, 376
cattle  4, 39, 71, 95, 305, 322-323, 325-327, 330-332, 335-336, 338-341, 358, 365-366, 370-371, 375, 377, 419, 421-422, 426, 451, 502-503

cavern  132, 134, 199, 204
centralization  242, 432
cesspit  35
charge  27, 121, 126, 245, 276, 281-283, 286, 290, 298, 310, 392, 406, 432, 435, 539
chemical treatment  120, 396
children  19, 51, 61, 136, 168, 262, 319, 339, 374, 491, 496, 503, 515, 554, 574, 582, 590, 595, 599
chloramine  515
chlorination  113
chlorine  113, 127, 406, 515
cholera  46, 103, 108-115, 132, 138, 139-142, 144-146, 152-154, 158-161, 163, 180, 190, 229, 260, 262, 265-266, 269, 296, 310, 312, 374, 382, 433, 464, 466-469, 490, 492-494, 498-499, 505, 517, 570
cistern / cisterns  24-25, 50-51, 53, 55, 57, 61, 72-74, 76-78, 82-83, 86, 100, 132, 136-139, 141-142, 144, 149-151, 157, 159, 198, 224, 422, 425
civic pride  536
civilian  70, 73, 75, 133, 134, 135, 136, 137, 138, 139, 140, 142, 144, 146, 149, 151, 346
civil engineer  172, 601, 605
civil servant / servants  124, 150
civil society  566
Clean Water Act (CWA)  512-513, 515, 522, 527
co-operative  436
combined system  148
comfort  80, 124, 481
commercialization  283, 291, 293
commodity  132, 134, 142, 144, 169, 312, 455, 456, 502, 527, 534, 541, 543, 544, 551, 555, 598
community control  413, 420
competition  127, 183, 242, 252, 262, 370, 522, 524
competitive bidding  252, 398
conduit  61, 71, 431
conduits  64, 73, 74, 85, 182, 431, 597
conflict  5, 175, 180, 210, 261, 276, 285, 299, 357, 369, 370, 375, 377, 379, 439, 457, 505, 539
consciousness  190, 250, 374, 398, 448
consensus-oriented  585
constraint  161, 198, 251, 252, 310, 585, 590
contemporaries  104, 106, 111, 122, 123
contracting out  249
convenience  59, 112, 303, 313

621

copper 49, 53, 54, 285, 448, 457, 514, 525
corporate influence 524
corruption 239, 252, 257, 354, 565, 566, 585, 590, 597
counterpoise lift 13, 14, 21, 29, 358
courtyard 34, 136
coverage 7, 287, 309, 310, 436, 437, 438, 439, 442, 569, 570, 571, 574
cowshed 38, 39, 95, 360, 365
cross contamination 138, 513, 517
cryptosporidiosis 498
customers 27, 101, 127, 144, 244, 397, 519, 521, 524, 526-528, 540, 590
decentralization 286, 294, 295, 438, 566
decision-making 254, 293, 377, 527, 586, 597, 598
defence 33, 70, 72, 536, 537
dehydration 80, 83, 138
democracy 194, 439, 454, 504, 546, 549, 564, 565, 566, 590, 595
democratization 439
desalination 143, 144, 146, 529
desert 73, 74, 75, 80, 458, 459
development paths 95, 237, 598
diarrhoea 52, 111, 491, 493, 567, 571, 578
disaster 77, 91, 497
diversity 7, 251, 377, 451, 459, 510, 564, 572, 573, 598
doctrine 70, 78
donkey 27, 101, 285
downstream 78, 79, 85, 391, 396, 484, 513, 517, 528
drilling 277, 322, 329, 331, 337, 358, 371, 375, 438, 502
dryland 356
dwelling / dwellings 50, 61, 71, 108, 135-139, 143, 175, 303, 383, 576, 579, 581-582
dysentery 46, 52, 61, 94, 111, 114, 151, 158-159, 260, 262, 296, 303, 492-494, 567
earthquake 58, 217
economic development 383, 389, 397, 449, 453, 522, 565, 586
elevated tank 361
elite 31, 104, 152, 193, 433, 435, 570
emigration 100, 104, 394
empower 300, 566, 578
Emscher tank 361
enforcement 121, 128, 246, 285, 295, 298, 308-309, 313, 351, 527, 569

English toilet 37
enterprising 88, 101
entrepreneur 2, 121, 244
environmental engineering 237
Environmental Protection Agency (EPA) 515-519, 524, 527, 530
environmental sanitation 286, 305, 306, 308, 309, 351
epidemiological studies 54
equality 40, 95, 127, 576, 578
equitable 4, 207, 210, 261, 585
erosion 199, 251, 325, 336, 351, 356, 451
esker 360, 361, 383
estate, estates 19, 95, 186, 236, 278, 301, 313, 352-353, 389, 600, 618
ethics 237
EU Directives 408
evaporation 203, 325, 331, 335, 339, 392
excavation 22, 42, 203, 333, 339
excreta 4, 40, 46, 255, 260, 278, 306, 309, 310, 388, 490, 582
executioner 35
expert 124, 332, 375, 389, 390, 394, 398, 480, 490, 577, 559, 560, 602
famine 62, 64, 153, 494
federation of municipalities 249
fee 35, 134, 158, 224, 245, 298, 351
fertilizer 27, 28, 38, 39, 466, 505
filtering 28, 30, 51, 55, 73, 81, 112, 113, 114, 160, 260
fire-:
    brigade 124, 186; fighting 19, 95, 124, 127, 144, 236; hydrant 361; insurance 119; protection 121; safety 360, 361
fishing 46, 483
fish tapeworm 46, 493
fissure 132
flood 85, 150, 152, 155, 156, 159, 188, 210, 529-530, 572
flotation 395
flushing 36, 37, 59, 74, 137, 141, 143, 144, 296, 348, 353
food resources 269, 476, 479, 482, 483-486, 488
foreign aid 565
fort 74, 80, 83, 166, 167, 302, 316, 317, 614
fountains 2-4, 11, 19, 58-59, 61, 72, 73, 93, 96, 100, 134, 157, 167-168, 180, 184, 187, 354, 431, 457, 570

Fourth World 491
fractured bedrock 207
free-riding 338
gallery 203
garbage 148, 581
garrison 73, 75, 87, 95, 132, 133, 134, 135, 137, 144, 149, 158, 159
gender 461, 576, 578
geology 198, 278, 323, 497
geopolitics 268, 448, 459
Germany Agency for Technical Corporation (GTZ) 290, 291, 314
germs 137
giardiasis 494, 498
gold rush 431
good social 387
government:
    central 240, 246, 247, 284, 482, 484, 488, 535, 545; federal 251, 350, 434, 438, 440, 504, 523, 525, 528; local 177, 252, 291, 299, 314, 350, 352, 365, 377, 522, 524, 566
hand pump 29, 285, 288, 289, 358
hazard 347, 388, 498
health hazard 490, 498
health officer 277, 285, 302
helminth 493
hookworm 493-494
horses 77-79, 83, 85, 87, 135, 150, 246, 260, 370, 404, 457, 503
hospital 136-139, 145, 149, 150, 152-153, 157-159, 245, 304, 305, 338, 387, 567
hot spring 53
house connection 120
house owner 357, 375
human development 510, 564, 565, 572
human waste 18, 296, 309, 479, 480, 481, 482, 516
human welfare 409
hydram 273
hydrological cycle xii, 5, 94
hydropower 408
ideology 250
infrastructure:
    aging 512, 513, 517, 521, 525, 590; basic 430, 431, 436, 442, 504; hidden 526
illegal connections 516
immigrant 168
immigration 138, 430, 433, 504, 527

immune system 148, 153
impurity 108, 111
Indian practices 449
indigenous-:
    development 203; knowledge 573; people 12, 413, 420, 423, 427, 476, 479, 505, 596
industry:
    forest 363, 389; leather 396; prepared food 365; sugar industry 431; textile industry 396-397
industrialization / industrialisation 4, 100, 104, 118, 123, 190, 261, 266-267, 382-384, 397, 438
industrial revolution 101, 236, 580
inequality 150, 160, 260
infant mortality 122, 127, 139, 190, 510, 566, 576, 577
infectious diseases 62, 111, 144, 468, 490, 498-499
infiltration 166, 167, 325, 408, 516
inspector / inspectors 78, 152, 169, 190, 284, 298, 302, 306, 316, 317, 432
institutions 104, 155, 146, 190, 204, 215, 237, 238, 245, 250, 287, 291, 294, 299, 344, 361, 397, 432, 440, 442, 480, 482, 485, 566, 573, 586, 591, 597
Integrated Water Resources Management 5
inter-municipal 118, 359, 363, 366, 376, 377
International Drinking Water Supply and Sanitation Decade 491, 571, 578
in situ repairs 525
iron 2, 21, 28, 34, 49, 53, 143, 168, 178, 223, 224, 226, 228, 229, 231, 232, 262, 263, 358, 382, 385, 408, 481, 509, 514, 516
irrigation 5, 14, 18, 20, 62, 125, 166, 198, 200, 205, 207, 211, 216, 219-220, 243, 261, 274, 277, 279, 316, 318, 335, 451-452, 498, 544, 547, 568, 570, 571, 596
joint venture 144, 243
karst 199, 204, 207, 210, 211, 387, 393, 397
knowledge 13, 14, 29, 35, 100, 120, 122, 124, 127, 207, 208, 215, 216, 339, 375, 377, 409, 516, 540, 541, 569, 573
law:
    Germanic 238; Roman 238; rule of 585
labour-intensive 322, 502
landowner / landowners 207, 459
landownership 454-456, 459
land use 148, 326, 356, 396, 527
latrine / latrines 18, 35, 74, 78-80, 85, 95, 100, 262, 277, 296, 298, 300-306, 308-310, 312-314, 317, 320, 481, 491, 569, 571, 595

laundry 75, 138, 204
lavatories 35, 74, 182
lead 14, 26, 53, 62, 63, 71, 75, 114, 120, 168, 262, 433, 496, 514-515, 517-518, 522, 525, 590, 596
leadership 101, 104, 435, 573
leakage 30, 144, 228, 289, 325, 394
lesson / lessons 5, 66, 129, 253, 255, 256-257, 342, 375, 440, 442, 504-505, 544, 573, 590
life expectancy 566, 568
limestone 24, 132, 134, 369, 387
livelihood 6, 12, 285, 365, 370, 453, 576
liver infection 136
livestock 39, 95, 101, 118, 166, 289, 322, 335, 337-338, 340, 373, 482, 547
loading / loadings 83, 356, 383, 391, 393, 598
loan / loans 119, 156, 159-160, 276, 281, 284, 290, 346-347, 350, 354, 361, 383, 513, 519, 524, 526, 571
lobbying 241, 523-524
local-:
 administration 155, 216; authority 251, 291, 484, 551; government 177, 252, 291, 299, 314, 350, 352, 365, 377, 522, 524, 566; material / materials 95, 215, 301, 573
Lübeck Act 27, 239
luxury 57
maintenance 64, 87, 119, 127, 148, 154, 176, 207, 214, 215, 216, 219, 261, 272, 281-286, 289, 292, 294, 301, 305, 307, 332, 337-339, 344, 346, 517, 521, 584, 597
malaria 46, 62, 64, 67, 94, 106, 114, 308, 340, 435, 498
manhole 516
manor / manors 35, 39, 357, 360, 619
manure 35, 38-39, 335, 482, 539
market 120, 151, 178, 200, 205-206, 214, 219, 240-241, 245, 250, 253, 255, 268, 274, 305, 316, 351, 358, 365, 402, 430, 455, 456, 460, 544, 546
market economy 255, 268, 402
media 390, 537-538, 542-543, 546-548, 550
medicine 46, 48, 50, 53, 65, 67, 90, 104, 106, 111, 115, 146, 152, 163, 194, 266, 387, 490
medieval 27, 31-32, 35, 38, 43, 74, 100-101, 268, 402, 564-565
mercury 494
meter / meters 19, 21-22, 24-25, 31, 112, 132, 134, 139, 142-143, 152, 155, 157, 159, 160, 222-224, 226, 228-229, 325, 357, 383-384, 386, 394, 397, 405, 407-408, 449, 451, 514, 550, 596
metropolitan 159, 172, 195, 436, 438, 579
miasma 106, 108, 110, 122, 123, 136
microbes 54, 110, 113-114, 260, 492, 596
microbiological studies 54, 596
Middle Ages 4, 30, 64, 67, 90, 238, 382, 403
migration 100, 195, 266, 392, 454, 458, 494
military 57, 70, 72-75, 77-79, 81, 85-91, 132-133, 135-140, 142, 144-146, 150-152, 154, 158-159, 178, 215-216, 260, 438-440, 448, 453-454, 455, 481, 504, 536
Millennium Development Goals (MDGs) 5, 569, 577-578
mining 52, 183-185, 371, 431, 455-457, 494, 534, 545
mismanagement 457
mobile sewer 374
modernization / modernisation 208, 266, 383, 433-434, 440, 454
morbidity 148, 154, 497, 510, 567, 570, 576, 577
mortality 61, 67, 122, 127, 136-142, 144-146, 148, 152, 154, 185, 190, 195, 236, 254, 262, 435, 437, 497, 510, 566-567, 570-571, 576-577
mosquito / mosquitoes 46, 62, 94, 114, 136-137, 340, 498
mule / mules 25-26, 150, 185
mummies 497
municipal 37, 40, 118-123, 156-157, 161, 166, 169, 185-190, 193-194, 215-217, 236, 238-241, 243-244, 247-250, 252-253, 255, 257, 261, 273-274, 297, 299-300, 319, 345, 359-363, 365-367, 375-378, 392, 394, 403, 406-407, 409, 433, 435, 439, 441, 445, 515-517, 522-523, 537-538, 543, 548, 553, 559-561
municipalization 239, 242
mutualisation 247
myth / myths 12, 48, 94
National Agency for Food and Drugs Administration 344, 351
National Water Master Plan 287, 290, 293, 316, 318
natural monopoly 239, 257, 457
navy 75, 132, 135, 139, 154-157, 260
neo-liberal 440, 454-455
networking 124, 598
night soil 27, 28, 296, 297, 319
non-core operations 118, 252
non-governmental 287, 374, 571
nutrient 40, 408, 515

624

oasis 74, 285, 451, 456, 457, 459-460
odour 39, 81, 219, 395, 492
on-site system 40, 375, 595
opposition 18, 125, 222, 440, 456, 526, 536, 543, 565
organic-:
    carbon 409; matter 106, 137, 157, 159, 406, 408
orthophosphate 514-515
outbreak 78, 142, 148, 190, 260, 312, 431, 466, 505
outskirts 200, 205, 206, 330, 345, 384, 502
outsourcing 118, 252
over-exploitation 209
ownership 120, 186, 237, 239, 241, 247, 251-252, 254-255, 284, 292, 338, 370, 393, 397, 440, 526, 548
oxygen depletion 363
ozonization 160, 406
packaged water 351-353
pail-closet 184, 189
Paleolithic times 382
pandemic 103, 138, 140, 152, 466, 494, 505
paradigm 237, 397
parasite 62, 491
participation 209, 219, 286, 298, 309, 312, 353, 354, 373, 400, 433, 436, 510, 548, 569, 585, 597
path dependence 6, 237, 252, 376
payment 27, 119, 120, 183, 244, 253, 322, 329, 337, 347, 502
performance 217, 229, 237, 251, 286, 287, 347, 553, 584, 602
pesticide / pesticides 482, 495
phosphorus 406, 515
pilgrimage 152
pipe:
    cast-iron 168, 226, 229, 262; ceramic 62; clay 23-24, 35, 47, 71; concrete 516; ductile iron 514; polyvinylchloride (PVC) 514; wooden 168, 222-223, 226-227, 229, 231, 244, 262, 357-358, 360, 379, 404-405
plague 177, 188, 190, 296, 297, 498
point source 484, 516
poison / poisons 85-88, 94, 97, 109-111
policy 6, 8, 156, 184, 195, 209, 219, 237-239, 250-257, 261, 272, 276, 278, 280-281, 283, 289-290, 292-295, 297-298, 308-309, 312, 318, 341, 379, 384, 400, 402, 406, 410, 432, 434-436, 438-439, 444-454, 502, 527, 540, 554, 566, 569, 584, 587, 595
poliomyelitis 492
pollution 4, 28, 37, 73, 112, 115, 118, 122, 130, 166, 167, 171-172, 219, 222, 251, 266, 281, 285, 289, 295, 304, 307, 319, 340, 363, 376-377, 382, 384, 387-390, 392, 397-400, 405, 469, 476, 479-480, 484-485, 488, 490-491, 512, 515, 529, 534, 598
population-:
    density 144, 368, 374, 402, 403, 468, 469, 472, 505, 579, 595; expansion 356; growth 4, 6, 13, 71, 138, 158, 174-175, 191, 266, 267, 278, 310, 356, 360, 382, 394, 431, 451, 476, 505, 532, 537, 565, 566, 576, 595
port 152, 160, 177, 180, 189, 193, 260, 273, 299, 308, 354, 432, 452, 456, 617
poverty 5, 294, 318, 283, 499, 564, 565, 571-573, 576, 581, 587, 591, 595, 597
poverty-:
    alleviation 573; reduction 565, 571, 573, 591
practise 598
precaution 481
precipitation 30, 132, 134-135, 143-144, 198, 325, 369, 449, 453, 567
prehistoric 497
pressure system 47, 122
preventive health care 118, 240
preventive maintenance 127
price 35, 135, 151, 186, 228, 243, 245, 348, 349, 384, 529, 542, 543, 544, 550, 552, 559, 561, 582
principle 48, 51, 125, 201, 203, 205-206, 249, 252-253, 292, 398, 437, 442, 448, 552, 555, 590
priority 5, 283, 299, 304, 387, 389, 391, 397, 409, 454, 525, 571, 596, 598
prison 32, 136, 151
private -:
    company / companies 183, 189, 192, 236, 238-239, 242-243, 247-249, 261, 274, 433, 435, 524, 526, 534; enterprise 176, 241, 247; entrepreneur 239; involvement 251-252, 433; operator / operators 192, 238, 239, 249, 251, 252, 253, 261, 262, 274, 346, 352, 430, 434, 441, 502, 598; sector 127, 240, 243-244, 251-253, 262, 273, 283, 291, 309, 312, 346, 348-349, 354, 398, 440, 455, 457, 510, 534, 538, 585
privatisation / privatization 42, 102, 129, 237, 241, 246-248, 250-251, 253, 254-256, 262, 283, 291, 314, 348, 354, 394, 439-440, 442, 449, 504, 505, 524-527, 529-530
privy 32, 112, 308-309
proactive 586, 591

profit / profits  27, 241, 247, 251, 370, 392, 436, 524, 555, 597-598
progress  37, 100, 101, 104, 109, 123, 134, 155, 195, 302, 305, 307, 314, 431, 436, 448-449, 484, 485, 553, 564, 569, 570, 574, 588
property owners  175, 396, 538
protosystem  121, 122, 124-125
protozoa  52, 492, 494, 517
public-private partnership  42, 251, 254, 256
public-:
  concern  389, 526; debate  390, 543; duty  433, 504; fountain  19, 58-59, 61, 157; health  35, 37, 40, 46, 54, 61, 64, 66, 94, 101, 104-106, 111, 114-115, 118, 123, 125, 129-130, 140, 144, 146, 186, 188, 190, 231, 236, 240, 256, 260, 266, 269, 277, 297- 298, 300, 302-303, 313, 348, 353, 402, 432-433, 437, 442, 481, 490-491, 494, 497, 499, 504-505, 510, 512, 515, 517, 522-523, 525, 528, 569, 590, 595; interest  295, 431, 524-525, 527, 590; pressure  398, 433, 504; response  533, 537, 550; sector reform  253, 256; tap  157; welfare  514; well  20, 31, 135, 139, 141, 180, 187, 238; works  42, 148, 179, 236, 243, 254, 256, 275, 277, 297-298, 315, 362
pump / pumps  14, 29, 34, 110, 112, 137, 154-155, 168-170, 177, 180, 184, 202, 209-210, 222-228, 231, 265, 274, 284-285, 288-289, 322, 332, 337-338, 358, 360, 371-375, 377, 385, 394, 404-405, 502-503, 513, 516, 534
purification  73, 81, 87, 112, 160, 185, 361, 362, 376, 408
purity  219
rainwater harvesting  569
Rand Water Users' Council  547
rate:
  increasing block  526
reconstruction  84, 118, 409, 574
recreational use  362
reel  14, 29, 34, 38
reform  106, 121, 123, 175, 177, 194, 208, 211, 236, 247, 253, 256, 257, 286, 291, 293-295, 309, 438-440, 454-455, 524, 536, 565-566
refugee / refugees  70, 153, 262, 494
refuse  35, 108-110, 113, 180, 183, 186, 275, 296-297, 300, 303, 309, 313, 315, 319, 347, 481-482
regulation / regulations  35, 122, 169, 180, 210, 215-217, 220, 236, 240, 261, 268-269, 294-295, 297, 299, 318, 344, 346, 351, 360, 382-383, 398, 400, 430, 439, 448, 459, 479, 481- 482, 512-513, 515-516, 523-524, 527-528, 586
rehabilitation  219, 288-289, 406, 521, 572, 584, 597
relief work  362
religious belief  12
religious institutions  204
renovation  134, 160, 219, 597
rent-seeking  295
repair  160, 216, 245, 285, 305, 510, 512, 515, 517, 521, 522, 525, 553
replacement  229, 288, 337, 352, 448, 464, 472, 496, 510, 512, 518-520, 525, 528-529, 590, 597
replicable  598
reservoir / reservoirs  77, 86, 102, 134, 139, 141, 143, 152, 154, 155, 156, 166, 168, 169, 170, 171, 172, 175, 176, 177, 178, 179, 180, 181, 183, 187, 188, 191, 193, 200, 203, 204, 209, 222, 226, 228-229, 242, 335, 349, 361, 385, 409, 537
retail  353, 502
revenue  169, 244, 281-282, 284-285, 291-292, 347, 521-522, 525-528, 542, 551, 586
reverse osmosis  143
rights:
  human  528, 565, 566, 572, 587; land  370, 454; water  169, 448, 455, 458, 460
Safe Drinking Water Act (SDWA)  512-513, 521-522
safe yield  120, 127
salt  137, 144, 150, 157, 159, 175, 219, 329, 464-465, 516
sanitary conditions  80, 138-139, 177, 180, 184, 269, 347, 434, 435, 464-466, 468, 469-470, 472, 505
sanitary-:
  engineering  237, 435, 490; inspector  190, 306; movement  100, 430, 433-435; policy  432, 434; reform  106, 236; system  161, 183
scarcity  136, 169, 183, 277, 301, 344, 349, 353-354, 413, 415, 417- 427, 451, 453, 457, 532, 553
schistosomiasis  62, 64, 94, 340, 493-494, 498
seafood  476, 485- 486
sealing  154
season:
  dry  136, 149, 151, 157, 159-160, 260, 277, 325-326, 338-339, 344, 349, 353, 370; rainy  134, 150, 154, 183, 325, 339, 353, 559
security  124, 302, 512-513, 578, 581-582

sedentary agricultural life  12
sediment  49, 55, 200, 204, 516
seepage  203, 339
self-help  268, 286, 288, 344, 569, 583
self-purification  408
separate sewer  23, 121, 596
separate system  148
settler  12, 41, 166-167, 193, 274, 280, 284, 313, 331, 336, 479, 480
settling tank  54, 64, 71, 74, 94
sewage disposal  170, 175, 183, 185, 247, 306
sewage ponds  485, 486
sewage treatment  304, 365, 366, 513, 516, 517, 605
sewer:
 brick  22; overflow  231
shaduf  13, 21
shaft / shafts  25, 29, 31, 180, 201-203, 224, 514, 538
shanty town  580
shellfish  483, 485, 494, 517, 518
siege / sieges  70, 72, 74, 77, 83, 85-86, 89, 133, 139, 142, 144
slaughter  306
slave  19, 167, 168, 433-434
slow sand filtration  122
sludge  105, 365-366, 386, 392-394, 408, 515
sluice / sluices  73, 81, 85-86, 88, 166
slum / slums  27, 67, 149, 157, 160-161, 177, 190, 194, 260, 388, 510, 576-577, 580-583, 586-587
smallpox  137, 139, 180, 190, 433, 570
social-:
 consensus  208; equity  528; good  291, 312, 502, 527, 598; responsibility  540; status  18, 61; unrest  434
socio-economic  100, 104, 268, 430, 435, 544, 557,-578
soda  49, 53
soldier / soldiers  15, 31, 70, 73, 77-80, 85-87, 95, 133, 139, 144, 153, 159
solid waste  38, 243-244, 266
specific water demand  532
sphere  239, 436, 527, 541, 549, 598
sponge  27, 80
sprinkler  159, 396, 547
squatter  510, 546, 580, 581, 587
sliding scale tariff  538, 552
stakeholder / stakeholders  219, 242, 352, 548, 586, 597

standpipes  373, 553
steam engine  222-223, 404, 464, 503
storage  70, 73, 84, 87, 135, 136, 139, 144, 170, 171, 176, 179, 180, 187, 198, 404, 480, 481, 492, 518, 541, 546
street improvement  175
structural change  266, 397
struggle / struggles  134, 382, 430, 434, 435, 439, 448, 516
subsidiary  237, 238
subsidy  283
subsistence economies  454
sulphur  49, 53, 454, 456
sustainability  40, 208, 210, 254-255, 257, 261, 283, 290-292, 389, 444, 453, 483, 488, 572, 574, 577
Swedish International Development Agency (SIDA)  286, 289, 306-308
syphilis  137
taboo  18, 95
tanker  268, 344, 346, 348-350, 353, 502, 537
tap / taps  59, 76, 157, 175, 202, 203, 209, 217, 274, 349, 352-354, 384, 387, 390, 496, 514, 515, 517, 536, 541, 560, 583
taste  48, 51, 64, 78, 394, 395, 492
technological choice  6, 595-596
tenant  18, 136, 177, 553
terrace  35
territory  214, 215, 216, 217, 220, 272, 342, 345, 370, 378, 394, 435, 448, 449, 454, 456, 459
tide  182, 539
toilet / toilets  14, 18-19, 22, 24, 27-28, 30-33, 35-41, 43, 47, 59-61, 67, 74, 79, 94-95, 100, 120, 144, 149, 172, 239, 296, 348, 353, 360, 373, 379, 466, 468-470, 472, 481-482, 505, 536, 553, 569, 582-583, 595
topology  132, 144
town council  27, 175, 182, 185-186, 239, 382, 387
toxic  389, 408, 410, 496-497
trade route  109, 322
trading station  160, 260
tradition  231, 237-238, 248, 252, 365, 375, 455
traditional technologies  564
trajectory / trajectories  268, 272, 430, 436
trans-boundary  382
transformation / transformations  150, 160, 214, 260-261, 437-439, 442
transnational corporations  522, 524
transparency  252, 566

transportation  61-62, 118, 150, 222, 273, 277, 514
trench / trenches  80, 83, 154, 296, 333, 337, 371
trenchless technologies  525
tuberculosis  190
tunnel / tunnels  73, 74, 143, 175, 177, 182, 187, 199-204, 207-208, 209, 211, 222, 223, 385, 451
typhoid fever  111, 114-115, 122, 125, 127, 151, 158-159, 185, 260, 266, 347, 349, 464, 490, 492-493
typology  237, 238
underground  123, 134, 198-200, 202-204, 207-210, 231, 333, 337, 348, 451-452, 469, 513, 534, 538, 591
upstream  85, 150, 155, 158-160, 166, 228, 245, 387, 394-395, 404, 455, 512
urbanization / urbanisation  6, 47, 61, 64, 70, 94, 100, 104, 118, 129, 183, 190, 255, 266, 382-384, 397, 434, 438, 454, 464, 532, 534, 541, 579, 580, 586, 587, 591
urban-:
    agglomeration  579; growth  151, 430, 433, 438, 504, 579, 586; history  5, 256, 268; locality  579; planning  118, 148, 442, 449; poor  148, 150-152, 158, 160, 260, 580-581, 586-587; segregation  189
utility / utilities  239, 241, 243, 247-253, 256, 261-262, 294, 277, 374, 376-377, 379, 434-436, 438-439, 513, 517, 519, 520-521, 524-528, 534, 536-537, 540, 542, 546, 555, 586, 590
vendor / vendors  4, 27, 101, 136, 259, 344, 345, 347-348, 352, 373, 431, 465-466, 492, 502, 505, 586
viability  155, 254, 291
victim / victims  152, 153, 160, 260, 537
vinegar  81
virus / viruses  52, 136, 137, 491-492, 493, 516-517
visionary  253
vulnerability  155, 413, 424, 433, 497, 504, 513, 576, 597
washhouse  170
washing place  22
water:
    boiling of  54, 64, 94, 113; carried  4, 150, 168, 178; control of  170, 251, 279, 370, 503; cool  53; domestic  59, 209, 214, 219, 243, 323, 541, 553; flowing  12, 20, 53, 75, 88, 480; ground  28-29, 38, 94, 120-122, 124-125, 127, 202, 217, 219, 325-326, 342, 356-357, 360-361, 366, 376, 385-386, 394, 397, 403, 405-408, 410, 455, 459, 498, 503, 515-516, 559, 569; healthy  71, 87, 476, 479, 480, 482; importance of  15, 46, 48, 94, 368, 541; impure  142; marshy  52, 63, 64, 94, 595; non-potable  141; non-revenue  586; odourless  595; piped  4, 37, 59, 105, 112, 246, 272-273, 283-284, 309, 338, 357, 360, 377, 390, 395, 492, 570, 571, 578; polluted  114, 260, 266, 397, 481, 494; potable  46, 60, 112, 135, 139, 141-144, 150-151, 266, 289, 345-346, 348, 352-354, 372, 437, 441, 490-492, 498, 534, 591; pure  12, 47- 48, 94, 142, 268, 344, 346-349, 351-354, 366-367, 479, 497, 560; rain  13, 22-23, 28, 31, 47, 63, 72-74, 136, 141-142, 180, 325, 339, 353, 387, 413, 415, 422-427, 431, 481, 569; running  27, 30, 47, 51, 61, 384, 480-481; sacred  479; salty  141, 150; scarcity of  136, 169, 183, 277, 344, 349; stagnant  136, 481; storm  181, 186-188, 231, 404, 484; surface  101, 112-113, 120, 122, 124, 126-127, 132, 144, 274, 300, 330, 376, 385-386, 394-395, 397, 407-408, 503, 515, 534, 568, 597; thermal  457; turbid  110; unfiltered  137; usages of  70, 88, 455; value of  553
water-:
    allocation  205, 206, 207, 211; board  177, 238, 540; bodies  118, 260, 325, 356, 388, 389, 392-393, 397, 408, 476, 591; bottling  456; carrier / carriers  4, 19, 20, 27, 141, 168, 238; charge / charges  121, 246; closet / closets  13, 22, 36, 37, 38, 40, 100, 111-113, 260, 569; 447; code  447, 455, 459; consumption  6, 39, 75, 95, 126, 130, 176, 229, 339, 353, 365-367, 391-394, 396-398, 406-407, 422, 535, 537-539, 541-542, 544, 547, 549, 551, 561, 590, 595; container  81, 83; distribution  4, 140, 142, 157, 161, 406, 409, 410; losses  217, 553; policy  219, 238, 252, 276, 278, 283, 290, 292-294, 527; pollution  118, 166, 266, 281, 304, 306-307, 363, 376-377, 382, 389, 392, 397, 399-400, 490-491, 512, 515, 598; power  226; quality  5, 112, 114, 124, 160, 214, 219, 246, 250, 307, 340, 363, 383, 389, 397, 400, 405, 408-409, 471-472, 494, 503, 514, 527-528, 578, 598; quantity  5, 125, 382, 598; reservoir / reservoirs  77, 177, 181, 361, 385; restrictions  175, 510, 532-540, 542, 544- 548, 551-552, 555, 561, 590; security  512-513; seller  167; shortage / shortages  72, 125, 150-151, 158, 166, 170, 175, 183, 261, 273, 277, 349, 353-354, 371, 384, 387, 459, 510, 537, 549; source / sources  7, 20, 54, 64, 71, 73-74, 77-78, 83, 85-86, 94, 104, 121, 180, 186, 191, 195, 208, 222, 236, 273, 282,

285, 288, 313, 322-323, 327, 328, 338-340, 360, 370, 383, 388, 406, 408, 456-457, 464, 493, 502, 503, 526, 528, 564, 572, 598; stagnation 468, 472, 505; table 198, 202, 207, 208, 209, 210, 513; tank / tanks 34, 74, 154, 156, 360, 365, 382, 385-387; tower / towers 74, 127, 386, 404, 523; treatment 120, 124, 219, 260, 266, 383, 385-386, 388, 390, 392, 395-397, 406, 437, 441, 465, 472, 490, 492, 505, 513-515, 523, 526-528, 596; user / users 214, 217, 219, 526, 561, 566, 568, 569; use purposes 5, 8; vein 357; vendor / vendors 27, 101, 259, 344-345, 347-348, 352, 373, 431, 465-466, 492, 502, 505, 586; wastage 533, 552; wheel / wheels 24-25, 228, 373, 383, 431

water-lifting 18

waterborne 46, 64, 94, 114, 177, 182, 186, 188-189, 260, 262, 269, 303-304, 306, 310, 314, 490, 491, 493, 498, 499, 505, 520, 521, 595, 596

waterhole / waterholes 277, 326

watershed 183, 191, 246, 601, 606

waterway / waterways 269, 390, 476, 479, 480, 482-485, 487-488, 513, 516

waterworks 40, 120-122, 124-125, 127, 151-152, 154, 176, 180-181, 183-185, 187-188, 237, 241-242, 244, 246, 250, 254, 345-346, 350, 360, 375, 385, 387-388, 390-392, 394, 399, 473, 502, 524

Water Supply and Sanitation Collaborative Council 5

well / wells:
    deep 20, 209, 465, 466, 505; shallow 183, 322, 338, 403, 431, 502; tube 14, 29, 497

wetland / wetlands 325

wholesale 303, 363, 525

windlass 13, 14, 29

windmill / windmills 14, 338, 360, 371, 373, 379

witcher 357

women 19, 168, 204, 338, 358, 359, 374, 503, 558, 566, 569, 572, 574, 578, 582, 590, 595

working class 112, 127, 150, 152, 158, 160, 177, 240, 260, 389, 392

World Health Organization (WHO) 115, 266, 269, 278, 280-282, 284-286, 306-307, 309, 319, 490-492, 494-500, 567, 569, 574, 587

World Water Development Report 4, 597

yellow fever 114, 132, 136-138, 144, 152, 160, 222, 231, 433

yoke 358